Geologica

The Origins of the Earth

Geologica

EARTH'S DYNAMIC FORCES

Chief Consultants
Dr Robert R. Coenraads and John I. Koivula

Geological Time • Supercontinents
Climate • Landforms • Animals • Plants

MILLENNIUM HOUSE

Map Symbols

~~	River
⌁	Lake
▨	Fjord
▨	Basin
▨	Desert
▨	Craton
～	Boundaries
⌐	Rift/Plate line
▲	Active volcano
△	Dormant volcano
▲	Mountain peak/Extinct volcano
⊔	Canyon/Gorge
⏢	Cave
●	Geyser/Hot spring
⌣	Impact crater
■	Volcanic remnant

Map Lettering

FRANCE	Country
MONTANA	State/Province
Sardegna	Island
Antarctic Peninsula	Peninsula
ANDES	Mountain range
Everest	Mountain peak
Grand Canyon	Featured location
Gibson Desert	Desert
Cape of Good Hope	Cape
Amazonas	River
Lake Victoria	Lake
ATLANTIC OCEAN	Ocean
Caspian Sea	Sea
Gulf of Carpentaria	Gulf

PUBLISHER	Gordon Cheers
ASSOCIATE PUBLISHER	Janet Parker
PROJECT MANAGER	Fiona Doig
ART DIRECTOR	Stan Lamond
PRODUCTION	Simone Russell
EDITORS	Loretta Barnard
	Fiona Doig
	Catherine Etteridge
	Heather Jackson
	Carol Jacobson
	Deborah Nixon
	Anne Savage
	Marie-Louise Taylor
CHIEF CONSULTANTS	Dr Robert R. Coenraads
	John I. Koivula
CONSULTANTS	David McGonigal
	Armstrong Osborne
	Robyn Stutchbury
COVER DESIGN	John Canty
DESIGNERS	Stan Lamond
	Avril Makula
PICTURE RESEARCH	Fiona Doig
	Catherine Etteridge
	Philippa Hutson
	Rebecca Jervis
	Oliver Laing
	David McGonigal
ILLUSTRATORS	Andrew Davies
	Glen Vause
CARTOGRAPHY	Sergio Boggio
	Ruth Coombes
	Alison Davies
	Mark Fairbairn
	Kim Farrington
	John Frith
	David Hosking
	Robin Hyatt
	Kevin Klein
	David Maltby
	Joe Nunn
	Alan Palfreyman
	Colin Reid
	Timothy Rideout
	Alan Smith
	Martin Smith
	Mary Spence
	Matthew Townsend
MAP RESEARCH	Jan Watson
INDEX	Diane Harriman
PUBLISHING ASSISTANTS	Bernard Roberts
	Liam Wilcox

Photographs are credited on page 523.

First published in 2007 by
Millennium House Pty Ltd
52 Bolwarra Rd, Elanora Heights
NSW, 2101, Australia

ISBN: 978-1-921209-72-7

Text © Millennium House Pty Ltd 2007
Maps © Millennium House Pty Ltd 2007

Reprinted 2009, 2010

Sales
For all sales, please contact:
Millennium House Pty Ltd
52 Bolwarra Rd, Elanora Heights
NSW, 2101, Australia
Ph: (612) 9970 6850   Fx: (612) 9970 8136
email rightsmanager@millenniumhouse.com.au

Authors
Millennium House would be happy to receive submissions from authors. Please send brief submissions to:
editor@millenniumhouse.com.au

Photographers and Illustrators
Millennium House would be happy to consider submissions from photographers or illustrators. Please send submissions to:
editor@millenniumhouse.com.au

Photographs on preliminary pages:

PAGE 1 These unique rock formations in Cappadocia, Turkey, are known as "fairy chimneys." They are formed from weather-worn tufa, a type of volcanic rock.

PAGES 2–3 An unusual fall of recent rain has caused fresh vegetation to spring up in the dunes in the Namib-Naukluft National Park, an ecological reserve in the Namib Desert, Namibia.

PAGES 4–5 The Gooseneck Bends of the San Juan River carve their way through Gooseneck State Park, Utah, USA.

PAGES 10–11 The mighty Iguazú Falls thunder across the border of three countries—Argentina, Brazil, and Paraguay. They flow over 19 major drops through a narrow 260-foot (80-m) deep canyon.

EARTH is an epic publishing feat never to be repeated, proudly created by Millennium House

Western Europe

Locator box indicates location of map within world

Locator box indicates location of map within region

Color-coded bars make identifying each continent easy

Each page is edged with silver gilding to help in the preservation of the book over time

Comprehensive labelling with current data

Specially created state-of-the-art full-color background relief

Colored relief bar indicating elevation

Feature boxes throughout the book focus on milestone events in a country's history or take an in-depth look at a unique feature about a country

Exquisite full-color images are used throughout, with more than 800 images in total

Scale bar and map scale information

A profile of each country provides a snapshot of vital information, including official name, population, area, capital city

A map of each country shows the major cities, highest point, and neighboring countries

Handy map reference information will take you straight to the corresponding map for each country

A locator map pinpoints the location of the country within the region

This limited edition atlas—the world's largest modern atlas—is bound by hand in beautiful custom-dyed leather with silver-gilded pages for preservation and silver-plated corners for protection. *Earth* establishes a new benchmark in the world of atlases, with highly detailed mapping and comprehensive country profiles. With 580 pages, including 355 maps of 194 countries, more than 800 images, 4 breathtaking gatefolds, and information on every country, territory and dependency in the world, *Earth* will appeal to book collectors, libraries and institutions, map lovers, and anyone wishing to enjoy now, and preserve for the future, a record of our world today.

Earth can be ordered from all good book stores or contact Millennium House for your nearest supplier

www.millenniumhouse.com.au

Contributors

Dr ROBERT R. COENRAADS

Robert Coenraads is a consultant geo-scientist, lecturer, and author of three books and over 30 scientific publications. His interest in geology and the natural world has heightened during his 30 years of field experience, including working with diamonds, sapphires, rubies, opals, and other beautiful gemstones. He has led geology, archeology, and natural history field trips to various corners of the globe, including visits to active volcanoes of the USA, Mexico, Chile, and the Pacific, and was present for the Mt St Helens eruption in 1980. As chief author and a chief consultant, he wrote the chapters on geological time; Earth's formation; volcanoes; volcanic remnants; mountain ranges; rift valleys and fault lines; geysers and hot springs; and fjords, glaciers, and icefields. Robert also researched the illustrations in *Geologica*.

JOHN I. KOIVULA

John I. Koivula has spent 45 years studying and photographing the microworld of minerals. Chief Gemologist at the Gemological Institute of America, he has published more than 800 articles and notes on inclusions and related topics, and is a contributor to several books. He is coauthor with Dr E. Gübelin of the *Photoatlas of Inclusions in Gemstones, Volumes 1, 2 and 3*, and the author of *The MicroWorld of Diamonds*. John holds university degrees in geology and chemistry; the gemological credentials GG, CG, FGA; was awarded a fellowship in the Royal Microscopical Society; and serves on the board of the International Gemmological Conference group (IGC). He is a member of several gemological societies and has won many awards. As a chief consultant on *Geologica*, John wrote the introduction and glossary.

Dr ARMSTRONG OSBORNE

Armstrong Osborne started caving in the scouts, soon becoming puzzled with how they formed. He studied geology and worked as a teacher, but the cave bug would not go away. He has been studying the caves of eastern Australia for 30 years and regularly visits central Europe to collaborate with colleagues to study some of the more unusual caves there that have similarities to Australian caves. Armstrong is involved in the management and conservation of caves and their geological features internationally, as a consultant and advisor, and as an expert witness in high-profile environmental cases. He supports his caving activities by teaching Science Education at the University of Sydney. Armstrong wrote part of the karst and caves chapter.

BELOW Zabriskie Point is in Death Valley National Park, California, in the USA. This arid landscape, fractured by a long history of earthquakes, lies in the Mojave Desert, and is one of the hottest and driest regions in North America.

DIANE ROBINSON

Diane Robinson is a freelance writer who has published much material on travel, the environment, and health in many publications including the *Sydney Morning Herald*, the travel magazine *Implosion*, and the books *Natural Disasters* and *Historica's Women*. She also works as a writer–editor for an environmental agency in Sydney, Australia. Slipping around on glaciers and mountains, trying not to fall into crevasses and abysses, and trudging up never-ending slopes while pursuing a fondness for trekking have provoked her interest in geology and fueled her passion for writing the canyons and gorges chapter. She is currently planning an extensive trip to visit some of the exotic locations featured in the book.

PHIL RODWELL

Phil Rodwell wrote and compiled the gazetteer section at the end of the book. He has been researching and editing general interest books on topics as varied as gardening, history, travel, motor touring, heritage architecture, DIY, Australiana, and cookery for 30 years. After prolonged exposure to innumerable images and thousands of words describing the world's breathtaking and dramatically varied physical forms he is a changed man. As a resolutely inner-urban dweller who has generally confined his working life and many travel experiences to cities, Phil is now committed to seeing as many of the wonders of the natural world, and of the world's amazing landforms as possible.

BARRY STONE

Barry Stone is a freelance writer, photographer, and a graduate of the Australian College of Journalism. He lives in a pastoral setting in the foothills of the Southern Highlands near Sydney with his wife who hates bugs and would love nothing more than to return to life in the suburbs, and their two boys—Jackson, aged 6, and Truman, aged 2. He has long been a contributor on travel and architectural themes to some of Australia's leading daily newspapers including the *Canberra Times* and the *Sun-Herald*, along with various in-flight magazines including those for the international airlines Qantas and Air New Zealand. Barry has also co-authored books on travel, history, the arts, and architecture.

ROBYN STUTCHBURY

Robyn Stutchbury is passionate about all things natural. After teaching biology for some years, she completed a degree in geology. This background, followed by a graduate qualification in science communication, gave her an excellent understanding of Earth and its processes and set her on the path of freelance writing. Robyn wrote the supercontinents and climate text and the rivers and coastal landforms chapters, revealing what a powerful force water is in shaping our planet. She also wrote for *Natural Disasters*, and co-authored *Exploring Nature in Lakes, Rivers and Creeks* as well as many articles for scientific publications with her husband, Noel Tait. Extensive travels have led Robyn to see many of the world's most beautiful landforms.

Contents

Foreword

Imagine being able to gaze at our Earth's entire 4.6 billion years of geologic history passing by in the space of 12 hours. On this accelerated timescale, we would be able to see the workings of our Earth's heat-driven internal convection engine. We would witness the continents torn apart and the pieces driven around like toy boats in a lake. Pushed across the globe, the continents are periodically rammed together again with such force that giant Himalaya-like mountain ranges are thrust miles into the air. As individuals, we can only witness a snapshot of this story. In reality, Earth's powerful engine moves no faster than our fingernails grow.

To us, even 12 hours is a long time to wait and watch Earth's story unfold. Using this timescale, it takes over 10½ hours before life evolves and diversifies, bringing us to the Cambrian "explosion" of life about 542 million years ago. The insignificance of humankind's entire existence is realized as we witness the first *Homo sapiens* appear on the planet just a little over one second before midnight. These early humans lived in the Great Rift Valley of Africa more than 130,000 years ago.

Geologica takes us on a journey of discovery from Earth's very beginnings, through geologic time, to the present day. We travel all the continents and oceans, visiting some of the most spectacular places of geological importance that the world has to offer—Earth's greatest mountain ranges, its fragile icecaps and glaciers, violent volcanic chains, empty deserts, and stunning coastlines. We explore the processes behind the scenery and rocks around us, and how they shape the way that life evolved.

Robert Coenraads

RIGHT The remains of the volcano Naboyatom lie on Kenya's Lake Turkana, which is located in the heart of the East African Rift Valley—the site of some of the earliest ancestral human fossils.

Introduction

Nature is alive with color and form. A vivid sunrise or the sudden appearance of a rainbow will almost always make one pause from whatever one is doing to take in the natural beauty. The ever-changing forms displayed by plumose wind-driven clouds can delight our imaginations with images of things that aren't really there. The variety in nature is as seemingly endless as a clear night sky dotted with a myriad flickering "stars."

Where does nature fit into our modern world? On a daily basis, most people rarely take note of nature's offerings. Throughout geological and recorded history, nature has shocked us with displays of raw power. Evidence suggests that the dinosaurs were pushed toward extinction 65 million years ago by the impact of an asteroid in the Yucatan Peninsula, Mexico, and its subsequent aftermath. The "lost city" of Pompeii was destroyed by a volcanic eruption in 79 CE, "freezing" people in place through incineration and burial in hot ash. There is not much left of the Indonesian island of Krakatau, another victim of volcanic violence recorded by those who witnessed this event in ships' logs and scientific journals, or heard its thunderous roar in other parts of the world.

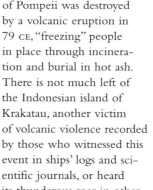

In more modern times, Mt St Helens in the Pacific northwest of the USA exploded with tremendous force. The eruption was heard more than 100 miles (160 km) away and the ash cloud looked like a black wall, blocking out sunlight as it traveled east, covering structures with a thick, heavy blanket, and choking internal combustion engines and causing them to fail.

Thousands died or were made homeless when an earthquake-generated tsunami (a seismic sea wave) devastated the western side of Southeast Asia and parts of the Indian subcontinent in 2004. When Katrina, a super-hurricane, made landfall in 2005 and the levees failed along the southern Gulf coast, New Orleans, and other populated areas along this part of the US coast experienced the power of nature in the worst possible way. To nature, these are simply interactions between various natural systems, but to those who lived through them to tell of their ordeal, these are wake-up calls that will never be forgotten.

Such wake-up calls from nature need not be violent and dramatic. Subtle clues to nature's ultimate control over our lives surround us. The power of nature and its relentless force can be experienced when we stumble on a cement-covered sidewalk where a tree root has lifted a slab, or when we see a single flower that has broken through the surface of a paved road in order to reach sunlight.

SPHERES OF INFLUENCE

Earth science or geoscience is the all-encompassing term that gathers together, under the one academic roof, all the scientific disciplines used to understand our planet and its interrelated geological processes. Geology, one of these earth sciences, is the study of the composition, history, natural forces, physical properties, and overall structure of the solid substances that make up our planet.

Some of the other closely related scientific fields of study that have an effect on geology include astronomy, biology, chemistry, meteorology, oceanography, and physics, though most of these are not actual earth sciences per se. These primary sciences are further divided into numerous closely related disciplines or areas of focused research. And, as our body of knowledge grows, these areas of study expand. The interconnected relationships of these different sciences to each other as well as their practical application to geology show us that our planet is not a solitary and isolated structure, but instead a dynamic and rather delicately balanced ecosystem that is influenced by all of the forces of nature in interactive combination.

The daily weather, continuous climate change, water and air quality, our consistent food supply, the health and welfare of all life on our planet (not just our own) … these things are all important to our basic survival as a species, and they are all related. In studying geology and the other related sciences, we must take into account not only the most obvious rock structures, such as mountain ranges, but also the interactions of the various layers or spheres of influence that create the whole planet we experience as Earth.

ABOVE The spectacular nature of the rugged volcanic landscape of Na Pali Coast State Park on the island of Kauai is best viewed from the air. Each island in the Hawaiian Islands is part of a chain of hotspot volcanoes, and the Big Island is home to the currently active volcano.

OPPOSITE Winding ribbons of dark moraine trail through Russell Glacier in Alaska's Wrangell–St Elias National Park and Preserve, USA, contrasting with the white ice. This is one of several large glaciers fed by the Bagley Icefield, which sits on the crest of the Chugach Range.

ABOVE An aerial view of the dunes near Sossusvlei, Namibia, reveals the impact a good rain year has on the landscape. In the heart of the Namib Desert, Sossusvlei has the world's highest sand dunes.

ABOVE An ancient ice sheet carved the plateau that is now cut through by the Chilcotin River in British Columbia, Canada. The surrounding region is home to old growth Douglas fir trees as well as a variety of mammals, including river otters, bears, wolves, cougars, and lynxes.

Earth's four interactive spheres that constantly influence each other are the atmosphere, the biosphere, the hydrosphere, and the geosphere. Or, on a more personal note, they are the air we breathe, the food we eat, the fresh water we drink, and the solid ground we live on.

These can be studied independently of each other, but they are all so closely and complexly related in such a significant way that scientists refer to them collectively as "the earth system," and their study has now evolved into a relatively new earth science known as earth system science. Powered by a combination of energy from the Sun, and heat, radiation, and pressure from Earth's interior, a change in any one of the spheres that form the earth system can produce changes in all the other spheres. Most critically, the whole earth system is delicately balanced, and what we collectively do now influences how we live in the near future.

ATMOSPHERE

Earth's atmosphere is the envelope of mixed gases surrounding our planet that is held in place by our gravitational field. It is subdivided into several different layers or zones that extend from the surface (the troposphere) to the exosphere, more than 300 miles (500 km) above Earth's surface.

Our atmosphere is composed primarily of nitrogen and oxygen, which account for more than 90 percent of the atmospheric content, particularly near the surface. The remainder is made up of other gases, such as minute amounts of carbon dioxide, ozone, and argon, and a few percent of water vapor as a representative of the hydrosphere. Concentrated in a layer called the troposphere, this is the life-sustaining mixture of gases that we call air. In addition to providing the "air" we breathe, the atmosphere also shields us from deadly ultra-violet solar radiation while at the same time stabilizing the planet's overall surface temperature, making the biosphere more comfortable for us.

The effects of the atmosphere are quite easily observed, and they are sometimes quite dramatic. Wind and rain sculpt the land over long periods of time. Freezing rain can break down mountains through a process known as frost heaving. Large hailstones can break windows. Windblown sand can remove paint from any surface. And the devastation caused by tornadoes and hurricanes is all too well known to many.

Another area of atmospheric concern is air pollution. Because controlling air pollution directly affects our world economy, it has proven to be a tough adversary. As a species we must decide if breathable air is more important than business profits. Stratospheric ozone depletion, higher and higher pollen counts, excessive carbon dioxide emissions—the pollutants are biological, chemical, and physical, and they all impact on the biosphere.

BIOSPHERE

The biosphere, the area in which we thrive, is constantly influenced by changes in the other three spheres for a logical reason. It is the area in which life exists, and it therefore incorporates portions of the other three spheres—the atmosphere, the geosphere, and the hydrosphere; or air, land, and water. Life is found in all of them, virtually everywhere on or near Earth's surface—from pole to pole, up into the atmosphere, down into the deepest recesses of the oceans, and even the crust; including some places previously thought of as uninhabitable or unable to support even the simplest microscopic forms of life.

Observations of both simple and complex life forms have helped to establish the relative thickness of the biosphere. Excluding manned space-flight, in the atmosphere, Rueppell's vultures (along with their parasites) are the champions. These birds can fly at incredible heights—one actually collided with an aircraft at an altitude of 37,000 feet (11,300 m). Some forms of micro-organisms can exist under very extreme conditions, stretching the biosphere 30,000 feet (9,000 m) below sea level. By comparison, without special equipment to protect us, we humans are far more limited in the extent of our own biosphere.

GEOSPHERE

The geosphere is most simply the dense solid-to-molten part of Earth. It is composed primarily of crystallized matter in the form of minerals and rocks on the comparatively thin external surface, as well as substances in both the molten and solid state in the deeper regions. The geosphere is divided into three basic zones or parts, described as the core, the mantle, and the crust, which are layered around each other in more or less concentric spheres.

From the inside out, the geosphere consists of the dense core with an inner solid component and an unusual outer layer that, from seismic data, appears to be in a liquid state. The core is the innermost region or nucleus of Earth's interior. The solid inner core is approximately 1,600 miles (2,600 km) in diameter. It is thought to be solid and composed primarily of elemental iron. The outer core, which is approximately 1,350 miles

BELOW An aerial view captures the lush greenery of the Amazon rainforest in Brazil, in the lower Amazon River floodplain. During the wet season, this vast river releases up to 32 million gallons (121,000 L) of water per second into the Atlantic Ocean—more than the volume of the six next greatest rivers combined.

WINDOWS INTO GEMS: MINERAL INCLUSIONS

Minute amounts of water trapped inside a mineral is part of the hydrosphere. Other fluids or solids, such as other minerals, may also become trapped. By studying these inclusions, scientists can learn much about the mineral that contains them. These exquisite photographs show close-up details of inclusions.

A. Under fluorescent light, this quartz from Baluchistan, Pakistan, contains petroleum, which shows as a chalky blue; under normal light it is usually yellow.

B. This polarized fossil resin from Boyaca, Colombia, contains inclusions of liquid and gas from an ancient biosphere and, most interestingly, a bird's feather.

C. The brassy irregular mass inclusion inside this quartz from Montana, USA, is chalcopyrite (copper ore). The copper was present when the quartz crystallized.

D. Diamonds' durability and transparency provide a window into the ancient rocky crust and Earth's molten interior. Most diamonds range between 1 billion and 4 billion years old. Mineral inclusions in diamonds, such as this one from Yakutia, Russia, containing pyrope, are fascinating geological time capsules.

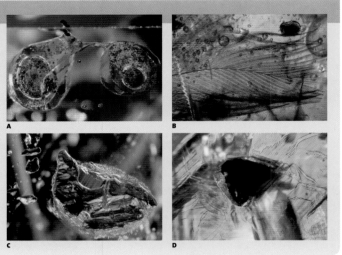

(2,170 km) thick, is thought to consist mainly of molten metal (possibly iron and nickel, as well as smaller amounts of other elements).

The mantle surrounds the core. Measuring approximately 1,800 miles (2,900 km) in thickness, it is the layer or "shell" that extends from beneath the crust all the way to the outer core. Geoscientists divide the mantle into two specific sections—the lower mantle or mesosphere, and the upper mantle or asthenosphere. Very little is actually known about the mesosphere. It certainly must have its own effects on the planet and interact with both the core and the upper mantle, but this interaction and the resulting impact still remain geological mysteries. The asthenosphere or upper mantle is where significant volumes of magma are formed that generate or drive tectonic activity and influence the shape of the solid crust or lithosphere on which we live in the delicately balanced biosphere.

BELOW The entrance to Paradise Harbour on the western side of the Antarctic Peninsula is riddled with chunks of ice which have calved off the numerous glaciers that flow down into the bay.

HYDROSPHERE

The hydrosphere is composed of all the water found in, on, or above our planet regardless of its state. This includes the oceans (which make up its overwhelming mass); inland seas, rivers, and lakes; ground water; all of the ice in its various forms (the cryosphere); and all of the water vapor (which comprises a small percentage of the gaseous atmosphere). Snow, sleet, rain, and fog … if it's composed of water, then its part of the hydrosphere. Even the water trapped as fluid inclusions in small voids in rocks and minerals is part of our hydrosphere. The hydrosphere even encompasses the water in plants and animals, including ourselves.

It is the presence of abundant water that makes Earth unique. In their ongoing search for life on other worlds, cosmologists and astronomers have yet to find a similar planet. Like a mottled blue marble suspended in space, feathered by clouds, as far as we know, as a planet, Earth even has a look all its own.

Constantly flowing water erodes away existing land and re-deposits it as sediments, which over time will ultimately solidify to form new rocks in a continuous rock-cycle. The hydrosphere is constantly interacting with the atmosphere and geosphere to reshape our world.

Pollution of the hydrosphere is as damaging as air pollution. It limits the availability of drinkable water. With a continuously growing population, the demand for drinking water will grow. While people may be able to get along for several days without solid food, it is not possible to do without water for an extended period of time. Fresh water may end up being one of the most precious resources we have. In some parts of the world, it already is. The hydrosphere is as delicately balanced to population as any of our other major resources. What we do to the hydrosphere will affect everyone.

EXTINCTION: THE INEVITABLE CONCLUSION

Applying the science of geochronology, using sophisticated age-dating techniques on micro-inclusions of zircon, geoscientists have established that Earth is about 4.6 to 5 billion years old. Scientists estimate that in a few more billion years, the primary source of all we recognize as living, our very own star, the Sun, will begin to die. Through nuclear fusion, the same solar process that now gives us life, our Sun will have exhausted its supply of core hydrogen and will begin to swell into a bloated but still very hot, dying star. In this process of unstable gigantism, the Sun's outer atmosphere will dramatically expand, engulfing the inner planets if they still exist—our own Earth will be one of the victims. In a rather dramatic form of global warming, the atmosphere will at first be swept away, then the cryosphere and the biosphere—and with it the hydrosphere, as all water on and in the remaining husk of the lifeless rocky geosphere is boiled off. Exit Earth.

John I. Koivula

BELOW The myriad channels of the Rapa River Delta in Sarek Nature Reserve wind through a region of Sweden that lies in the high Arctic. The lagoons of this delta are an important habitat for a number of waterbirds.

Earth's Formation

Earth started life as a pile of debris that began forming in the Sun's spinning proto-planetary disk. As proto-Earth orbited, it cleared a swath by accreting numerous smaller bodies. It grew larger and hotter, and then melted. There were no surface rocks at this time, only a thin atmosphere of hydrogen and helium bleeding into space.

RIGHT Highly prized for the attractive gemstone peridot, the abundant mineral olivine forms deep beneath Earth's crust in the mantle. A significant component of magma, it can be brought to the surface by volcanic activity.

BELOW The intensive tectonic activity that occurred at the beginning of the Archean Era would have been unlike anything known today. Frequent and violent volcanism would have occurred as the first continental crust began to form.

This was the time known as the Hadean Eon, stretching from Earth's beginning, 4.6 billion years ago, to about 3.8 billion years ago. The name comes from the Greek word for 'hell,' aptly chosen to describe this phase when Earth was a heavily bombarded inferno. It is also sometimes called the "rockless eon," as there was no hard crust. Eventually, Earth became massive enough, and its gravitational field strong enough, to hold an atmosphere in place. It was a heavy poisonous mix of carbon dioxide, ammonia, methane, nitrogen, and water vapor. A thin crust began to form toward the end of the Hadean Eon and, even at these temperatures, water vapor was able to condense under high atmospheric pressure. Torrential rain poured into boiling oceans. There was no free oxygen around at this time, as it immediately became bound to hydrogen or other elements.

EARTH BEGINS TO DIFFERENTIATE INTO LAYERS

During this early stage, Earth's own strengthening gravity pulled the heaviest elements, mainly metals such as iron, toward its center while the lighter ones, such as oxygen and silicon, rose upward toward its surface. Finally, Earth differentiated itself into three distinct layers of different density—the core, mantle, and crust.

At Earth's center sits a dense iron core. It is divided into the inner core, which is solid, and an outer core, which is liquid. It is only the incredibly high pressures toward Earth's center that maintain the inner core in solid form, otherwise it too would be molten. The mantle surrounds the core and is made mainly of the elements magnesium, silicon, and oxygen, which are combined as the mineral olivine. At times, fragments of the mantle, including the occasional diamonds that also form there, are carried to the surface by rapidly rising magma. A thin brittle low-density crust, akin to the cracked shell of a hard-boiled egg, lies on top of the mantle. The crust is made up almost entirely of the lightest elements—silicon, aluminum, and oxygen, combined mainly as the minerals feldspar and quartz.

THE ARCHEAN EON— WHEN EARTH'S EARLIEST ROCKS BEGAN TO FORM

Earth was still a very active place during the Archean Eon, which followed the Hadean Eon, and lasted from 3.8 billion years ago to 2.5 billion years ago. The surface was very hot and continents were not yet forming— probably because heavy meteor bombardment was

still shattering the crust, and vigorous plate tectonic activity was recycling the pieces at a very great rate. By the end of the Archean Eon, tectonic activity had slowed to near present-day levels.

The few traces of Earth's very first rocks that still remain today are found in the most ancient centers of the continents (known as continental shields) of Africa, Australia, Greenland, and North America. The oldest mineral is a zircon crystal, dated at 4.4 billion years old, but it was no longer part of a rock. It was found as a recycled grain in much younger rocks. The oldest dated

ABOVE Steam rises from a hot spring in Iceland, geologically one of the youngest lands. Subjected to constant geothermal activity and volcanism, it may resemble the young Earth.

complete rock formations were found in the Isua Greenstone Complex, in Greenland. These are 3.8 billion years old and are mainly altered metamorphic and igneous rocks. Traces of organic carbon have been found in some of these rocks indicating that they were once sedimentary rocks, and that primitive life had already begun to evolve on Earth by this time.

FIRST LIFE ON EARTH

Although the Archean Era atmosphere lacked free oxygen, temperatures had dropped to near normal levels, and there were now vast oceans. About 3.5 billion years ago, primitive cyanobacteria—simple prokaryote cells with no nucleus —appeared in the oceans. Increasing rapidly in numbers, they began to grow hard mound-like structures called stromatolites by depositing concentric layers of calcium carbonate. These simple bacteria built huge reef complexes that became the first limestone rocks. Although these stromatolite reefs have been largely replaced by modern coral reefs, a living example can still be seen at Shark Bay in Western Australia.

THE PROTEROZOIC EON—OXYGEN BUILDS IN THE ATMOSPHERE

The Proterozoic is the last eon before complex life literally exploded into abundance on Earth at the beginning of the Phanerozoic Eon. The Proterozoic Eon started at the end of the Archean Eon, 2.5 billion years ago, and finished 542 million years before the present. These first three eons, the Hadean, the Archean, and the Proterozoic, are also informally known as Precambrian Era.

From about 2.3 billion years ago onward, oxygen that was being produced by photosynthesis finally began to build up to a significant level in the atmosphere. This was only able to take place after all of the elements that readily bond with oxygen, such as sulfur and iron, had been fully satisfied. This oxygen increase is recorded in the world's banded iron formations—beautiful red and silver striped sedimentary rocks that also contain most of the world's iron ore reserves. Their tree-ring-like bands show that oxygen levels rose and fell in seasonal cycles. It is thought that the

free oxygen poisoned the bacteria that produced it, before exhausting itself by oxidizing all the available iron. The cycle then began again with bacteria numbers increasing and oxygen levels building up in the atmosphere. These unusual banded sedimentary rocks ceased to be deposited more recently than about 1.9 billion years ago, indicating that, by then, atmospheric oxygen levels were permanently high.

EARTH TURNS COLD

The Proterozoic Eon rocks have recorded a more detailed picture of Earth's geologic history as they are far more abundant, and less distorted by metamorphism and alteration, than those of the previous eons. These rocks reveal that continents were rapidly growing by accretion, and splitting and reforming in supercontinent cycles. Sediments were also being laid down in shallow seas and the first glaciations were taking place.

There were at least four major Ice Ages during the Proterozoic Eon. These start with the Huronian from 2.4 to 2.1 billion years ago, and culminate with the Cryogenian (Sturtian–Marinoan) from 850 to 635 million years ago. This stage is also sometimes known as "Snowball Earth."

PROTEROZOIC LIFE BECOMES MORE ADVANCED

The rising oxygen levels created by the cyanobacteria also created an ozone layer that blocked out harmful ultraviolet radiation from the Sun, making it safe for more complex life to evolve.

It appears that the first complex cell may have evolved by means of a symbiotic relationship, called *endosymbiosis*. This theory envisages that larger prokaryote cells may have engulfed smaller highly oxidizing prokaryote cells, which became mitochondria (the cell's internal powerhouse).

Prokaryotes may have also engulfed photosynthetic prokaryote cells, which evolved to become chloroplasts (the plant cell's source of food). Thus a

more complex type of life form was born, the eukaryotic cell. Single-celled eukaryotes, such as acritarchs, evolved and bloomed in the Proterozoic Eon oceans, while the prokaryotic stromatolites continued to diversify, peaking in their abundance about 1.2 billion years ago.

The Proterozoic Eon comes to an end at the close of the Ediacaran Period, 542 million years ago. By this time, an abundance of bizarre-shaped multicellular organisms had evolved. However, being soft bodied, very little trace remains of them in the fossil record. At this time the very first of the shelled animals, *Cloudinia*, appeared, marking

a critical point in the fossil record. The Cambrian Period of the Phanerozoic Eon then opens with the evolution of most of the major animal groups, including numerous organisms with hard parts such as supporting frameworks, teeth-like structures and protective shells.

Some of the first Cambrian Period life forms included calcareous archeocyathid sponges and early arthropods (trilobites). One of the best-known localities where Cambrian life has been fossilized in abundance is the World Heritage-listed Burgess Shale site in British Columbia, Canada. Here, some unusual life forms have been unearthed.

BELOW Four major Ice Ages are known to have occurred on Earth since it first formed. The last Ice Age first commenced during the Tertiary some 15 million years ago, peaking during the Quaternary, which began 1.8 million years ago. Today, the ice caps of Antarctica (pictured) and the Arctic are all that remain of the ice sheets that covered up to 30 percent of the planet.

Geological Time

The vastness of geological time, from the formation of Earth 4.6 to 5 billion years ago to the present day, is a difficult concept to portray. We have no means to visualize such immensity of time. A human may reach the age of 100 years old if they are lucky. Europeans only reached the Americas a little over 500 years ago.

RIGHT Fossils were the original and most vital clues for setting the original geological time scale. Different types of fossils found at different layers reveal the relative ages of the rocks in which they are buried. Little is known about the earlier stages of the time scale, largely owing to a lack of fossils. This *Ornithomimus* dinosaur lived during the late Cretaceous Period.

There are various ways of representing geological time by comparing it with something familiar that we can visualize, such as a 12-hour clock face. Sometimes teachers use a 1,000-sheet roll of toilet paper stretched out in a long corridor as a representation of our 4.6 billion-year-old Earth, to teach their students about geological time. Using 920 sheets—practically the entire roll—each sheet represents 5 million years.

WALKING THROUGH GEOLOGICAL TIME

Starting at the beginning of Earth time, we walk past 85 percent of the length of the roll before we can mark on the first appearance of simple soft marine organisms at the beginning of the Ediacaran Period, about 635 million years ago. A little bit further on, there is a virtual explosion of life forms at the beginning of the Phanerozoic Eon and the Cambrian Period, some 542 million years ago. The evolution of teeth and other hard parts for attack, along with shells for defense, marked the beginning of an evolutionary "arms race."

Fossils become common in the stratigraphic record from this time onward—and it is usually only this part of the time scale that is shown in geology textbooks. Marking on all of the following geological periods, and the arrival and departure of key animal and plant types, students realize they are very quickly arriving at the end of the roll. The dinosaurs, together with numerous other animals and plants, die out at the end of the Cretaceous Period in a massive extinction event just 13 sheets from the end (65 million years ago).

The ancestral horse, which was about the size of a fox, appears a mere 10 sheets from the end. Whales, bats, and monkeys appear five sheets from the end, and our own genus, *Homo*, appears only a little less than half a sheet before the end. The earliest anatomically modern humans (*Homo sapiens*) appear in the last one-fortieth of the last sheet (130,000 years ago), so our written history comes down to no more than the tangle of fibers at the very end of the last sheet.

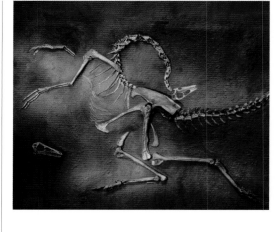

ABOVE The first primitive wingless insects appeared during the Devonian Period. Modern insects such as butterflies, dragonflies, ants, termites, and grasshoppers emerged in the Cretaceous Period as the newly evolved flowering plants proliferated, providing an abundance of food in the form of nectar. Insects now make up over 90 percent of all known animal species.

RELATIVE VERSUS ABSOLUTE TIME

Despite the development of a complex time scale relating the age of one region to another, no one had any idea about the real age of Earth in absolute numbers of years until "geological clocks" were discovered within different rocks. These "rock clocks" are based on the decay of radioactive elements (such as uranium) locked inside crystals of common minerals (such as zircon) that are found in many rocks. As we know the exact rate at which radioactive elements decay, we can determine the age of the minerals containing them simply by measuring the ratio of the parent element (say uranium) to its daughter decay product (say lead). The older the rock, the greater amounts of daughter elements it contains. The oldest known zircon crystal formed 4.4 billion years ago.

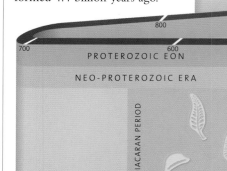

900

800

700   600

PROTEROZOIC EON

NEO-PROTEROZOIC ERA

EDIACARAN PERIOD

CAMBRIAN PERIOD

630   542

DIVIDING GEOLOGICAL TIME

The time scale shows how we divide geological time into *eons*, and smaller subdivisions called *eras* and *periods*. These are based on the appearance and disappearance of important life forms. At first the periods were arbitrary divisions, often named for regions of the United Kingdom where certain fossil types had been found. Boundaries were placed between one rock unit and another with different fossils. These fossil types were tracked over Europe and then linked with the rest of the world. It was only realized much later that many of these period boundaries represented a major, often worldwide, catastrophe, known as a *mass extinction*, in which most of life on Earth was killed, providing the opportunity for entirely different species to evolve.

ABOVE Any rock strata reveals a piece of geological history. The rocks in Andrew Gordon Bay on Baffin Island in Nunavut, Canada, tell a unique story of glacial and tectonic events. This island has metamorphic bedrock dating back to the Precambrian, with areas of overlying sedimentary rock laid down as recently as the Quaternary Period, thus encapsulating the bulk of geological time. Much has been revealed about Ice Ages in these rocks.

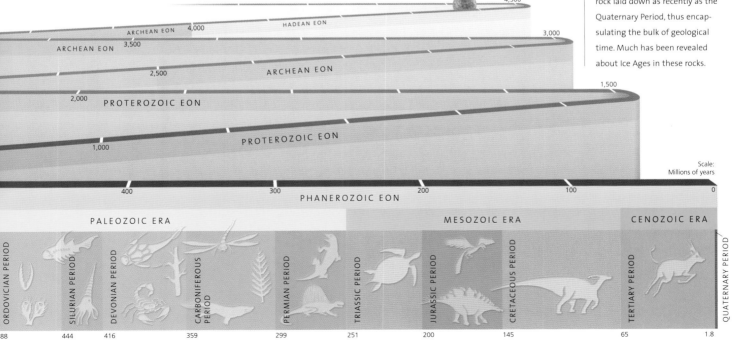

Precambrian Time

Precambrian time is an informal term covering Earth's first three eons of existence; the Hadean, the Archean, and the Proterozoic. These eons extend from Earth's mysterious birth, 4.6 to 5 billion years ago, to the beginning of the Cambrian Period, 542 million years ago, and collectively they cover the very beginnings of life.

Life had to bide its time as Earth differentiated itself into a dense core of iron surrounded by a mantle of lower density during the Hadean or "rockless" Eon. Cooling eventually led to the formation of a thin and highly mobile crust. Then, following the meteorite bombardment of the Archean Eon, life first appeared in the early oceans. It is thought that Earth's early atmospheric gases—such as ammonia, hydrogen sulfide, and hydrogen—combined with carbon dioxide to form the first simple organic compounds. Some scientists speculate that life may have been brought to Earth by the incoming meteorites.

FIRST LIFE ARISES WITHOUT OXYGEN

The simplest of single-celled prokaryotes (cells with no nucleus) are the anaerobic ("without oxygen") methanogens, which grow by using hydrogen for energy and carbon dioxide as their carbon source. They produce methane, or marsh gas, during this process of anaerobic respiration, and still survive today in places where there is little oxygen, such as swamps, hot springs, hydrothermal vents, and within the digestive tracts of animals.

About 3.5 billion years ago, primitive cyanobacteria (simple prokaryote cells) began photosynthesizing sunlight and producing oxygen. Increasing rapidly in numbers, these simple bacteria gathered in crude mat-like colonies, and began building rounded mounds called stromatolites. Depositing thin layers of calcium carbonate, they built huge reefs that became Earth's first limestone rocks, today found in Western Australia and South Africa. Although superceded by modern corals, a living stromatolite reef can still be seen at Shark Bay in Western Australia.

The story of atmospheric oxygen build-up is recorded by the layers of red jasper and metallic hematite (iron oxide) in the world's banded-iron formations. These rocks tell of the seasonal variations in oxygen levels as the oxygen-producing bacteria alternately flourished, then killed itself off by producing too much toxic oxygen. Excess

RIGHT It is unknown how Earth appeared during the Precambrian, but the first part of the time period was devoid of both rocks and life. Around the time the first rocks appeared, the first life evolved in the seas. After photosynthesis began, they began creating an oxygenated atmosphere that would later support all the complex forms of life on Earth.

oxygen combined readily with iron to form these hematite layers. Oxygen finally reached stable levels about 1.9 billion years ago. Other evidence for oxygen in the atmosphere includes the appearance of terrestrial Red Beds at about this time. These sandstones and siltstones are colored red by iron oxide that does not occur in older rocks.

An ozone layer formed, blocking out harmful solar radiation, and protists began to evolve. These are simple cells with a nucleus, but are classified as neither plant nor animal. Colonies of these simple cells saw different forms of algae flourish in the early oceans. Fossils from this time, however, are extremely rare because their lack of hard parts means they are not easily preserved. Eventually, more complex single cells evolved (eukaryotes, such as acritarchs) opening the way for the development of multicellular plants and animals of the late Proterozoic Eon.

THE SUPERCONTINENTAL MOTHERLAND

Rodinia was the first supercontinent; it was a lifeless expanse surrounded by the Mirova Ocean (Russian for "global sea"). Lack of fossil evidence makes it difficult to look any further back in time than about 1.1 billion years ago when Rodinia formed. The Proterozoic Eon saw this supercontinent break apart about 750 million years ago, with the Panthalassic Ocean (Greek for "all the seas")

ABOVE Along the shores of Shark Bay in Western Australia are some curious reef-like formations. These are stromatolites, living structures built by photosynthetic bacteria that were among the very first forms of life. Fossil stromatolites 3.5 billion years old have been found in central Western Australia.

forming between its spreading fragments. These fragments included the proto-Americas, Scandinavia, Siberia, Australia, and Antarctica. Icehouse conditions developed as these landmasses traveled out to the polar regions. This caused more extensive glaciations than at any other period in Earth's history, a time known as "Snowball Earth."

By the end of the Proterozoic Eon, the continental fragments came together to form a new supercontinent in the Southern Hemisphere, Pannotia (meaning "all southern"). The collision resulted in the Pan-African mountain-building event. The short-lived Pannotia supercontinent began to break up about 600 million years ago.

LEFT Seaweed is a type of algae, which arose in the Precambrian. There are as many as 40,000 species of algae, varying from microscopic to giant kelp forests. The process of photosynthesis—the conversion of sunlight into food—is believed to have started during the Archean Era as long as 3.8 billion years ago. Algae were the earliest plants to appear.

Ediacaran Period

The Ediacaran Period was a time when bizarre-shaped soft-bodied creatures dwelt on the sea floor, just prior to the explosion of skeletal life forms in the Cambrian Period. The Ediacaran is the last period of the Proterozoic Eon, and the International Commission on Stratigraphy only officially adopted the name in 2004.

The Ediacaran Period goes from 650 million years to 542 million years before present and is named after the type-fossil locality at Ediacara Hills in the Flinders Ranges, north of Adelaide in South Australia. Fossil sites from this time are extraordinarily rare because all of the organisms were soft-bodied and as such, are rarely preserved.

THE DISCOVERY OF THE EDIACARAN FOSSILS
Ediacaran fossils from Namibia were first described as far back as the 1930s, though their great age was not fully appreciated until 1946 when geologist Reg Sprigg discovered the abundant fossil "jelly-fish" site at Ediacara Hills. The discovery aroused international interest in this unusual fauna.

It was initially thought that the creatures were primitive versions of later animals, as the creatures from Ediacara Hills were fossilized in coarse sandstone, and so their fine features were not very well preserved. Guy Narbonne's later discovery of fossils at Mistaken Point in Newfoundland, Canada, allowed much finer details of the fauna to be seen and studied. At Mistaken Point, an entire seafloor community was preserved in situ beneath a layer of fine ash from a 565-million-year-old volcanic eruption. Similar fossils have been uncovered at about 30 widespread localities including England, Canada, Russia, Siberia, and the USA. The Ediacaran Period is often referred to as the Vendian Period owing to similar fossil types found west of the Ural Mountains in Russia. It appears that Ediacaran fauna were common in the oceans of this period, living on the continental shelves and slopes.

THE SOFT-BODIED EDIACARAN LIFE FORMS

Soft-bodied multicellular animal fossils from the rocks of this time were quite different from anything that lived during later periods. They had no shells or hard parts and disappeared mysteriously from the fossil record at the end of the period. This has led some scientists to speculate that this fauna may have been wiped out by an extinction event, possibly related to the major "Snowball Earth" glaciation that took place at the close of this period.

There are well over 50 distinct genera of Ediacaran organisms worldwide, but most are so peculiar in their appearance that they

cannot be easily placed into any group of living or extinct animals. Scientists have variously grouped the more unusual organisms in a separate class, phylum, or kingdom. Several have even proposed that some of the Ediacaran organisms belonged to kingdoms other than that of the animal kingdom, such as lichens, algae, or colonial protists. Controversy still reigns as to where they really belong.

The earliest Ediacaran fossils are of frond-like organisms that attached to the sea floor by holdfasts. They date back 575 million years. The largest of these was a flattened, oval-shaped, segmented creature called *Dickinsonia* that grew up to 3 feet (1 m) in length. Many of the Ediacaran creatures appear to be immobile "sack-like" or "air-mattress-like" shapes—even very large single-cells. These animals appear to have taken in food by direct slow absorption through their body walls, because there is no evidence of an advanced digestive system, such as a mouth, gut, and anus. They also lacked legs or jointed limbs. With a general absence of predators, it appears that there was no need for movement or the ability to quickly process nutrients.

LIFE BECOMES MORE DIFFICULT

Although many Ediacaran fossils are difficult to classify, about 560 million years ago, fossils such as *Kimberella, Parvancorina,* and *Spriggina* appear on the scene. Some of these resemble trilobites, and may be early examples of mollusks and arthropods. Feeding trails left in soft sediments prove that at least some creatures were becoming mobile.

Lines of fossilized fecal pellets have been discovered which show that some animals had one-way guts, and the ability to graze systematically for food. It is perhaps the evolution of significant predators that triggered the disappearance of the Ediacaran life forms, and the subsequent Cambrian explosion of different—and better adapted—life.

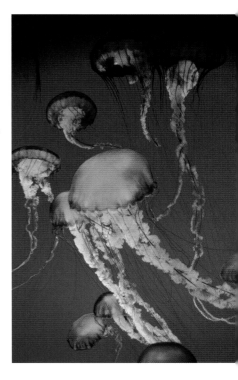

ABOVE The first primitive animals with a nervous system, the metazoans, evolved during the Ediacaran Period. Among the most ancient forms of metazoan life that still exist today are the cnidarians—the jellyfish and corals. However, Ediacaran life was distinctly different to anything still existing today and there are simply no living equivalents.

LEFT Wilpena Pound is a massive curved natural syncline in southern Australia. It contains the highest peaks of the Flinders Ranges. These ranges are also the site of the Ediacaran Hills, formed during the Ediacaran Period. This fossil site lies in the northern part of the ranges, around 400 miles (650 km) from Adelaide, the capital city of South Australia.

LEFT The bizarre and since unseen life forms that arose during the Ediacaran Period led to some unusual theories. One idea was that they were lichen, not animals, as their fossils were too well preserved to be soft-bodied organisms. Lichen would resist being compacted, explaining how fossils up to several feet across remained intact.

Cambrian Period

The Cambrian marks the opening of the Phanerozoic Eon and is one of the most remarkable periods in the evolution of life on Earth. Numerous varieties of hard-shelled animals appeared in the oceans for the first time. In fact, almost every animal phyla were in existence by the close of the period, 488 million years ago.

RIGHT A fossil *Saukia* trilobite, which first appeared in the late Cambrian Period. Trilobites were among the first arthropods (insects, crustaceans, and spiders) to evolve, and they had a massive radiation and abundance. Trilobites flourished during the Cambrian and Ordovician periods but became extinct by the end of the Permian Period.

BELOW Yoho National Park in British Columbia, Canada, is home to the fossil site Burgess Shale. It has yielded numerous fossils from the Cambrian, including many that went extinct by the end of this period. These delicate animals were preserved in seafloor sediments and later uplifted into the Rocky Mountains.

At the beginning of the Cambrian Period, 542 million years ago, the massive Proterozoic Eon supercontinent Rodinia was now starting to break apart into continental fragments.

Young shallow ocean floor and melting ice helped push Cambrian sea levels up over the edges of the continents, and all at once, the skeletal remains of life became abundant everywhere.

Hard shells are much more easily preserved in the fossil record than those of the rarely preserved soft-bodied life forms found from previous eons. It is perhaps because these types of animals made such a grand appearance in the fossil record that their sudden appearance in Cambrian strata led people to at first believe life originated during this time. Cambrian rocks were originally studied in Wales, so the Roman name for Wales, "Cambria," was adopted for this period.

THE GREAT CAMBRIAN "ARMS RACE"

The soft-bodied animals of the Ediacaran Period were swamped by the great "arms race" that was taking place around them. It was only when these immobile sacks of highly concentrated protein became food targets for other animals that the evolutionary race finally sped up.

Animals that could move were better advantaged than those that couldn't—but moving also required more energy and more fuel.

The ability of a cell to remove dissolved ions from water and secrete them as a hard compound, such as calcium carbonate (calcite) and silicon dioxide (quartz), became a great tactical advantage. Hard parts that could be used to scrape against the cell walls of other animals and rupture them had to be alternately matched by hard protective coverings or exoskeletons for protection from attack.

Improvements in defense, such as thicker shells and spikes, then had to be matched by more sophisticated methods of attack. Stinging cells in the cnidaria (jellyfish) became more sophisticated for both defense and attack. Sensitivity to light and shadows developed. Trilobites were among the first arthropods to develop proper eyesight, which assisted in their search for food. They had compound eyes with lenses of transparent calcium carbonate—an adaptation of their exoskeleton.

LIFE ON THE CAMBRIAN SEA FLOOR

Giant reefs built by primitive calcareous sponges (archaeocyathids) teemed with gastropod mollusks, echinoderms, graptolites, and more than 100 families of trilobites.

The Burgess Shale fossil site in British Columbia, Canada, has yielded much valuable information about life on the Cambrian sea floor. Worm-like *Pikaia*, the earliest known representative of the phylum Chordata (animals that have a nerve chord), swam through the water-filtering particles as it went. Strange spiny, 14-legged *Hallucigenia* and armor-plated *Halkieria* foraged on the sea bottom, together with the trilobites.

One of the largest Cambrian animals was *Anomalocaris*, a segmented arthropod-like predator, which reached lengths of over 6 feet (2 m), a staggering size for that time. It had grasping arms and a ring of strong sharp teeth that were capable of crushing the exoskeleton of primitive arthropods, such as *Marella*. All of these strange creatures were extinct by the end of the Cambrian Period. The trace fossil *Climactichnites* appeared in the late Cambrian, about 510 million years ago. These are possibly the tracks created by a slug-like creature foraging about for food in the wet sand, and indicate that certain animals may have already begun their transition from the marine environment to land by this time.

THE CAMBRIAN MASS EXTINCTION

It is not clear what disaster befell life in the marine communities of the continental shelf at the end of the Cambrian Period, but many species went extinct at this time. It is thought that ice sheets building up on the barren polar continents at this time and glaciers pushing seaward may have significantly reduced the ocean temperatures and their oxygen content. Entire species that were unable to adapt quickly enough to these changes would have died out.

The resultant dropping sea levels would have also exposed the continental shelves, reducing the availability of shallow-water habitats, and driven the Cambrian communities into increasingly restricted areas toward the continental slopes.

Whatever the exact cause, the fossil record tells us that the numbers of different species of trilobites, archaeocyathids (reef-building sponges), and brachiopods were severely depleted by this mass extinction event.

ABOVE Peripatus or velvet worms (Onychophorans) first made their appearance during the Cambrian Period. Although worm-like in appearance, they are more closely related to the arthropods, though they do have an ancestral link between these two groups. Despite being such ancient animals, modern-day forms have barely changed. Out of around just 200 known species, 74 occur in Australia, and most have Gondwanan ancestry.

Ordovician Period

The Ordovician Period was a time of great biodiversity, stretching from 488 million years to 444 million years ago. Life that survived the Cambrian Period mass extinction rebuilt as it was rafted away on the continental shelves of the separating continents. Life now had the opportunity to evolve independently into new forms.

RIGHT Mineral salt deposits and sulfur vividly color the crater of Dallol Volcano in Ethiopia, a scene perhaps reminiscient of the Ordovician Period. During this time, enormous quantities of mineral salts were deposited in evaporitic basins during times when the sea level fell.

BELOW During the Ordovician Period, enormous coral reefs thrived in the warm seas. By its end, the continents had shifted to the polar regions, and the seas cooled. During this glacial episode, global temperatures plummeted to their lowest level, resulting in a mass extinction that extinguished many of the tropical species of that time.

The name "Ordovician" comes from Ordovices, one of the Celtic tribes living in the British Isles before the Roman invasion in the first century. They inhabited a region in Wales where rock strata of this period occur. Geologist Charles Lapworth described these stratigraphic sequences in 1879.

Ordovician rocks are typified by thick beds of carbonate rocks, such as limestone and dolomite, which accumulated in shallow subtidal and inter-tidal environments. Interestingly, Mt Everest is also composed of Ordovician strata, and it is often Ordovician carbonates that are crushed for cement production. As the continental fragments moved apart, life diversified and specialized to its new climate and living conditions. Sea levels continually rose throughout most of the Ordovician Period, drowning much of the continental land surface.

LIFE FLOURISHES AND DIVERSIFIES

The appearance of planktonic graptolites (tiny colonial animals living in interconnected cup-shaped skeletons) marks the lower boundary of the Ordovician. They evolved and diversified into various forms during this period with some of the graptolite species quickly becoming extinct. When found as fossils, graptolites are very useful for accurately dating the rocks (often to within a million years) as well as for estimating ocean depth and temperature of the time.

Ordovician reef ecosystems were still dominated by red and green algae and stromatoporoids (colonial sponges); however, rugose and tabulate corals were rapidly increasing in number. Blastoids (stemmed echinoderms), bryozoans (lacy fan-like colonies of interconnected individuals), crinoids (sea lilies), as well as many kinds of brachiopods (shelled filter-feeders), bivalves (clams), and gastropods (snails) appeared in the geological record for the first time during the Ordovician Period.

Trilobites (segmented armored arthropods) took on numerous and distinctively location-dependent forms. They survived through to the end of the Permian Period before finally becoming extinct. Eurypterids (sea scorpions) and intelligent nautiloid cephalopods such as *Cameroceras* and *Treptoceras* (free-swimming, squid-like creatures with straight, cone-shaped shells) were the new predators of the Ordovician oceans. One variety, *Endoceras*, developed shells that grew up to 13 feet (3.5 m) long and 10 inches (25 cm) in diameter. They replaced *Anomalocaris* as king of the oceans.

Some of the oldest vertebrate chordates, the ostracoderms, also appeared at this time. They were jawless, armored fish with large bony shields protecting the head, and plate-like scales covering the rest of the body. Abundant teeth (conodonts) from unknown soft-bodied chordates are also commonly found in strata from this period, and they too can be used for dating rocks.

ORDOVICIAN MASS EXTINCTION

Although there were several minor extinction events during the Ordovician, the period ended with a catastrophic mass extinction event that caused the demise of about 60 percent of all marine genera and 25 percent of all families. The worst hit were the brachiopods, bryzoans, conodonts, graptolites, nautiloids, reef-building corals, and trilobites. This worldwide mass extinction, believed to have been caused by a glaciation event, led to a complete collapse of the reef ecosystem.

The largest supercontinental fragment at the time, Gondwana, moving over the South Pole during the late Ordovician, is believed to have

triggered the glaciation event. The outcomes were identical to the extinction event that terminated the Cambrian—ice sheets built on Gondwana caused a drop in sea temperature and oxygen levels, a fall in sea level, and a loss of shallow continental shelf habitat. Glaciation continued and became extensive, and global climate became colder and the seas choked with icebergs. As sea levels fell, massive salt and other evaporate mineral deposits formed as huge seawater lakes, left trapped in enormous land-locked basins, slowly evaporated. Australia's Lake Eyre basin is a classic example. Today such sites are excellent playas for petroleum exploration.

RIGHT Crinoids flourished during the Ordovician Period. A living crinoid is on the left and a fossil crinoid from the Ordovician Period is on the right. These filter feeders live on the sea floor.

Silurian Period

During the Silurian Period (444 million to 416 million years ago) sea levels rose as the massive end-of-Ordovician glaciers melted. The climate stabilized, allowing marine life to recover from the Ordovician extinction event and continue its diversification in the tropical seas. Primitive fungi and vascular plants appeared for the first time and arthropods began their advance onto the land.

BELOW The soft glow of iridescent algae on Turnagain Arm in Alaska, USA, could possibly resemble a Silurian landscape, where vast new mountain ranges were forming after the collision of several landmasses.

The Silurian was named after a Celtic tribe known as the Silures, who lived south of the Ordovices tribe in southern Wales. Here, back in the 1830s, geologist Sir Roderick Murchison described a system of strata. Then, in 1878, one of geology's founding fathers, Charles Lapworth, formally defined the Ordovician–Silurian boundary using the first appearance of the graptolite microfossil *Akidograptus ascensus,* which was found in deep-water black shale exposures at Dob's Lin near Moffat in southern Scotland.

By the Silurian Period, most of the Rodinian supercontinent had become fragmented. Gondwana remained positioned over the South Pole, while the other continental fragments migrated toward the equatorial region and began to collide and assemble into a large single continent known as Euramerica. These collisions thrust up the enormous Caledonian Mountain Range that ran from the eastern USA (the Appalachians), across Scotland, Ireland, England, Wales, Greenland, and western Norway, which were all then joined.

There was a vast north polar ocean and, despite Gondwana's high latitude position, the ice caps and glaciers melted. Greenhouse conditions prevailed, sending warm shallow seas back over the continents, forming a great unconformity. At the end of the Silurian, sea levels fell again, creating another series of evaporite basins and causing another minor extinction event.

LIFE REBOUNDS AND DIVERSIFIES

Despite their setback in the Ordovician extinction, the tabulate and rugose corals, and the calcareous sponges, recovered and continued to build huge limestone reefs throughout the period. There was also a widespread diversification of crinoids and a prolific expansion in the number of brachiopod types. Brachiopods became the most common hard-shelled organisms, making up about 80 percent of hard-shelled life. Trilobites, however, failed to recover and their numbers remained low. The sea scorpion predator, *Eurypterus*, continued to prowl the shallow seas, with some reaching the enormous length of several yards. Other marine fossils commonly found in Silurian strata include conodonts, mollusks, and stromatoporoids.

FISHES CONTINUE TO EVOLVE

The jawless cartilaginous fish that appeared in the Ordovician Period spread widely and rapidly through the Silurian oceans. Also for the first time, jawless fish, eurypterids, and xiphosurids (horseshoe crabs) began to colonize fresh and brackish water environments. The first bony fish, belonging to the Osteichthyes class, appeared and began to diversify rapidly. Known as acanthodians, they were covered with bony scales and developed movable jaws, which appear to have evolved from the supports of the front two or three gill arches. These bony fish remained in very low numbers until their proliferation later in the Devonian Period.

LIFE TAKES A FOOTHOLD ON LAND

Most Silurian plants belonged to the genus *Cooksonia*. These branching-stemmed plants with no leaves and no vascular tissue managed to gain a foothold on the otherwise barren land, perhaps as early as the Ordovician. These first plants were most likely to have initially colonized damp pockets adjacent to the shoreline in the moist regions near the equator. They probably evolved from marine green algae living in coastal rock pools.

One of the most significant advances to occur during the Silurian was that plants with a primitive but fully developed vascular system made their first appearance. These had true stems containing connective structures for transferring water and food between the leaves and roots. The evidence for this advance comes from the Silurian rocks of Australia where the fossil lycophyte *Baragwanathia* was found. The lycophytes are a division of plants that includes the club mosses and spike mosses.

At some time during the early Silurian, the first arachnids (spiders) and myriapods (centipedes) appeared. Aquatic arthropods, such as the eurypterids, followed the mosses and may have spent at least some part of their life cycle on land. Scorpions probably branched off as a lineage from the eurypterid stock at about this time.

ABOVE The Appalachian Mountains are the North American remains of a massive Silurian mountain range that existed after the supercontinent Eumerica was formed when Laurentia and Baltica collided. The other remnants are found in Europe.

RIGHT The first fungi appeared during the Silurian. Unlike plants, they germinate by emitting spores. Fungi have their own kingdom, which has over 100,000 known species of mushrooms, molds, mildews, stinkhorns, puffballs, truffles, and others. They vary considerably in size, from the microscopic to a soil fungus covering 30 acres (12 ha).

BELOW Mosses were among the very first plants to colonize the land. These non-vascular plants belong to the bryophytes group, which arose early in the Silurian. They most likely evolved from onshore algae that emerged from the oceans.

Carboniferous Period

The Carboniferous Period is known as the "Great Age of Plants." Massive tall evergreen forests and dense swamps inhabited by giant insects covered the continents. Atmospheric oxygen levels were incredibly high and living growth rates vigorous. Reptiles made their first appearance during this period, venturing further from their watery origins.

RIGHT First appearing in the Devonian Period, during the Carboniferous, the ammonites—a now-extinct group related to octopus, squid, and cuttlefish—thrived. These ocean-dwelling animals evolved more and more highly complex shapes and forms and were extraordinarily abundant during this period. Fossils ranging from the size of snails to some resembling large truck tires have been unearthed.

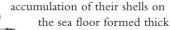

The Carboniferous Period extended from 329 million years to 299 million years ago, and was named for the thick and widespread coal deposits that formed in the dense forests of Western Europe and the United Kingdom during that time. It was the first geological period established, being recommended in a paper published in 1822 by English geologists William Conybeare and William Phillips. In North America, the Carboniferous is divided (at 318 million years) into an older sub-period, the Mississippian, and a younger sub-period, the Pennsylvanian. During the Carboniferous, the enormous continents of Euramerica and Gondwana continued their slow drift toward one another.

LIFE REBUILDS IN THE SEAS

Life flourished in the seas following its demise in the mass extinction at the end of the Devonian Period. The ray-finned bony fishes diversified enormously at the expense of the other fishes. The ammonoids, blastoids, brachiopods, bryozoans, and crinoids rebuilt their species numbers. Fusilinids (single-celled protists with a calcium carbonate shell) became abundant in the oceans. They were so numerous that the accumulation of their shells on the sea floor formed thick limestone strata. As such, the fusilinids became excellent guide fossils of the time, telling researchers whether they were looking at Carboniferous or Permian rocks. The trilobite, however, was not faring so well. Only one order, the Proetida, remained out of the previous nine trilobite orders.

GIGANTISM IN THE GREAT CARBONIFEROUS FORESTS

Life on land flourished considerably during the Carboniferous Period. The moist and swampy conditions encouraged the lycophytes (plants that use spores for reproduction) to diversify. The early insects from the Devonian had since evolved wings and continued to grow in size and fill many different environmental niches.

The burial of organically produced carbon in the swamps is believed to have pushed atmospheric oxygen levels to a peak of about 35 percent some 300 million years ago (as compared with today's 21 percent). This is thought to have been responsible for the huge increase in size of some insects and amphibians. The lack of predators around at this time may also have played an important role in allowing these insects to become so big.

The forests were alive with giant arthropods, such as mayflies, spiders, and scorpions. *Arthropleura* was a millipede-like creature that measured up to 10 feet (3 m) in length. The largest insect that ever lived was the giant dragonfly, *Meganeuropsis permiana*. It had a wingspan of over 2 feet (75 cm). These large species peaked at the end of the Carboniferous, coinciding with the highest oxygen levels, but had disappeared by the middle to late Permian Period, some 50 million years later.

LEFT Ferns and tree ferns flourished during the Carboniferous Period, along with other vascular plants, mainly the clubmosses and horsetails. Forests thrived, and with so many plants competing for sunlight they grew to heights of up to 150 feet (45 m).

THE IMPORTANCE OF THE AMNIOTIC EGG

Reptiles appeared in the Carboniferous, evolving from the Devonian amphibians, but competition from other amphibians meant that their numbers remained low until the end of the period.

However, despite being best known for being a time of vast forests and subsequent coal deposits, one of the greatest evolutionary achievements of the Carboniferous Period was the development of the amniotic egg. This type of egg has a tough water-impermeable membrane or a shell to prevent it from drying out. Such eggs could now be laid safely on dry land, allowing further colonization of the continents by four-legged animals. Reproduction was now possible on land without the need for water.

COAL AND THE CARBONIFEROUS

Much coal was laid down in the swamps of the Carboniferous where dense stands of spore-bearing lycopod trees grew to more than 100 feet (30 m) tall. Plants use the Sun's energy and carbon dioxide from the atmosphere to build their carbon-rich living tissue and release the excess oxygen into the atmosphere. When plants die and fall into swamps, thick deposits of vegetation can eventually build up. The carbon is prevented from oxidizing in the dank waters and the layers of dead vegetation are slowly compressed into solid coal.

ABOVE During the Carboniferous Period, swamps were widespread and the lush plant life that dominated the land at this time was to eventually be buried in these swamps and slowly fossilize into coal. This created many of the world's major coal-deposits that are mined today.

Permian Period

The Permian Period extends from about 299 million to 251 million years ago. On land, drier and more variable weather conditions prevailed, brought about by the coming together of the supercontinent Pangea. Amphibian numbers declined with the loss of their swamp habitat and reptiles became the dominant land animals. Coral reefs proliferated in the Permian seas.

British geologist Sir Roderick Murchison named the Permian Period after the ancient kingdom of Permia, near the present-day city of Perm in the Ural Mountains in Russia. He described the extensive sedimentary exposures overlying the Carboniferous rocks during a visit to the region in the 1840s. The Permian is the last geological period of the Paleozoic Era, which ended with the most savage of Earth's mass extinction events.

BELOW This almost intact fossil of a dicynodont was discovered in fossil beds in South Africa. It is a type of therapsid—a mammal-like reptile that evolved during the Permian. Therapsids did not survive the mass extinction at the end of this period.

AN ABUNDANCE OF LIFE ON LAND AND IN THE SEA

The biodiversity of plants, arthropods, and amphibians continued for some time into the Permian Period. However, the drying climate that occurred during the mid Permian reduced the extent of the giant swamp forests. This in turn forced the water-loving lycopods to shrink in size. The seed-ferns, conifers, and the ginkgos now began to fill the forests.

Amphibians decreased in both size and numbers while the reptiles diversified and began to occupy the new habitats. Their scaly skin and development of the amniotic egg enabled them to adapt more easily to the changes.

The reptiles evolved into pelycosaurs, such as the huge sail-backed *Dimetrodon*, and the therapsids (mammal-like reptiles). The therapsids may have even been warm blooded and had fur; they soon began out-competing and displacing the pelycosaurs. Near the very end of the Permian Period, the reptiles were joined by the first archosaurs—they were ancestors of the Triassic dinosaurs.

The Permian oceans teemed with life forms very similar to those of the Carboniferous Period. Great limestone reefs were built by organisms like algae, bryozoa, corals, foraminifers, sponges, and stromatolites, and provided shelter for ammonoids, brachiopods, gastropods, echinoderms, nautiloids, and fish. Trilobites continued their decline until they became extinct at the end of this period.

Ammonoids diversified and the shapes of their coiled shells and internal buoyancy chambers were becoming increasingly complex. Today, the numerous shapes of fossils of these long-extinct animals are used as a guide to the age of the strata in which they are found. Sharks ruled the oceans as the dominant predator.

FORMATION OF PANGEA, SUPERCONTINENT

By the early Permian, a supercontinent known as Pangea (which means "all Earth") had assembled, which included all of the continents that we know today. Pangea practically stretched from pole to pole and was surrounded by a single sea called Panthalassa (meaning "all oceans"). The giant landmass was shaped like a giant letter "C" which enclosed a smaller sea known as the Tethys (named after the Greek sea goddess Tethys). As Pangea came together, there would have been a drop in sea level and a decline in warm shallow shelf environments. The interior of this vast continent would have turned dry with great diurnal and seasonal temperature fluctuations owing to the increased distance from the moderating effects of the sea—and this was perhaps also a contributing factor to the extinction of many life forms at the end of the Permian Period.

THE MASSIVE PERMIAN EXTINCTION

The Permian Period ended with the worst catastrophe ever suffered by life on Earth. Some 90 to 95 percent of marine species (53 percent of marine families) became extinct along with 70 percent of all terrestrial species, including plants, insects, and vertebrate animals. The coral reefs that had been rebuilding since the Devonian Period extinction, and their marine communities, were devastated. Life forms that were wiped out forever include the blastoids, fusulinid foraminifera, rugose corals, tabulate corals, and trilobites.

On land, the reign of the therapsids ended. Around the world, fossil-rich "death beds" mark the end of the Permian Period. The land and seas were once again practically empty, ready for the

new surge of Mesozoic Era life. Severe global climatic changes probably caused this devastating extinction. Drying of the land shows in the geological record as thick strata of desert sand dunes and evaporites. However, the exact reasons for this are still not clear. Gas and ash injected into the atmosphere from volcanic eruptions in Siberia and China may have been responsible. Glaciation on Gondwana, similar to that which caused the earlier Ordovician Period and Devonian Period mass extinctions, is also a possibility.

ABOVE The first fossils of leaves are known from the Permian Period, and one of the primary trees to thrive during this period was the ginkgo. Today there is just one species, *Ginkgo biloba*, which survived only in certain mountainous regions of China.

LEFT The Stone Forest in the Yunnan Province of China is part of the Lunnan karst. This Permian landform is one of the world's foremost examples of karst formation; the outcrops are comprised of nearly pure limestone sculpted into some bizarre and unusual formations.

BELOW During the Permian, all lands were joined up into the supercontinent Pangea. The western parts and center of this vast continent were extremely arid, which enabled arid-adapted reptiles to begin to dominate.

Triassic Period

The Triassic Period, stretching from 251 million to 200 million years ago, heralded the beginning of the Mesozoic Era. It was a time of rebirth, following the crushing extinction that wiped out nearly all life on Earth and terminated the Paleozoic Era. It took 150 million years for life to rebuild its former diversity.

ABOVE *Herrerasaurus* was one of the first known carnivorous dinosaurs. This skeleton was found in late Triassic Period fossil beds at Ischigualasto in Argentina. There were very few dinosaurs at this time and most were diminutive; *Herrerasaurus* was up to 15 feet (5 m) in height.

BELOW Early on in the Triassic, forests were dominated by cycads, conifers, and ginkgos, which thrived in the northern regions. Toward the end of the period, however, conifers such as these modern-day monkey puzzle trees (*Araucaria araucana*) were more representative.

Friedrich Von Alberti named the Triassic Period in 1834 for the three distinctive strata, the Trias (Latin, meaning three), which are found throughout Germany and northwestern Europe and belong to this time. These are red sandstone, overlain by chalk, which in turn are overlain by black shale.

PANGEA BEGINS TO BREAK UP

At the start of the Triassic, the world was dominated by the giant supercontinent Pangea that lay straddling the equator. Being such an immense piece of land, Pangea controlled the global climate. Monsoonal conditions prevailed along its coastline, which was dominated by forests of conifers and cycads. Meanwhile, Pangea's enormous interior, away from the moderating effects of the ocean, was hot and dry. Thick evaporite deposits were laid down as great salt flats in its vast inland basins.

By about the mid Triassic, convection currents in the mantle beneath the supercontinent had started to tear it apart. Initially Pangea broke into two—Gondwana (South America, Africa, India, Antarctica, and Australia) in the south, and Laurasia (North America and Eurasia) in the north.

There was much geological activity associated with Pangea's breakup. Extensive basalt outpouring along the rift zone occurred, plus subduction of the ocean floor along the leading edges of the spreading plate fragments. In the mid to late Triassic, a long volcanic mountain chain arose along the west coast of the Americas, extending from Alaska to Chile. This was the beginnings of the Andes and the North American Cordillera.

REPTILES DOMINATE THE LAND, SEAS, AND SKIES

The reptile survivors of the Permian Period extinction went on to dominate the Triassic landscape. The archosaurs—reptilian ancestors of crocodilians, dinosaurs, and birds—quickly diversified. In the arid conditions, they out-competed the warm-blooded mammal-like therapsids. Interestingly, the arid conditions are thought to have driven many reptiles back into the sea. New semi-aquatic and aquatic groups appeared on the scene, including phytosaurs (semi-aquatic crocodile-like reptiles), placodonts (mollusk-eating marine reptiles), proto-crocodiles, and turtles. Early plesiosaurs, ichthyosaurs, and nothosaurs evolved, with their legs becoming more and more flipper-like. They went on to dominate the Jurassic seas.

Reptiles also adapted their limbs for flight. Thin membranes of skin that stretched between the fingers, arms, and bodies evolved, allowing them to take to the skies as the pterosaurs.

A few mammal-like reptiles, such as the barrel-chested dog-like *Lysrosaurus*, survived from the Permian Period and managed to spread widely early in the Triassic, and the first true mammals had evolved by the end of the period. However, these mammals, such as the nocturnal rodent-like Megazostrodon, remained small and obscure until the end of the Mesozoic Era.

Dinosaurs appeared by the mid Triassic and, by becoming faster and more efficient, they eliminated most of the larger reptiles. Although many of these early dinosaurs became

extinct at the end of the period, a few key survivors went on to dominate the Jurassic Period.

In the seas, the first of the modern scleractinian corals appeared in the niche which was previously occupied by the rugose and tabulate corals. The first belemnites—relatives of the modern-day cuttlefish—appeared, and the ammonites continued to evolve into more complex forms.

EXTINCTION ENDS THE TRIASSIC

Life suffered another setback at the end of the Triassic Period. It is not clear what precipitated this extinction event, though significant climatic changes are again implicated. Massive volcanic eruptions around 208 to 213 million years ago associated with the fragmenting Pangean supercontinent may have been responsible. The meteor that created the 210-million-year-old Manicouagan impact crater in Quebec, Canada, has also been suggested as a possible trigger.

Around 35 percent of all animal families died out, including most of the remaining therapsids, the non-dinosaurian reptiles, the early marine reptiles, and all of the remaining large amphibians. The mammals, ichthyosaurs, plesiosaurs, and the more adaptive dinosaurs survived. The Triassic extinction allowed the dinosaurs to become more abundant and diversify into many unoccupied niches during the Jurassic Period.

LEFT The trees of Arizona's Petrified Forest National Park were alive in the Triassic Period. This now-arid region was then a forested, muddy, tropical floodplain grazed by armored reptiles such as *Desmatosuchus* and gigantic herbivorous mammal-like reptiles like *Placerias*.

BELOW With the supercontinent Pangea lying across the equator during the Triassic Period, climatic conditions were monsoonal on the coasts but hot and dry in its vast interior. Here, numerous salt basins formed, leaving behind some of the world's significant evaporite deposits. This soda lake is in California, USA.

Jurassic Period

The Jurassic Period lasted from 200 million to 146 million years ago, marking the rise of the dinosaurs as the new rulers of the land. Spreading Pangean continental fragments carried their parcels of life away from one another into new environments where they adapted and evolved in isolation, forcing a new burst of terrestrial biodiversity.

ABOVE In 2006, the discovery of a beaver-like fossil, *Castorocauda lutrasimilis,* in northeastern China rewrote the history books on mammalian evolution. It is the oldest swimming mammal ever discovered and also revealed that furred mammals occurred 40 million years earlier than earlier fossil finds had shown.

The Jurassic Period was named after the Jura Mountains, which lie along the border between France and Switzerland. It was here that French mineralogist, chemist, and naturalist Alexandre Brongniart first studied the extensive marine limestone exposures of this age. The period was popularized by the success of the movie *Jurassic Park,* even though several of its "stars" did not actually evolve until the Cretaceous Period.

PANGEA'S BREAKUP LEADS TO WARM SEAS AND ABUNDANT LIFE

The Triassic breakup of the supercontinent Pangea—which began as the long, narrow proto-Atlantic seaway between Laurasia and Gondwana—carried on. It was accompanied by extensive basaltic volcanism. The large continents, in turn, fractured into smaller continental pieces, surrounded by numerous narrow seaways.

This meant that the enormous climatic variations of the previous period were moderated. The huge inland deserts and salt basins disappeared and were replaced by temperate and subtropical forests. The warming climate caused the polar ice caps to melt and the sea level to rise.

RIGHT This extraordinarily well preserved specimen of *Archaeopteryx lithographica* was found in Solnhofen, Germany, in 1876. Remarkably, its feathers were preserved, including its flight feathers. Being nearly identical to other maniraptoran dinosaurs such as *Velociraptor,* it was clearly bird-like and the numerous *Archaeopteryx* fossils found at this site revealed an evolutionary path from dinosaurs to birds.

Flooding of the continental shelves allowed marine life to diversify, and plankton thrived in the warm shallow waters. Dead plankton accumulated on the bottom in huge numbers, forming today's petroleum fields in the North Sea and the Gulf of Mexico. Scleractinian coral reefs proliferated, abounding with ammonites, belemnites, bivalves, bryzoans, gastropods, and sponges. The top marine reptile predators were the ichthyosaurs, pliosaurs, long-necked plesiosaurs, and crocodiles.

JURASSIC DINOSAURS

On land, dinosaurs became both diverse and abundant. They differed from other reptiles in the structure of their pelvis. Dinosaur hip sockets allowed the hind limbs to extend beneath the body, giving them a fully erect posture and making them faster and more efficient. Most other reptiles have limbs that project sideways from their body, giving them a sprawling posture like a crocodile.

The sauropods were immense plant-eating dinosaurs, and the largest land animals of all times. These include *Apatosaurus* (formerly *Brontosaurus,* with four massive legs and an incredibly long neck balanced by a long tail—from North America), *Brachiosaurus* (found in North America and Africa, with forelimbs distinctly longer than hind limbs), *Camarasaurus* (the most common of the large sauropods from North America), and *Diplodocus* (with a long whip-like tail). *Brachiosaurus* reached lengths of over 80 feet (25 m) and was considered the largest dinosaur until the discovery of a more massive plant-eating dinosaur in South America.

Argentinosaurus appeared during the Cretaceous Period, reaching lengths of 115 feet (35 m) and weighing up to 100 tons. The sauropods roamed the Jurassic forests, grazing on ferns, cycads, or conifers, while keeping a watchful eye out for the predatory theropods, such as *Allosaurus* (with sharp slashing claws on its small forelimbs, strong jaws, and serrated teeth), *Ceratosaurus* (with a short horn on its snout, large eyes, four-fingered slashing claws, and sharp teeth), and *Megalosaurus* (sharp three-fingered claws, strong neck, and powerful jaws). These smaller meat-eaters were more intelligent and may have hunted in packs.

ARCHAEOPTERYX AND PRIMITIVE BIRDS EVOLVE

Late in the Jurassic Period, small nimble coelurosaur dinosaurs appeared in the fossil record. They had feathers for insulation and may have been warm-blooded. These probably evolved into the first birds, such as demonstrated by the well-known *Archaeopteryx*. This "missing link" fossil was about 1½ feet (0.5 m) long, with feathers and other features similar to those of modern-day birds. However, it also had small teeth as well as a long bony tail, similar to other dinosaurs of the time. The first complete *Archaeopteryx* specimen was found in a limestone quarry near Solnhofen, Germany. Its description in 1862 followed only two years after Charles Darwin's book, *On the Origin of Species*, and sparked a massive public debate on evolution that continues to this day. However, these feathered forms remained relatively obscure as the pterosaurs continued to rule the skies until the close of the Mesozoic Era.

ABOVE A ribbon of black dolerite runs through the dry valleys in Antarctica, stretching for about 1,900 miles (3,000 km) along the Transantarctic Mountains. These intrusions are associated with the early stages of the breakup of the supercontinent Gondwana during the Jurassic Period. Corresponding dolerite cliffs in Tasmania are where the final hinge snapped, separating Australia from Antarctica.

Cretaceous Period

The Cretaceous Period runs from 146 million years to 65 million years before present, and brings the Mesozoic Era to a close. It is a period when dinosaurs continued their rule of the land, and the giant reptiles dominated the sea. Angiosperms (flowering plants) increased in abundance and gradually displaced the long-established gymnosperms (non-flowering plants).

RIGHT This dragonfly fossil was found near Sihetun, a village in Liaoning Province in northeastern China. This site has become a fossil-hunting paradise as the quarries here have produced rare and spectacular fossils of birds complete with feathers.

BELOW The Cretaceous Period was named for the chalk cliffs along England's southern coast. Global temperatures were high and the seas warm; supercontinents were rifting apart, and in the newly forming seas, vast quantities of remains deposited on the sea floor eventually fossilized into this limestone.

The Cretaceous takes its name from the Latin *creta* meaning "chalk," for the thick extensive chalk beds of Europe and the United Kingdom (including the well-known White Cliffs of Dover). Belgian geologist Jean d'Omalius d'Halloy defined the Cretaceous in 1822, based on strata from the Paris Basin.

The Cretaceous climate was uniformly warm and the poles were free of ice. Vigorous tectonic activity and young, high, mid-oceanic mountain ranges associated with continental rifting pushed the shallow tropical seas up over large areas of the continental crust.

These conditions were very favorable for life in the oceans. Calcium carbonate plates around the algae cell walls, known as coccoliths, accumulated on the sea floor in greater numbers than during any other period, and are responsible for the famous chalk beds. Widespread black-shale deposits also indicate that the oceans were calm and abounded with life. The continents continued to move toward their current configuration, and differences between the flora and fauna of the northern and southern continents continued to increase. Warm-climate plants thrived in Alaska and Greenland at this time, and dinosaurs roamed to within 15 degrees of the Cretaceous South Pole.

MARINE MONSTERS AND FLOWERS

The warm Cretaceous oceans continued to harbor an abundance of life as they had done during the Jurassic Period. The great marine reptiles—the ichthyosaurs, plesiosaurs, and mosasaurs—hunted

fish, ammonites, belemnites, and other mollusks. Rays, modern sharks, and ray-finned (teleost) fishes became common. By the close of the Cretaceous, teleost fish dominated both ocean and freshwater habitats. Other curious Cretaceous animals included the straight-shelled ammonite, *Baculites*, and the toothed Hesperornithean birds. These birds were largely flightless, but would swim and dive in the sea to catch their prey. Echinoderms, such as sea urchins and sea stars, and diatoms (single-celled plankton with siliceous rather than carbonaceous shells) also proliferated in the Cretaceous seas.

A particularly significant evolutionary step was the appearance of the angiosperms, or flowering plants, around 125 million years ago. By the close of the Cretaceous, numerous recognizable flowering forms had evolved. Angiosperms arose in co-evolution with the insects, and their diversification led to the appearance of the ants, aphids, bees, butterflies, termites, and wasps as the Cretaceous Period progressed. Magnolia trees become predominant near the end of the Cretaceous, their flowers being fertilized by bees.

Mammals still remained small and insignificant as the ruling dinosaurs had now attained their peak diversity. The best-known of the dinosaurs—*Tyrannosaurus rex*, *Triceratops*, *Velociraptor*, and *Spinosaurus* —roamed the continents at this time. At Liaoning (the early Cretaceous Chaomidianzi formation in China), coelurosaur dinosaurs have been found with hair-like feathers. Pterosaurs were commonplace, but faced increasing competition from the birds as the Cretaceous progressed.

THE CRETACEOUS MASS EXTINCTION

The last major mass extinction on Earth marked the end of the Cretaceous Period and concluded the Mesozoic Era. In total, about 50 percent of all species, and about 25 percent of known families, disappeared. The dinosaurs, the marine reptiles, and the flying pterosaurs all ended their lengthy rule. Marine life, especially the foraminifers and ammonites, was the hardest hit, while land plants were the least affected. Flowering plants, amphibians, crocodilians, lizards, and snakes were unaffected, and mammals went on to occupy many of the niches vacated by the dinosaurs.

Anomalous amounts of the metal iridium found along the boundary layer between the Cretaceous and the Tertiary periods—often referred to as the K–T boundary—implies that a natural disaster of climate-changing proportions was responsible. The two candidates are the Chicxulub meteorite, with an estimated diameter of 65 miles (100 km) that struck the Yucatan Peninsula, Mexico, or the massive basalt outpourings of the Deccan Traps in India. Both of these events occurred at about that time.

ABOVE The first flowering plants appeared during the Cretaceous Period. Among the earliest known flowers were water lilies and magnolias. Redolent of sweet nectar, flowers were a rich food supply for the abundance of insects that soon followed.

Tertiary Period

The Tertiary Period heralds the dawn of the Cenozoic Era and the age of mammals. It stretches from 65 million years to 1.8 million years before present. Open woodland and grassland plains supporting large grazing animal populations replaced the forests as the climate cooled and became more seasonal. The first primates arose.

The Cenozoic (sometimes called the Cainozoic) means "era of new life" and sees the continents move into their present positions. It follows on from the massive Chicxulub meteorite impact in the Yucatan Peninsula, Mexico, and extensive outpourings of basalt in the Deccan Traps, India, that most likely ended the Mesozoic Era.

Giovanni Arduino first introduced the name Tertiary in 1760, when he divided geological time into three—the primary (primitive), secondary, and tertiary periods. A fourth, or quaternary, period was added later. In 1828, Charles Lyell built on this framework for his own, more detailed, classification system. The primary became the Paleozoic

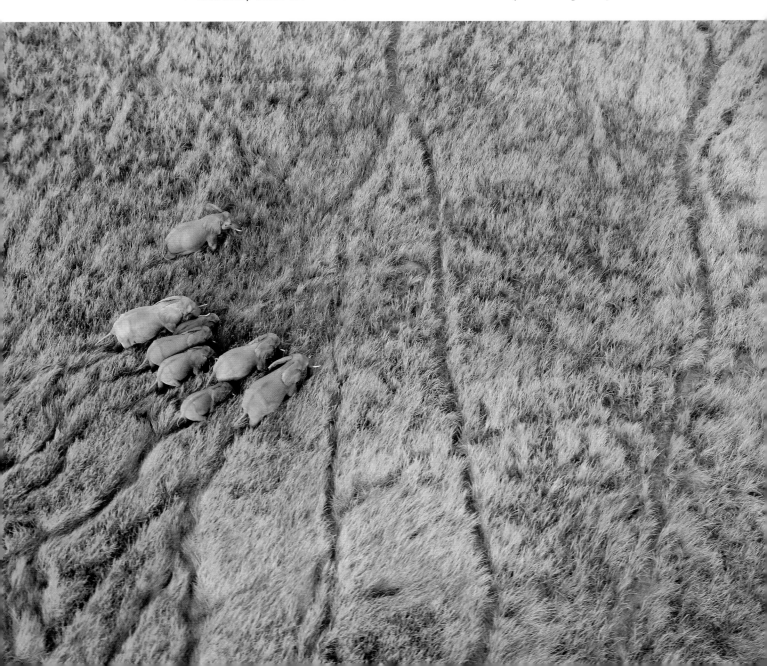

Era and the secondary became the Mesozoic Era, and each was broken into shorter periods. The International Commission on Stratigraphy does not use the term Tertiary. The time interval is instead represented by the Paleogene Period (65 to 23 million years ago) and the Neogene Period (23 to 1.8 million years ago). These terms come from Greek, meaning "ancient born" and "new born" or "a new start," respectively.

RISE OF THE MAMMALS

After hiding in the shadows of the forests for millions of years, the mammals adapted to the vacant niches left by the dinosaurs and diversified into several branches during the Paleocene Epoch. Today, these groups are represented by the egg-laying monotremes (platypus and echidnas), the pouched marsupials (such as kangaroos and wombats), and the placentals (such as monkeys, bat, and cows). The rodent-like multituberculates became extinct. The placentals became the most dominant group, outcompeting the other groups whenever they have come into contact.

The grasses appeared and have continued to expand right through to the present. They represented an amazing new renewable food source for the herbivores. As grasses grow from their roots,

RIGHT Sculpted into some bizarre shapes by the erosional forces of the waves, this marine porcelanite and chert on the Marin Headlands at Ano Nuevo State Reserve in California dates back to the Miocene Epoch in the Tertiary Period.

rather than the tip of the plant, they do not die when their tops are grazed off. Grazing herbivores adapted to life on the open plains and evolved strong grinding teeth and powerful stomachs suited to their diet of grass.

Many other symbiotic relationships evolved as plants and trees continued to diversify. During the Oligocene Epoch, the ancestors of modern elephants and rhinoceros in Africa grew to a large size, and the first primates appeared (the suborder Anthropoidea, which later branched into monkeys, apes, and humans). A major expansion of grasslands occurred in the cooler and drier climate of the Miocene Epoch, leading to selection of the ruminant herbivores (with compartmentalized stomachs suited to grass), such as antelope and cattle. These closely resembled today's forms. Different predators evolved to take advantage of the grazing herds, including bears, wolves, saber-tooth cats, and pack-hunting dogs.

During the Pliocene Epoch, the global climate became cool and dry. Tropical rain forests shrank back to just a band around the equator, allowing savanna, grassland, desert, deciduous and coniferous forest, and tundra environments to expand into their present configuration.

ABOVE This fossilized sassafras leaf was found in fossil deposits in Washington State, USA, and is dated to the Eocene Epoch of the Tertiary Period. Sassafras belongs to the laurel family.

LEFT As the flowering plants expanded into new niches, by the end of the Tertiary Period, grasses were flourishing. They, in turn, set the scene for the expansion of grazing mammals, which took up the gap left by the extinction of the dinosaurs. They would evolve into the more familiar animals such as elephants, deer, and horses.

EPOCHS OF THE TERTIARY

The Tertiary Period is divided into five epochs, each with its own important marker events:

PALEOCENE (65 to 56 million years). An early increase in fern numbers occurs, perhaps suggesting recovery from Chicxulub meteorite-impact wildfires. Open woodlands replace the previously dense forests as the climate now cools.

EOCENE (56 to 34 million years). The first modern mammals emerge at this time. The end of this epoch is marked by an extinction event called Grande Coupure ("Great Break" in continuity), related to impacts in USA and Siberia.

OLIGOCENE (34 to 23 million years). Antarctica breaks free from South America and begins to cool rapidly. Baleen and toothed whales appear. Riverbank grasses expand into new habitats.

MIOCENE (23 to 5.3 million years). Grasses continue to diversify, co-evolving with large grazing herbivores, including ruminants (cows, deer, giraffes, and their relatives). All modern mammal families were present by this time.

PLIOCENE (5.3 to 1.8 million years). The continents are now close to their present positions. Antarctica is completely covered with permanent ice. The Americas join, the Panama isthmus linking the two continents, leading to the extinction of South America's unique marsupial fauna. The circum-equatorial current is blocked, leading to the cooling of the oceans.

Quaternary Period

The Quaternary Period is the most recent and shortest period of geological time, beginning about 1.8 million years ago and bringing us to the present day. The continents are all essentially in their modern-day positions. Ice ages come and go, and the first close human ancestors appear, as well as today's modern animal forms.

RIGHT This fossil tooth along with a skull were excavated from a construction site in 2005 in Beijing, China. Experts believe that they were from a Paleoloxodon, also known as "straight-tusked elephant," a kind of vertebrate that went extinct before the Quaternary Ice Age.

BELOW During the Pleistocene Epoch, the first part of the Quaternary Period, global temperatures dropped and Earth was plunged into another Ice Age. Ice covered both polar caps, moving out over the adjacent seas and continents, and plants and animals had to adapt to the colder conditions.

The Quaternary Period (meaning "fourth") was proposed by Jules Desnoyers in 1829 to include sediments of France's Seine Basin that lay on top of the Tertiary (or "third") Period rocks. The Quaternary is sometimes known as "the age of humans," and roughly covers the time span of recent glaciations, including the last glacial retreat. The International Commission on Stratigraphy is still in the process of deciding whether to place the base of the Quaternary Period and Pleistocene Epoch at 1.8 million years or 2.6 million years before present.

SOME VERY COLD TIMES

The climate of the Pleistocene Epoch was one of periodic repeated glaciations. Alpine glaciers extended out from the mountain ranges and continental glaciers pushed equator-ward from the polar regions to as far as 40 degrees latitude. Large areas of North America, Europe, Asia, South America, New Zealand, and Tasmania were covered with ice, plus all Earth's major mountain ranges. The drop in temperature profoundly affected life on Earth. Animals, such as bison, musk ox, mammoth, reindeer, and rhinoceros, developed woolly coats for insulation against the cold. The hominids also evolved during this epoch.

Massive sea level falls accompanied the glacial maxima, causing interesting geographic changes. The straits of Bosphorus and Skagerrak emerged, turning the Black Sea and Baltic Sea into freshwater lakes. Britain and Europe were linked by a land bridge. The Bering Strait closed, forming a land bridge between Asia and North America. Today's northern American lakes (including the Great Lakes), Hudson Bay, and the Baltic Sea–Gulf of Bothnia are all depressions left behind after loading by the last glacial ice. These temporary features will shrink and drain as the land continues to rise isostatically.

HUMANS SPREAD OVER THE PLANET

The intelligent genus, *Homo,* appeared around 2 million years ago, and the first anatomically

EPOCHS OF THE QUATERNARY

The Quaternary Period is broken into two epochs:

PLEISTOCENE (1.8 million years to 11,500 years ago). The name of the epoch means "most new," and it covers the most recent episodes of glaciations when up to 30 percent of Earth's surface was covered by ice. Sea levels were very low. Rivers were more powerful and huge lakes covered the continents as ice blocked their outlets. The end of the epoch corresponds with the close of the Paleolithic age used in archeology.

HOLOCENE (11,500 years ago to the present). The name comes from Greek, meaning "wholly new." It starts with the last retreat of the Pleistocene Epoch glaciers, so it is considered the present interglacial period within the current Ice Age. Human civilization developed entirely within the Holocene Epoch. Continental areas above about 40 degrees north latitude are still rising today following melting of the Pleistocene ice.

modern human, *Homo sapiens*, at least 130,000 years ago. Human impact was most marked during the Holocene Epoch, when nomadic hunter-gatherer based groups began to change into sedentary agricultural communities. Development of a stable source of food meant that permanent cities could grow in size and civilizations develop. Civilization was accompanied by rampant population growth.

At the beginning of the Holocene Epoch there was a major worldwide extinction of large mammals, such as the saber-tooth cats, mammoths, mastodons, and the glyptodonts (giant armadillos).

In North America, horses, camels, and cheetahs all became extinct, and this has been attributed to the arrival of the ancestors of Amerindians. In fact, the proliferation of hominids on every continent corresponds with a significant drop in the number of mammal, flightless bird, and reptile species. At the present time, the combination of human population growth, agriculture, forestry, fishing, and industrial pollution is creating severe habitat loss for most other species. Extinction of living species is occurring today at a rate at least equivalent to that of any previous mass extinction event.

ABOVE Much of the landscape in the Yosemite Valley was sculpted by extensive glaciation that occurred during the Pleistocene Epoch in the Quaternary Period. The large areas of granitic rocks, such as these at Cathedral Rocks and Bridal Veil Falls, were originally laid down much earlier, during extensive volcanism from 20 million to 5 million years ago.

Tectonics: Earth's Powerful Driving Force

Plate tectonics provides our most up-to-date understanding of how Earth works. The theory proposed in 1912 to explain the amazing jigsaw fit of the continents is now a measurable certainty. Today, anyone with an accurate global positioning system can monitor the slow movement of their home across the surface of the globe.

In 1912, German climatologist Alfred Wegener studied the pattern of scratch marks (striations) left behind by Permian Period glaciers on various continents. He noticed that, if those continents were all once joined together in a tight cluster around the pole, the otherwise random striations all pointed outward in a radial pattern from the South Pole. Furthermore, this would also explain the perfect jigsaw fit of the Atlantic coastlines of Africa and the Americas. It all seemed so logical, but Wegener could not explain how or why such

a movement of the continents would have happened. Then the following year, physicist Arthur Homes realized that the heat escaping from Earth's central furnace must be driving a series of convection currents in the mantle, much like what happens in a pot of soup boiling on the stove.

THE MOVING CONTINENTS

It is the slow-moving mantle convection currents that push the tectonic plates around the globe—at about the rate at which a fingernail grows. For decades, this concept was considered to be preposterous by the wider scientific community. Then in the 1960s, echo sounders, developed from ex-World War II submarine-tracking equipment, brought about an explosion in understanding of the topography of the ocean floors. Mid-oceanic ridges—huge submarine mountain ranges—were discovered and mapped.

More information came to light to support the theory when the patterns of Earth's magnetic field fluctuations recorded in the ocean floor basalts were found to be perfectly symmetric about the ridges. Samples of basalt collected from the ocean floor were also found to be youngest closest to the ridge and progressively older toward the edges of the continents. It was realized Earth's crust is like the thin brittle shell of an egg. The plates grow and move apart at the mid-oceanic spreading ridges then eventually sink and disappear into the deep ocean trenches at a steady rate. The continents themselves are thicker and lighter areas of the

PLATE TECTONICS IN ACTION

The illustration below shows how convection currents rising and falling in Earth's interior drive the movement of its surface plates creating all of the landforms around us.

A. COLLISION When light thick continents collide, their edges buckle and fold, pushing up tall mountain ranges, such as the Himalayas and the Alps.

B. HOTSPOTS Enormous volcanic islands grow on the sea floor above deep mantle hotspots until they are moved from their source by the continual movement of the tectonic plate. As each volcano dies and sinks, new islands grow.

C. SUBDUCTION VOLCANIC ISLANDS Denser and thinner ocean crust pushes beneath lighter crust, such as younger ocean. The descending and melting oceanic plate gives rise to magma that rises to form arcs of volcanic islands, such as Indonesia and the Philippines.

D. MID-OCEANIC RIDGE New basaltic crust forms along the mid-oceanic ridges. Here, magma intrudes into the widening cracks and solidifies as the tectonic plates move apart.

E. SUBDUCTION VOLCANIC MOUNTAIN RANGES Continental crust easily overrides oceanic crust, which is forced to descend and melt. Molten magma rises to form chains of volcanoes like those of the Andes Mountains.

F. RIFTING Rifting occurs in the middle of large continents sitting on top of rising convection plumes. The land is put under tension, which causes it to crack and move apart.

A   B   C

crust that float in the mantle at a higher level than the surrounding sea floor crust, and hence make dry land. Once created, the continental crust, like the froth on boiling soup, is never dragged down and recycled into the depths of Earth's mantle.

FROM SUPERCONTINENT TO SUPERCONTINENT

It is now understood that the continents move in enormous supercycles lasting hundreds of millions of years. Supercontinents form when all of Earth's landmasses are joined together in one huge continent. But they don't stay together as a single landmass for long before it becomes unstable. A supercontinent acts like a gigantic thermal blanket, trapping Earth's internal heat underneath. This causes enormous plumes of rising mantle material to push upward against the supercontinent's base near its center. The convection currents gain strength and eventually they rip the continent apart into jigsaw pieces.

We can only see dimly back about 1.1 billion years into Earth's continental history. At this time, a supercontinent existed, called Rodinia (which is Russian for "the motherland") surrounded by the Mirova Ocean—Russian for "global sea." Earlier supercontinents probably existed before Rodinia, but the lack of fossils means that these cannot be pieced together. Rodinia broke apart about 750 million years ago. Its pieces reassembled to form Pannotia about 600 million years ago, which in turn broke up 50 million years later. The continental fragments came together again about 275 million years ago to form Earth's last huge supercontinent known as Pangea. Our present continents are all fragments moving outward from the breakup of Pangea. At a future time, all of today's continental fragments will come crashing together to form a new supercontinent.

Biodiversity, too, reflects the cycle of supercontinental formation and breakup, with the number of different species becoming more abundant at times when the continents are fragmented, and less

when the landmasses are joined. In recent geological time, a different species of big cat has evolved on each of four major continents—the lion in Africa, the tiger in India and Asia, and the panther in South America. These were all the same species at the time of Pangea's breakup.

ABOVE The same rifting process that causes these volcanoes to rip Iceland apart along the Mid-Atlantic Ridge caused ancient oceans to open and supercontinents to split apart and separate.

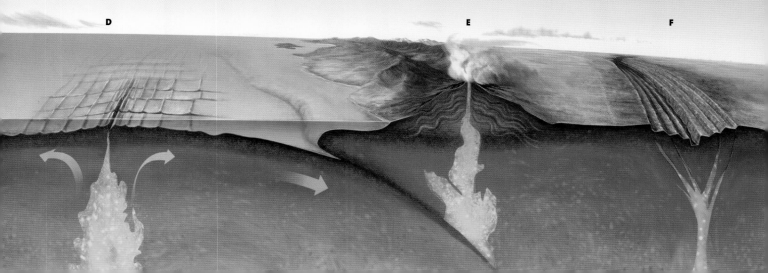

D   E   F

ABOVE The vibrant scarlet crazed pattern on the shores of Lake Natron in Tanzania is created by mineral salts and micro-organisms with a high rate of evaporation. The lake lies in the heart of the East African Rift Valley, which is slowly tearing Africa apart. Alkaline lakes are common in rift valleys.

SPREADING PLATES AND NEW OCEANS

Amazing landforms develop when a continent begins to break up. Initially a dome develops, creating a vast upland, before huge tensional cracks appear and open. Blocks of crust slide down along these faults creating deep flat-bottomed valleys, sometimes with their floors well below sea level. The great East African Rift and Lake Baikal are examples of where rifting is happening today. Basalt magma from Earth's mantle wells into the opening cracks, in places forming large volcanoes.

Year by year, the rift slowly widens, with each valley wall eventually destined to become the shoreline of a distant continent. A very interesting event occurs when the land-locked rift valley finally breaches the edge of the separating contin-ent and is flooded by the sea, either gradually in stages, or as a raging torrent. The narrow seaway continues to grow into a giant ocean, like today's Atlantic Ocean, with the shape of the initial crack preserved in the shape of the mid-oceanic spread-ing ridge and the shape of the continental edges.

appropriately named, the Pacific "Ring of Fire." Therefore, as the continents separate, the widening of the Atlantic Ocean is being balanced by the closing of the Pacific Ocean along zones of crustal destruction. The term "subduction zones" is used to describe how one plate, usually the oldest or most dense, slides beneath the other along an inclined plane. Numerous and often strong earthquakes mark this dipping subduction plane—sometimes also known as a megathrust.

Their source locations (foci) can be mapped to a depth of about 440 miles (700 km), at which point they die out, indicating that at this depth the down-going plate has begun to soften and melt. The molten products of the descending plate rise as huge buoyant bodies of andesitic magma and are responsible for the lines of active volcanoes that lie inland from, and parallel to, the deep trenches.

GIANT COLLISION MOUNTAIN RANGES

Earth's crust cannot so easily disappear when two plates of buoyant continental crust are being pushed together. In this case, neither plate is able to descend beneath the other. The result is a collision of gigantic proportions, with the edges of both continents crushing together, folding and buckling upward. The Himalayas are the best example of a collision mountain range forming today—a result of India ramming into the Asian continent. This commenced about 50 million years ago, after the last remaining part of the ancient sea, the Tethys, was swallowed up when the two continental edges first touched. Today, some of the Himalayan peaks reach over 5 miles (8 km) in elevation and are rising at a rate of over 1 inch (25 mm) per year. As India is far too light and thick to subduct beneath Asia, the subduction process will eventually jam up and stop. Stress will continue to build, however, causing the subduction zone to eventually "jump" to a new locality—most probably southward off the coast of India—and continue anew.

ABOVE Stratovolcanoes such as Karymsky volcano in Kamchatka occur along subduction margins where the subducted plate is "swallowed" by the continental plate, pushing up volcanoes. It lies on one of many subduction zones ringing the Pacific Ocean, in the Pacific "Ring of Fire."

SUBDUCTING PLATES AND RINGS OF FIRE

It was ultimately realized by scientists that if the planet is not to expand like a great balloon, then the crust must disappear at a rate equal to that which is being created at the spreading ridges. The crust is being destroyed along lines of deep submarine trenches that ring the globe—it descends into these trenches and is melted.

Most of Earth's trenches occur in a line around the edge of the Pacific Ocean. It is a zone marked by active volcanoes and many violent earthquakes,

ABOVE When the collision of the Indo-Australian Plate with the Eurasian Plate formed the Himalayas, the intensity of this impact forced the high level of folding of the Sulaiman and Makran Ranges along the western border of Pakistan.

SUPERCONTINENTS AND TECTONIC PROCESSES

The super cycles involve a sequence of tectonic regimes. During a supercontinent breakup, rifting environments predominate, followed by the development of passive plate margins and seafloor spreading. When the continents amalgamate, collisional environments become dominant. Such collisions are accompanied by mountain-building activity, such as the formation of South America's Andes, and volcanic activity around island arcs, such as that around the northern Pacific Rim, where the Pacific Plate is colliding with the Eurasian Plate. The Himalayas are still being up-lifted as the Indo–Australian Plate dives beneath the Eurasian Plate during this current supercontinent cycle phase. With Australia and Africa now moving steadily northward on their respective plates, we might be witnessing the next phase of supercontinent formation—and that might be only 50 million years away.

RIGHT When a large continental landmass straddles a pole, as does Antarctica today, it may be one of the significant contributing factors that leads toward the generation of icehouse planet conditions. Here, sea ice extends out into the Weddell Sea.

RODINIA, THE FIRST KNOWN SUPERCONTINENT

At present there is strong evidence for the existence of only four of the past supercontinents. Those that might have occurred earlier than Rodinia are difficult to determine because of lack of evidence from suitable fossils and other data. There is sufficient evidence to suggest that one of the oldest known, Rodinia, formed between 1.3 and 1.1 billion years ago. Rodinia, which was made up of almost all Earth's continental crust of the time, included Laurentia (most of which later would become North America), which formed the core of the supercontinent, together with what were to become Australia–East Antarctica along the western margin and Baltica–Amazonia along the eastern margin. Between these colliding masses was a third craton that would eventually become north Central Africa.

During the time of Rodinia, the climate was generally cold with icehouse conditions persisting from 850 to 630 million years ago. Evidence for this comes from the extensive deposits of glacially formed rocks of the time. Some scientists believe that back then Earth resembled a giant snowball. Rodinia finally rifted into eight smaller continents over the period from 830 to 745 million years ago. The opening that occurred along the eastern margin saw the formation of the initial Pacific Ocean and that which occurred along the western margin would become the Atlantic Ocean.

THE SUPERCONTINENT GONDWANA

About 750 million years ago, large parts of today's southern hemisphere continents formed one of the fragments from the breakup of Rodinia. These were grouped around a core landmass consisting of Australia, India, and East Antarctica. Around 520 million years ago, fragments carrying Africa and South America that had moved eastward around the globe docked with the core landmass to form a new supercontinent called Gondwana.

Altogether, it had taken almost 230 million years for the Rodinian fragments to amalgamate into the final assemblage of Gondwana—made up of the modern landmasses of South America, Africa (including Madagascar), India, Antarctica, Australia, and New Zealand. The eastern edge of the Precambrian rocks today marks the boundary, known as the Tasman Line, of what would have been the coastline of Gondwana.

BIODIVERSITY AND SUPERCONTINENT CYCLES

Apart from the increase in biodiversity that is imposed on life from isolation caused by fragmented continents and the decreased biodiversity associated with the formation of supercontinents, there are other factors that affect biodiversity on this tectonic scale. For example, when continents and oceans occur along north–south axes, there is greater isolation and thus more diversity than when they occur along east–west axes. This is because changes in climatic conditions over latitude would interrupt communication routes and so cause isolation. Consequently, zones that are oriented east–west would experience less isolation and thus reduced diversification and rate of evolution.

Isolation also has the effect of allowing the independent evolution of similar traits in unrelated organisms, for example, Australia's marsupial dog-like thylacine and a true canine such as the golden jackal, a placental mammal. This is known as convergent evolution, which results from organisms adapting to similar environments by independently evolving similar traits. Divergent evolution can also result from continental breakup. In this case, structures of common origin can evolve into very different forms over time. For example, the vertebrate limb evolved through a common ancestor, but has diverged in both its structure and its function.

Whale fins resulted from divergent evolution. Whales evolved from land mammals, possibly from an ancestor of the hippopotamus and cow, so their fins evolved from legs.

THE SUPERCONTINENT PANGEA

Between 350 and 255 million years ago, the land-masses forming the present-day northern hemisphere continents assembled to the north of western Gondwana. These included Europe, Russia, North America, Siberia, and Greenland, together known as Laurasia. They eventually collided with the western section of Gondwana that contained Africa and South America to form the supercontinent Pangea, which means "all lands."

By this time, life had evolved on land, and the supercontinent of Pangea, which stretched almost from pole to pole, gave the opportunity for wide-scale movement of life. Aquatic life also dispersed along the rivers, swamps, and inland seas that characterized Pangea. The lungfish, whose family evolved during Pangean times, is now found only in Australia, Africa, and South America, all of which formed the Gondwanan portion of Pangea.

The sea level was low over the time of the formation of Pangea as a super-continent, but rose rapidly during the Ordovician and Cretaceous periods, when rifting drove the continents apart. About 240 million years ago, Pangea began to break apart to form Laurasia to the north and Gondwana to the south, opening the North Atlantic Ocean and the Mediterranean Sea.

When eastern North America and northwestern Africa broke apart, the early Atlantic Ocean was formed. At that time, Greenland, as part of North America, was still joined to Europe but their separation began about 60 million years ago and Laurasia finally divided into the continents—Laurentia (North America) and Eurasia including Baltica, Siberia, Kazakhstania, and parts of China (but without India and Arabia).

GONDWANA BREAKS UP

Gondwana remained intact until 180 million years ago when it started to break up in the Jurassic Period to form the continents of today's atlas. Africa, western China, and India were the first to detach from the remnants of Gondwana.

Australia was one of the last to separate. It started to move north when it split from Antarctica about 90 million years ago. However, it was India that took the record for tectonic "speed." It broke free from Gondwana about 135 million years ago,

BELOW Armadillos are ancient placental mammals that gestate their young in utero and evolved largely on Laurasian continents. The mammals that evolved on Gondwanan continents were primarily marsupials—which produce immature young and gestate them in a pouch.

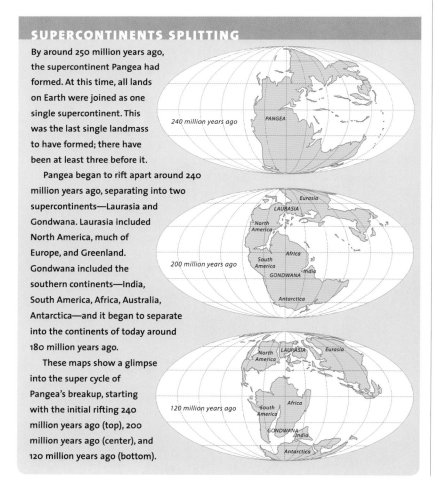

SUPERCONTINENTS SPLITTING

By around 250 million years ago, the supercontinent Pangea had formed. At this time, all lands on Earth were joined as one single supercontinent. This was the last single landmass to have formed; there have been at least three before it.

Pangea began to rift apart around 240 million years ago, separating into two supercontinents—Laurasia and Gondwana. Laurasia included North America, much of Europe, and Greenland. Gondwana included the southern continents—India, South America, Africa, Australia, Antarctica—and it began to separate into the continents of today around 180 million years ago.

These maps show a glimpse into the super cycle of Pangea's breakup, starting with the initial rifting 240 million years ago (top), 200 million years ago (center), and 120 million years ago (bottom).

240 million years ago — PANGEA

200 million years ago — Eurasia, LAURASIA, North America, Africa, South America, GONDWANA, India, Antarctica

120 million years ago — North America, LAURASIA, Eurasia, South America, Africa, GONDWANA, India, Antarctica

moving with a velocity of about 4 inches (10 cm) per year and crashing into the Eurasian Plate. The force of that collision produced the highest mountain range on Earth—the Himalayas. The force was such that the uplift continues today.

When Gondwana finally broke up into the landmasses we know today, these newly formed continents took with them species of the flora and fauna of the time. Many of the southern continents of today share living representatives of these. For example, the Antarctic Beech tree (*Nothofagus* species), which probably evolved in the Late Cretaceous Period, is found living on the former Gondwanan landmasses of New Guinea, New Calendonia, New Zealand, southern South America and south-eastern Australia. It is not found in Africa or India, which means that *Nothofagus* probably evolved after these two continents separated from the rest of Gondwana.

BELOW An aerial view of water, sediments, and rocks in Djibouti, Africa, near the Red Sea. This is at the opening of the East African Rift Valley near the "triple junction," where three major tectonic plates are now tearing East Africa away from the rest of Africa.

RIFTING AND MAGMATIC PROVINCES

Fragmentation or rifting of continental crust is characterized by widespread magmatic activity when massive volumes of magma (molten rock) from within the mantle pour out over the land as lava. It is thought that plumes of magma rise through the mantle and heat overlying crust, which eventually melts and releases the magma.

One of Earth's most extensive flood basalt provinces has been associated with the breakup of Pangea during the late Triassic through to early Jurassic periods. It covers some 4,250,000 square miles (11 million km²) over areas within central Brazil, western Africa, Iberia, and France and North America, all once together as part of Pangea.

This massive outpouring of magma occurred along the mid-oceanic ridge where it dissects Iceland.

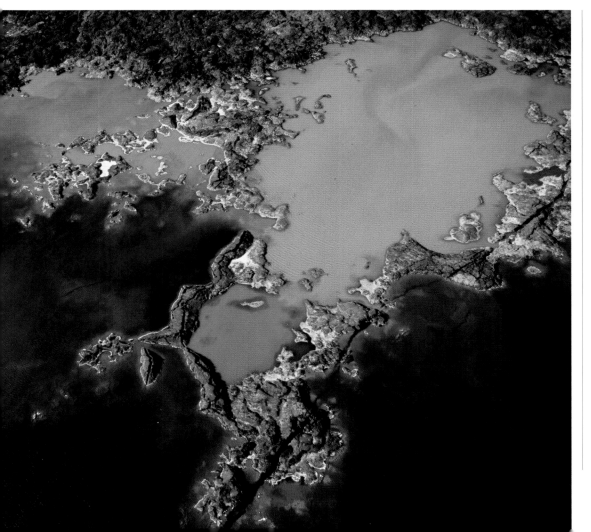

ABOVE *Nothofagus* leaves displaying their autumnal colors in Tasmania, Australia. These deciduous beech trees are found in most Gondwanan continents, showing their common ancestry and thus providing a clue that these continents were all once joined together.

Climate: Past to Present

Earth is continually evolving. From the time it formed, there have been ongoing changes to every feature—structure, atmosphere, oceans, continents, climate, and life. Once the planet cooled sufficiently for the crust and oceans to form, cyclic tectonic processes—over hundreds of millions of years—assembled and then rifted apart supercontinents, changing the distribution of land and sea.

These changes to the crust would have greatly influenced the climatic and biological conditions, with climate varying between periods of icehouse-driven glaciation and more extensive greenhouse-driven warm tropical times. Climate, in this context, means looking at global trends over geologic time. These cycles of glaciation can be understood through paleoclimatology, the study of ancient climates, which uncovers evidence for the many climatic changes buried in the geologic record.

EVIDENCE FOR PAST CLIMATES

Science has discovered a vast number of different clues to ancient climates. Some idea of past temperatures and weather patterns have been found in sea-floor sediments, which also reveal how much ice existed in past times. Ancient atmospheric gas bubbles in cores drilled from polar ice caps can be examined to show the atmospheric composition of the time as well as offer clues to past temperatures. Cycles of drought and rain are recorded in both living and fossil tree rings and coral, as well as in cave formations.

EARTH'S ICY PAST REVEALED

Over geological time there have been at least four major glaciations when Earth was an icehouse—covered in massive polar ice sheets. The oldest dates back to the early Proterozoic Era, around 2.3 billion years ago. There was a second icehouse period during the late Proterozoic Era, 780 million years ago, and another, between the Permian and Carboniferous periods, which started about 330 million years ago. The most recent began during the Tertiary Period, some 15 million years ago. Since each of the first three major glacial periods lasted 75 million years or more and the most recent started only 15 million years ago, it seems likely that today, although the climate is warming artificially, we are still under the influence of the icehouse conditions of the Tertiary Period.

When the roughly 600 million years for all four icehouse periods are expressed as a percentage of Earth's history, it is just 13 percent. So icehouse periods are "minor." They are the exception rather than the rule. Warmer climates with no permanent icecaps at the poles are apparently the norm. The warming that occurred between each of the icehouse periods of would have been from the effect of greenhouse gases that caused the icecaps to melt and sea levels to rise significantly.

WHAT CAUSES CLIMATE TO CHANGE?

The configuration of the continents and oceans over cycles of millions of years greatly affects climate, but where the continents were positioned

ABOVE Trees record annual climatic variations in tree rings. When trees become fossilized, as did an entire forest in Petrified Forest National Park in Arizona, USA, they can yield clues to ancient climates. This forest existed during the late Triassic Period, and fossils here have been dated at more than 200 million years old.

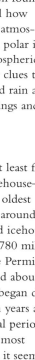

in relation to each other, the circulation of the ocean currents around them, and at what latitudes they lay were all instrumental. For example, when Australia finally separated from Antarctica about 45 million years ago, ocean currents started to flow between the two continents for the first time. This had the effect of producing cooler and drier climates on both continents.

When continents are assembling into super-continents, collision of the plates causes the crust to buckle through compressional forces, pushing up extensive mountain chains. These are often of an altitude that can interfere with the jet stream and so produce the cooling effect required to trigger the beginning of an icehouse cycle.

Further, when landmasses are concentrated near the polar regions, there appears to be an increased chance for snow and ice to accumulate. But this is not always the case; there have been past warm periods when polar landmasses were home to deciduous forests. So, the causes of the four major periods of glaciation are not fully understood.

ABOVE During periods of intense global cooling, large parts of Earth have been covered by ice-fields and glaciers that extended outward from the polar ice caps. These icefields were responsible for forming the world's fjords—drowned U-shaped valleys that were carved by glaciers.

LEFT The climate in steamy tropical rainforests is very similar to that experienced throughout most of Earth's history: for long periods of time the planet was a warm and humid place. However, the variation in the global mean temperature is really only about 18°F (10°C) between periods of extensive glaciation and times of intense warming.

CLIMATE CHANGE OVER SHORTER CYCLES

There are also climatic changes over lesser periods of time. From years to decades, Earth has cycles of drought and rain, and over periods of hundreds to hundreds of thousands of years, there is historical evidence of minor glitches like the Little Ice Age. This was possibly tied to the changing alignments of the Earth, Moon, and Sun, which affect tidal pull, thus altering ocean currents. However, some believe it might have been linked to reduced sunspot activity.

Some climatic cycles over periods of thousands of years have been shown to relate to variations in Earth's orbit around the Sun. These have the capacity to alter the amount of solar radiation reaching the Earth and seem to work in recurring time intervals of 23,000, 41,000, 100,000, and 400,000 years. The cycles were calculated by a Serbian mathematician, Milutin Milankovich in the 1920s, and now bear the name Milankovich Variations. They match many of the glacial–interglacial cycles of the Pliocene–Pleistocene Ice Age over the past 2.5 million years.

PRESENT-DAY CLIMATES

"Climates" is more generally used to describe the average weather in a region over a period of at least 30 years—periods of mere decades. The climate that affects us from day to day is influenced by a number of factors. Climate changes with latitude or distance from the equator, and this is because of the amount of solar radiation received. It is much hotter where the Sun's rays strike Earth's surface at 90° than it is at the poles, where the angle is very much less. Seasonal movements of air masses within the atmosphere also play a major role. These, in turn, influence two important climatic factors—air temperature and precipitation such as rain, hail, or snow.

Seasons result from the annual variation in solar radiation reaching Earth's surface due to the tilt of its axis (23.5°) as it rotates in its orbit around the Sun. When the Northern hemisphere tilts toward the Sun, it is winter in the Southern hemisphere and summer in the North. This is reversed when the Southern hemisphere is tilted toward the Sun.

ABOVE The amount of dust and gases entering the atmosphere from volcanic activity can be enormous, sufficient to cool the atmosphere. Large historic eruptions that lowered global temperatures for many years include Mt Tambora (1815), Krakatau (1883), and Mt Pinatubo (1991).

Variation in solar radiation could be a possible culprit. For example, just small changes can tip the balance between snows completely melting over successive summers to snow remaining throughout the seasons. We understand cycles of solar activity over very short-term periods, but not for millions of years.

Other factors might include variations in volcanic eruptions and Earth's orbit, which can alter the amount of solar radiation received. These variations are cyclic and tie in well with the known glacial-interglacial cycles during the Pliocene–Pleistocene Ice Age of the last 2.5 million years.

RIGHT As the seasons change, landscapes alter dramatically as the plants and animals adapt to variations. Deciduous trees shed their leaves in a burst of autumnal glory because it is energetically "expensive" for them to maintain their growth during the shorter and darker winter days.

CLIMATIC ZONES

It is possible to divide Earth's surface into climatic zones that broadly follow changes in latitude. These climatic regions coincide to a certain extent with patterns of soils and vegetation. It is the climate that determines what plants will grow in a region and what animals will live there. Collectively, climate, plants, and animals create a biome.

A. TROPICAL CLIMATES are generally characterized by high temperatures and rainfall throughout the year. These are the regions of broad-leafed evergreen rainforests and savannas.

B. ARID AND SEMI-ARID CLIMATES are characterized by little rain and a huge daily temperature range. These include the semi-arid or steppe, and the arid or desert regions of the world.

C. HUMID MIDDLE-LATITUDE CLIMATES are characterized by warm dry summers and cool wet winters. They include the Mediterranean climate and both the warm and mild moist temperate climates. In middle latitude climates, variations in the type of land and bodies of water play a large part. Plants have adapted to the extreme difference in rainfall and temperature between winter and summer seasons.

D. CONTINENTAL OR COLD TEMPERATE MOIST CLIMATES are generally characteristic of the interior regions of large landmasses. Total precipitation is not very high and seasonal temperatures vary widely.

E. POLAR CLIMATES are characterized by permanent ice and tundra. Only about four months of the year have temperatures above freezing.

A. Being near the equator, the climate at San Rafael Falls in Ecuador's Cayambe-Coca Reserve is tropical.

B. Oases are pockets of vegetation sometimes found in arid areas, such as this stretch of sandy desert in Morocco.

C. The islands of Greece have a Mediterranean climate, featuring cool wet winters and hot dry summers.

D. Cradle Mountain–Lake St Clair National Park, in Tasmania has a cold temperate climate, with snow in winter.

E. Greenland, like all lands that lie across the Arctic Circle, has a polar climate and its plant life is largely tundra.

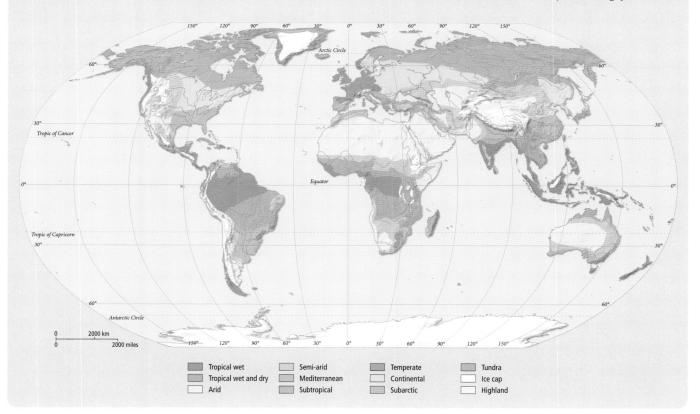

Tropical wet
Tropical wet and dry
Arid

Semi-arid
Mediterranean
Subtropical

Temperate
Continental
Subarctic

Tundra
Ice cap
Highland

BELOW Viewed from space, the long stretch of the Andes—the massive cordillera of the South American continent—is clearly visible. The altitude of these mountains, combined with a westerly wind off the Pacific Ocean, creates a rainshadow on the eastern side of the mountains. This means little or no rain falls there, creating ultra dry deserts such as the Atacama.

Weather is the regional day-to-day experience of climate. There can be weather extremes, such as severe storms, hailstorms, hurricanes, droughts, and floods. Although distance from the equator plays an important role in a country's climate, not all countries at the same latitude will experience the same weather patterns. Some countries, such as parts of India, might have very seasonal monsoons where summer brings warm wet weather with dry sunny winters; others at the same latitude might be arid, such as the Sahara Desert. Still others, like the United Kingdom, have very changeable weather year-round. North America has cycles of very extreme weather.

Altitude also influences weather. Mountain ranges can increase or decrease rainfall depending on the moisture content of prevailing winds. The climate of some coastal regions varies greatly from that found inland, such as the warm to cool temperate climate of Australia's eastern coast compared with the hot arid and semi-arid interior regions.

OSCILLATIONS IN WEATHER PATTERNS

On the scale of decades, there appear to be regular cycles of drought and flooding. In the southern hemisphere, these seem to be related to changes or oscillations in sea-surface temperatures and their interaction with the atmosphere. These episodes

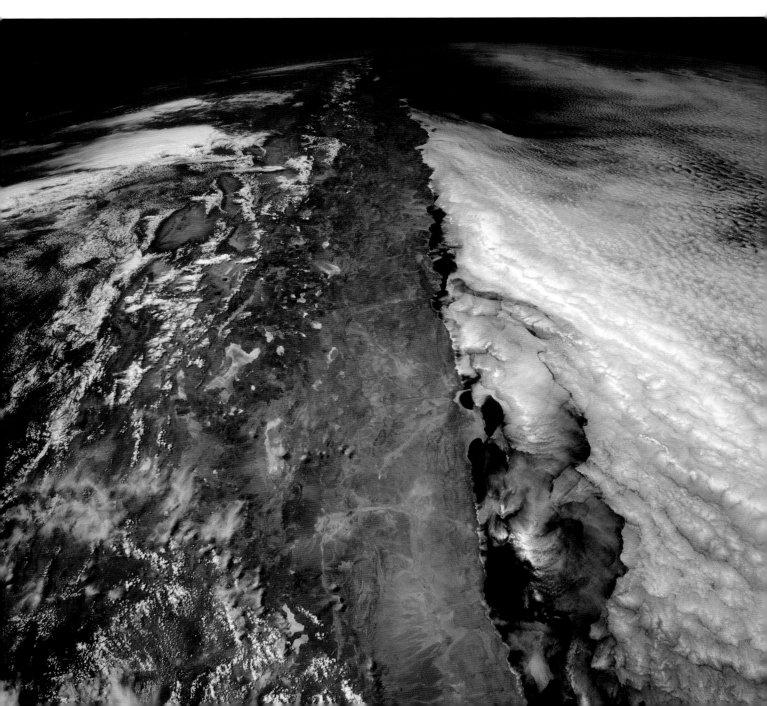

BELOW Viewed from space, the long stretch of the Andes—the massive cordillera of the South American continent—is clearly visible. The altitude of these mountains, combined with a westerly wind off the Pacific Ocean, creates a rainshadow on the eastern side of the mountains. This means little or no rain falls there, creating ultra dry deserts such as the Atacama.

are known as El Niño, when sea-surface temperatures are exceptionally warm across the tropical eastern Pacific. Periods of sea-surface temperatures below average have become known as La Niña episodes. Together they operate as the El Niño–Southern Oscillation (ENSO), at cycles of around three to five years, and can have disastrous impacts on weather on either side of the Pacific Ocean.

The northern hemisphere also has a climatic pattern that seems linked to variations in solar radiation. This affects the climate in the atmosphere during winter and at sea level it becomes a large-scale fluctuation, the North Atlantic Oscillation, which affects how air flows over the Atlantic Ocean. High solar activity brings drier weather, while low activity brings wetter weather.

OUR FUTURE CLIMATE AND PREDICTING CLIMATE CHANGE

Relative stability of climate has persisted throughout 10,000 years of the Holocene Epoch, which started at the end of the last Ice Age. The very first modern humans, *Homo sapiens,* are thought to have appeared during the Pleistocene Epoch, around 200,000 years ago, and, over that time, they would have survived several bleak glacial cycles.

Evidence is clear that the climate is now changing at an alarming rate. We use records of long-term climate in order to understand how climate has changed in the past and how it might change in the future, as global warming is indicated. It is now agreed that this is the result of the buildup of greenhouse gases in the atmosphere, and that these gases are the products of human industrialization.

The greenhouse effect is a natural phenomenon that has occurred throughout the planet's history. It is what keeps Earth's surface warm enough to be inhabited by life. It results from certain greenhouse gases in the atmosphere, such as carbon dioxide and water vapor, absorbing solar energy and radiating it back to the surface. If this did not occur, this energy would be reflected back out into space, in which case Earth's surface would be about 60°F (33°C) colder.

The volume of greenhouse gases in the atmosphere has varied considerably over geologic time and this variation has been the cause of the fluctuations from icehouse glaciation to greenhouse warm tropical climates. What we are currently facing is the rapidly accelerating addition of greenhouse gases, such as carbon dioxide, into the atmosphere from human activities. This significantly increases average global temperature. This is not a measure of how hot or cold it is from day to day, but the overall average temperature at the planet's surface.

In considering the average global temperature as we would the average temperature of the human body, then a 1° or 2° rise in body temperature would probably be caused by fever at the onset of an illness. So, too, is it with our planet. Even a single degree rise is a significant increase. The resulting global warming can cause dramatic changes to climate—polar ice caps and glaciers melt, and weather patterns change so that there is more severe weather with greater frequency. There are also climate changes where desert areas receive greater rainfall and regions of higher rainfall become more arid.

Earth's climate has never been stable. Climate change has been the norm. What is different about the predicted climate changes is that never before has there been a single species of animal in such vast numbers with an unprecedented power to alter the environment. *Homo sapiens* is in plague proportions. If we are to survive, then we must change our ways. It is not Earth we have to save; we need to save ourselves and the biosphere on which we depend. The Earth has a 4.6 to 5 billion year history, during which it, along with various life forms, has survived many dramatic changes and it will continue to do so. The question is: will we?

ABOVE Most glaciers have been retreating for a century—a significant indicator that the global temperature is rising. The loss in Europe is 30 to 40 percent, in New Zealand, 25 percent, and 60 percent of the ice on Africa's Mt Kilimanjaro and Mt Kenya has gone.

FALSE SECURITY

We have been lulled into a false sense of security when it comes to climate. For thousands of years now, climate has remained relatively stable. There have been minor glitches, such as the Little Ice Age that saw Earth cool significantly for almost 400 years, from the thirteenth to the seventeenth centuries. However, from the time of the earliest written history 5,000 years ago, there has been little variation. But all of that is about to change.

LEFT With global warming comes a rise in ocean temperatures, which may threaten many ecosystems. On coral reefs, rising sea temperature can contribute to coral bleaching, which will kill the reef if the temperature does not revert fast enough for coral regrowth. During 2002, over 60 percent of Australia's Great Barrier Reef suffered bleaching.

The Atlantic Ocean

SOUTH ATLANTIC OCEAN

SOUTH AMERICA

Peru–Chile Trench

Chile Basin

Tropic of Capricorn

Pernambuco Plain

Chain Fracture Zone

Ascension Fracture Zone
Ascension

St Helena Fracture Zone
St Helena

Stocks Seamount

Brazil Basin

Vitória Seamount

Ilhas Martin Vaz
Ilha da Trindade

Mid Atlantic Ridge

Santos Plateau

Rio Grande Rise

Rio Grande Fracture Zone

Argentine Basin

Argentine Rise

Argentine Plain

Falkland Islands (Islas Malvinas)

Falkland Plateau

Falkland Escarpment

Tierra del Fuego

Drake Passage

Scotia Ridge

Yaghan Basin

Scotia Ridge

South Shetland Trough

Antarctic Peninsula

Scotia Sea

South Orkney Islands

Shag Rocks

South Georgia

South Sandwich Islands

South Sandwich Trench

Scotia Ridge

American–Antarctic Ridge

Atlantic–Indian Ridge

Agulhas Ridge

Shona Ridge

Bouvetøya

Discovery Seamount

Gough Island

Tristan da Cunha

Cape Basin

Namibia Plain

Vema Seamount

Angola Basin

Walvis Ridge

Mornington Plain

Southeast Pacific Basin

Antarctic Circle

500 km
500 miles

N

15°S
30°S
45°S
60°S

90°W 75°W 60°W 45°W 30°W 0° 30°E 45°E 60°E

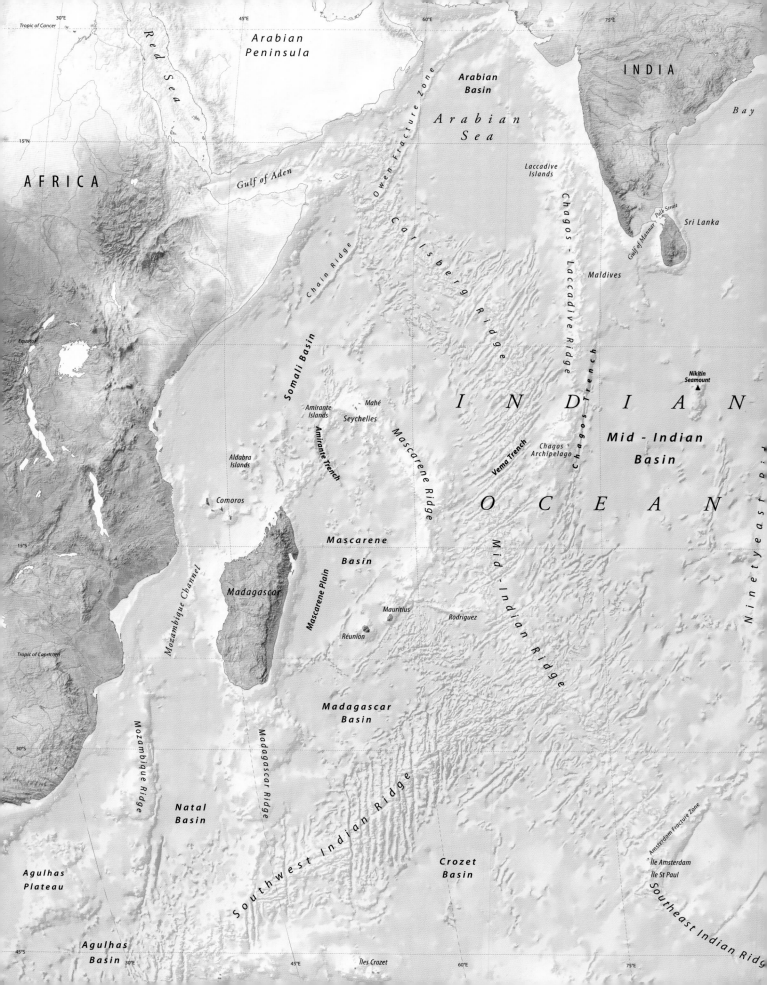

ASIA

Andaman Basin

Gulf of Thailand

South China Sea

Luzon Strait

Central Basin

Ryukyu Trench

Oki-Daito Ridge

Bonin Trench

Philippine Basin

West Mariana Basin

South Honshu Ridge

Philippines

Philippine Trench

Palawan Trough

Sulu Sea

Philippine Sea

Challenger Deep -10,923m

Mariana Trench

East Mariana Basin

Palau Trench

Yap Trench

Kyushu-Palau Ridge

Sunda Shelf

Strait of Malacca

Sumatera

Celebes Sea

Borneo

West Caroline Basin

East Caroline Basin

Cocos Basin

Makasar Strait

Sulawesi

Seram Sea

Bismarck Sea

Maluccca Sea

Teluk Bone

Banda Sea

Weber Basin

New Guinea

Investigator Ridge

Java Sea

Java Trench (Sunda Trench)

Jawa

Flores Sea

Java Ridge -7,125m

Arafura Sea
Arafura Shelf

Gulf of Papua

Cocos Islands

Christmas Island

Timor Sea

Joseph Bonaparte Gulf

Gulf of Carpentaria

Coral Sea

North Australian Basin

Great Barrier Reef

West Australian Basin

Exmouth Plateau

Wallaby Plateau

Cuvier Plateau

East Indiaman Ridge

Broken Ridge

Perth Basin

AUSTRALIA

Naturaliste Plateau

Great Australian Bight

Diamantina Deep -6,602 m

Diamantina Trench

South Australian Basin

Bass Strait

N

The Indian Ocean

Tasmania

Tasman Basin

0   500 km
0   500 miles

Hudson Bay

NORTH AMERICA

ATLANTIC OCEAN

Gulf of Mexico

Caribbean Sea

SOUTH AMERICA

Kodiak Island
▲ Patton Seamount
Comstock Seamount
▲ Pratt Seamount
▲ Welker Seamount
Alexander Archipelago
Queen Charlotte Islands
Endeavour Seamount ▲

Cascadia Basin

Tufts Abyssal Plain

Mendocino Fracture Zone

North east

Murray Fracture Zone

Patton Escarpment

Cedros Trench

Molokai Fracture Zone

Kaua'i
O'ahu
Maui
Hawai'ian Islands
Hawai'i

Clarion Fracture Zone

Pacific Basin

Islas Revillagigedo

Middle America Trench

Guatemala Basin

Tabuaeran

PACIFIC

Clipperton Islands

Colon Ridge

Cocos Ridge

Panama Basin

Kiritimati

Clipperton Fracture Zone

Galapagos Fracture Zone

Carnegie Ridge

Islas Galápagos

OCEAN

Malden
Starbuck

Gallego Rise

Galapagos Rise

Grijalva Ridge

Rakahanga

Vostok
Rebecca
Nuku Hiva
Îles Marquises
Hiva Oa
Marquesas Fracture Zone

Bauer Basin

Peru Basin

Penrhyn Basin
Flint
Tiki Basin
Puka Puka
Tuamotu Fracture Zone

Yupanqui Basin

East Pacific Rise

Nazca Ridge

Islands
Archipel de la Société
Tahiti
Archipel des Tuamotu
Raroia

Austral Fracture Zone

Manuae

Maria
Îles du Duc de Gloucester
Îles Gambier
Ducie

Chile Basin

Isla San Felix

Tubuai
Îles Australes
▲ President Thiers Seamount
Pitcairn
Pitcairn Islands
Sala y Gomez Ridge
Isla de Pascua
Isla Sala y Gómez

Isla San Ambrosio

Rapa
Marotiri

Islas Juan Fernández

Southwest Pacific Basin

Roggeveen Basin

Challenger Fracture Zone

N

Agassiz Fracture Zone

East Pacific Rise

Mocha Fracture Zone

The Pacific Ocean

Guafo Fracture Zone

0   1000 km
0   1000 miles

Menard Fracture Zone

Mornington Abyssal Plain

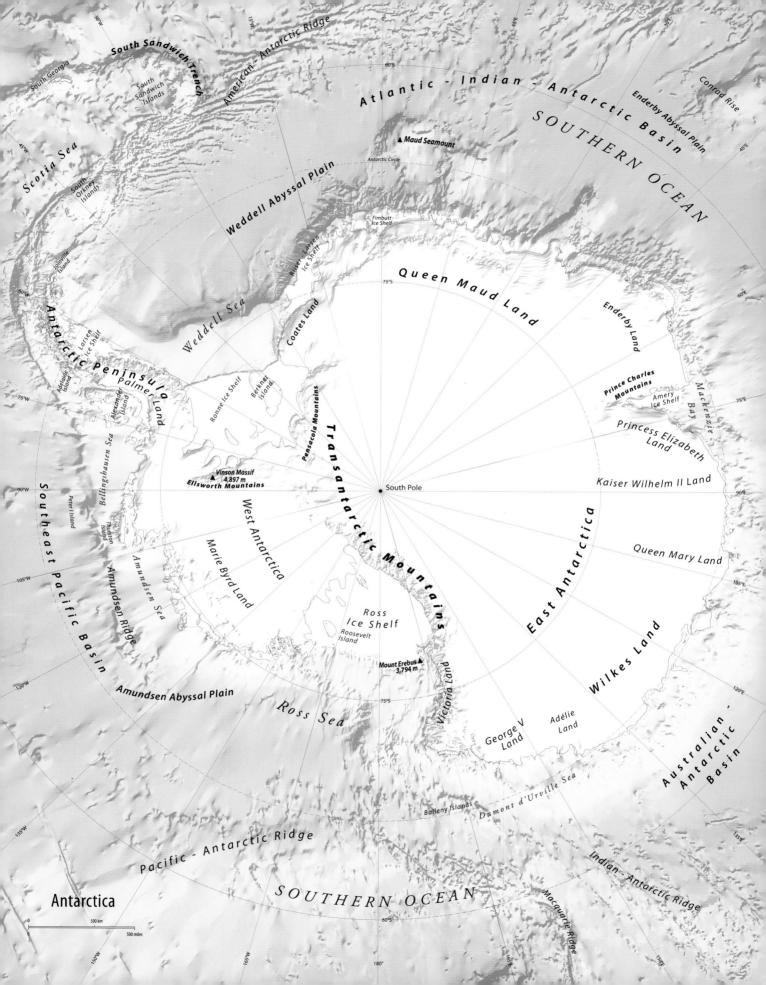

South Georgia

South Sandwich Trench

South Sandwich Islands

American - Antarctic Ridge

Atlantic - Indian - Antarctic Basin

Conrad Rise

Enderby Abyssal Plain

SOUTHERN OCEAN

▲ Maud Seamount

Antarctic Circle

Scotia Sea

South Orkney Islands

Weddell Abyssal Plain

Fimbutt Ice Shelf

Riiser-Larsen Ice Shelf

Queen Maud Land

Enderby Land

Joinville Island

Weddell Sea

Coats Land

75°S

60°S

60°E

Larsen Ice Shelf

Antarctic Peninsula

Palmer Land

Adelaide Island

Alexander Island

Ronne Ice Shelf

Berkner Island

Penacola Mountains

Prince Charles Mountains

Amery Ice Shelf

Mackenzie Bay

75°W

Princess Elizabeth Land

Bellingshausen Sea

Thurston Island

▲ **Vinson Massif**
4,897 m
Ellsworth Mountains

Transantarctic Mountains

● South Pole

Kaiser Wilhelm II Land

90°E

Peter I Island

90°W

Southeast Pacific Basin

West Antarctica

Marie Byrd Land

East Antarctica

Queen Mary Land

Amundsen Ridge

Amundsen Sea

105°E

Ross Ice Shelf

Roosevelt Island

105°W

Wilkes Land

120°W

Amundsen Abyssal Plain

Ross Sea

Mount Erebus ▲
3,794 m

Victoria Land

George V Land

Adélie Land

Australian - Antarctic Basin

120°E

75°S

Balleny Islands

Dumont d'Urville Sea

135°W

Pacific - Antarctic Ridge

SOUTHERN OCEAN

Indian - Antarctic Ridge

Macquarie Ridge

125°E

Antarctica

60°S

0   500 km

0   500 miles

150°W

165°W

180°

165°E

150°E

PACIFIC OCEAN

Gulf of Alaska

Bering Sea

Sea of Okhotsk

Norton Sound

Bering Strait

Arctic Circle

Chukchi
Sea

NORTH AMERICA

ASIA

Ostrov
Vrangelya

East Siberian
Sea

Beaufort Sea

ARCTIC
OCEAN

Novosibirskiye
Ostrova

Yanskiy
Zaliv

ARCTIC
OCEAN

Chukchi
Abyssal
Plain

Canadian Abyssal Plain

Canada Basin

Northwind Ridge

Plateau
Chukchi

Mendeleyev Ridge

Laptev Sea

Victoria
Island

Severnaya
Zemlya

Lomonosov Ridge

Makarov Basin

Vozonin Trough

Alpha Ridge

North Pole

Amundsen Basin

Central Kara Rise

Ellesmere Island

– 4,346 m ▼

Arctic Mid-Ocean Ridge

Nansen Basin

Franz Josef
Land

Kara Sea

Baffin Island

Baffin
Bay

Novaya Zemlya

Barents
Sea

Greenland

Yermak
Plateau

Spitsbergen

Pechorskoye More

Davis Strait

Greenland Sea

Boreas Abyssal Plain

Greenland Fracture Zone

White Sea

Jan Mayer Fracture Zone

Mohns Ridge

Denmark Strait

Jan Mayen Ridge

Norwegian
Basin

Voring
Plateau

Icelandic
Plateau

Norwegian
Sea

Irminger Basin

Iceland

The Arctic

Faroe-Iceland Ridge

Faroe
Islands

EUROPE

0   500 km
0   500 miles

Iceland Basin

Volcanoes are one of Earth's most fascinating phenomena. They are direct proof that only a short distance below our planet's familiar surface exists molten rock (called magma) at high temperatures and pressures, awaiting the opportunity to burst forth, either in flows or explosions, wherever a weakness in the solid crust presents itself.

ABOVE Hawaii's Big Island is a classic example of a shield volcano. This type of volcano results from basaltic magma breaking through Earth's crust.

Volcanoes are found in greatest abundance at the boundaries between Earth's tectonic plates, such as all along the west coast of North and South America, throughout the Indonesian archipelago, and even underwater, right down the middle of the Atlantic Ocean. The type of volcano that will form depends on whether the plates are pulling apart or coming together.

RIGHT Shishaldin volcano on Unimak Island, Alaska, has the typical symmetrical cone shape of a stratovolcano. It is one of the most active volcanoes in the Aleutian Arc, and was last active on February 17, 2004.

At spreading margins, the plates move apart and magma is able to well up passively in between and solidify. Rift volcanoes result from this process. At subduction margins, one plate is forced beneath the other where it is heated to melting point. Here, molten bodies of

magma known as "plutons" rise through the overlying plate, often under pressure, to form Earth's most violent and dangerous volcanoes. These are referred to as composite volcanoes or stratovolcanoes.

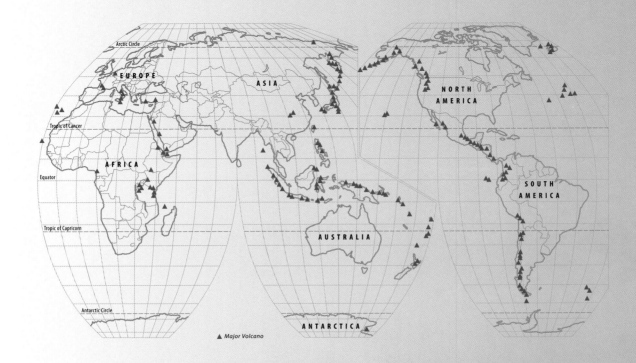

▲ *Major Volcano*

THE DIFFERENT VOLCANO TYPES

There are three types of volcano whose form is usually determined by their plate tectonic setting: rift volcanoes along spreading margins; composite volcanoes or stratovolcanoes along subduction margins; and shield volcanoes above intra-plate hotspots.

Rift volcanoes are the most abundant type of volcano on Earth, though we rarely see them. They lie mostly out of sight at the bottom of the oceans, forming a line of continuous eruption thousands of miles long between spreading plates. There are only a few places on Earth where these mid-oceanic rift volcanoes are visible on land, such as in Iceland, where the rift rises out of the ocean and cuts straight across the middle of the island. The people of Iceland live with the constant threat of volcanic activity, but also utilize the abundant heat from the rift volcanoes for heating and electricity.

Composite volcanoes or stratovolcanoes grow at closely spaced intervals along all of Earth's subduction margins, such that it is often possible to view several of them in a line from a single vantage point. A good place to find them in abundance is along the edges of the continents encircling the plates of the Pacific Ocean, hence the name "Ring of Fire." Another good place is the Indonesian archipelago. Deep trenches on the sea floor, such as the Japan

Trench, Peru–Chile Trench, or the Java Trench, running parallel to the chains of volcanic peaks, mark the line along which plates are being forced beneath others and consumed. The viscous and highly gas-charged molten rock (magma) that forms the volcanoes is of intermediate composition (andesitic), as it is the product of the descending and melting plate, as well as the seawater and ocean sediments that have been dragged down with it. The magma tends to erupt explosively, as ash and various-sized pieces of broken rock (collectively known as tephra), and as lava flows. Over time, the

ABOVE The people of Iceland live with the constant threat of activity from mid-oceanic rift volcanoes. Iceland is sited over the boundary of the Eurasian and North American plates.

LEFT Part of Iceland's Dyngjufjöll mountain chain, Askja caldera rises 4,974 ft (1,156 m) above sea level. This vast volcanic crater was formed in 1875 as a result of a devastating eruption of the crater in front of it.

TYPES OF VOLCANOES

The four main volcano types are shown below. A. Shield volcanoes punch up lava through weak spots in Earth's crust, such as the volcano on Hawaii's Big Island. B. Stratovolcanoes (also called composite volcanoes) erupt along plate edges, such as those found all around the Pacific Ocean rim. C. Rift volcanoes are most commonly found on the sea floor, but they also occur on land in Iceland. D. Cinder cone volcanoes, such as Mt Paricutín, in Mexico.

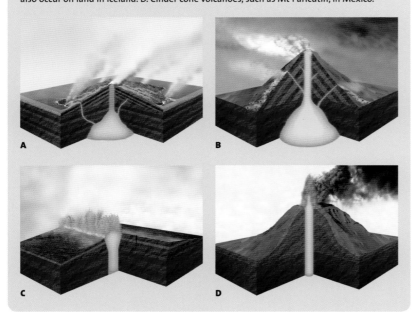

A

B

C

D

of Hawaii is a classic example that stands 5 miles (8 km) tall, from ocean floor to summit. Activity ceases when the moving plate severs the shield volcano from its feeder pipe, stopping the lava flows. The immense weight of the shield volcano is so great that it begins to depress the sea floor and slowly sinks. Finally, after a few more million years the volcano disappears from sight beneath the waves. This mechanism is responsible for the many oceanic island chains, such as the Hawaiian and Tahitian chains in the Pacific. An active volcano sits at the head of the chain above the hotspot with a line of increasingly older volcanic islands and submerged seamounts heading away from that point in the direction of plate movement. Shield volcanoes erupt gently and do not often pose a threat to those living on them.

LIFE CYCLE OF AN EXPLOSIVE STRATOVOLCANO

Stratovolcanoes, with their majestic classical shape, are by far the most dangerous type of volcano found on Earth. Their immensely destructive cycle, which terminates with a cataclysmic explosion followed by a rebuilding phase, is an incredibly long period of time in human terms. As such, their danger only becomes apparent every few centuries or even after a much longer time.

During its brief explosive phase, a single volcano can wipe out generations of human development that has occurred around it since the last eruption. The volcano explodes with a massive venting of gas, steam, and tephra high into the atmosphere (an eruption column or plume). The plume collapses under its own weight as the bulk of destructive hot debris surges (pyroclastic flows) radiate outward from the volcano.

Such destruction took place in minutes when the 1902 explosive eruption of Mt Pelée totally obliterated Saint-Pierre, a city of 29,000 on the island of Martinique. The explosion that destroyed the summit of Mt St Helens in 1980 was even larger, but only 57 lives were lost due to the volcano's remote location and the successful management of the situation by the US Geological Survey.

The general populace living within the vicinity of a volcano is usually unaware of the unseen danger. Today, Indonesian farmers work the fertile soils around lakes within the craters of their nation's numerous active volcanoes, as new volcanic edifices grow noisily nearby. Entire Japanese cities are spreading around the flanks of Mt Fuji, which has not erupted since 1707. Even Mexico City, one of Earth's largest urban populations, is located just 45 miles (70 km) from Popocatepetl volcano. This highly active volcano has had 15 eruptions since 1519 and is currently on Yellow Alert.

alternating tephra falls and lava flows build up a distinctive, steep-sided, often symmetrical cone known as a composite cone or stratovolcano.

Shield volcanoes occur above specific "hot spots" in Earth's mantle where basaltic magma is able to find its way to the surface through a weakness in the crust. Flow after flow of runny low-viscosity lava builds up an enormous broad, flat, shield-like volcanic edifice, with the entire process taking about one or two million years. The Big Island

BELOW Mt Bromo in eastern Java has a small pyroclastic cone and a broad caldera. On June 8, 2004, at least two tourists were killed and several people were injured when this stratovolcano erupted.

SUPERVOLCANIC ERUPTIONS
AND THE GLOBAL CONSEQUENCES

Supervolcanic explosions are felt around the world. They have global climate-changing consequences and have even resulted in mass extinctions of life, such as occurred at the end of the Permian Period 248 million years ago. The aftermath landscape of a supervolcanic explosion is completely unrecognizable, with the entire volcanic cone blown away leaving a huge caldera. The giant Yellowstone caldera, with its hot pools, springs, and geysers, is an example of the remains of a supervolcano that exploded three times in the last two million years. Although not in the supervolcano class, the largest explosion to occur in more recent times was that of Tambora volcano, Indonesia, in 1815. The most deadly explosion in human history, 92,000 deaths resulted from the blast, accompanying tsunami, and subsequent disease and starvation. In August 1883, the Indonesian archipelago was again hit when

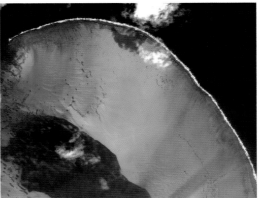

Krakatau volcano exploded in the Sunda Straits, killing 36,400 people. Spectacular red sunsets were seen all over the world as a result of the vast amounts of ash that were ejected into the atmosphere. The fine airborne particles reflected some of the Sun's energy back into space, resulting in a

ABOVE Mt St Helens and the area around it were dramatically changed within minutes by the 1980 eruption. On May 18, 1980, an earthquake measuring 5.1 on the Richter scale shook the strato-volcano, which spewed out gray ash that soon fell over Washington, USA, and beyond.

LEFT This Pacific Ocean atoll was once a shield volcano that eventually sank following millions of years of gentle lava eruptions. The large amounts of magma produced by shield volcanoes is rich in magnesium and iron.

drop in global temperature by as much as 2.2°F (1.2°C) in the year following the eruption.

The cycle begins anew when a new volcano starts to slowly grow within the old caldera. Anak Krakatau ("child of Krakatau") was first spotted when Javanese fishermen observed plumes of steam and ash shooting from the sea on December 29, 1927. Since then it has grown into a 100-foot (300-m) tall conical-shaped island. Another well-known example is the Wizard Island volcanic cone in the Crater Lake caldera in Oregon, USA. These small cones will continue to grow into majestic adult cones, such as Mt Fuji, and over time will eventually overlap the older caldera structures from the earlier explosive episode. At this point they are ready to explode anew and repeat the cycle.

PREDICTING VOLCANIC ERUPTIONS

Prior to an eruption, pressure builds up beneath the volcano as magma rises into its subterranean chamber. Numerous micro-earthquakes accompany the upward movement of the magma, as cracks and fractures open and the mountain begins to swell.

While all of this is imperceptible to the human eye and ear, it is well within the capability of sensitive seismometers and tilt meters that can be arrayed on dangerous volcanoes. Although the exact hour, day, or even week of an impending eruption may not be exactly known, many lives may be saved with such a system of continuous monitoring and data collection in place.

An example of how effective this can be was evidenced by the modern-day eruption of Mt St Helens. Warning signs were received early, and good communication between scientific organizations and civil authorities meant that systems to restrict entry to the area were put swiftly in place. Amid mounting protests from the logging companies, whose livelihoods depended on access to the tree-covered slopes of the volcano, and from residents wishing to return home, the cordon remained firmly in place.

As a result of these safety measures, on the morning of May 18, 1982, when Mt St Helens finally exploded, only 57 lives were lost. For many of Earth's volcanoes, however, no such monitoring or data collection systems exist, and there can be no such advance warning.

Safe air travel is threatened when clouds of fine, glassy, abrasive volcanic material in the atmosphere cross the flight paths of jet aircraft. To help air traffic controllers and pilots plan their flight paths, an international aviation color-code alert system has

been put in place in accordance with International Civil Aviation Organization (ICAO) procedures to describe the status of active volcanoes:

1. Green Alert: Active volcano is at normal non-eruptive background level.

2. Yellow Alert: Signs of elevated unrest above known background levels. Elevated seismic activity.

3. Orange Alert: Accelerating intense unrest. An eruption is likely within hours to days. Small ash eruptions are expected or confirmed. Seismic disturbance is recorded on local seismic stations, but not at more distant stations.

4. Red Alert: Eruption is in progress. A plume (of specified height) is present or is likely to rise over 25,000 feet (8,000 m) above sea level. Strong seismic disturbance is recorded on all local stations, and commonly on more distant stations.

MEASURING THE SIZE OF EXPLOSIVE VOLCANIC ERUPTIONS

The size of a volcanic eruption can be measured and ranked using a number of different methods. One is to measure the height of the eruption column, another the volume of material erupted. Also the radial distance of thrown fragments versus their size can be calculated, along with the amount of fine ash injected into the atmosphere and duration of the eruption. All of these measurements determine, directly or indirectly, the amount of energy released by the eruption.

The Volcanic Explosivity Index (VEI) is based on several of these variables and is used to rank the size of an explosive eruption on a scale from 0 to 8. A VEI of 0 denotes a non-explosive eruption, regardless of volume of lava flows erupted. One example is the continuous outpourings of lava from Kilauea, Hawaii. Eruptions with a VEI of 5 or higher are "very large" explosive events, and occur on average about once every 20 years. Since the year 1500, there have only been 15 eruptions of VEI 5 (including Mt St Helens, 1980), four eruptions of VEI 6 (including Krakatau, 1883), and one of VEI 7 (Tambora, 1815).

The VEI has been determined for more than 5,000 eruptions that have occurred in the course of last 10,000 years; however, all of these eruptions pale into insignificance compared with the hypothetical maximum VEI of 8. These include giant caldera-forming eruptions from volcanic systems such as Long Valley Caldera (California), Yellowstone Caldera (Wyoming), and Valles Caldera (New Mexico). If a VEI 8 event does occur, it will eject over 240 cubic miles (1,000 km³) of material in a cataclysmic explosive eruption that will last more than 12 hours and produce a plume over 15 miles (25 km) in height.

DANGEROUS VOLCANOES

There are numerous active volcanoes that are approaching their explosive phase. Some of those that pose the greatest risk to nearby population centers have been declared as "decade volcanoes" by the International Association of Volcanology and Chemistry of the Earth's Interior (IAVCEI) and earmarked for careful monitoring and study. These are Avachinsky-Koryaksky (Kamchatka), Colima (Mexico), Etna (Italy), Galeras (Colombia), Mauna Loa (Hawaii), Merapi (Indonesia), Nyiragongo (Democratic Republic of the Congo), Sakura-jima (Japan), Santa Maria (Guatemala), Santorini (Greece), Taal (Philippines), Teide (Canary Islands), Ulawun (Papua New Guinea), Unzen (Japan), and Vesuvius (Italy).

The Estación Volcanológica de Canarias has drawn up an eruption-risk map for Teide volcano.

LEFT Volcanologists carry out their work on Sicily's Mt Etna during an eruption. In March, April, and May 2007, Mt Etna put on a spectacular show of volcanic activity.

OPPOSITE The small cinder cone volcano Eldfell, on Iceland's Heimaey Island, was formed in 1973. Local people made good use of the fallout from the eruption that created Eldfell: tephra was used as landfill and to lengthen the airport runway.

Volcanoes of the Northwestern Pacific Rim

The Aleutian Islands describe an elegant volcanic arc across the northern Pacific Ocean linking Russia's Kamchatka Peninsula to the Alaska Peninsula. The chain of active volcanoes follows the Alaska Range then loops sharply south around Anchorage and joins those of the North American Cordillera. The broken line of active and dormant peaks runs south along the western seaboard of North America into Mexico.

ABOVE American black bears (*Ursus americanus*) are widely distributed but they are particularly common in Alaska and the Cascade Ranges.

RIGHT Sheer 100-foot (30 m) cliffs of eroded volcanic ash, carved out by the Ukak River, rise before Mt Katolinat in Katmai National Park, Alaska. The nearby Plinian eruption of Novarupta in 1912—lasting three days—covered the entire region with up to 700 feet (200 m) of volcanic debris.

BELOW Mt Rainier is one of the dormant Cascade Ranges volcanoes. Despite only suggestions of historic activity, it is considered one of the most dangerous volcanoes as any activity could collapse its delicate sides, which are built of ancient mudflows.

This region of the world is known as the Northwestern Pacific Rim and it is part of the circum-Pacific "Ring of Fire," a zone of intense earthquake and volcanic activity. The region's volcanoes are born of andesite magma and explosive ignimbrites, produced by melting of the Pacific Ocean seafloor as it is pushed beneath the North American Plate.

CONTINUOUS VOLCANIC ACTIVITY
Volcanic activity in this region is more or less continuous as attested by historical accounts and, before them, Native American legends. In the Aleutian Arc of Alaska, there are over 40 volcanic centers that have been historically active, resulting in hundreds of eruption reports. Written records began in 1760, with the eruption of Kasatochi taking place in that year.

On some of the islands, debris avalanches from collapsing volcanic domes cause tsunamis when they enter the sea. The 1883 eruption of Augustine generated a tsunami, which measured 30 feet (10 m) at Port Graham, Alaska. Another volcano, Pavlof, is particularly interesting as most of its eruptions tend to occur between September and December. This highly sensitive volcano appears to be triggered by minor seasonal stress variations,

such as snow and ice buildup or even changes in atmospheric pressure. Recently, the most active volcanoes in the Aleutian Arc have been Augustine, Mt Cleveland, and Shishaldin.

In Canada, the Iskut volcanic field, the Tseax River cones, and the Mt Edziza complex all form part of the northern cordilleran volcanic province. According to a legend of the Nisga'a First Nations people, lava from the Tseax cinder cones flowed 14 miles (23 km) to the Nass River, where it blocked the river's flow, forming a natural dam. Today, outcrops of hardened lava along the edge

of the Nass River show evidence of having flowed over wet ground, supporting the Nisga'a legend.

In the USA's Cascade Ranges, Mt Baker, Glacier Peak, Mt Rainier, Mt St Helens, Mt Hood, Mt Shasta, and Lassen Peak have all shown varying degrees of unrest since historical record-keeping began. The well-publicized explosive eruption of Mt St Helens in May 1980 has made it the best known of these volcanoes.

NOVARUPTA: THE LARGEST TWENTIETH-CENTURY ERUPTION

In 1912, Novarupta, on the Alaskan Peninsula, exploded violently releasing more than 7 cubic miles (30 km^3) of ash and lava in about 60 hours. Fine ash darkened the sky over much of the northern hemisphere. It was reported that in nearby Kodiak during the following two days, a lantern held at arm's length was not visible, and in far-off Vancouver that the resulting acid rain caused washing to disintegrate on clotheslines. The eruption (volcanic explosivity index 6) was 10 times more powerful than Mt St Helens. Pyroclastic flows surged into the adjacent valleys of Knife Creek and the Ukak River, traveling up to 15 miles (24 km), and filling them with as much as 700 feet (200 m) of hot volcanic debris.

A National Geographic Expedition led by Robert Griggs in 1916 witnessed this valley still roaring with thousands of steam fumaroles created by boiling water trapped beneath the hot ignimbrite blanket. The amazing spectacle inspired Griggs to name the place, "The Valley of Ten Thousand Smokes." It took until the 1930s for the fumarolic activity to die away. The eruption emptied the magma chamber beneath the original Katmai volcanic cluster causing the ground to collapse into a caldera that has since been filled with

a lake. Although Novarupta exploded with a volcanic explosivity index of 6, the same as that of the massive eruption of Krakatau in Indonesia in 1883, there were no human casualties here due to the extreme remoteness of the volcano.

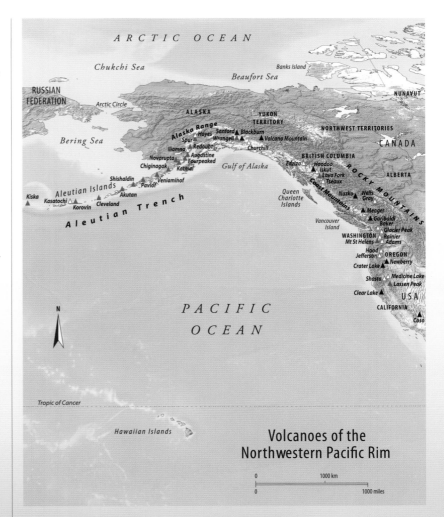

Volcanoes of the Northwestern Pacific Rim

MT ST HELENS, UNITED STATES OF AMERICA

Following its spectacular eruption on May 18, 1980, Mt St Helens has become the world's most studied volcano. The US Geological Survey has set up a permanent observatory on Johnston Ridge, located 5 miles (8 km) from the volcano. It is named for geologist David Johnston who lost his life on the ridge that fateful morning.

OPPOSITE The lava dome in the blown-out crater of Mt St Helens releases volcanic gases. Dormant since 1857, the force of the 1980 eruption blasted the top off the mountain, reducing its height from 9,647 feet (2,950 m) to 8,335 feet (2,549 m).

The status of the mountain is constantly being monitored via a network of video cameras, seismometers, and other instruments covering its slopes, and its explosive fury is by no means spent. Minute earthquakes signify the movement of new magma, while swelling of the mountain indicates pressure buildup within its subterranean magma chambers, heralding the possibility of another eruption. These changes in shape of the volcano are mapped continuously on centimeter-scale accuracy by Earth-orbiting satellites, so hazards can be assessed well before they pose a threat to life.

THE YOUNGEST, MOST ACTIVE VOLCANO IN THE USA

Louwala–Clough or "smoking peak" was the name for Mt St Helens used by the Pacific Northwest Native Americans. Its cone has grown up entirely within the last 2,200 years and the oldest volcanic ash deposits date back only 40,000 years, making it the youngest major volcano in the Cascade Ranges. Apart from intermittent activity between 1831 and 1857, the volcano had been peaceful. Visitors came by the thousands, lured by the mountain's reputation of tranquility, to fish in the clear-water lakes or hike on the forested slopes of its graceful conical peak.

However, all of that changed in May 1980 when the upper 1,300 feet (397 m) of the mountain's summit was blown away, leaving a massive horseshoe-shaped crater. The sudden pressure release due to an avalanche allowed superheated groundwater around the volcano's magma chamber to flash to steam. It blasted laterally northward, out of the avalanche scar, and formed a hot, destructive pyroclastic surge of ash and debris. Immediately north of the volcano, tens of thousands of acres (hectares) of productive forest were destroyed. Giant trees were uprooted and laid down like matchsticks, while at greater dis-

CURRENT STATUS OF THE MOUNTAIN

Mt St Helens is currently, in 2007, on Alert Level 2 (Aviation Code: Orange). A new lava dome is slowly growing within the volcano's crater, accompanied by minor seismic activity and emissions of ash, steam, and volcanic gases. All of the other Cascade Range volcanoes are presently at background levels of seismic activity and pose no immediate threat.

tances they were stripped bare. Within minutes, an enormous column of gas and ash rose from the summit accompanied by massive lightning bolts. The ash column rose some 12 miles (19 km) into the atmosphere before flattening into a giant umbrella shape. Then it collapsed on itself, raining down huge quantities of ash, pumice, and rocks. The eruption lasted nine hours.

FAR-REACHING HAZARDS

The enormous ash cloud was pushed quickly eastward over the United States by strong high-level winds, blocking the morning sun. It turned day into night, grounding all forms of transport and forcing people to wear masks. Fine airborne ash took three days to cross the continent and 17 days to circle the globe.

In total, some 57 people lost their lives, including scientists working on the mountain, campers, a few logging company workers, and Harry Truman, a man who stubbornly refused to leave his lodge on Spirit Lake. The death toll might have been much higher were it not for the US Geological Survey's 20-mile (32-km) cordon around the volcano that restricted access to all but essential personnel. As well as the human toll, it is estimated that the eruption killed 5,000 black-tailed deer, 1,500 elk, 200 black bears, plus unknown numbers of mountain lions, bobcats, rodents, birds, fish, and insects. Complete rebuilding of the local food chain may take many decades.

It is not clear whether Mt St Helens will blow itself apart in another major eruption in the near future, or sometime later, after it has rebuilt a new majestic conical top.

ABOVE Spirit Lake was littered with dead trees blasted by the 1980 eruption of Mt St Helens. The eruption created the largest recorded landslide, which thrust the lake's water into 600-foot-high (183-m) waves and wiped out an entire forest.

Volcanoes of Mexico and Central America

The volcanoes of Mexico and Central America form a chain connecting with those of the Andes to the south and the North American Cordillera to the north. The chain of large active volcanoes is paralleled offshore by the deep Middle American Trench, into which the floor of the Pacific Ocean (the Cocos Plate) is disappearing.

BELOW Glowing cinders and lava spout out of Arenal in Parque Nacional Volcan Arenal, Costa Rica. On July 29, 1968, a massive eruption blew the west flank off Arenal after a 400-year dormancy. It is considered one of the world's most active volcanoes, being active almost daily.

The ocean floor is being overridden, or subducted, by the American and Caribbean plates along a line of massive overthrusting marked by a descending plane of earthquakes. Its destruction, by melting, generates large bodies of molten magma that buoyantly rise through the overlying crust and source the region's volcanoes. The region is well known for its large explosive eruptions.

MEXICO'S NEW VOLCANOES

Most interesting are the "new volcanoes" found in central Mexico. Paricutin is the famous volcano that grew from a crack in a cornfield from 1943 until 1952. The eruptions produced a large cinder cone and slow-moving lava flows. No lives were lost, but the volcano destroyed agricultural lands and livestock. The town of Paricutin was buried over a period of two years, leaving only the cathedral's spires rising from the lava.

ACTIVE AND DANGEROUS

Popocatepetl, "The Smoking Mountain," is Mexico's most potentially hazardous volcano. It has produced three major Plinian eruptions since the Holocene Epoch, the last being about 800 CE. Each time, thick pyroclastic flows and lahars swept

Obsidian is a glassy looking igneous rock formed when flows of silica-rich lava cool so quickly that individual mineral crystals do not have time to form and grow. It is commonly black but may be brown or red, and some varieties display a beautiful iridescence in the sunlight. This may appear as a golden or silver sheen or as bands of rainbow color. This natural volcanic glass was prized by native American cultures as it could be worked into razor-sharp implements such as spearheads and knives, or highly polished to use as a mirror, or used in other artifacts.

into the surrounding basins, which are presently inhabited by some 20 million people. The giant stratovolcano has recently become active again after five decades of quiet.

Colima is the most active volcano in Mexico, having erupted violently several times over the last 450 years. About 4,000 years ago, it produced a cataclysmic avalanche much larger than that of Mt St Helens, USA. Today nearly 300,000 people live within 25 miles (40 km) of the volcano. Colima's explosive eruptions, the last being in 1913, periodically destroy the summit, leaving a crater. Lava dome growth then rebuilds the summit prior to the next explosion. Mexican authorities remain on standby to evacuate surrounding villages in the event of an eruption.

El Chichon is a small andesitic tuff cone in Mexico. It awoke in 1982, when three very explosive eruptions destroyed its summit lava dome and created a new crater, which now contains an acidic crater lake. Pyroclastic surges destroyed villages around the volcano, killing more than 2,000 people. There was major destruction of crops and livestock and tens of thousands of residents were forced to flee their homes. The eruption injected large volumes of sulfur dioxide aerosols into the atmosphere causing brilliant sunsets for several years and temporarily decreasing solar radiation reaching Earth's surface.

Fuego is one of Guatemala's most active stratovolcanoes, located dangerously close to the city of Antigua. The modern volcano was born from the dramatic collapse of the ancestral Meseta Volcano about 8,500 years ago. Frequent eruptions of ash, lava, and pyroclastic flows have been recorded at Fuego since the arrival of the Spanish in 1524.

Santa María Volcano in Guatemala is best renowned for its devastating Plinian eruption of 1902 that killed over 5,000 people. The eruption was quite unexpected and followed a more-than 500-year period of quiescence. This stratovolcano remains a hazard as the massive lava dome, which is growing from the 1902 crater, is unstable and

could perhaps collapse, as it did in 1922, causing further deaths from pyroclastic flows and lahars.

The San Cristobal volcanic complex in Nicaragua was the site of a catastrophic landslide and lahar in 1998 that destroyed two towns, killing over 2,500 people. Instability of the slopes of the volcano resulted from the torrential rains of Hurricane Mitch, which also caused numerous other slope failures throughout Central America.

Arenal is Costa Rica's youngest as well as its most active stratovolcano. The symmetric andesitic cone has been growing for at least 7,000 years—built by successive lava flows alternating with periods of major explosive eruptions at several-hundred-year intervals. Arenal's most recent explosive phase began in 1968 accompanied by occasional pyroclastic flows.

ABOVE Costa Rica has 11 volcanoes, seven of which are considered active. The active volcanoes are: Arenal (pictured), Barva, Irazu, Miravalles, Poás, Rincón de la Vieja, and Turrialba. Inactive volcanoes are: Cerro Tilaran Orosí, Poás, Platanar, and Tenorio.

TOP LEFT Born in 1770, Izalco Volcano in El Salvador was so active until 1958 it was dubbed "the lighthouse of the Pacific." Apart from a very small eruption in 1966, it has since remained dormant.

Volcanoes of the Caribbean

The beautiful islands of the Antilles lie in a great arc, like a giant necklace, stretching across the Caribbean from Puerto Rico to Venezuela. These island jewels are all active volcanoes lining the eastern edge of the Caribbean Plate. They are being built from rising magma created by the melting of westward-moving Atlantic oceanic crust as it plunges beneath the Caribbean Plate.

Seventeen volcanoes have been active in the West Indies during the last 10,000 years, including one island yet to be born, Kick-'em-Jenny, a submarine volcano. Most of the volcanoes are stratovolcanoes, which commonly produce lava domes, hence most of their eruptions are explosive events that generate pyroclastic flows.

THE REGION'S DEADLIEST ERUPTION: MT PELÉE

On May 8, 1902, pressure building within Mt Pelée for over a year finally blasted out the plug of solidified magma blocking the volcano's throat. An eruption column rose from the summit followed by a blast from a vent in its side. The pyroclastic cloud, dense with ash and superheated gases, incinerated the entire city of St Pierre's 29,000 residents in less than three minutes. Witnesses on ships in the harbor who were lucky enough to escape with their lives documented the short but deadly event.

In a striking comparison, only 1,600 people perished in the almost-simultaneous May 6 eruption of La Soufrière, which occurred on nearby St Vincent Island. Although both volcanoes erupted with the same explosive force (volcanic explosivity index 4), the difference in the death toll can be attributed largely to the fact that most of the people living in the region around La Soufrière were evacuated in a timely manner.

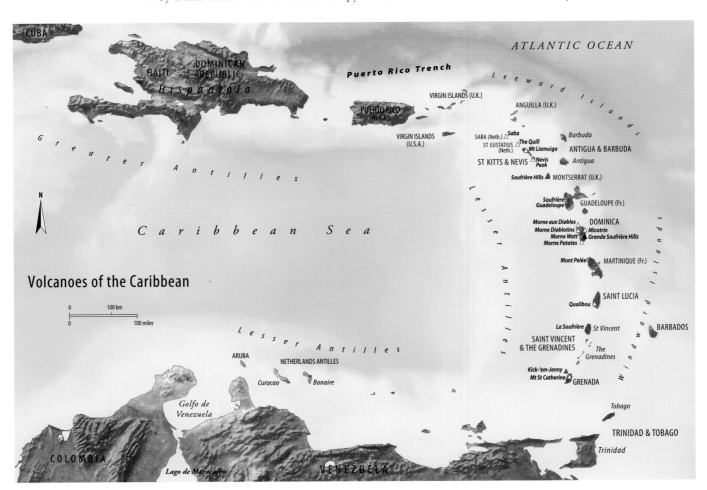

Volcanoes of the Caribbean

RIGHT The rugged peak of Gros Piton juts 2,619 feet (799 m) above sea level, near its smaller twin, Petit Piton. Both are the cores of volcanoes on the west coast of St Lucia. This island formed 50 million years ago when the Atlantic Plate thrust beneath the Caribbean Plate, creating the Lesser Antilles.

THE CARIBBEAN'S ACTIVE ISLAND VOLCANOES

La Soufrière, St Vincent, erupted explosively in 1718, 1812, 1902, 1971, and 1979. A heavy ash fall during the 1812 eruption killed 75 people, and in 1902 the northern third of the island was devastated within minutes by a pyroclastic flow. Before its last eruption in April 1979, there were about 24 hours of precursor activity. This advance warning was heeded and about 17,000 people were successfully evacuated, with no fatalities. La Soufrière is a highly active volcano and will continue to pose a threat to those living in its shadow. The populace should be prepared for an explosive eruption about once every century.

Soufrière, Guadeloupe, has erupted seven times since historical records began in 1690. The last eruption was in July 1976, with a central vent explosion and radial fissure eruptions. Pyroclastic flows and mudflows damaged land and property, causing severe economic disruption. Basse-Terre, the island's capital city, sprawling immediately below the volcano, had to be evacuated. The volcano has undergone several cycles of cone build-ing, explosion, and caldera collapse, beginning with the construction of the Grand Découverte volcano about 200,000 years ago. It was followed by growth and destruction of the Carmichaël Volcano then the Amic Volcano. The presently active Soufrière Volcano grew within the Amic crater, following collapse of the previous edifice in about 7490 BCE.

Soufrière Hills Volcano on Montserrat is also active. Soufrière means sulfur in French, explaining why there are three active volcanoes in the Antilles bearing that name.

Microearthquake activity at Morne Patates and Micotrin on Dominica, Liamuiga on St Kitts, Saba on the Netherlands Antilles, Mt Pelée on Martinique, and Qualibou on St Lucia indicate that these volcanoes will one day erupt again.

A DEADLY POLITICAL LESSON

The story of Mt Pelée's eruption at St Pierre is one to learn from—it tells about the dire consequences of fail-ing to heed the ample warning signs of an impending volcanic eruption. Despite ashfalls, mudflows, and explo-sions, local authorities were actively encouraging the residents to remain in St Pierre. Ironically, it was impor-tant to the politicians that everyone remained in town to vote in an election scheduled for May 10, 1902. The French Governor, Louis Mouttet, even attempted to stop the general panic by traveling to St Pierre to personally reassure the populace. He was killed by the eruption along with the rest. The election was never held.

St Pierre lay in ruins after the catastrophic eruption of Mt Pelée in 1902. The only land survivor was a prisoner in a cell.

ABOVE The rugged west coast of St Vincent is lined with exposed cliffs of volcanic rocks shrouded in vegetation. Lying in the southern part of the Lesser Antilles, the island is entirely vol-canic in origin and built principally of lavas of basaltic and andesitic composition. The island is young in geological terms, with all of its rocks being less than 3 million years old.

SOUFRIÈRE HILLS, MONTSERRAT

Perhaps one of the most interesting of the Antilles volcanoes is Soufrière Hills on Montserrat in the Caribbean Sea. It began to reawaken in July 1995. By 1997, the intensity and threat of its activity had forced over half the island's population to evacuate, leaving Montserrat's economy in ruins.

BELOW Houses were destroyed in the village of Dryers, about 2 miles (3 km) from English's Crater, in Montserrat, after the June 25, 1997, pyroclastic flow. Another eruption, the strongest, occurred on August 5 and almost razed the capital, Plymouth.

The volcano lay dormant since the island was first colonized in 1632. By 1995, Montserrat's population had grown to about 11,000, and Plymouth was its attractive colonial capital.

The volcanic island was born about 100,000 years ago, and its rocks tell a story of five major episodes of lava-dome growth and collapse during the last 30,000 years. Each involved considerable explosive activity, but none of that was apparent in the emerald isle's now-serene verdant peak. There were significant earthquake swarms and associated fumarolic outbursts in 1897–98, 1933–37, and 1966–67, but those were quickly forgotten. Plymouth continued to grow and prosper atop the relatively flat pyroclastic-flow deposits of the previous eruptions.

THE PRESENT ERUPTION BEGINS

For about 400 years, a dome had been growing quietly in English's Crater, which is hidden atop the peak just 2½ miles (4 km) east of the capital. On July 18, 1995 all that changed, with earthquake swarms and steam explosions caused by rapid heating of the groundwater by rising magma. The magma reached the surface by mid-November 1995, swelling the lava dome. In April 1996, pyroclastic flows, avalanching from the dome, filled the Tar River valley on the volcano's eastern flank, and by May they began building a delta into the sea on the east coast.

THE LAVA DOME EXPANDS RAPIDLY

The dome increased rapidly in size over the following months, repeatedly collapsing and rebuilding. Eventually it completely filled up English's Crater, overtopping the crater walls that had been protecting Plymouth to the west and the airport to the north. A safety zone was established but, on June 25, 1997, a major pyroclastic flow tore down the volcano's northeastern flank. It destroyed seven villages and killed at least 20 people who had not yet evacuated.

On August 3, 1997, another flow of heated rock, gases, and ash cascaded down the volcano, reaching Plymouth and setting fire to its buildings, including many of Plymouth's stately Georgian-style homes. Further eruptions began to bury the town beneath drifts of ash. With ash deposits of 4 feet (1.2 m) thick, Montserrat's citizens compared their capital to a modern-day Pompeii.

After scientists warned of the possibility of a "massive, cataclysmic" eruption of the volcano, its sovereign nation Britain decided to organize a voluntary evacuation. Islanders were offered £2,500,

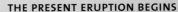

ROCK FEATURE: PUMICE

Pumice is a light-colored, silica-rich igneous rock formed during explosive volcanic eruptions such as the one at Soufrière Hills. It forms when highly gas-charged lava cools so quickly that the gas does not have time to escape. The result is a light frothy looking rock that is so full of bubbles it can float on water. If the gas pressure in the lava is high enough, the bubbles will explode, shattering the cooling pumice into myriad sharp glassy fragments known as volcanic ash. Huge rafts of pumice floating on the ocean often signal the presence of an erupting submarine volcano.

about six months' average wages, to assist with their relocation to Britain or elsewhere in the Caribbean—a difficult decision.

ANOTHER DEVASTATING BLAST

A violent eruption on August 6, 1997, left over 80 percent of Plymouth destroyed and even showered the northern end of the island with debris. Plumes of ash from this eruption reached heights of 40,000 feet (12,000 m) and dropped ash on the islands of Nevis and Antigua, which are over 25 miles (40 km) away. By early September, pyroclastic flows continued to press farther to the north, crowding the remaining 5,000 residents into shelters at the northern tip of the island.

On September 21, 1997, a large pyroclastic flow buried the airport runway and set the terminal on fire. After the flow reached the sea, the water heated to uninhabitable temperatures and many fish and lobsters were found washed up on shore. Thick ash also fell across the entire island. Since 1998, the British government and the United Nations have been working to replace the lost housing, hospitals, and the airport, in a major plan to rebuild on the northern part of the island.

During 2006, the lava-dome has continued to grow, with incandescent rockfalls visible at night, coursing down the sides of the dome. Two-thirds of the island is still considered unsafe, but the island's economy is slowly rebuilding in the north.

TOP An aerial view of Montserrat reveals the extent of the damage from the pyroclastic flows created by multiple eruptions in 1997. Access to the island is restricted due to increasing volcanic activity since January 2007.

ABOVE LEFT The courthouse in Plymouth was one of many buildings buried by the eruptions in 1997. Evacuation of essential services commenced on June 25, before the volcano was to destroy the city.

Volcanoes of South America

South America's volcanoes nestle within the mountainous Andes ranges along the continent's west coast. Offshore, a subduction trench, known as the Peru–Chile Trench, runs parallel to the mountains. It is into this deep slot that the eastern edge of the Pacific Ocean crust—the Nazca Plate—descends to its ultimate destruction beneath South America, thrusting up volcanoes all along the continent's western seaboard.

OPPOSITE Orsono Volcano in the Lake District of southern Chile last erupted in 1869. However, it is considered one of the most active in the southern Andes, having erupted explosively and producing many pyroclastic flows over the last 14,000 years.

BELOW The glacier-topped Villarrica Volcano in Argentina and Lanin Volcano in central Chile lie perpendicularly across from the Andes near the Chilean–Argentinian border. Villarrica was last known to be active in 2006 but Lanin has not erupted since 560 CE

Partial melting of the seafloor slab, and its overlying seafloor sediments, generates great buoyant bodies of molten magma—plutons of volatile andesitic magma that rise through the overlying crust along channelways of least resistance to feed the volcanoes.

The volcanic activity occurs in three discontinuous strips separated by two distinct volcanically inactive zones, one in Peru and the other in central northern Chile. Earthquakes occurring along the entire length of the subducting slab allow geophysicists to calculate the descent angle of its downward passage. Interestingly, the zones with no volcanoes correspond to regions where subduction is at a shallower angle, so there is insufficient melting to generate magma to initiate volcanism.

South America has more volcanoes than any other region in the world. Most of these are large glacially clad stratovolcanoes. Nevados Ojos del Salado in Chile is the world's highest volcano, with an elevation of 22,595 feet (6,891 m). The region's earliest recorded eruption was El Misti, near Peru's second-largest city, Arequipa, sometime after 1438. Apart from intermittent fumaroles, El Misti was most recently active in 1948 when explosions formed a scoria-rimmed crater. The volcano is considered a potential hazard being only 11 miles (18 km) from Arequipa.

SOUTH AMERICA'S DANGEROUS SENTINELS

Lascar Volcano, near Toconao, Chile, is one of the most active volcanoes in the central Andes. Fumarolic activity on the peak is continuous, interspersed with lava-dome growth and lava eruption. Occasionally, Vulcanian explosions throw ash several thousand yards (meters) above the crater.

In 1987, the small village of Talabre, situated in a steep-walled valley that lies just 8 miles (12 km) from the volcano, and significantly at risk from mud and pyroclastic flows, was abandoned and rebuilt on higher ground.

Galeras Volcano in Colombia has erupted more than 20 times since the 1500s. Volcanic activity has increased in frequency in recent decades, including the January 1993 eruption that killed nine people on a scientific expedition to the volcano's summit. In July 2006, explosion plumes blanketed surrounding villages in ash causing authorities to evacuate about 10,000 people closest to the volcano. Pasto, a city of 400,000 inhabitants, just 5 miles (8 km) to the southeast, is dangerously close.

Ubinas in Peru has been intermittently active, with persistent steam and fumarolic activity. Heightened explosive activity during May and June in 2006, with ash plumes reaching altitudes of 26,000 feet (7,900 m), prompted the issue of an orange alert. The volcano is considered to be a hazard for the many small villages built in a deep valley draining the southeastern side of the volcano. During the 2006 alert, approximately 550 families were evacuated.

SOME SERIOUS ERUPTIONS

Tungurahua, "the black giant," is one of Ecuador's most active volcanoes, having erupted historically at least 17 times. In 1995, it forced the temporary evacuation of more than 25,000 inhabitants of Baños and its surrounding villages. On July 14, 2006, Tungurahua exploded dramatically with falling incandescent rocks and pyroclastic flows destroying crops and killing farm animals on the volcano's flanks. It is estimated that more than 1 million people were affected. Most villages were safely evacuated in time, although their buildings were destroyed. There were 60 fatalities and an estimated 4,000 people were displaced. These villages will be permanently relocated to land paid for by the Ecuadorian government.

Key to Volcanoes
1. La Cumbre
2. Darwin
3. Wolf
4. Alcedo
5. Sierra Negra
6. Mojanda
7. Imbabura
8. Cayambé
9. Chachani

Major Volcanoes of South America

Perhaps the region's most tragic eruption was that of Nevado del Ruiz in Colombia in 1985. It was not a large eruption, but sufficient to melt glacial ice and snow at the volcano's summit, triggering a massive mudflow. The vibrant town of Armero was buried within minutes and about 27,000 people lost their lives.

Volcanoes of Europe

Europe's volcanic chain stretches from Spain to the Caucasus. It runs through southern Italy (the Calabrian Arc), the Greek Islands (the Hellenic Arc), and Cyprus (the Cyprean Arc). Most of its volcanoes are inactive, but several, such as Italy's Mt Etna and Mt Vesuvius, still pose a continuous threat to their surrounding populations.

RIGHT Stromboli Volcano is one of Italy's Aeolian Islands. Italy's second most active volcano, its minor eruptions made it a tourist attraction. More violent eruptions starting in February 2007 opened two new fissures and those in March sent lava flows into the sea. Its last major eruption was in December 2002.

BELOW Vesuvius looms behind Italy's city of Naples. This tranquil scene belies the fact that its population is at threat if there were an imminent eruption. The prevailing winds put Naples at high risk. The current evacuation plan to remove 60,000 residents will take 72 hours to carry out, and is inadequate if a highly explosive eruption were to occur.

These volcanoes lie along the collision line that occurs between the Eurasian Plate and the northward-moving and subducting African Plate. Both the Aegean and Anatolian (Turkish) micro-plates in the Mediterranean push the subduction boundary into a sweeping recumbent curve around the Adriatic Sea due to the strong westerly component to their movement.

EARLIEST RECORDS, FIERY GODS
The region has a tradition of record-keeping going back thousands of years and is the cradle of modern volcanology. The word volcano comes from Italy's island of Vulcano, the legendary fiery forge of Vulcan, the Roman god of fire. The earliest record of volcanic activity is an Anatolian wall painting of a cinder cone erupting in 6200 BCE, and record-keeping of Mt Etna's many eruptions began in 1500 BCE. Pliny the Younger witnessed and described the catastrophic eruption of Vesuvius in 79 CE, which buried Pompeii. The world's first volcano observatory was built in 1845 to keep an eye on Vesuvius.

STROMBOLI'S HISTORY OF VIOLENCE
Stromboli Island, in the Calabrian Arc, is one of the most active volcanoes on Earth, having been in almost continuous eruption for over 2,000 years. Several small gas explosions occur each hour, throwing incandescent blobs of lava above the

crater rim, while lava flows are less frequent. The term Strombolian is used to describe this type of volcanic eruption elsewhere.

Violent eruptions are rare at Stromboli, but in 1919, four people were killed and several homes destroyed by huge blocks ejected by a series of powerful blasts. Pyroclastic flows killed four people in 1930, and in 1986 a biologist on the crater rim was killed by a falling block.

VESUVIUS: A THREAT TO A VAST METROPOLIS

Mt Vesuvius is perhaps Earth's most potentially dangerous volcano owing to its proximity to a vast metropolis. The population of Naples, in its immediate vicinity, has grown to over 3 million, with some 600,000 people living within the immediate area of the cone or red zone. The red zone is the area that would be destroyed in an eruption with a volcanic explosivity index of 4. That is an order of magnitude smaller than the Mt St Helens eruption of 1980, which killed only 57 people because of its relative isolation.

Mt Vesuvius is best known for its catastrophic eruption of 79 CE that buried the towns of Herculaneum, Stabiae, and Pompeii beneath layers of cinders, ash, and mud. About 2,000 people died in Pompeii alone when a dense cloud of hot ash and gas surged down the mountainside. This eruption was the first in history to be documented in detail by Pliny the Younger. Volcanologists now use the term Plinian to describe explosive volcanic eruptions that generate high-altitude eruption columns and blanket large areas with ash.

SANTORINI AND THE LEGEND OF ATLANTIS

In 1640 BCE, the Minoan civilization, one of Europe's first high-cultures, was destroyed by a volcanic eruption. The explosion (volcanic explosivity index 6) literally blew apart Santorini Island in the middle of the Hellenic chain, and probably gave rise to the legend of the sunken city of Atlantis. About 7 cubic miles (30 km³) of material was ejected, with the Plinian column reaching a height of 23 miles (37 km). The volcano collapsed into its submarine magma chamber leaving only a ring of land—now the holiday isles of Thera, Therasia, and Aspronisi. In 197 BCE, Santorini's new cone began growing from the sea. It was the first documentation of such an event. While Santorini has been dormant since 1950, this does not rule out the possibility of a future eruption, which would threaten the Island's 10,000-strong population. Other potentially active Hellenic Arc volcanoes are Milos, Kos, Nisyros, and Yali.

The entire island of Santorini is a volcanic caldera, which features plunging cliffs dropping almost vertically into the sea. Santorini's Volcano Observatory keeps checks on the level of activity.

MT ETNA, ITALY

Sicilians do not regard Mt Etna as being particularly dangerous, referring to it as their "good volcano." Thousands live on its flanks; its fertile volcanic soils are perfect for agriculture, with the mountain slopes a green patchwork of countless vineyards, olive groves, and orchards. Catania, Sicily's largest city, with its population of over 1 million, sprawls across the lower slopes of the volcano.

Despite its apparent tranquility, Mt Etna can be very destructive. It is Europe's largest and most famous volcano, with the longest historic record of eruptions. The first was recorded about 1500 BCE when a large explosion caused its eastern flank to catastrophically collapse in an enormous landslide. Diodore of Sicily described the event, which reads similarly to the well-known Mt St Helens eruption of 1980.

MT ETNA'S DESTRUCTIVE PAST

One of the most dramatic eruptions of Mt Etna began on March 11, 1669, when a fissure opened on the volcano's south flank. Activity slowly moved down the flank with the biggest vent opening up near the town of Nicolosi. Powerful explosions began to build the Mt Rossi cinder cone and exude voluminous lava. Sections of Nicolosi and nearby villages were quickly incinerated. Fortunately, most villagers had abandoned their homes because of increasingly strong earthquakes in the two months prior to the eruption. Lava flowed down the volcano's flank, overwhelming numerous hamlets, villages, and towns in its path. Refugees fled ahead of its advancing front to Catania and were housed there.

In early April, the lava reached the city wall on the western side of Catania and began piling up against it. The flows were initially deflected to the south by the wall and began filling the city's harbor. Lava kept rising until eventually, on April 30, the pressure grew too strong and a section of the wall collapsed allowing lava to enter the city. It destroyed monasteries and other buildings around the church of San Nicolò l'Arena. Residents rallied and quickly built walls across the roads leading down to the lower part of the city in an attempt to stop the flows. Although much destruction ensued in the upper part of the city, the strategy worked and widespread damage was avoided. The eruption finally ceased on June 11.

The most destructive eruption of the twentieth century began on November 2, 1929. Explosions and plumes began near the volcano's summit, propagating along opening fissures down the eastern flank. On November 4, the lowermost fissure opened and lava began pouring into a ravine above the town of Mascali. The next day there was an orderly evacuation of Mascali's 2,000

ABOVE A fiery new cone on Mt Etna glows in Sicily's night sky in 2002. For 24 days during the summer of 2002, Etna gave its most dazzling show in a decade. It again spouted lava fountains in May 2007 after a gash opened. It is the highest active volcano in Europe, reaching 10,910 feet (3,328 m).

RIGHT Being a highly active volcano, Mt Etna is one of 16 volcanoes around the world listed as a decade volcano. These are intensely active volcanoes, often near large populations, that have exhibited highly destructive eruptions in the past and are considered worthy of intensive study.

inhabitants, before the lava flow advanced slowly through the entire town, consuming its buildings and cultivated land. The cathedral was the last building to be buried. By November 7, Mascali was gone. The eruption stopped on November 20. No one was killed, and residents had time to save their lives and household goods, even removing roof tiles in front of the slowly advancing lava.

THE FUTURE OF MT ETNA

The volcano continues to be active with some of its more recent eruptions being quite explosive. Monitoring systems on the volcano, capable of tracking sub-surface movements of magma, now warn civil protection authorities days to months in advance of an impending eruption. Micro-earthquakes indicate that, since 1994, magma has been rising from a chamber 4 to 9 miles (6 to 15 km) beneath the volcano and is filling a shallower chamber 2 to 3 miles (3 to 5 km) below the surface. The buildup in pressure in the shallow storage area appears to have triggered the eruptions that took place in 2001 and 2002. Volcanologists are concerned that this explosive pattern of behavior will continue. Flank eruptions are occurring and are expected to continue at intervals ranging from one to three years, some of which might be much more voluminous and explosive than in the past.

LEFT Closeup of the sulfurous deposits found on Mt Etna. Steaming fumeroles on volcanoes release gases including noxious sulfur dioxide, which precipitates into sulfur deposits. Along with this gas, copious amounts of carbon dioxide are released that have altered the atmosphere for years following a major eruption.

Volcanoes of the Western Pacific Rim

The Western Pacific Rim is made up of Kamchatka Peninsula, the Kuril Islands, Japan, the Ryukyu Islands, the Philippines, and the Izu-Mariana islands. They are all arcuate chains of volcanic islands that form the western edge of the "Ring of Fire." The term arc was coined to describe these active volcanic island archipelagoes.

OPPOSITE Black sand beaches line the eastern shores of Iwo Jima (sulfur island), a volcanic island in the Magellan archipelago south of Japan. It is known as Invasion Beach, as it was the site of a major battle during World War II. The 550-foot-tall (168-m) volcanic vent, Mt Suribachi, is in the background.

BELOW Colonies of filamentous bacteria thrive in a hot spring of a volcano on the Western Pacific Rim. These heat-loving bacteria live in the vent waters of an active volcano, which range from 122° to 180°F (50 to 80°C).

These volcanic arcs are separated from the Asian mainland by marginal seas—the Sea of Okhotsk, the Sea of Japan, the East China Sea, and the South China Sea. Offshore, to the east of the arcs, lie the deep narrow Japan, Kuril, and Mariana trenches. The deepest point in the ocean is recorded at 36,160 feet (11,029 m) in the Mariana Trench. The trenches mark the line along which the Pacific Ocean crust is disappearing beneath the island archipelagos, dragged downward by convection currents in Earth's mantle. The descending ocean crust melts to form giant bodies of magma that rise to feed the arcs' continuous volcanism.

Many islands are single volcanoes with the larger landmasses, such as Japan's island of Honshu, built up from coalesced cones. In the distant future, these islands will be crushed against the Asian mainland as the Pacific continues to close.

JAPAN'S REVERED VOLCANOES

Mt Fuji is the highest volcano in Japan, standing 12,388 feet (3,778 m) above sea level, and revered by the people for its perfectly conical symmetry. The present cone began to grow some 11,000 years ago, forming from alternating periods of lava flows and explosive eruptions. Fuji is currently quiet, its last eruption being in 1707 when it threw ash over Tokyo from a new crater on its east flank. Aso is Japan's most active volcano, with more than 165 eruptions since its first documented eruption in 710 CE. However, it was 6,300 years ago that Kikai produced one of Earth's largest Holocene Epoch explosive volcanic blasts. Based on the extent and thickness of the ash and tephra deposits surrounding the volcano, it was determined to have erupted with a volcanic explosivity index of 7.

Unzen is a large volcano made of several overlapping lava domes located dangerously close to

RIGHT The north face of Mt Fuji at dawn, Japan's most revered volcano. History records it has erupted 17 times since 81 CE. Its last eruption in 1707–08 was one of the largest recorded. It lasted 16 days, blew out a flank, and produced an ash column 6 miles (10 km) wide. Low level seismic activity around 2002 has led to increased surveillance.

the city of Nagasaki. In 1792, a major collapse of one of the lava domes resulted in a pyroclastic avalanche and tsunami that killed 14,524 people. Most were killed by the tsunami. Unzen lay dormant for 198 years after its 1792 eruption, before coming to life again on November 17, 1990. Continuous tremors beneath the volcano indicated the movement of new magma and its volcanic dome began to grow over the years from 1991 to 1994. During this time, some 10,000 pyroclastic flows, generated by the collapse of lava blocks along the edge of the dome, were recorded on Unzen. On June 3, 1991, 43 people, including volcanologists Maurice and Katia Krafft and Harry Glicken, were killed by one of these pyroclastic flows. Unzen, along with Sakura-jima near Kagoshima, is a decade volcano, earmarked for detailed observation and research due to their threat to nearby large population centers.

TOKYO'S GREAT EARTHQUAKE THREAT

The region around Tokyo is particularly active seismically as here the "Ring of Fire" divides into two around the Philippine Sea. The Izu-Mariana Islands form the easternmost or outer arc, while south Japan, the Philippines, and the Ryukyu Islands form the westernmost or inner arc. The stresses in the crust here are immense, and the stored energy is periodically released in the form of catastrophic earthquakes. The Great Kanto Earthquake of September 1, 1923, completely destroyed the city of Yokohama and 70 percent of Tokyo, killing 143,000 people. September 1 is remembered nationally in Japan as Disaster Prevention Day and much effort is spent in preparing residents for the next big earthquake.

THE IZU-MARIANAS ARC

The Izu-Marianas Arc currently contains 42 volcanoes. Twenty of these are submarine, still growing toward the ocean surface. Anatahan became active in March 1990, resulting in the evacuation of the small island's 23 residents. By April, its crater lake had boiled away. Sarigan Volcano has also shown recent signs of unrest with hundreds of microearthquakes being recorded during August 2005.

Volcanoes of the Western Pacific Rim

KAMCHATKA'S LINE OF GIANTS

With over 100 active volcanoes and many inactive, Russia's Kamchatka Peninsula in its far southeast is one of the most volcanically active places on Earth. The volcanoes here form a belt stretching 430 miles (700 km), with about 30 volcanoes having erupted recently. Presently, the most active volcanoes on Kamchatka are Klyuchevskoy, Shiveluch, Karymsky, and Bezymianny.

Karymsky is the most active of these volcanoes. The 6,100-year-old volcano erupted 23 times dur-

ing the twentieth century, with the last being in 1996. Eruptions beneath Karymsky Lake, 4 miles (6 km) from the volcano, generated tsunamis and turned the lake waters acid, killing its freshwater salmon population. It is currently on orange alert.

Klyuchevskoy is Eurasia's largest active volcano, with its perfectly symmetrical cone reaching a height of 15,575 feet (4,750 m). This in an 8,000-year-old giant that became active again in 1697 and since then eruptions have been occurring about every five years or less. Its activity has not

ABOVE Volcanic ash rises above Klyuchevskoy in Kamchatka during its major eruption in October 1994. This northeasterly view shows the plume from the newly erupted volcano. It was photographed by NASA astronauts 115 nautical miles above Earth in the space shuttle *Endeavour*.

ROCK FEATURE: ANDESITE

Andesite is one of the most common igneous rocks found in subduction zones such as the Pacific "Ring of fire." It has cooled from the intermediate-composition lavas erupted by stratovolcanoes. It sometimes contains well-formed crystals of feldspar within its very fine-grained or glassy matrix. The andesite line was recognized, marking a major change in regional geology from the basaltic volcanic rocks of the Central Pacific Basin to the andesitic rocks around its continental margin. Around the Pacific margin, the andesite line follows the deep subduction trenches.

yet threatened the inhabitants of Kluchi, a town that lies 20 miles (30 km) away from the volcano.

Shiveluch is Kamchatka's northernmost active volcano, standing 10,771 feet (3,285 m). Written records of its activity, dating back to 1739, describe large Plinian eruptions in 1854, 1964, and most recently in May 2001. Due to its frequent and violent explosive eruptions, Shiveluch is a potential hazard for the nearby towns of Kliuchi and Ust'-Kamchatsk, and also for aviation routes between the USA and northern Asia. It is on orange alert.

THE PHILIPPINES' DEADLIEST VOLCANOES

Philippine volcanic eruptions, such those of Taal, Mayon, and Pinatubo, have had a higher-than-usual impact. Lahars (mudflows) are common due to the heavy rainfall in this typhoon-plagued archipelago, and secondary mudflows remain a hazard long after the volcanic activity ceases.

Beautifully symmetric Mayon is the Philippines' most active and notoriously destructive strato-volcano. Although it has erupted about 50 times since 1616, many of its residents still farm its fertile slopes. Legazpi City, population 157,000, lies only 9 miles (15 km) to the southeast of the volcano. Its most violent manifestation was in 1814, when more than 1,200 people were killed. The entire town of Cagsawa was buried by a mudflow. Mayon last erupted in 2001, forcing the evacuation of about 50,000 people. There is a 4-mile (6-km) permanent danger zone around the volcano's crater, and in 2006 the Philippine volcano observatory warned that an explosive eruption could occur at any time.

Pinatubo Volcano awoke from 450 years of quiescence on April 2, 1991. As the situation worsened, Philippine authorities established a 12-mile (20-km) exclusion zone around the volcano, resulting in the evacuation of 58,000 people. On June 15, two explosions occurred at dawn, accompanied by incandescent pyroclastic flows. This violent climactic eruption (volcanic explosivity index 6) ejected a 22-mile (35-km) high ash column. The volcano's summit was blasted away, creating a new caldera $1\frac{3}{5}$ miles (2.5 km) across.

The eruption of Pinatubo coincided with the arrival of Tropical Storm Yunya. Ash in the air mixed with the storm's water vapor, causing heavy black-ash rain to fall across the island of Luzon. The dense wet mixture built up quickly, causing the roofs of many buildings to collapse on people. In total, 722 people died in the volcanic eruption and associated mudflows. The area will take many decades to recover. Bare ash deposits on the volcano's flanks are seasonally remobilized by monsoon and typhoon rains to form giant mudflows.

Mt Pinatubo caused the greatest stratospheric disturbance since the eruption of Krakatau in 1883. Suspended fine ash spread around the globe in two weeks and within a year had covered the planet, causing global temperatures to fall by about 1°F (0.5°C). Pinatubo also ejected some 20 million tons (22 million tonnes) of sulfur dioxide into the atmosphere. The sulfur combined with atmospheric water and oxygen to form sulfuric acid, which fell as acid rain. Spectacular sunsets were witnessed around the world for several years following the eruption, and during 1992 and 1993, the ozone hole that lies over Antarctica was much larger than normal.

BELOW Steller's sea-eagle (*Haliaeetus pelagicus*) inhabits seashores of eastern Asia. Their breeding ground is in Kamchatka, where they build nests up to 6 feet (1.8 m) wide. Also common in the peninsula are brown bears, salmon, and sea otters.

BOTTOM View of three volcanoes in Russia's Kamchatka Peninsula: Bezymianny (front), Kamen (center), and Klyuchevskoy (background). Once thought to be dormant, Bezymianny exploded in 1956. An eruption in December 2006 destroyed its lava dome.

Volcanoes of Indonesia

Indonesia's abundant earthquakes and its belt of active subduction volcanoes, including the highly explosive volcanoes Merapi, Krakatau, and Tambora, make it tectonically speaking, one of the most dangerous places on Earth to live. Sweeping volcanic ash off the floors is a daily chore for local residents.

ABOVE A flowering vine establishes new life on the barren ash of Anak Krakatau's flanks. The ash produced by volcanoes can act like a type of nutrient-rich mulch. Seeds can arrive in many ways; birds drop them or they may be blown in by the wind.

The Indonesian island archipelago is a sweeping curve made up of more than 13,000 islands. It is built from rising bodies of magma that became melted during the continuous down-thrusting of the Indo–Australian Plate beneath the Eurasian Plate. The subduction takes place in the deep Sunda Trench along the southwestern edge of the island archipelago at a rate of about 2½ inches (6.5 cm) per year. Vast bursts of earthquake energy are released as the seafloor crust shudders, jerks, and cracks on its downward journey to its ultimate destruction by melting.

Indonesia has a large number of historically active volcanoes (76) and, when combined with Japan, these two regions together produce around a third of the world's explosive volcanic eruptions. Indonesia's eruptions produce the highest number of fatalities, damage to farmland, tsunamis, mudflows, and pyroclastic flows.

TAMBORA, HISTORY'S LARGEST ERUPTION

Tambora Volcano had been in repose for some 5,000 years when on April 5, 1815, several thunder-like blasts, heard at a distance of 870 miles (1,400 km), heralded its reawakening. A still-larger eruption occurred on April 10 to 11, described as "three columns of fire rising to a great height," which ejected about 12 cubic miles (50 km^3) of material. The eruption, one of the largest in 10,000 years (volcanic explosivity index 7) killed 92,000 people, and blew the top off the stratovolcano, leaving a deep summit caldera. Earthquakes from the blast were felt up to 300 miles (500 km) away.

The massive quantities of fine volcanic ash that Tambora ejected into the stratosphere—some reports suggest it was as much as 100 cubic miles (400 km^3)—spread quickly. It lowered global temperatures by as much as 5°F (3°C), with the northern hemisphere being particularly affected. In parts of Europe and in North America, 1816 was known as "the year without a summer," with significant crop failures resulting in widespread famine. A new cone, Doro Afi Toi, has since started to grow within Tambora's summit caldera.

KRAKATAU'S VOLCANIC CYCLE OF FURY

On the morning of August 27, 1883, Krakatau in the Sunda Straits erupted, killing 36,400 people from the eruption and subsequent tsunami. The five enormous explosions that took place were the loudest sounds in recorded history. Black clouds of searing ash and gas surged outward from the collapsing eruption columns, obliterating all life on the nearby islands. The accompanying tsunamis, reaching heights of 130 feet (40 m), were even more devastating, wiping out 165 villages without a trace. Floating pumice clogged the Sunda Straits and minute particles of ash circled the equator within 13 days, causing exotic green or blue halos around the sun and moon. Three months later, the ash had spread to higher latitudes, creating vivid red sunsets. Global temperatures lowered by as much as 2.2°F (1.2°C) in the following year and did not return to normal until 1888.

KRAKATAU'S CHILD IS BORN

The eruption of 1883 blew the top 4 cubic miles (18 km³) off the summit of Krakatau, which originally stood 2,720 feet (830 m) above sea level. This left behind only ocean—about 820 feet (250 m) deep. Two-thirds of the original 5½ by 3 mile (9 by 5 km) island had vanished, believed collapsed into Krakatau's submarine caldera, emptied by the powerful eruption (volcanic explosivity index 6).

On December 29, 1927, anglers from Java noticed plumes of steam and debris shooting from the sea. They were witnessing the birth of Anak Krakatau, which translates to "child of Krakatau." Since then a cone has emerged from the sea, built up from layer upon layer of ash deposits and andesitic lava flows. Since the 1950s it has continued to grow at an average rate of 5 inches (13 cm) per week to the point where in 2006 it was over 980 feet (300 m) high. Anak still has some way to grow before it reaches the height of its original parent volcano and is ready to complete the next explosive cycle.

MERAPI, THE MOUNTAIN OF FIRE

The top of Merapi has no vegetation because erupted ash often falls there, and because of the *nuées ardentes* resulting from crumbling of the summit lava dome. Dense vegetation covers the volcano's lower flanks, where many farmers live. The volcanic ash makes rich soil for growing crops, but it is a dangerous place to live. A particularly devastating eruption took place in 1930, which killed 1,300 people. Another, in 1976, killed 28 people and left 1,176 people homeless. In 1979, heavy rainfall turned lahar deposits into mudflows that

LEFT Like a dot-to-dot puzzle, volcanoes line the edge of Indonesia's islands. This archipelago nation sits on one of the world's most active subduction zones, the sinking tectonic plate pushing up volcanoes that caused the world's biggest eruptions.

ABOVE Anak Krakatau erupting in May 1994, during one of its many eruptive periods since 1883, with ash-ejecting explosions every 20 minutes or so. Blocks, ash, and lava were hurled out over several months.

BELOW Mt Bromo is an active volcano in the Bromo-Tengger-Semeru National Park, Java, in Indonesia. It has erupted more than 60 times since 1767—an eruption in 1966 killed 39 people and another in 1994 killed two tourists walking on the rim. It is currently on orange alert.

THAILAND

South China Sea

M A L A Y S I A

BRUNEI

Awu

Pulau Sague

Kerangetang

Simeulue

SINGAPORE

Tongkoko
Mahawu
Soputan

Nias

*Sumatera
(Sumatra)*

Borneo

Colo

Molucca

Equator

Tandikat   *Merapi*
          Talang
       Kerinci

*Sulawesi
(Celebes)*

Macassar Strait

Kaba

B

Dempo

Java Sea

I N D O N E

Krakatau Tangkuban
          Parahu
       Gede   Cereme
     Guntur   Slamet
  Papandayan Galunggung Dieng
          Sundoro Merapi
                   Kelut   Tengger
                Semeru Raung Ijen Batur *Bali* Rinjani Tambora Sangeang *Flores Sea* Batu
                 Api   Api   Tara
                                                        Agung   Leteboleng *Flores* Sirung
                                                                                    Iya Egon Iliwerung
                                                                                 Lewotobi

*Jawa
(Java)*

Lombok *Sumbawa*

Sumba

Tir

Volcanoes of Indonesia

INDIAN OCEAN

0 ___ 200 km
0 ___ 200 miles

surged 12 miles (20 km) down the flank of the volcano. Eighty people were killed. Then, in 1994, a dome collapse sent pyroclastic flows down the volcano's southern side, killing 43 people. Today, 50,000 people live on the volcano's flanks, and the city of Yogyakarta (population 3 million) lies only 22 miles (35 km) to the south.

MERAPI'S IMMINENT EXPLOSION

Because of Merapi's extremely violent history and close proximity to Yogyakarta, it was designated a decade volcano as part of the International Decade for Natural Disaster Reduction (1990–2000). It is subject to increased surveillance by the Volcanological Survey of Indonesia.

Early in April 2006, the volcano showed increasing seismicity, and a bulge was detected in its cone. By April 23, smoke plumes and numerous micro-earthquakes signaled subterranean movement of magma. By May, 17,000 people were evacuated from the area; then, during a week of intermittent lava flows, all residents were ordered

On the morning of May 27, 2006, a devastating magnitude 6.3 earthquake struck the densely populated region on the southern flank of Merapi. During the 57-second earthquake, tens of thousands of poorly constructed buildings collapsed, killing over 6,000 people and injuring seven times that number. The town of Bantul, closest to the epicenter, was worst hit, with 80 percent of its buildings destroyed. In Yogyakarta many of the city's buildings suffered severe structural damage. In total, some 67,000 houses were destroyed, leaving up to 200,000 people homeless. With the memory of the 2004 earthquake and tsunami fresh in their minds, coastal residents fled inland as soon as they felt the shaking. Fortunately no tsunami eventuated due to the location of the earthquake's epicenter near the volcano rather than in the ocean. Two aftershocks of magnitude 4.8 and 4.6 followed several hours later, causing further panic among the survivors.

An aerial photo taken on May 31, 2006, reveals the extent of structures damaged or destroyed in the Klaten district in Yogyakarta. The epicenter was about 10 miles (16 km) southwest of the city.

off the mountain. Volcanic activity had begun to calm by May 16, and villagers started to return to their villages to tend their livestock and crops. But then something unexpected happened: a large earthquake struck nearby. Micro-seismic activity jumped to three times its pre-quake level and Merapi became active again.

EARTHQUAKE SPECULATION

Government response to the earthquake was terribly slow, with survivors facing a lack of clean water, electricity, food, shelter, and medical supplies. Relief funding was also inadequate, and even six months after the disaster 40 percent of all those who lost their housing still had insufficient shelter. To address the problem of providing urgent shelter for approximately 50,000 needy families, the government adopted a "roof-first" strategy so that most could have at least some shelter for the rainy season.

Increased volcanic activity during June was blamed on the earthquake. Scientists believed that

the earthquake may have cracked the magma dome growing at Merapi's summit. There were grave concerns that it could collapse, spewing out massive quantities of volcanic rock and lava. However, by the end of June, the alert level was downgraded. That means an eruption was no longer imminent, but it doesn't mean one will not occur in the near future.

LEFT Local tourists visit Kaliadem Resort area on June 17, 2006 where two search-and-rescue members were trapped in a bunker and baked to death two days after a cloud of ash and hot gas descended from Merapi. This volcano has the unfortunate distinction of producing more *nuées ardentes* or pyroclastic flows than any other volcano.

OPPOSITE Merapi on June 11, 2006, spewing lava and ash. Merapi, meaning fire mountain, has erupted 68 times since 1548 and produced many deadly eruptions, including a massive pyroclastic flow in 1930 that killed 1,400 people.

Volcanoes of the Southwestern Pacific Rim

The relentless push of the Pacific Ocean seafloor westward and beneath the Australian Plate edge has created a long boomerang-shaped trench, the Tonga–Kermadec Trench, and an archipelago of little-studied volcanic islands and submarine volcanoes. The trench runs north from New Zealand's Taupo Volcanic Zone then hooks to the west around the donut-shaped Tongan island of Niuafo'ou, a volcano with a large lake-filled caldera.

BELOW Ruapehu's crater lake in the Tongariro volcanic complex can indicate changes in the activity of the volcano's magma chamber. An increase in the amount of gas released alters the color of the lake. The striking blue-green means there is no activity; gray indicates a change.

In New Zealand, most volcanism during the last 1.6 million years has occurred in the Taupo Volcanic Zone (TVZ), a deep elongate volcano-tectonic depression running across the North Island. It includes the active Ruapehu, Tongariro–Ngauruhoe, and White Island cone volcanoes, and the Okataina and Taupo calderas.

Ruapehu, rising 9,175 feet (2,798 m), is a complex andesite stratovolcano constructed during at least four cone-building episodes. The Tongariro volcanic complex, located northeast of Ruapehu, is composed of over a dozen composite andesitic cones built over a period of 275,000 years. The youngest cone, the symmetrical, steep-sided Ngauruhoe, has grown into the highest peak of the complex in just 2,500 years. Ngauruhoe and Ruapehu have been New Zealand's most active volcanoes during historical time.

A CITY ON A VOLCANIC FIELD

Auckland, New Zealand's largest city, is built on an active volcanic field created over the last 150,000 years, with its youngest volcano Rangitoto rising above sea level in about 1400 CE. Rangitoto forms a near-perfect cone at the entrance to Waitemata Harbour. Eruptions from numerous small basalt volcanoes have created an alluring landscape of cones, craters, and lava flows in the downtown area. One example is the city lookout atop Maungarei (Mt Wellington), the field's largest scoria cone. They are mostly short-lived volcanoes, formed during a single eruption, or series of eruptions lasting perhaps a year or two. The field's next eruption will probably occur in a new location.

THE TONGAN ISLANDS

The Tonga group is made up of about 14 active volcanoes. Over half of these are submarine volcanoes that will someday grow into new islands. Metis Volcano first rose above sea level in 1851, but the temporary island was a soft cinder cone that was quickly destroyed by ocean waves. Home Reef Volcano built a short-lived island in 1984 and more recently, in August 2006, a new island appeared accompanied by huge rafts of floating pumice. Locally, these temporary islands are known as *fonuafo'ou*, which loosely translates as "jack-in-the-box," and are a navigation hazard to shipping.

Niuafo'ou is the northernmost island in the Tonga group. It is an active volcanic-rim island with a population of approximately 750 residents. In 1853, explosive eruptions destroyed the village of 'Ahau, killing 25 people. Lava flows from the

1912 and 1929 eruptions destroyed the village of Futu and closed the harbor. Other eruptions occurred in 1935, 1936, 1943, and 1946. The 1946 eruption was so violent that Niuafo'ou's inhabitants were evacuated to another island and not allowed to return until 1958.

Tonga's most active volcano, Tofua is probably best renowned as the site of the mutiny on the *Bounty* in April 1789, during which the crew set Captain Bligh adrift and took command of his ship. Tofua, with a population of about 50, is a ring of land surrounding the summit caldera of a steep-sided submarine cone. A new cone, Lofia, is growing in the northern edge of the lake-filled caldera. In 1958–59, explosive eruptions from Lofia forced most islanders to evacuate for a year or more.

ABOVE The birth of a new volcanic island near Tonga on August 11, 2006, was witnessed by the crew of the yacht *Maiken*. The skipper reported the sea turned to rock, and they found themselves sailing through a raft of pumice, gradually thickening and behaving like concrete. The next day, sailing toward the source, they watched as the fledgling volcano emerged from the sea. It may be the first time anyone has seen such an event.

ISLAND OF LEGENDS

Uninhabited White Island, off the coast of New Zealand's North Island, is the emergent tip of a submarine volcano. Its intermittent steam and tephra eruptions have become an integral part of Maori legends. Sulfur was mined on the island until 1914, when a portion of the crater wall collapsed, burying the mine and miners' quarters. The landslide killed 12 people. Mining resumed again for a short time during the 1920s. Ash plumes from White Island are often seen from the shores of Bay of Plenty. Despite this harsh environment, the island hosts several bird species including a gannet colony.

LEFT New life in deep-sea vents was discovered on a five-month voyage by researchers from Hawaii's Undersea Research Laboratory. Previously unexplored submarine volcanoes in the Southwest Pacific yielded an array of unknown life forms that live in these gaseous high-temperature conditions, including an unusual tube worm that is possibly new to science.

Oceanic Hotspots

Fixed hotspots, located deep in the mantle, spout lava through cracks and weaknesses in the ocean crust to form the most intriguing of volcanic islands. Flow upon flow of basaltic lava eventually builds giant shield volcanoes—broad, smooth, gently sloping edifices that grow from the seafloor. They can reach several miles above sea level in less than a million years.

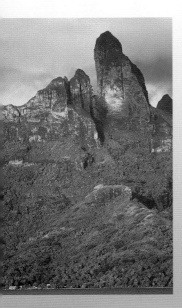

ABOVE Bora Bora is an island in French Polynesia. At its center are the remnants of an extinct hotspot volcano that created its highest peak, Mt Otemanu.

Tectonic forces pushing Earth's crust over the hotspots, like cloth passing over a giant sewing machine, slowly move the currently active volcano off the hotspot and, in turn, a new part of the seafloor is punctured by rising magma, beginning the island-building process anew.

AN ISLAND IS DISCONNECTED, FOLLOWED BY A SLOW DANCE WITH DEATH

Once disconnected from its hotspot source, the volcanic island is doomed to die over the course of another several million years. Ravaged by the forces of weathering and erosion, the island is slowly washed away and leveled flat. Furthermore, the immense weight of the edifice causes an isostatic readjustment of the seafloor and the island slowly disappears beneath the waves.

Amazingly, in the tropics, coral reefs growing around the flanks of the volcano continue to grow upward as the island ultimately sinks. This occurs very slowly but eventually, when the volcano has become completely submerged, all that is left is a ring of coral surrounding a flat lagoon. This is known as an oceanic atoll.

CHARLES DARWIN AND THE VOYAGE OF THE *BEAGLE*

Charles Darwin did not know anything about plate tectonics, as the concept was only proven during the twentieth century, but he was intrigued by the consistent differences between volcanoes within the island chains.

Sailing along the Society Island chain in 1835, Darwin noticed how the islands became progressively older to the northwest, and he prepared a series of sketches depicting these changes. In the southeast was Mehita, a young-looking volcano with its edges only slightly eroded and a small fringing reef of coral. To the northwest, Darwin studied Tahiti, which he recognized as a more worn-looking volcano surrounded by a barrier reef. Moving further along the chain Darwin drew Bora Bora and Maupiti, noting that they were still older-looking, until he finally sketched an atoll, with no remaining volcanic central peak. Coral can only grow in warm waters, so the ultimate fate of an island carried out of the tropics by the plate-tectonic conveyor belt is to sink and become a submerged reef.

HOTSPOT CHAINS ON LAND

Hotspot volcanoes can erupt through continental crust as well as in the ocean. One example is the line of volcanic remnants running along the eastern Australian highlands. This inland line, which was built by the East Australian hotspot, parallels the Lord Howe and Tasmantid oceanic hotspot chains located off Australia's east coast. These hotspots have been building volcanoes since the Australian Plate began its northward journey after breakup from Antarctica some 95 million years ago. The East Australian hotspot is currently located in Bass Strait, north of Tasmania. Another classic example of a sub-continental hotspot is Yellowstone, located beneath Yellowstone National Park in northwestern Wyoming, USA. It is responsible for the park's abundant geothermal activity and spectacular geysers.

CHAINS OF OCEANIC ISLANDS

The world's oceans have dozens of these parallel chains, each with its active volcanic hotspot lying beneath the youngest island. In the Pacific, one of the longest is the Hawaiian–Emperor chain. The chain's parent hotspot is currently erupting on the southeastern corner of the Big Island of Hawaii.

Other well-known hotspots in the Pacific are the Pitcairn hotspot at the head of the Tuamotu–Line Island chain and the Macdonald hotspot at the head of the Austral-Cook Island chain. Atlantic Ocean hotspots include the New England hotspot, responsible for the seamount chain east of New York, and the Cape Verde hotspot west of Dakar. In the Indian Ocean, the Reunion hotspot is presently building Piton de la Fournaise shield volcano on the eastern side of Reunion Island. A 530,000-year-old volcano, it is one of the most active in the world, and in an almost-continuous state of eruption. Older islands in this chain include Mauritius, the Laccadive Islands, the Maldives, and Chagos Archipelago (a group of seven atolls).

BELOW An aerial view of Huahine Island shows the crater that was created by the original volcano. An extinct hotspot volcano, it is one in a chain of volcanoes in French Polynesia that began forming 6.5 million years ago then moving in a northwesterly direction.

RIGHT A gray heron (*Ardea cinerea*) builds its nest on a remote island in the Indian Ocean. Hotspot islands formed by volcanoes, like other remote oceanic islands, are vital stops or breeding grounds between continents for many migrating birds.

ABOVE Piton de la Fournaise on Reunion Island is a hotspot volcano in the South Indian Ocean near Mauritius. Resembling Hawaii's volcano, it has erupted more than 153 times since 1640.

Volcanoes of Antarctica

Most of the volcanoes on the Antarctic mainland appear to be related to a giant rift valley that is growing between East Antarctica and West Antarctica. The floor of the rift, or the graben, is hidden well below sea level, and covered by the thick West Antarctic Ice Sheet.

The rift valley is bounded by the uplifted edge of East Antarctica, the Transantarctic Mountains, and Marie Byrd Land in West Antarctica. Volcanic activity, such as Mt Erebus—currently erupting on Ross Island—occurs along the edges of the rift. Here, deep faults caused by stretching of the crust appear to be providing channelways to the surface for magma rising from the mantle. More recently, however, evidence of active volcanism hidden beneath the ice has been discovered.

THE HIGHLY ACTIVE MT EREBUS

Mt Erebus, discovered by Captain James Ross in 1841, is Earth's southernmost historically active volcano. Mt Erebus is less than 1 million years old and together with two dormant basaltic shield volcanoes, Mt Terror (between 0.82 and 1.75 million years old) and Mt Bird (between 3.8 and 4.6 million years old), they have built the roughly triangular-shaped Ross Island.

Mt Erebus is the tallest volcano on the ice-covered island with an elevation of 12,444 feet (3,795 m). Within its large summit rim, a lake of red molten lava slowly convects in an oval crater measuring 820 feet by 330 feet (250 m by 100 m). Upwelling at the edge of the crater, the lava flows slowly across the lake before disappearing into an opening at the other edge. Degassing Strombolian explosions erupt from the lake almost daily, showering the crater rim with feldspar crystals, glass, and bombs. There are occasional more violent episodes, such as one that threatened a party of New Zealand scientists climbing into the crater in 1974. Again in 1984, large bombs were ejected up to a mile (1.6 km) from the crater.

Activity on Mt Erebus was first observed by Shackleton's expedition in March 1908. The volcano's eruptions and lava movements are currently

ABOVE Weddell seals (*Leptonychotes weddellii*) live in the sea and ice surrounding Antarctica. The species has been studied in Erebus Bay since 1968. The seals chew breathing holes through fast ice with their canine teeth. Their eyes are adapted to the low light conditions here.

RIGHT The Ross Ice Shelf juts out in front of Mt Erebus, a combined shield- and stratovolcano featuring a rare lava lake in its crater. Antarctica's most active volcano, it has shown continual activity since 1972. Hot gases from fumeroles sculpt ice towers on its flanks and its "warm" ice caves are thought to be a possible source of undiscovered life.

ACTIVE VOLCANO ERUPTS BENEATH ANTARCTICA'S ICE

In February 1993, while flying over the West Antarctica Ice Sheet about 300 miles (480 km) inland from where it joins the Ross Sea Ice Shelf, American scientists Donald Blankenship and Robin Bell noticed a round depression in the ice. It was about 160 feet (50 m) deep and 4 miles (6.5 km) wide. Immediately they realized that only an active volcano could melt that volume of ice. Later, Bell told *Discover* magazine, "It's so hard to make things on Earth that are round that aren't volcanoes." Their suspicions were proved correct when they returned with radar equipment to penetrate the mile-thick ice cover, and discovered a cone-shaped mountain 2,100-foot (640-m) high and 4-miles (6.4-km) wide at its base. They now believe that the volcano may be just one of several, and that the meltwater created by the heat is lubricating the base of the ice sheet. This is allowing the overlying glacial ice to push toward the Ross Sea, at speeds up to 2,500 feet (760 m) per year.

monitored continually by a network of instruments known as the Mount Erebus Volcano Observatory (MEVO) linked to McMurdo Station in Antarctic US Territory. Core drilling into the flanks of Mt Erebus have revealed several layers of ice interbedded with lava flows, telling of previous eruptions of lava onto glacial ice.

Other volcanoes that mark the edge of the Transantarctic rift system include Coulman Island, a 7.2-million-year-old complex of overlapping shield volcanoes, and Mt Discovery, which grew between 5.3 and 1.8 million years ago.

VOLCANOES OF THE ANTARCTIC PENINSULA

Another group of volcanoes found on the northern Antarctic Peninsula have a different origin. These volcanoes, together with those of the South Sandwich Isles, have been formed by subduction as the Scotia Plate pushes eastward and southward over the Antarctic Plate. The subducting and melting Antarctic Plate is feeding these explosive stratovolcanoes.

Other volcanoes also currently considered active are Penguin Island and Deception Island in the South Shetland group.

ROCK FEATURE: DOLERITE

Dolerite is a dark gray to black silica-poor igneous rock. It is composed of medium-sized crystals of white feldspar and dark pyroxene and olivine that are just visible to the naked eye. They vary in size between those of fine-grained quickly cooled basalt and coarse-grained slowly cooled gabbro. Dolerite occurs as intrusive volcanic necks, dykes, and sills, or as thick lava flows, and is often associated with rifting continents. This rock takes a high polish and is used for bench and table tops, building facings, and monuments.

ABOVE Antarctic pearlwort (*Colobanthus quitensis*) is one of only two species of flowering plants found along the edges of continental Antarctica. The other species is Antarctic hairgrass (*Deschampsia antarctica*).

TOP Tiny Paulet Island, a volcanic island at the tip of the Antarctic Peninsula, is largely ice-free owing to residual heat. Its status is unknown, but there are signs of volcanic activity during the last 1,000 years. It has a large colony of adelie penguins.

DECEPTION ISLAND

Deception Island is an intriguing active volcanic island in the South Shetlands, a group of 11 major islands and a few smaller ones, located off the northern tip of the Antarctic Peninsula. It is a near-circular volcanic rim island with a diameter of about 9 miles (14 km) enclosing a flooded caldera around 4 miles (7 km) in diameter. The caldera forms a natural sheltered harbor, Port Foster, and is known as one of the safest anchorages in Antarctica.

The name Deception Island came into being, however, as ships must first pass through its narrow and difficult-to-find entrance, Neptunes Bellows, just 750 feet (230 m) wide on the southeast side of the island, and avoid the treacherous submerged Ravn Rock on the way through.

ACTIVE AND DANGEROUS

Deception Island and nearby Penguin Island, also part of the South Shetlands, are active strato-volcanoes fed by magma bodies melting from the Antarctic oceanic plate as it descends beneath the overriding Scotia Plate. Deception's caldera probably formed during a major eruption over 10,000 years ago, when the volcano collapsed into its own empty magma chamber following the explosive ejection of at least 7 cubic miles (30 km^3) of magma. Evidence of the volcano's more-recent explosive past has been preserved on the slopes of Mt Pond, where a glacier contains alternating layers of snow and ash. Historic eruptions have been mostly explosive outbursts originating from the ring fractures around the edge of the caldera, rather than lava flows. These have made the island a particularly hostile uninhabitable place.

GAINING A TOEHOLD ON THE ISLAND

In 1906, whalers increasingly began to use the island's safe harbor, starting with the arrival of the Norwegian–Chilean factory ship, *Gobernador Bories*. By 1914, the number of factory ships had grown to 13 vessels. A station was built at Whalers Bay to boil down and extract additional oil from the whale carcasses. It operated until 1931 when whale oil prices plummeted during the Great Depression.

In 1944, the British Antarctic Survey established a permanent base at Whalers Bay, but was forced to evacuate during a volcanic eruption on

December 5, 1967. The British returned again in 1968, but they finally had to abandon the base when it was destroyed by another violent eruption that occurred on February 23, 1969 with a volcanic explosivity index of 3.

In 1955, Chile established its own station at Pendulum Cove, Pedro Aguirre Cerda. Later it set up another station, Gutierrez Vargas, to increase Chilean presence on what it claimed to be its territory. Both were destroyed, along with the British station, during the 1969 eruption. Presently there are only two scientific summer stations still in use

ABOVE A hot outdoor bath in a chilly climate in Pendulum Bay is enjoyed by a group of tourists who arrived in Deception Island on a cruise ship. The island was closed to tourists following a fuel spill in January 2007 after a cruise ship ran aground.

on Deception Island—Spain's Gabriel de Castilla Station, and Argentina's Decepción Station.

Tourist cruises have often visited Deception Island to enjoy its unusual volcanic beauty, its history, and abundant wildlife. A popular highlight is Pendulum Bay, where a hot bath can be enjoyed in pits dug into the black sand beach beside the chilly harbor shore. Here hot water, heated by a recent shallow magma intrusion, is continually rising to the surface along fractures. Those that are brave then plunge into the icy harbor before returning to the hot pool.

DECEPTION ISLAND'S ABUNDANT MARINE AND SEABIRD LIFE

The nutrient-rich Antarctic waters around the South Shetland Islands sustain substantial fish populations that in turn support major seabird, penguin, and seal colonies. During summer, they can seem to occupy every available patch of ground around the coast. Sea mammals include an abundance of seals (Antarctic fur, crabeater, Weddell, leopard, and southern elephant), and whales (orca

and humpback). Deception Island supports the world's largest colony of chinstrap penguins (*Pygoscelis antarctica*) at Baily Head, on the outer coast, with over 100,000 breeding pairs. Other seabirds include marconi penguins, brown skuas, kelp gulls, cape petrels, Wilson's storm-petrels, Antarctic terns, and snowy sheathbills. With flora, however, the picture is less abundant. Deception Island's polar climate supports only mosses and lichen, several of which have not been recorded anywhere else in the Antarctic.

LEFT Named for its once-coveted skin, the Antarctic fur seal (*Arctocephalus gazella*) breeds mainly on subantarctic islands. Their diet may include krill, fish, and sometimes squid. They can dive to depths up to several hundred yards or meters and can maintain a constant core body temperature in the icy waters.

BELOW A birdseye view of Port Foster in Deception Island reveals the dramatic flooded caldera. Though most of the volcano's activity occurred in the seventeenth and eighteenth centuries, and it is now calm, the seafloor of Port Foster is rising, marking it as a restless caldera.

As soon as a volcano ceases to erupt and grow, the forces of erosion rapidly gain a foothold. Rivers and streams carve into its unstable flanks, reducing them to sea level. Soft volcanic ash is removed first while harder parts of the volcano, like solid lava and magma, are more resistant. These interesting volcanic remnants are often spectacular and popular tourist attractions, and afford geologists a first-hand opportunity to study the internal workings of a volcano. Numerous volcanic remnants are protected within the boundaries of national and state parks.

The volcanic cone itself is very susceptible to erosion, and can be substantially reduced by the rivers and streams flowing down its sides within the period of a million years. The "Achilles heel" of the cone results from the thin layers of volcanic soil or ash that often separate its hard lava flows. Ash and soil is soft and susceptible to being washed out from in between the harder flows. Once the resistant lava flows become undermined, they break away easily. Usually the most resistant to erosion is a volcano's internal plumbing system, or lava feeder pipes that are filled with solidified magma. Sometimes these are the only remaining evidence of an ancient volcanic complex.

DANGEROUS CALDERA LAKES

An early, and often catastrophic, event in the life
of a volcanic cone occurs when erosion breaches
its crater or caldera wall, thereby allowing its lake
waters to drain away into surrounding river systems. An Aleut village at Cape Tanak on Umnak
Island, Alaska, was destroyed in 1870 by the sudden
draining of the Okmok caldera lake. Nowadays,
geologists and civil authorities watch such situa-
tions closely so that the potential hazards can be
addressed in time. One such case reached crisis
point ten years after the 1991 eruption of Mt
Pinatubo in the Philippines. The level of the lake
forming in the volcano's new summit caldera was
rising steadily and by 2001 the lake had reached
dangerous levels. Authorities were worried that the
caldera walls could burst and flood communities
surrounding the volcano with wet volcanic debris,
and had to quickly design and construct a canal.
The project was completed safely. About a quarter
of the lake was drained into the Maraunot River,
relieving pressure on the upper caldera walls.

PLUGS, NECKS, PIPES, SPIRES, AND TOWERS

A volcanic plug, or neck, is the term used for the
often nearly circular and vertical feeder pipe of a
volcano that has been filled with solidified magma.

The descriptive words "spire" or "tower" are often
used in the name of such a topographic feature because these words most appropriately describe it.
Devils Tower is a classic example that rises above
the Wyoming Plains, USA. Spectacular volcanic
plugs are located near the village of Rhumsiki in
the Mandara Mountains of Cameroon, and have
created one of the most beautiful landscapes in
the world. The largest of these plugs is Kapsiki
Peak, a spire standing 4,016 feet (1,224 m) tall.

In many older countries with long and varied
histories, these natural features have often played

ABOVE Mt Pinatubo is a stratovolcano on the island of Luzon in
the Philippines. In 2001, authorities feared that its caldera lake
would burst its banks, causing
widespread flooding.

LEFT At Borobudur in Indonesia,
a wide basalt plug has been
transformed into a Buddhist
temple. Three round platforms
sit atop six square platforms.

an important role in the architectural and historical development of a region. Being generally difficult to climb and therefore protected, volcanic plugs have become the sites of buildings such as castles, churches, monasteries, or temples. Many of the castles of the British Isles are located on isolated, craggy plugs, making them invincible to attack. Edinburgh Castle and Stirling Castle in Scotland are well-known examples. The name "puy" comes from the volcanic plugs of the Avergne district in France's Central Massif, which stick out as steep, pillar-like hills. The twelfth-century St-Michel-d'Aiguilhe chapelle, in the town of Le Puy-en-Velay, is built atop a 280-foot (85-m) high volcanic rock pinnacle. Towering above a more modern part of this city, a bronze statue of the Virgin Mary stands on another 500-foot (152-m) tall bare rock plug.

As sites for places of worship, the tops of volcanic plugs demand a certain degree of dedication or penance from devotees trying to reach them. Near Mandalay in Myanmar there is a shrine dedicated to animal spirits known as "Nats" at the top of an extinct volcano, Mt Popa. The summit of the plug is reached by climbing 777 steps.

One of the most extensive Buddhist shrines is the temple of Borobudur in Java, Indonesia, Here, among the rice fields of the Kedu Plain, a broad plug of basalt has been turned into an ornately decorated stepped pyramid 115 feet (35 m) high and with a base 394 feet (120 m) square. It was built during the Sailendra dynasty in the ninth century, and took about 80 years to construct.

VOLCANIC DIKES: NATURAL GREAT WALLS OF ROCK

Volcanic dikes are cracks or fractures, discordant with the surrounding geologic structure, which have opened as a result of stresses in the ground

BELOW The three monoliths known as Torres Del Paine were carved from a huge granite laccolith by the forces of ancient glacial ice. The peaks are located in a remote part of the Andes Mountains in southern Chile.

created by a heaving and erupting volcano. When a volcano alternately expands and contracts during an eruption, numerous stress fractures develop in the surrounding ground. High pressures within the volcano often drive molten magma into these fractures where it cools and hardens. Within larger dikes, a variation in crystal size can usually be seen from the edges to the center. The volcanic rock at the edges is cooled more quickly due to heat loss into the surrounding rocks, whereas cooling is slower and the crystals have more time to grow toward the center of the dike.

Dikes are generally almost vertical and radiate outward from the center of the volcano (these are called radial dikes), or they may form concentric rings that dip inward toward the volcanic center (ring dikes). They also tend to become thinner and more spaced out with distance from the volcanic center. Dikes are of great assistance to geologists

FRACTIONAL CRYSTALLIZATION

Larger igneous bodies (thick sills, laccoliths, lopoliths, batholiths) may become differentiated as cooling and crystallization takes place over a very long time. In magmas containing many different elements, certain minerals will crystallize out, first at higher temperatures followed progressively by others as the temperature falls. This process is known as "fractional crystallization." The early forming crystals, such as the iron- and magnesium-rich silicate minerals olivine and pyroxene, will sink to the bottom of the igneous body and accumulate there because they are heavier than the liquid in which they grew. Consequently, later-crystallizing minerals are more abundant near the top of the sill. Metal sulfide and oxide-rich liquids will also segregate out from the magma into immiscible blobs, producing thick layers of valuable minerals. The nickel deposits of the Sudbury lopolith in Canada formed in this way. South Africa's igneous Bushveld Complex contains the world's largest reserves of chromium, and the platinum group elements, as well as vast quantities of iron, tin, titanium, and vanadium.

making maps of ancient volcanic terrains. By carefully measuring the orientation and angle of the dikes, the positions of the old volcanic centers can be determined quite accurately.

Dikes often form the most amazing of volcanic remnants, as they are usually more resistant than the other rocks of the volcano and, like volcanic plugs, can be left standing proudly above the surrounding landscape. Dikes can form spectacular natural walls that are miles long and hundreds of feet high, but perhaps only several feet wide. One of the most visited examples is the 30-million-year-old Shiprock volcanic plug and its radial dikes protruding from the floor of the New Mexico desert in the United States. Shiprock is probably the world's finest example of the exhumed remains

BELOW Stirling Castle was built on a volcanic plug in the Midland Valley Sills, which underlies a large part of central Scotland. The castle sits 250 feet (75 m) above the surrounding plain.

Volcanic Remnants of North America

Driven westward by the opening Atlantic Ocean, the North American Plate has been steadily swallowing the adjacent Pacific Ocean. This has caused extensive volcanic activity along the American West Coast over the last 200 million years. Erosion and glaciation of the continent has exposed many of these volcanic features, and today they are key points of attraction in many of North America's national parks.

A belt of granite, known as the Sierra Nevada Batholith, stretches for 400 miles (650 km) along the United States' Western Cordillera, and forms much of California's spectacularly scenic Sierra Nevada Mountain Range. From the Triassic Period (150 million years ago) through to the Cretaceous Period (80 million years ago), granitic magma rose buoyantly from the subducting and melting oceanic plate, emplacing itself in the semi-solid state as hundreds of individual plutons. The batholith originally intruded deep below the surface, but it has since been exposed by erosion, revealing many interesting details about the inner workings of the magma chambers of subduction zone volcanoes.

HALF DOME, YOSEMITE NATIONAL PARK

Half Dome is an exposed granite dome or peri-cline in Yosemite National Park and is one of the park's most recognizable landmarks. Glacial action is responsible for carving this dome, which rises 4,737 feet (1,444 m) above its U-shaped valley floor. It is possible to ascend Half Dome's vertical northwest face via a number of technical routes; however, most visitors climb the more rounded but still steep eastern side of the dome. It is not an ascension for the faint-hearted, as it involves a cir-cuitous hike of 8½ miles (13.5 km) with the final 900 feet (275 m) of the ascent made between a pair of steel cable hand railings on posts. The large flat summit of the dome offers spectacular views of the surrounding landscape. Other spectacular glacially carved granite domes in Yosemite National Park include El Capitan, a vertical 3,000-foot (1,000-m) rock formation that is very popular with rock climbers. Another, Sentinel Dome, is a gently sloping dome accessed via a relatively easy 2¼-mile (3.5-km) round trip hike starting at its base.

DEVILS POSTPILE NATIONAL MONUMENT

The Devils Postpile formation, located near Mammoth Mountain in eastern California, is the remnant of a thick basalt flow that erupted into

BELOW The Wawona Tunnel View Overlook provides a spectacular view of the glaciated landscape of Yosemite Valley in Yosemite National Park, which includes the geological features known as El Capitan, Three Brothers, Half Dome, and Sentinel Dome.

the already-glaciated valley of the Middle Fork of the San Joaquin River less than 100,000 years ago. In the vicinity of the Postpile, the lava was most likely over 400 feet (120 m) thick, though much of it has been removed by subsequent erosion. Devils Postpile is a spectacular example of columnar basalt formed by inward cooling from the bottom and the top of the lava flow. Perfect vertical and curved fanning shrinkage columns have been exposed on the side of the volcanic remnant, with an impressive jumble of loose columns forming an apron at its base. The top of the formation is a perfect polygonal pavement, like a giant's tiled floor, that has been polished smooth by glacial action. Numerous parallel striations, or scratch marks left by rocks carried in the glacier, indicate the direction of ice movement. The columns are mostly six-sided and may reach 60 feet (18 m) in length. Their average diameter is 2 feet (60 cm), the thickest being 3 feet (1 m) wide.

An early proposal to build a hydroelectric dam required Devils Postpile to be blasted into the river; however, the formation was granted permanent protection in 1911 when President William Howard Taft declared the area to be a United States National Monument.

MORRO ROCK, CALIFORNIA

Morro Rock, sometimes called the "Gibraltar of the Pacific," stands out prominently, offshore from the town of San Luis Obispo, California, at the entrance to Morro Bay. The Rock stands 576 feet (176 m) tall and is one of a line of eroded volcanic plugs known as the "Nine Sisters." The other eight "sisters" are Black Hill, Cabrillo Peak, Hollister Peak, Cerro Romauldo, Chumash Peak, Bishop Peak, San Luis Mountain, and Islay Hill. These volcanoes finished erupting approximately 20 to 25 million years ago, leaving the last rhyodacitic magma to freeze solid in their extinct volcanic

ABOVE These giant sequoias *(Sequoiadendron giganteum)* are part of Mariposa Sequoia Grove in the lower montane vegetation zone of Yosemite National Park. The combination of nutrient-rich ash and mineral soil provides ideal growing conditions for these magnificent trees.

ABOVE LEFT The granite pericline called Half Dome dominates Yosemite Valley in California's Sierra Nevada. Glacially polished granite is common in the valley, providing further evidence of glaciation taking place here.

BLACK TUSK, BRITISH COLUMBIA

The Black Tusk is a prominent spire of dark volcanic rock that stands 7,608 feet (2,319 m) above sea level in Garibaldi Provincial Park, British Columbia, Canada. It is perhaps the best-known mountain in the Garibaldi Ranges. Its pointed black peak is an immediately recognizable landmark visible from all directions. The Black Tusk is the remnant feeder pipe of a cinder-rich strato-volcano that erupted about 1.2 million years ago. The loose cinder has eroded, leaving only the more-resistant lava plug. Black Tusk is just one of many volcanic peaks of the Pacific Ranges and Cascade Range that lie in a line along the coast from British Columbia to California, and were formed by the subduction of the Pacific Plate under the North American Plate. As the descending oceanic crust melts, rising bodies of molten magma fuel the numerous volcanic eruptions.

Mt Garibaldi, which towers behind the town of Squamish, British Columbia, is part of a complex of much younger volcanoes built by explosive eruptions. It is relatively young, at only 15 to 20 thousand years old. Lava flows in the ancient valley dammed the glacial meltwater streams, leading to the formation of Garibaldi Lake.

NEW EDDYSTONE ROCK, ALASKA

New Eddystone Rock is part of an eroded ancient volcanic vent in Misty Fjords National Monument, south of Ketchikan, Alaska. This jagged basalt tower, which is located at the entrance of Rudyard Bay, was first noted in the journals of Captain George Vancouver. He was exploring the region around the Alaska Panhandle in 1793 and named the spectacular 236-foot (72-m) pillar "New Eddystone Rock" after a lighthouse south of England.

The tower is the remains of magma hardened in a volcanic dike. It was the conduit for repeated eruptions of flow basalts that took place less than 5 million years ago, the remains of which can be found on several islands surrounding New Eddystone Rock. The surrounding lavas cooled from their upper and lower surfaces, forming vertical polygonal columns, in contrast to those of New Eddystone Rock, which are more haphazard and broken. A later glacial advance removed most of the lava flows, leaving just New Eddystone Rock and the other small islands.

Misty Fjords National Monument was created on December 1, 1978; it is a wilderness covering approximately 3,500 square miles (9,000 km^2) of mostly western hemlock, spruce, and western red cedar forest within Tongass National Forest. Its abundant wildlife includes grizzly bears, black bears, salmon, whales, mountain goats, and deer.

TOP The eroded volcanic plug called Morro Rock is reflected in Morro Bay. The plug is one of the "Nine Sisters" volcanic peaks, which separate the Los Osos and Chorro valleys.

ABOVE The abrupt peak Black Tusk is part of the Garibaldi volcanic belt. It is prominent from the Sea-to-Sky Highway south of Whistler, Canada.

feeder pipes. Erosion of the softer surrounding rock has since exposed the resistant plugs.

Morro Rock was first sighted in 1542 by the Portuguese explorer, Juan Rodriguez Cabrillo. He called it "El Morro," which means crown-shaped hill. The hill was quarried from 1889 until 1963 to build the breakwaters of Morro Bay and Port San Luis Harbor. In 1968, Morro Rock was protected as a state landmark and declared a reserve for the peregrine falcon and other bird species. Visitors are no longer allowed to climb the rock but it can be viewed from nearby trails and viewing lookouts.

PALISADES SILL, NEW YORK AND NEW JERSEY

On the eastern side of the continent, the massive Palisades Sill is evidence of the tectonic activity associated with North America's Triassic Period breakup from Europe 200 million years ago. The Palisades Sill is a 1,000-foot (300-m) thick dolerite (diabase) intrusion that was injected, possibly in several pulses, into the sandstones of the Stockton Formation within the Newark Basin. It extends beneath parts of the states of New York and New Jersey, and its most significant outcrop forms the cliffs known as The Palisades. These rise 300 feet (100 m) above the western bank of the Hudson River, and are exposed along the river for a distance of approximately 50 miles (80 km).

Because of its thickness, crystals growing in the slowest cooling central part of the dolerite sill had time to either settle out under gravity or be differentiated during horizontal flow. There is an olivine-rich layer approximately 30 feet (10 m) above the lower contact.

STONE MOUNTAIN, GEORGIA

Stone Mountain is a granite dome in Atlanta, Georgia. It dates back before the Atlantic Ocean opening to a time 300 million years ago, when the Appalachian Mountains were forming during the collision of Africa and North America. The granite formed from melting of metamorphosed sedimentary rocks under the intense pressures and temperatures associated with the collision. It is an

Volcanic Remnants of North America

enormous monolith, rising 825 feet (250 m) above the surrounding plain with a summit elevation of 1,683 feet (513 m) above sea level. The mountain is famous for the huge carvings on its north face: Stonewall Jackson, Robert E. Lee, and Jefferson Davis in a group on horseback.

BELOW Stone Mountain may be the largest exposed granite dome in North America. Made up of light- to medium-grained gray granite, it is located in the Piedmont province of Georgia.

SHIPROCK, NEW MEXICO

Shiprock, New Mexico, is one of North America's most spectacular volcanic remnants. It is an exposed volcanic throat, a massive jagged pillar of igneous rock that protrudes 1,700 feet (500 m) above the surrounding plains, which themselves lie at an elevation of 5,500 feet (1,700 m) above sea level. Radiating from the central vent are three wing-like walls, or dikes, of resistant rock.

ABOVE The bald-headed eagle or America eagle *(Haliaeetus leucocephalus)* is the national symbol of the United States. It can be seen soaring over Shiprock looking for prey.

Shiprock is part of the Navajo Volcanic Field that erupted some 25 to 30 million years ago over an area of about 7,700 square miles (20,000 km²) where the four corners of Arizona, New Mexico, Colorado, and Utah meet. Highly gas-charged magma coming from Earth's mantle exploded onto the surface, creating numerous diatremes, vents, tuff pipes, and dikes. The violent nature of the eruptions is clearly evident in the sheer abundance of fragmented volcanic rock (breccia and tuff) in this area. Shiprock itself is composed of fractured volcanic breccia that has been injected with black dikes of an igneous rock called minette.

RARE ROCKS AND AN EXPLOSIVE PAST

Minette is a type of lamprophyre, which is a rare potassium-rich and silica-poor intrusive igneous rock believed to have come directly from small amounts of melting within Earth's mantle. The name lamprophyre is derived from the Greek *lampros*, meaning bright, and *porphyry*, meaning large crystals, which describes the appearance of the rock. It contains large, shiny, well-shaped crystals of biotite or phlogopite mica, sometimes with amphibole and pyroxene, set in a fine matrix of alkali feldspar and other minerals, such as diopside, iron and titanium oxides, apatite, quartz, amphibole, and glass. Lamprophyres occur as dikes and small intrusions of magma thought to be derived from deep volatile-driven melting in a subduction zone setting. As with all lamprophyres, minette is prone to rapid weathering.

Gas trapped in volatile-rich magma expands and becomes explosive as it rises toward Earth's surface, where the confining pressure is lower. As the gas begins to expand, it propels the magma even faster toward the surface, where the pressure drops even further, creating a violent chain reaction that has

THE NAVAJO LEGEND OF SHIPROCK

The Navajo (Diné) name for Shiprock is Tse'Bit'Ai, meaning "rock with wings." It refers to the massive central volcanic spire and the thin, wing-like, radiating dikes. According to Navajo legend, the rock was once a great bird that carried the ancestral people of the Navajos to their lands in present-day New Mexico. Far away to the northwest, the Navajo ancestors had crossed a narrow sea (perhaps referring to their crossing of the Bering Strait) to escape from a war-like tribe that was pursuing them. When the tribal shamans prayed to the Great Spirit for help, the ground rose beneath their feet suddenly and took the shape of a giant bird. The bird flew south throughout the night and at the end of the next day it finally landed at sunset where Shiprock now stands.

been likened to a shotgun being fired into the air. It is thus estimated that the exposed breccia in Shiprock formed approximately 2,500–3,000 feet (750–1,000 m) below Earth's surface. Following the main explosion, sheet-like dikes of fluid minette magma were injected into numerous expansion cracks in the Shiprock volcanic vent itself, and also radially outward into the rocks surrounding the volcanic neck.

THE VOLCANIC FIELD REVEALED

Millions of years of erosion have slowly stripped away the softer rocks of the volcanic cone. The host rocks forming the plains surrounding the intrusions are also soft, being late Cretaceous Period marine-origin mancos shale. As a result, erosion has been able to gradually lower the land surface by thousands of feet. Thus, the more resistant rocks of Shiprock and the other volcanic vents of the Navajo Volcanic Field have become exposed. The hard skeletal remains of the central vent and the radial vertical dikes of Shiprock have allowed geologists unique access to rarely seen deeper levels of extinct volcanoes.

CLIMBING SHIPROCK

A Sierra Club party that included David Brower first climbed Shiprock's precipitous walls in 1939, ending its standing as one of North America's great unsolved climbing problems. There are now at least seven routes to the top, all of which are technically difficult. Owing to the risks involved, and the peak's importance in Navajo religion and culture, permission to climb the rock is no longer being given.

BELOW Shiprock is a classic example of a volcanic throat or plug. The rugged exposed rocks are a great teaching resource for geologists and volcanologists.

Volcanic Remnants of Mexico, and Central and South America

One of South America's most spectacular volcanic remnants would have to be Brazil's Sugar Loaf (Pão de Açucar). This granitic gneiss dome rises near vertically from Guanabana Bay, forming one of the well-known backdrops to Rio de Janeiro's cityscape. It rises to 1,300 feet (400 m) above sea level and, as its name suggests, resembles a giant cone of cane sugar concentrate.

ABOVE The southern caracara (*Caracara plancus*) is one of more than 100 bird species that can be seen in Torres del Paine National Park. Some of these birds are classified as endangered species.

BELOW The three granite monoliths known as Torres del Paine are best viewed from Lake Pehoe. UNESCO declared the Torres del Paine National Park in Chile a Biosphere Reserve for its numerous ridges, glaciers, crags, waterfalls, rivers, lakes, and lagoons.

Sugar Loaf is just one of several monolithic domes that rise abruptly from the Atlantic Ocean around Rio de Janeiro. Corcovado Mountain is renowned for its amazing views and the giant statue of Jesus atop its peak. Morro da Babilônia and Morro da Urca are very well known in rock-climbing circles. Topsail Rock (Pedra da Gávea) is the tallest dome in the city, rising to 2,762 feet (842 m) above sea level. Visitors can travel to its summit by cable car. Rio de Janeiro's morro parks host numerous well-traveled climbing tracks of varying grade, and preserve remnants of Atlantic rainforest.

The strongly metamorphosed and re-melted rocks of the Rio de Janeiro area belong to the Ribeira Metamorphic Belt. They formed under intense heat and pressure during the final collisions and assembly of the Gondwana continent that took place around 550 million years ago. Basaltic dikes then cut these older rocks about 130 million years ago during the breakup of the Pangean supercontinent in the early Cretaceous Period.

TORRES DEL PAINE, CHILE

The Cordillera del Paine is a beautiful but remote segment of the Andes Mountains in southern Chile. The breathtakingly spectacular Torres del Paine (Towers of Paine) are the three best-known peaks—the South Tower, Central Tower, and North Tower of Paine. They are monoliths carved from an enormous granite laccolith by the forces of ancient glacial ice. The peaks lie within Torres del Paine National Park, which was declared a Biosphere Reserve by UNESCO in 1978. The park is very popular with hikers wanting to see the three stark chiseled peaks. Visitors can take a day trip to the peaks or they can hike the park's eight- to nine-day circle route.

MESETA DE SOMUNCURA, ARGENTINA

Meseta de Somuncura is an isolated plateau of flood basalt located in the Rio Negro province of Patagonia in Argentina. The name Somuncura comes from the indigenous Araucanian language meaning "rock that sounds or speaks." Most of the basaltic lava flows were extruded during the late Oligocene Period about 25 million years ago, with the youngest flows erupting only 5 million years ago. Erosion has carved the edges of the plateau into abrupt wall-like cliffs, and cut channels into the top of the plateau. Individual hills rise around

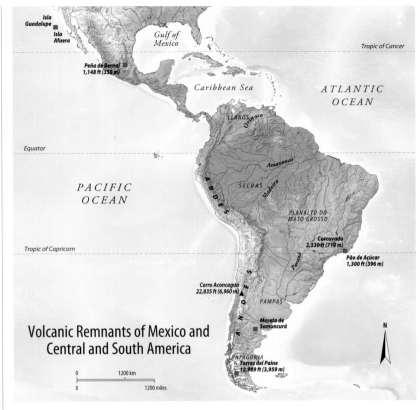

Volcanic Remnants of Mexico and Central and South America

6,200 feet (1,900 m) above sea level and are surrounded by low forest alternating with depressions that are filled by spring-fed lagoons. The basalt flows that make up the cliffs along the northwest edge of the plateau near Sierra Colorada have eroded into a series of rock staircases. The sparse population of this rocky desert lives off cattle ranching, with the main center being the small town of Valcheta. It is locally known as "the oasis of Río Negro" because its neat green farms appear in such stark contrast to the regular plateau landscape.

PENA DE BERNAL, MEXICO

San Sebastian Bernal is a tidy little town of approximately 5,000 inhabitants in the Mexican state of Queretaro. The town is popular with tourists as it is spread around the base of an impressive pyramid-like, rocky peak known as the Pena de Bernal. Visitors hike from the town to a small chapel about halfway up the peak. Some say that the peak's natural "cosmic energy" imparts an "extraordinary longevity" to those living near it.

Pena de Bernal, standing 1,148 feet (350 m) tall, is often referred to as one of the world's largest monoliths. This peak is the resistant remnant of

a porphyritic intrusion that occurred during the Jurassic Period, some 100 million years ago.

ISLA GUADALUPE AND ISLA AFUERA, MEXICO

Rugged Isla Guadalupe, offshore from Baja California, is the eroded remnant of two overlapping shield volcanoes. Only a handful of people live on the 22-mile (35-km) long, bare, rocky island. Isla Afuera, a sheer-sided volcanic plug off the southern end of the main island, is very popular with scuba divers. It is a haven for marine life, such as the northern elephant seal and the Guadalupe fur seal.

ABOVE LEFT Rio de Janeiro's landmark dome called Sugar Loaf is made up of granitic gneiss. This medium- to coarse-grained rock can be gray or pink. Feldspar and quartz create granular lighter streaks in the gneiss.

Volcanic Remnants of Europe

Europe's volcanic remnants include numerous spectacular castle- and cathedral-topped volcanic plugs or eroded batholiths—chosen originally for their impenetrability and isolation rather than their beauty. Strange columnar pavements, cliffs, and caves have also shaped human thought and legend long before their true geologic origin was surmised. These European sites have become valuable tourist drawcards.

ABOVE The 540-foot (165-m) tall basalt column Basiluzzo Island, in the Panarea Archipelago, is one of Sicily's Cyclopean Isles.

BELOW Volcanic Lanzarote Island is a UNESCO Biosphere Reserve. Although it is one of the oldest of the Canary Islands, the remnants of the last eruptions, which occurred in the eighteenth and nineteenth centuries, are still visible in the landscape..

Fingal's Cave, on the uninhabited Inner Hebridean island of Staffa, Scotland, is the legendary home of the giant Fingal. It is a sea cave formed within a series of Tertiary Period basalt lava flows that have cooled to create striking hexagonal columns. The 250-foot (75-m) long and 70-foot (20-m) high cave can be viewed from the sea or entered via a small path along its side. Sir Joseph Banks discovered Fingal's Cave in 1772 while studying the island's natural history. The melodic echoes of the water within the cave later became the inspiration for Mendelssohn's "Hebridean Overture." The cave's Gaelic name, Uamh-Binn, means "cave of melody," and Staffa means "island of pillars." Although Fingal's Cave is considered the classic site for basalt columns, there are numerous localities throughout Europe.

EUROPE'S BASALT COLUMNS AND SICILY'S CYCLOPEAN ISLES

The Atlantic islands, such as the Azores, Canaries, and Madeira, have remnant flows and plugs displaying columns including radiating patterns called "basalt flowers." Germany is home to several spectacular column localities, including Eifel and Siebengebirge near Bonn, Parkstein in Upper Palatinate in Bavaria; the Kassel-Marburg area; and also Scheibenberg and Pöhlberg in Saxony. Garni Gorge, east of Yerevan in Armenia, is also worth visiting. Near Garni's first-century CE temple, the Goght River has exposed basalt columns in the walls of a narrow gorge. Spectacular basalt columns are also common in Iceland.

The Cyclopean Isles of Sicily are legendary for their cooling columns formed in stacked basalt flows from Mt Etna. These tiny Mediterranean islands off Sicily's east coast take their name from the mythical Greek one-eyed giants, the Cyclopes. Inspired by the perfectly interlocking basalt columns found there, the Cyclopean construction method is also named after the islands. Buildings of this style, such as the Early Bronze Age structures in Sardinia, are made of closely fitting blocks without the use of mortar or other binding material.

CASTLES AND CATHEDRALS

Stirling Castle is another fortress that has played a key role in Scotland's history since its origins as a Stone Age settlement. The castle is built on an outcrop of the Midland Valley Sill, a 300-foot

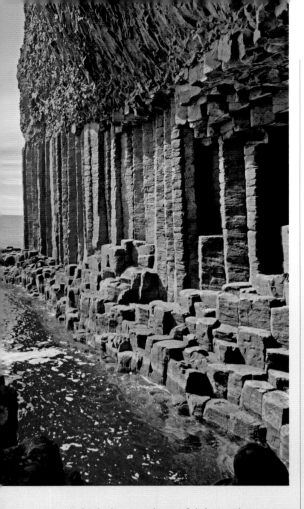

A CASTLE ON A PLUG

Edinburgh Castle (pictured below) is visible from all directions, towering 260 feet (80 m) above the cityscape of Edinburgh, Scotland. The royal castle is built on the summit of Castle Rock, a volcanic plug that has been occupied since 900 BCE. Over the centuries, Edinburgh Castle was modified and developed as a defensive fortress, to the point where its walls now appear to grow straight upward from the vertical cliffs of the rock of the same color. The castle's might has been tested on at least 13 occasions.

Nearby Arthur's Seat along with Castle Rock are the remnants of a volcano that erupted during the lower Carboniferous Period, approximately 350 million years ago. Glacial action has exposed the internal structure of the volcano including its ash cone, plugs, sills, and dikes. Castle Rock plug was carved into a classic "crag and tail" feature by glaciers flowing from the west. The ice pushed over and around the hard volcanic plug forming the "crag," which in turn protected and preserved a "tail" of softer sedimentary rocks that is now known as the Royal Mile. The "tail" is in fact longer than a mile, by 107 yards (98 m).

(100-m) thick dipping sheet of dolerite that intruded along the bedding planes of Carboniferous Period sediments and coal beds around 300 million years ago. The Midland Valley Sill underlies a large part of central Scotland.

Duomo di Orvieto is a massive fourteenth-century cathedral dominating the town of Orvieto in Umbria, northern Italy. The building was commissioned by Pope Urban IV and started in 1290.

It features a rose window and a golden facade with three huge bronze doors, and is built of alternating layers of white travertine and gray basalt. In the ninth century BCE, the Etruscans founded Orvieto on a small steep-sided mesa—a volcanic remnant left by the erosion of Quaternary Period ignimbrite layers that erupted from the Vulsini Volcano. The mesa sits on a bed of soft marine clay, so its cliffed edges are very prone to collapse.

ABOVE LEFT Fingal's Cave on Scotland's Isle of Staffa clearly shows the basalt columns that formed here millions of years ago. The site is very similar to Ireland's Giants Causeway, and is in fact part of the same lava flow.

GIANT'S CAUSEWAY, IRELAND

Giant's Causeway is a peninsula of tightly packed basalt columns jutting out into the sea in County Antrim, Northern Ireland. The perfectly formed flat-topped columns step down in a wide natural staircase that ultimately descends beneath the waves. This natural wonder is recorded in almost every geology textbook, and is so unusual that UNESCO declared the area a World Heritage Site in 1986.

BELOW Giant's Causeway resulted from the rapid cooling and contraction of lava flows. This phenomenon is also seen in many other places around the world, including Iceland, New Zealand, and Germany.

The causeway was formed about 65 million years ago when the Antrim chalk beds were intruded by highly fluid basalt. The lava flooded onto the surface, forming an extensive flat plateau. The cooling flows began to contract and develop tension fractures on their top and bottom surfaces, which grew into the distinctive columns that can be seen today. Visitors can walk around the causeway and see columnar formations with imaginative names such as the Camel's Hump, the Giant's Boot, the Giant's Eyes, the Giant's Granny, the Harp, the Honeycomb, the King and his Nobles, the Pipe Organ, the Shepherd's Steps, and the Wishing Well.

Approximately 40,000 polygonal stone columns have been counted at Giant's Causeway. They are on average hexagonal, with some having four, five, seven, or eight sides. The tallest columns are about 36 feet (12 m) high. A matching causeway is on the Scottish side of the North Channel on the Isle of Staffa, which was once part of the same ancient lava field. Staffa's well-known Fingal's Cave has formed in similar vertical basalt formations.

THE LEGEND OF GIANT'S CAUSEWAY

According to legend, the causeway once extended all the way to Scotland. It was built by an Irish giant by the name of Fionn mac Cumhaill who was challenged to a contest of strength by his foe, the Scottish giant, Benandonner. Fionn decided to build the staircase so his non-swimming foe could cross the channel. After laboring for weeks on the causeway, Fionn fell asleep. When Fionn failed to arrive in Scotland, Benandonner, who was much bigger, came across looking for him. In order to protect Fionn, his wife Oonagh covered him with a blanket and put a bonnet on his head, pretending that Fionn was her baby son. When Benandonner saw the "baby's" size, he reasoned that the supposed father, Fionn, must be huge. Benandonner fled back to Scotland in terror, pulling up the stones of the causeway so Fionn could not come after him.

Historic reports of Giant's Causeway date back to 1692, when the Bishop of Derry and Chevalier de la Tocnaye made the rugged journey to the site on horseback, and could barely believe what they saw. In 1693, Sir Richard Bulkeley of Dublin's Trinity College announced the discovery in a paper presented to the Royal Society.

Giant's Causeway became known internationally after Dublin artist Susanna Drury popularized it through an award-winning series of watercolor paintings in 1739. In 1765, engravings of her paintings illustrated an entry on Giant's Causeway in the *French Encyclopedia,* in which French geologist Nicolas Desmarest suggested that the columns were volcanic in origin. The popularity of Giant's Causeway grew when access became easier after completion of the coast road in the 1830s.

ABOVE During the time Giant's Causeway's stone columns were being formed, the area would have been experiencing the hot and humid conditions of an equatorial climate.

LEFT Dunluce Castle sits atop a basalt outcrop in County Antrim, Northern Ireland. Legend has it that some of the men who died when the *Girona* ran aground at Lacada Point were buried in St Cuthbert's cemetery not far from the castle.

THE *GIRONA* DISASTER

The treacherous waters of the Irish coastline are the scene of many a tragic story. The *Girona* was a ship in the Spanish Armada designed to carry 121 sailors, 186 soldiers, 50 bronze cannons, and 224 rowers. The year was 1588 and the Spanish Armada had run into difficulties. Its fleet was scattered around the west coast of Ireland. Gales and extreme weather eventually took their toll on the *Santa Maria Encoronada,* the *Santiago,* and the *Duquesa Santa Ana.* The conditions forced the surviving men ashore where they marched northward in search of other ships in the fleet. They discovered three other ships foundered by the gales at Killybegs—the *Girona,* the *Lavia,* and the *San Juan.*

After repairing the *Girona,* Commander Don Alonso decided to take all the surviving crews onboard and sail northward to Scotland and then back to Spain. The *Girona* set out from Killybegs with an estimated 1,200 men, and all their accumulated valuables and cannons. Once again, they encountered gale-force winds that damaged the rudder and blew the heavily overloaded ship toward the coast.

Even the desperate strength of the 224 rowers was not enough to keep the ship offshore. The *Girona* finally struck Lacada Point (in full view of Giant's Causeway) on October 28, 1588. Only a handful of men survived the tragic shipwreck.

Volcanic Remnants of Africa

Africa's volcanic remnants include rugged escarpments of flood basalt dating to the break-up of the Gondwana supercontinent into Africa and South America. Other features are great dikes and sills; huge domes of intrusive granite now exposed as impressive monoliths; beautifully columned volcanic plugs; and the enigmatic, but extraordinarily rich, ultramafic igneous Bushveld Complex.

ABOVE Clouds settle in the gullies between the basalt ramparts of the Drakensberg Range in the province of KwaZulu-Natal, South Africa. The range is known to the Zulus as Ukhahlamba, or "Barrier of Spears."

The igneous Bushveld Complex is located north of Pretoria in South Africa. The saucer-shaped igneous intrusion or lopolith covers an area of more than 25,000 square miles (66,000 km²) and reaches a thickness of 6 miles (9 km) in places. The intrusions are much lower in silica (ultramafic) than normal volcanic rocks and the crystallizing minerals have separated into layers of different density. This geological wonder contains some of the richest ore deposits on Earth.

It accounts for almost all of the world's primary platinum production and contains most of the world's reserves of the minerals andalusite, chromium, fluorspar, platinum, and vanadium, as well as vast quantities of iron, tin, and titanium.

The exact origins of the 2,000-million-year-old Bushveld Complex remain a mystery. One theory proposes that the lopolith formed from localized melting of Earth's mantle by a deep-seated hot mantle plume beneath southern Africa. Another theory suggests that the lopolith is the result of a large meteorite strike.

THE DRAKENSBERG PLATEAU, SOUTH AFRICA

The main ramparts of the Drakensberg Range were lifted to heights of over 10,000 feet (3,000 m) as the Atlantic Ocean began to open. Vast volumes of basaltic lavas of the Karoo Formation poured out of fissures onto the surrounding landscape to build the Drakensberg Plateau approximately 180 million years ago. These basalt flows match similar aged flood basalts found in South America. Individual flows can be followed for more than 20 miles (32 km). Their total thickness exceeds 4,300 feet (1,300 m) although, due to erosion, only a small remnant of the once mighty continental basalt plateau still exists.

AHAGGAR (HOGGAR) MOUNTAINS, ALGERIA

The Ahaggar Mountains are an elevated highland plateau in the central Sahara region of southern Algeria. In the mountains near the oasis town of Tamanrasset, the remains of ancient volcanic necks can be seen rising from the rocky desert floor. The Iharen basalt plug is one fine example displaying beautiful vertical cooling columns and an apron of debris around its base.

THE GREAT DIKE OF ZIMBABWE

Zimbabwe's Great Dike is a vertical volcanic dike that runs north–south, straight through the center of Zimbabwe for a distance of about 320 miles (515 km). It is 1 to 4 miles (3 to 12 km) wide and stands out like a giant natural wall, which is highly visible on satellite images. The dike is tallest in the north, rising 1,500 feet (460 m) above the Umvukwe Plateau. It formed about 2,500 million years ago, when molten magma rose from a depth of approximately 120 miles (200 km) in the mantle and intruded into Archean-age crust. Like the nearby Bushveld Complex, it is an ultramafic layered intrusion hosting vast metal and mineral ore riches—including gold, silver, chromium, platinum, iron, nickel, tin, mica, and asbestos—which support Zimbabwe's economy. Explorer Carl Mauch first discovered the Great Dike in 1867, but it took half half a century longer to realize the existence of its hidden mineral treasures.

VOLCANIC MONOLITHS

The Brandberg Massif is a cone-shaped granite monolith visible at great distances rising from the

---

ROCK FEATURE: BASALT

Basalt is a dark gray to black silica-poor igneous rock. It is composed of tiny crystals of feldspar, pyroxene, and olivine invisible to the naked eye. The rock is sometimes dotted with small holes (vesicles) caused by expanding gas in the lava. These can be filled with secondary minerals such as zeolite or calcite. Basalt occurs mainly as lava flows, sometimes very extensive, and is often associated with rifting continents. The rock takes a high polish and is used for bench tops and building facings, or crushed for road base.

ABOVE Remnants of ancient volcanic necks dot the desert landscape of the Ahaggar Mountains in Algeria. At 9,573 feet (2,918 m), Mt Tahat is the highest peak in these mountains.

flat Namib gravel plains. Its highest point and Namibia's tallest mountain is Königstein (German for "king's stone") at 8,441 feet (2,573 m) above sea level. The Brandberg granite batholith rose slowly through Earth's crust about 120 million years ago, stopping about 6 miles (10 km) beneath the surface, where it slowly cooled and crystallized. Later erosion stripped away the surrounding rocks, exposing the granite dome. Its brilliant pink-red color, which often glows strongly at sunrise and sunset, gives the batholith its name—Brandberg means "fire mountain" in German.

Another impressive granite monolith is Aso Rock on the outskirts of Abuja, Nigeria's capital city. This 1,300-foot (400-m) tall rock, whose caves once provided refuge for the Aso people against their hostile neighbors, has become the symbol of Nigeria's seat of power. Ben Amera in Mauritania, near the border with Western Sahara, is another significant African monolith. It is arguably the second largest monolith in the world.

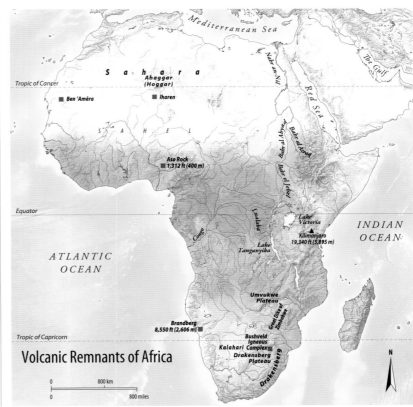

Volcanic Remnants of Africa

NGORONGORO CRATER

Ngorongoro Crater is truly unique and one of Africa's natural wonders. The giant unbroken natural wall of Ngorongoro encircles a flat floor of some 100 square miles (260 km²). The plains within have their own permanent water supply and so are home to thousands of animals, which are often quite rare elsewhere.

Ngorongoro, in Tanzania, was once a classic conical volcano, perhaps originally standing taller than neighboring Mt Kilimanjaro. About 2 million years ago, the volcano exploded, destroying its summit in a violent eruption probably similar to that of the Krakatau eruption of 1883 in Indonesia. The volcano's explosive self-destruction blanketed the surrounding area with enormous quantities of volcanic debris and ash. The spent volcano then collapsed on itself to form the present day caldera, and is now extinct. The circular structure is said be the world's largest unbroken caldera.

NGORONGORO'S DIVERSE
AND DENSE ANIMAL POPULATIONS

Today, the caldera forms a rich natural basin with streams flowing inward from its up to 2,000-foot (600-m) high walls. Seasonal swamps and wetlands form in the southeastern and the northwestern areas of the basin floor, providing habitat for hippopotamuses and many water birds. Acacia forests in the southwest are home to baboons, monkeys, and elephants. In the east, buffalo herds graze on the longer grasses growing there. Buffalo numbers have been increasing since human land use in the caldera dropped off, thereby lessening competition on the grassland. The short grass plains of the central region support herds of Thompson's gazelle, wildebeest, and zebra. The walls of Ngorongoro caldera also protect one of the two remaining herds of black rhinoceros (the other herd being in the Serengeti). The basin hosts nearly every species of wildlife found in East Africa, and an estimated total of some 25,000 individual animals.

Ngorongoro's dense grazing populations in turn support a number of predatory species and scavengers including the lion, leopard, cheetah, spotted hyena, serval (African wild cat), ratel (honey badger), jackal, and the bat-eared fox. Over the past few years, revegetation has returned the area's grasslands to their natural state, leading to increases in cheetah numbers.

NGORONGORO ERUPTIONS
HELP DATE HUMAN FOSSILS

The proximity of the Olduvai Gorge and Laetoli archeological sites to one another as well as to Ngorongoro means that they were both covered in ash each time the volcano erupted. These debris blankets, therefore, link the fossil ages at both sites. The Laetoli fossil beds record the early eruptive history of Ngorongoro beginning about 4.3 million years ago and overlap with the deposition of the Olduvai fossil beds, which record the volcano's later eruptions and demise 2 million years ago.

LAKE MAGADI

Toward the center of the caldera lies Lake Magadi, which supports thousands of migratory flamingoes and other wading birds. The lake was much larger during the Pleistocene Epoch, as evidenced by the freshwater fish fossils preserved in the ancient lake sediments found at higher levels in the caldera.

BELOW Not far from Ngorongoro Crater and lying within the Ngorongoro Conservation Area is the volcano Oldoinyo Lengai (steaming in the foreground), along with Empakaai Crater, one of a series of collapsed volcanoes that line the rim of the East African Rift Valley. Its crater is filled by a very deep wide lake.

Today, Lake Magadi has shrunk and become highly saline due to its internal drainage and high evaporation rate. Its waters support only a single species of cichlid fish, *Alcolapia grahami*. The extraordinarily high sodium and carbonate concentration comes from weathering of the volcanic ash and from abundant alkaline hot spring activity around the lake's edges. During the rainy season, a thin layer of briny water fills the lake pan. This evaporates quickly during the dry season, leaving a vast deposit of white sodium carbonate (trona) that dries out and cracks into large polygonal shapes.

Residents of the small town of Magadi, on the lake's eastern shore, mine the trona deposits, which are up to 130 feet (40 m) thick in places. The trona is crushed and washed in the Magadi Soda Factory to produce refined soda for a range of industrial uses, including leather tanning and the manufacture of glass, detergent, and paper products.

A WORLD HERITAGE SITE

Ngorongoro Crater, and nearby Olduvai Gorge, are protected within Ngorongoro Conservation Area (NCA), which was established in 1959 and

covers an area of 3,186 square miles (8,288 km²). The NCA is unique in that is designated as multi-use, meaning that limited human activities can take place within the park boundaries, while the wildlife enjoys protection status. The high altitude here creates a malaria-free microclimate. In 1979, the area was inscribed as a World Heritage Site for both its geological and ecological value.

ABOVE Thick fog fills Ngorongoro Crater in the Crater Highlands area of Tanzania, which is part of the East African Rift Valley. This rift valley was formed when tectonic forces began to tear apart Earth's crust.

LEFT Lake Magadi is home to many wading birds, including the graceful flamingoes that feed on algae growing in the shallow water. The jackal is one of the predatory species found in Ngorongoro.

Volcanic Remnants of Australia

Ancient volcanic complexes and basaltic lava fields that once erupted along lines of weakness parallel to the east Australian highlands have left behind dozens of spectacular remnants. Many of these volcanic centers formed as the Australian continent moved northward over a series of stationary mantle hotspots or plumes following the continent's breakup with Antarctica.

ABOVE The koala *(Phascolarctos cinereus)* is an Australian marsupial. Koalas are found in a range of habitats, including eucalypt forests and woodlands, both of which exist in Warrumbungle National Park.

Like the Hawaiian Islands and the other volcanic chains of the Pacific Ocean, the ages of the larger Australian volcanoes and Tasman Sea mounts become progressively younger toward the south. From their ages, a rate of northward movement for the Australian continent has been calculated at approximately $2\frac{1}{3}$ inches (6 cm) per year. Many of these eroded volcanoes are located in Australia's most often visited national parks. Warrumbungle National Park near Coonabarabran, New South Wales, includes the impressive trachyte plugs and dikes of a volcano that was active about 13 to 17 million years ago. Belougery Spire, Belougery Split Rock, Cater Bluff, and Bluff Mountain were all once part of the volcano's ancient plumbing system. The Breadknife, which is also part of the eroded volcano, is Australia's finest example of an exposed dike. This impressive wall stands 300 feet (100 m) high, but is only 10 feet (3 m) wide.

GLASSHOUSE MOUNTAINS

Glasshouse Mountains National Park, 40 miles (70 km) north of Brisbane, Queensland, contains 10 remnant plugs, including Mt Tibrogargan, Mt Beerburrum, Mt Beerwah, and Mt Coonowrin. These rhyolite and trachyte plugs are the remains of a volcanic complex that erupted between 25 and 27 million years ago. Captain James Cook named the mountains in 1770, after spotting them while sailing up the eastern coast of Australia; he thought they resembled glasshouses.

THE TWEED VOLCANO

The Tweed Volcano is an impressive eroded basaltic shield volcano located inland from Tweed Heads in northern New South Wales. It is clearly visible in satellite images, measuring approximately 60 miles (100 km) across its base and centered on the prominent 3,796-foot (1,157-m) tall Mt Warning volcanic plug, which last carried magma to the surface some 20 million years ago. Though the top of the volcano has been eroded away, the angles of its lower slopes indicate that it once stood around 6,500 feet (2,000 m) taller than at present. Mt Warning and the surrounding rainforests in the volcano's eroded caldera are within Mt Warning National Park. In 1975, these protected areas were listed as World Heritage Sites.

A TROVE OF GEMSTONE RICHES

Sapphires are found in many of the streams and rivers that flow from the east Australian alkali basaltic volcanic provinces. Initial highly explosive eruptions brought with them a wealth of sapphires from Earth's mantle, and they have formed some of the world's richest gem fields. Two of these volcanic centers are home to Australia's best-known sapphire mining districts, the New England Gem Field near Inverell, New South Wales, and the Central Queensland Gem Fields centered on the township of Sapphire. During the 1970s and 1980s, mining excavations in New England's Kings Plains valley dug up ground so rich that it was visibly blue with sapphires.

BALD ROCK GRANITE DOME

Bald Rock near Tenterfield, New South Wales, is a prominent granite dome rising 650 feet (200 m) above the surrounding landscape. It is one of the many plutons that form part of the New England granite belt. The molten granite magma rose up through Earth's crust about 220 million years ago and cooled slowly below the surface. Uplift and erosion have since exposed the granite, which is more resistant to weathering than the surrounding metamorphic and sedimentary rocks. The result is a landscape of rounded granite boulders, some of which are enormous and precariously stacked on top of one another. Bald Rock and the surrounding region are protected in Bald Rock and Girraween national parks. Other impressive granite landscapes can be seen in Wilsons Promontory National Park in Victoria, and Devils Marbles Conservation Reserve in the Northern Territory.

VOLCANICALLY ACTIVE NEIGHBOR

Unlike Australia, neighboring New Zealand is currently volcanically active. But some older remnants exist. Mt Cargill forms part of a massive eroded shield volcano that was last active some 10 million years ago. The breached caldera of the volcano now forms Otago Harbor in New Zealand. Mt Cargill forms the backdrop for the city of Dunedin and those making the popular walk to the top of the 2,237-foot (682-m) peak are rewarded with fine views of Dunedin, its harbor, and the surrounding Otago Peninsula. On the walk through the cloud forest, basalt columns in the lava flows are very attractively exposed at a locality known as the Organ Pipes.

ABOVE Located near Tennant Creek in the Northern Territory, Devils Marbles is a collection of rocks of varying sizes and shapes. Over many millions of years, a single, large, volcanic, granite mass of rock eroded to leave these balancing boulders.

BELOW The Warrumbungles landscape in New South Wales was shaped by volcanic activity and erosion over millions of years. The bedrock is referred to as Pilliga sandstone.

LORD HOWE ISLAND

Lord Howe Island is the tiny 7 mile by 2 mile (11 km by 2.5 km) remnant of a once-mighty basaltic shield volcano that rose out of the waters of the Tasman Sea about 7 million years ago. Today, less than a mere 2.5 percent of the original island remains, the rest having fallen to the forces of erosion.

BELOW The topography of Lord Howe Island is dominated by two peaks: Mt Lidgbird (left) and Mt Gower (right), both of which are made up of caldera-filled, horizontal lava flows. Mt Lidgbird's basalt volcanic rock is the youngest in the island area. Columnar jointing can be seen in several of the lava flows that make up this rock.

Initially, flow after flow of basalt lava erupted onto the floor of the Tasman Sea some 370 miles (600 km) off the east coast of Australia, and built up the broad volcanic edifice known as a "shield volcano." It took some 9,800 feet (3,000 m) of lava to reach sea level and then it built above sea level for at least a further 3,300 feet (1,000 m). At that time, Lord Howe probably looked much like the Big Island of Hawaii does today.

Eventually, the Lord Howe volcano was disconnected from its deep lava hotspot source by the relentless northward movement of the Australian Plate, and it ceased to grow. It joined a line of older extinct volcanoes on a "conveyor belt" moving north at a rate of about $2\frac{1}{3}$ inches (6 cm) per year. Middleton Reef, 120 miles (200 km) further to the north on the same hotspot chain, heralds the ultimate fate of Lord Howe Island. Today, there is no volcano left, and Middleton Reef is a small uninhabited ring of coral, exposed only at low tide.

TALLEST SEA STACK IN THE WORLD
A younger volcano erupted 14 miles (23 km) to the south of Lord Howe Island. Being smaller, it

eroded more rapidly; so even less is left of that volcano. Today all that remains of Balls Pyramid is a steep pinnacle of rock measuring only 3,600 feet by 1,300 feet (1,100 m by 400 m), and rising to an impressive 1,808 feet (551 m). It is listed in the *Guinness Book of Records* as the world's tallest sea stack. Unprotected by a coral reef, it is being continually smashed by ocean waves around its base, so does not have much time left, geologically speaking. On calm days, Balls Pyramid is often viewed by boatloads of visitors from Lord Howe Island interested in the seabird colonies covering its spectacular jagged spire.

A WORLD HERITAGE SITE

Although little remains of Lord Howe Island's once-extensive layer-cake of lava flows, explosive breccias, and volcanic dikes, these geologic features are responsible for the breathtaking scenery seen on the island today. The combination of this landscape with its unique evolved communities of plant, seabird, and marine life led to Lord Howe

EXTINCT INSECTS REDISCOVERED

The craggy and remote Balls Pyramid spire was found to be the last remaining home of the believed-to-be-extinct Lord Howe Island phasmid. A geologist, Dr David Roots, discovered this bizarre-looking stick insect in 1964, during an expedition to the pyramid while leading a group of climbers and Boy Scouts. Since that time, several pairs of phasmids have been collected and are breeding successfully in captivity. They are now ready for the day when feral rats, believed to be responsible for their extinction, are finally eradicated from Lord Howe Island, and they can be safely reintroduced into the wild.

Island and adjacent islets being granted World Heritage listing in 1982. Native rainforest, from which the ubiquitous Kentia indoor palm (*Howea forsteriana*) originated, covers over two-thirds of the island. Goats and pigs, originally introduced as a source of food for passing whalers, have been removed. Cats are now banned, with the last pet dying of old age in 2006, and programs are in place to control the rat population.

As a result of these initiatives, the Lord Howe woodhen, a flightless bird that is considered to be one of the rarest birds in the world, was brought back from the brink of extinction. Ninety-two chicks were reared in a captive-breeding program from just a handful of birds remaining on the summit of Mt Gower, and released into the wild.

Only 400 tourists can visit Lord Howe Island at any one time. The 350 permanent residents of Lord Howe live with a number of government restrictions on how they can employ themselves, and what they can build or grow—a small price to pay for a life in paradise.

ABOVE A Spanish dancer nudibranch swims in the Pacific Ocean near Lord Howe Island, which boasts the world's southernmost true coral reef. A tropical current brings warm waters and an interesting variety of marine life to the reef.

LEFT The sea stack called Balls Pyramid, just south of Lord Howe Island, is a good example of the almost final phase of destruction of a volcanic island. Balls Pyramid is part of the World Heritage Site, as are Admiralty Islands and Mutton Bird Islands.

LAVA TUBES

Networks of tube-like lava caves radiating away from old eruptive vents are common remnants around basaltic volcanoes. Many of the tunnels are small and flat, but there are some striking examples that can be entered walking upright and followed for miles, such as the Mt St Helens lava tubes in Oregon, United States, and the Undara lava tubes in Queensland, Australia.

OPPOSITE As red-hot lava cascades into the ocean in Hawaii Volcanoes National Park in the United States, fume clouds or plumes are created. These fumes contain volcanic glass and hydrochloric acid, and are a serious health hazard to anyone who happens to be in the vicinity.

BELOW The Undara lava tubes are remnants of basalt lava flows that poured onto the Atherton Tablelands from a crater near Mt Surprise in northern Queensland, around 190,000 years ago. So far more than 60 lava tubes and arches (collapsed tubes) have been discovered at Undara.

Basaltic lava leaves its vent at temperatures above 2,200°F (1,200°C). A lava tube begins to develop when lava flowing downhill forms a cooled skin of rock on its surface. The originally bright red lava progressively becomes darker red, and then black as the skin hardens. It thickens and then freezes into place against the valley walls while the still-molten lava beneath continues to flow unimpeded. This skin insulates the underlying lava, allowing it to flow for great distances down the flanks of a volcano without solidifying. Spectacular examples of active lava tubes spilling their contents directly into the sea, as waterfalls of incandescent lava, can be seen in Hawaii Volcanoes National Park on the southeastern side of Kilauea Volcano on the Big Island of Hawaii.

HORNITOS ON THE ROOF
Hornitos are small steep-sided spatter cones found on the roofs of lava tubes. They form when lava is

forced out of a hole or crack in the solid roof of an active lava tube, which results in lava spattering out around the edges of the opening. This builds up into a chimney-like structure (hence the Spanish name, which means "little oven") with a central opening to the interior of the tube. Skylight Cave in central Oregon provides a beautiful example, being illuminated by daylight coming in through three hornitos in its ceiling.

EMPTYING THE LAVA TUBE
When a volcanic vent ceases erupting, it acts like a tap turning off a giant water hose. New lava no longer enters the tube and the remainder of the lava drains out of the lower end. The result is an empty lava tube that is perfect for exploring. Lava tubes often feature lava stalactites hanging from the ceiling and lines of former lava flow levels along the sides of the tunnel. Over time, thin areas of the ceiling will collapse forming "skylight sinkholes" that allow light, water, and eventually vegetation into the tunnel. It is through these delightful pockets of sunken garden that access can be gained to the lava tube.

LAVA TUBES AROUND THE WORLD
The Mt St Helens lava tubes formed when the volcano erupted fluid lava, known as the Cave Basalt Flow, from a series of flank fissures about 1,950 years ago. Of the 64 identified lava tubes, the best-known and longest is Ape Cave. It is 12,800 feet (3,900 m) in length, with its ceiling sometimes vaulting to more than 23 feet (7 m) tall.

In places the tube looks like an abandoned railway tunnel. Most of the tubes have pools of water in their lower ends, such as Lake Cave. Little Red River Cave has a small stream running through it.

In Iceland there are lava tubes, such as Surtshellir, near Reykjavik, that are even bigger than those at Mt St Helens. During the seventeenth century, Surtshellir was used as a hideout by a gang of bandits who would come out of hiding to steal crops and livestock from the surrounding farms.

Thurston lava tube on the Big Island of Hawaii formed about 350 to 500 years ago during the eruption of a large vent, Ai-laau, on the east side of Kilauea Volcano. After the eruption ceased, the vent collapsed, forming the Kilauea Iki crater and two small pit craters. A short trail descends to the lushly vegetated floor of one of the small pit craters from where the tube can be entered. The flat-floored tunnel is well lit with electrical lights and can be followed for 400 feet (120 m). It was named after local newspaper publisher Lorrin Thurston, who played a vital part in the establishment of Hawaii Volcanoes National Park, which became the United States' thirteenth national park in 1916.

ABOVE The 'i'iwi *(Vestiaria coccinea)* is one of six Hawaiian honeycreeper species living in Hawaii Volcanoes National Park. This scarlet songbird can make a variety of sounds from metallic squeaks to clear flute-like calls.

LEFT Trees and rainforest vegetation thrive within a lava tube in Hawaii Volcanoes National Park. Thurston lava tube, also known as Nahuku, is a good example of this phenomenon.

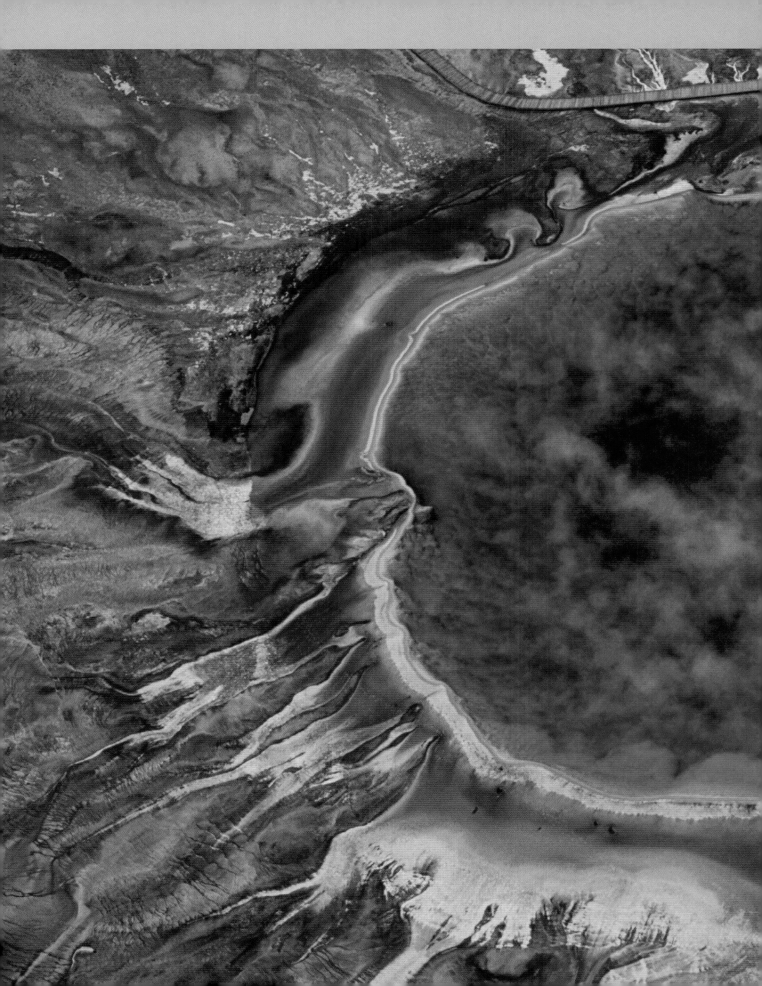

GEYSERS AND HOT SPRINGS

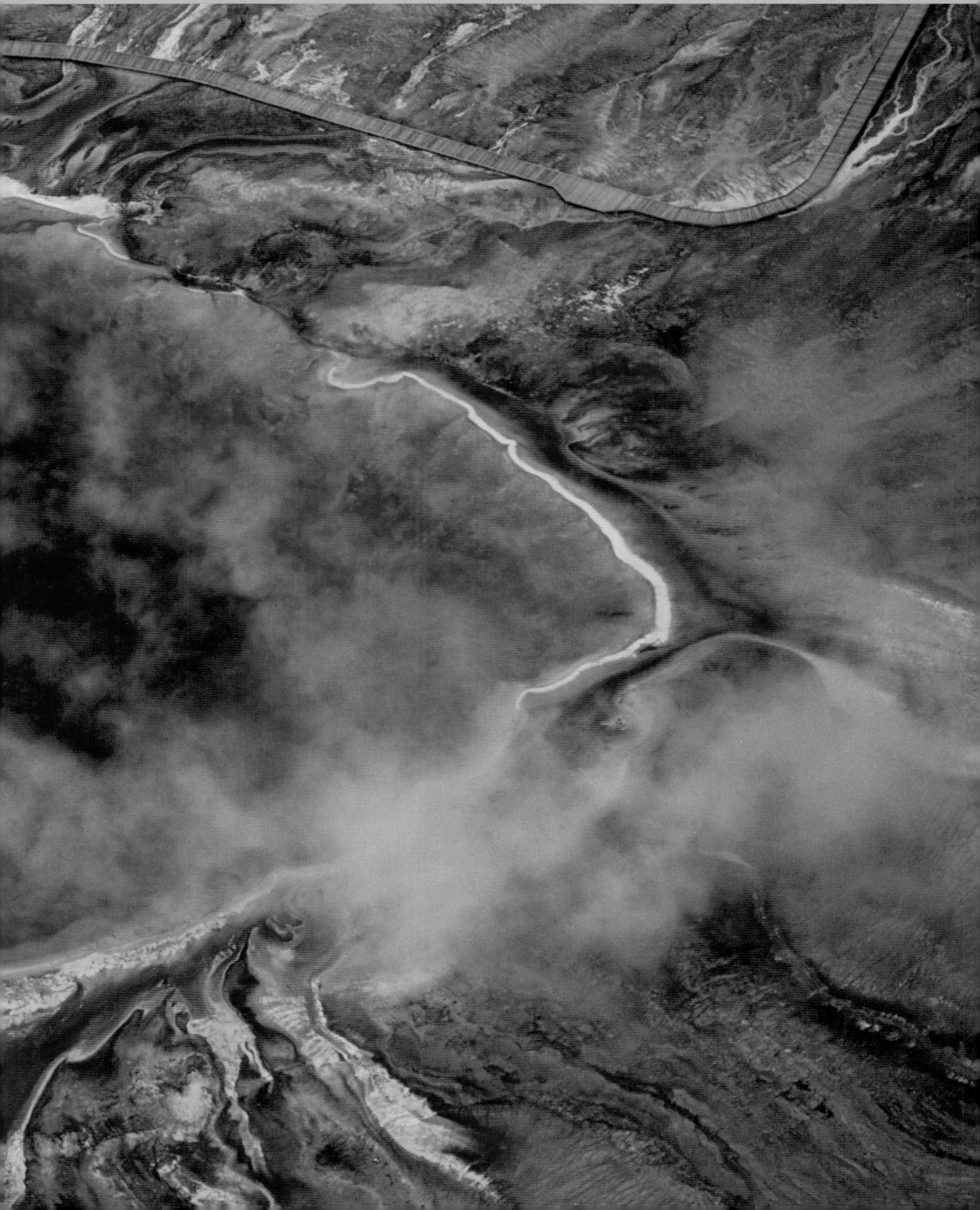

E voking a mood of mystery, suppressed energy, and even danger, geothermal areas have long attracted awe-inspired onlookers. Boiling pools of strangely colored water, thermal springs with supposed curative powers, pots of bubbling mud, and sudden noisy spouting geysers are typical of such areas.

The attraction and mystery of geothermal areas was even greater before modern science understood the process, as seen in the thirteenth century when Europeans would flock to Iceland's Great Geysir, believing it to be a supernatural phenomenon. After geothermal activity was better understood, it was realized that Earth's subterranean heat could be harnessed to provide us with a source of energy. However, experience has shown that we must be moderate in our consumption, and mindful of the ramifications of significantly altering a natural process.

Geothermal zones are defined as those regions where the geothermal gradient, or the increase in temperature with depth, in the top few miles of Earth's crust is significantly higher than normal. Groundwater percolating downward through cracks, faults, and channelways in these areas is heated to boiling point, and manifests itself in several different spectacular ways.

ABOVE Great Fountain Geyser in Yellowstone National Park, USA, puts on a spectacular display at sunset. Fountain geysers erupt from pools of water.

RIGHT The vivid blue, mineral-rich, geothermal waters in Iceland's hot springs are said to be helpful in the treatment of some skin conditions. This hot spring is at Hveravellir.

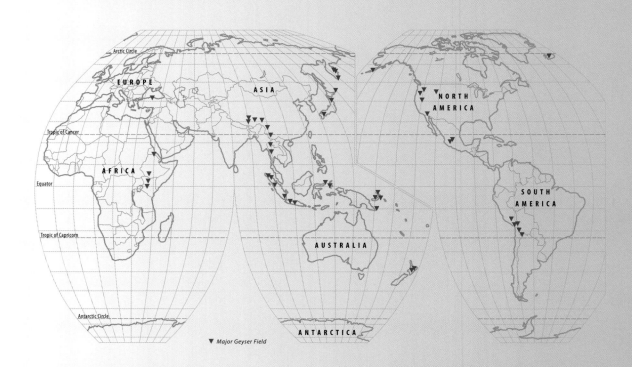

▼ *Major Geyser Field*

DISTINCT GEOTHERMAL
BELTS ENCIRCLE THE EARTH

Areas of high geothermal heat flow occur in distinct linear zones, which correspond to major boundaries between the tectonic plates. Earth's main geothermal belts are found along the mid-oceanic ridges and rift valleys, where the crust is young and thin, and where heat can easily escape from Earth's interior. These areas also coincide with Earth's major subduction zones where melting and volcanic activity is most common, and with collision margins where heat is being generated by intense and ongoing crushing and folding. Localized geothermal fields are associated with isolated mantle hotspots responsible for building giant shield volcanoes such as Mauna Loa, Hawaii.

SPREADING RIDGES AND BLACK SMOKERS

The spreading ridge geothermal zones are mostly located underneath the ocean where new basaltic ocean crust is being formed by the line of rift volcanoes at the mid-oceanic ridges, and also in rift valleys. Underwater geysers, or hydrothermal vents, in the deep oceans are visible only by using submersible craft. As the hot mineral-rich waters issue forth from these fissures and cool in the near-freezing waters of the ocean bottom, tiny

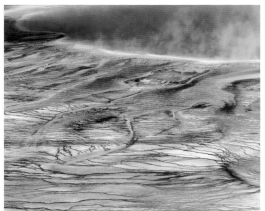

black crystals of metal sulfides precipitate instantly as billowing, thick, black clouds. Hence these vents are known as "black smokers" and sometimes build spectacular complexes of tall black chimneys. The first black smoker vent was only discovered in 1977.

Even in total darkness and at an average depth of about 7,000 feet (2,000 m), marine life has evolved near the black smokers, taking advantage of the geothermal heat. The unusual marine creatures include tubeworms and huge clams around the Pacific Ocean vents, while eyeless shrimp have been found at vents in the Atlantic Ocean.

ABOVE At Pamukkale in Turkey, calcite-rich waters flow down from hot springs in a cliff above the plain, resulting in a landscape of white terraced basins. Over-commercialization almost destroyed this natural wonder; hotels and pools were built on site, and the water stopped flowing. A restoration project is underway to reverse the damage.

LEFT Yellowstone National Park, USA, boasts an impressive 500 geysers, which occur in nine basins within the park. Pictured here is the Midway Geyser Basin.

ABOVE Archeologists have discovered numerous Roman bath complexes. The Romans utilized geothermal water for its health benefits as well as for heating their homes.

OPPOSITE A herd of elks graze through the snow around an erupting geyser in Yellowstone National Park, USA. They are the most common animals here and many remain in those parts of the park during winter where the snowfall is not too high.

BELOW Iceland's position atop the Mid-Atlantic Ridge gives it access to abundant, cheap, clean geothermal power. This plant is in the Krafla Thermal Area.

Iceland's geothermal and geyser fields lie on the spreading Mid-Atlantic Ridge where it rises from the sea and cuts across the island.

HEAT RINGS AROUND THE PACIFIC

The circum-Pacific geothermal zone, running through the Philippines, Japan, Kamchatka, the Aleutians, North and South America, Tonga, and New Zealand, is supplied by heat from the volcanically active subduction zones along which the Pacific Ocean basin is being swallowed underneath the surrounding plates. The opening of the Atlantic Ocean is driving the closure of the Pacific Ocean. Outside the Pacific, other geothermal belts fuelled by subduction heat include the Indonesian, Antilles, and South Sandwich island arcs.

HEAT GENERATED FROM THE COLLISION MOUNTAIN RANGES

The Mediterranean–Himalayan geothermal belt follows the line of closure of the ancient Tethys Sea, where the Indo-Australian, African, and Arabian tectonic plates began to crunch into the southern edge of the Eurasian tectonic plate beginning approximately 50 million years ago. The collision, which is still going on today, is building some of the world's mightiest mountain ranges and generating immense amounts of geothermal heat.

HISTORY OF GEOTHERMAL ENERGY

Humans have utilized geothermal energy in a variety of ways over time. Archeological evidence in North America shows that Paleo-Indians began forming settlements near hot springs more than 10,000 years ago. They appear to have used the mineral springs as a source of warmth, cleansing, and healing. The Romans used geothermal water to heat homes, and for treatment of injuries and illnesses in their bathhouses. In other geothermal

WHY ADD SOAP TO A GEYSER?

Surfactants are sometimes added to a geyser to make it erupt "on cue" for an expectant audience! A surfactant or "surface active agent" such as soap breaks down the surface tension of water, allowing it to better cover or "wet" a surface. In a near-boiling geyser this enhanced surface contact allows heat to transfer more effectively from the hot wall rocks into the water, bringing it quickly to the boil. Boiling decreases the pressure on the water below, which in turn boils. The reaction travels quickly downward, setting off the geyser. These are human-induced eruptions. Inducing eruptions of Iceland's Great Geysir has been discontinued because the forced eruptions and surfactants were damaging the geyser's plumbing system.

areas, such as New Zealand, Indonesia, Papua New Guinea, and Iceland, native peoples used geothermal steam vents or boiling pools to cook their food.

More recently, geothermal energy has been used to produce clean energy to replace dangerous or polluting non-renewable energy sources such as uranium or coal. The first steam-based geothermal power plant opened in Larderello, Italy, in 1904, and remains in operation today. Geothermal power plants harness the steam that reaches Earth's surface to spin turbines, which in turn drive generators that make electricity. In dry areas, energy can also be extracted from the hot rocks by pumping water down specially drilled boreholes. The superheated water returning to the surface at temperatures greater than 360°F (182°C) flashes into steam and is used to drive the turbines. The biggest users of geothermal power are the USA, the Philippines, Italy, Mexico, Indonesia, Japan, and New Zealand. Colder countries that are situated in geothermal zones, such as Iceland and Japan, also take advantage of direct geothermal heating by using piped hot water to heat homes and other buildings.

GEYSERS: A RARE PHENOMENON

The spectacular phenomenon of columns of boiling water spouting into the air at periodic intervals is extremely rare due to the specific set of geological conditions required for geysers to occur. There are only about 1,000 active geysers on Earth, with most of these occurring at just two sites. North America's Yellowstone National Park contains the lion's share with around 500 geysers, and there are up to 200 in Dolina Geizerov on the Kamchatka Peninsula, Russia. The world's third-largest geyser field, El Tatio, Chile, contains just 38 geysers.

Three important geologic ingredients are needed in order for geysers to occur. These are a shallow source of volcanic heat, an abundant groundwater

supply, and a unique natural underground plumbing system. The plumbing system is critical because it must be watertight and pressure-tight, much like a pressure cooker, in order to spurt out a jet of boiling water. Silica-rich volcanic rocks appear to be the best hosts. Rhyolite lavas and ignimbrites host most of the world's geysers. Silica dissolved from these rocks by the hot water is precipitated as mineral deposits called siliceous sinter, or geyserite, along the inside of the plumbing systems. Over time, these deposits form a watertight seal and strengthen the channel walls, enabling the geyser to work. Geysers are classified into two types: fountain geysers, which erupt from pools of water, usually as a series of sudden powerful bursts; and cone geysers, which erupt from cones or mounds of geyserite, usually in steady jets that can last anywhere from a few seconds to several minutes. Yellowstone National Park in Wyoming, USA, contains both types: Fountain Geyser in the Lower Geyser Basin is an example of a fountain geyser; and Castle Geyser and Old Faithful in the Upper Geyser Basin are both cone geysers.

HOW A GEYSER WORKS

Groundwater, which has been superheated by contact with hot rocks, circulates into the geyser's plumbing system from below and meets cooler water flowing in from the surface. They mix and fill the plumbing system, and are gradually heated from below. This process may take minutes for small geysers, or hours, even days, for larger geysers.

EVOLUTION OF A GEYSER SPOUT

Geysers such as Old Faithful, in North America's much-visited Yellowstone National Park, spout on an extremely regular basis. Following an eruption, surface water and groundwater drains back into the plumbing system: A. The geyser's empty plumbing system starts to fill with cold water, which runs back down into the plumbing system and begins to heat up. B. The water in the now-full plumbing system heats to above boiling point, and bubbles of steam start to form and rise up through the water. The system is still stable at this point. C. When sufficient gas bubbles amalgamate to make a "plunger" of high pressure gas, the system becomes unstable. The expanding gas then drives the overlying plug of water upward and out in the form of a spectacular geyser spout.

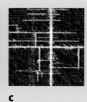

A   B   C

The water temperature reaches well above boiling point within the geyser's plumbing system, but it does not boil because of the pressure of the overlying water column. Eventually the water at depth becomes hot enough and starts to boil. Steam bubbles begin to rise through the water column in increasing quantity until they amalgamate and physically push the overlying column of water upward. At the surface, this is visible as a sudden outpouring of water or a rising dome of water sometimes called a "blue bubble." At this moment, the loss of water causes the system to become unstable. Due to the reduction in pressure (which equals a drop in boiling point temperature) the superheated water in the plumbing system is suddenly able to boil vigorously. Within seconds there is a sudden "flashing" of water to steam accompanied by a huge volume increase, so violent that it ejects the overlying water column into the air. Steam then continues to roar from the vent until either the water is used up or the temperature within the plumbing system drops below boiling.

The cycle of filling, heating, and boiling is then repeated, leading to another eruption. The duration of eruptions and times between successive eruptions will vary greatly from geyser to geyser; Strokkur in Iceland's Geysir geothermal field erupts for a few seconds every few minutes, while Grand Geyser in the United States erupts for up to 10 minutes every 8 to 12 hours.

GEYSERS UNDER THREAT

Since the advent of geothermal power plants in the twentieth century, it was often noted that the extraction of steam and the taking of groundwater to run the plant caused nearby geysers to cease being active. Nowhere has this been

LEFT Hot springs and geysers steam and spout at Lake Bogoria National Park in Kenya. Lake Bogoria waters are rich in the algae that attracts large numbers of lesser and greater flamingoes seasonally, and also turns the lake shores pink.

more clearly illustrated than in New Zealand. Of five major New Zealand geyser fields in existence a century ago (Whakarewarewa, Rotomahana, Orakeikorako, Wairakei, and Spa), Whakarewarewa is the only one with a significant number of geysers that are still active. Fewer than 15 geysers are still erupting out of more than 130 known to have been active throughout New Zealand in 1950.

Another growing threat to geysers comes from mining activity, because thermal areas are often very rich in valuable minerals, such as gold, precipitated by the hot groundwater. Extraction may even require removing the geyser plumbing itself. A dramatic example was the cessation of South America's second-largest geyser field (Puchuldiza, Chile) in May 2003, which was caused by mineral exploration being carried out nearby.

HOT POOLS, FUMAROLES, AND BOILING MUD POTS

Even if all of the essential elements for geysers are not present in a geothermal area, other geothermal phenomenon such as fumaroles (solfataras), hot springs, boiling springs, and mud pots can still occur. Fumaroles are steam vents that form when small quantities of water come in contact with an intense volcanic heat source. Fumaroles are particularly active after rain.

Hot pools and boiling pools, in turn, form where there is abundant groundwater present. The bright colors often seen in thermal pools, such as the Grand Prismatic Spring, Yellowstone National Park, are produced by thermophilic (heat-loving) prokaryotes (bacteria) that live in the hot waters despite the harsh conditions.

If the surrounding rock is soft and friable, or if there is much soil, then the water will turn into a thick muddy slurry and form mud pots.

Hot springs frequently have a very high soluble mineral content, containing everything from lithium, sodium, and calcium to iron and even radium. Because of the proven medical benefits of some of these springs, plus a good dose of folklore, many have become popular locations for tourist resorts and rehabilitation clinics.

COLD-WATER GEYSERS

Cold-water geysers are similar to normal geysers except that they are driven by carbon dioxide gas rather than steam. They mostly occur when boreholes are drilled into confined gas- and water-filled sedimentary rocks (artesian aquifers), and as such are rarely found in a natural setting. There are only two natural carbon dioxide driven, cold-water geysers in relatively undisturbed settings. These are Cold Water Geyser in Yellowstone National Park, and the Salton Sea Geyser in California. Erupting cold-water geysers are similar in appearance to steam-driven geysers, although the carbon dioxide-laden water often appears more white and frothy.

ABOVE Japanese women enjoy the benefits of an outdoor hot spring. Some springs can still be found in remote and pristine locations in Japan, where *onsen* bathing is an important ritual.

Geothermal Areas of Iceland

Iceland has numerous high-temperature geothermal areas located within the island's zone of active rifting and volcanism. Here, subsurface temperatures rapidly rise to about 460°F (240°C) within a depth of about 3,000 feet (1,000 m). The best known of these is the Geysir geothermal field in the Haukadalur Valley, which covers an area of about one square mile (3 km²).

Geological studies of the thermal rock formations indicate that the Geysir field has been active for at least 10,000 years. Its famous hot springs and geysers are created by groundwater that has been heated while circulating through narrow channels in the hotter rocks at depth. The water temperature determines the nature of thermal activity. Normal hot springs occur at temperatures up to boiling point, 212°F (100°C), while above that the springs erupt into geysers. The geysers in the Geysir field are the Great Geysir (the gusher), Strokkur (the churn), Sodi (the sod), Smithur (the carpenter), Fata (the bucket), Otherrishola (the rainmaker), Litli Geysir (the little Geysir), and Litli Strokkur (the little Strokkur).

THE GEYSER FIELD COMES TO LIFE

In June 2000, a series of large earthquakes awoke the Great Geysir from its dormancy and also caused the Otherrishola and Fata geysers to become active again, with Fata sometimes erupting twice a day. Interestingly, eruptions of Otherrishola are sometimes triggered by certain weather events,

THE MUD POTS OF THYKKUHVERIR

In the southern part of the Geysir geothermal area there are viscous hot springs or boiling mud pots. These are often at a temperature of around 158°F (70°C) to 176°F (80°C) and are formed by steam rising through groundwater-saturated soils. During dry weather, when the water level in the ground drops, these bubbling mud pots can change into ordinary steaming fumaroles.

if the air pressure becomes low enough. Several new hot springs were born during the earthquakes and the amount of water flowing from others increased. The hot springs Konungshver (the king's hot spring) and Blesi (the blazer) began boiling vigorously. North of Geysir are fumaroles, a type of phenomenon where only steam and gas escapes from the geothermal system. Bright yellow native sulfur can sometimes be seen crystallizing around the sides of the steam vents.

STROKKUR—THE MAIN ATTRACTION

The geyser Strokkur is now the thermal field's main tourist attraction. It erupts about every eight minutes, beginning with a beautiful blue bubble when the still pool of water is lifted and domed outward by rising and expanding steam. Once the steam reaches the surface, the water explodes upward to heights of 100 to 130 feet (30 to 40 m). Strokkur was formed following an earthquake in 1789 and was active until another earthquake struck in 1896, shutting it down. Its channel was thoroughly cleaned in 1963 and it has been operating reliably since.

GEOTHERMAL POWER IN ICELAND

As Iceland sits atop the Mid-Atlantic Ridge, where basalt lava is being continually emplaced, the country has access to abundant cheap geothermal energy. To generate steam, water is pumped down as deep as $1\frac{1}{4}$ miles (2 km) in boreholes, where it passes through the hot basalt rocks. The steam is used primarily for heating. Even the sidewalks in Iceland's two largest towns, Reykjavik and

Iceland's Geothermal Areas

Geysers in Haukadalur Valley
Konungshver · Great Geysir
Blesi · Fata
Otherrishola · Strokkur
Litli Strokkur · Litli Geysir
Thykkuhverir · Smithur
Sodi

CLOCKWISE FROM TOP The eruption phase of Strokkur Geyser commences with a dome of water bubbling up from the cone. The eruption always has three bursts and lasts about eight minutes, with the water spout reaching up to 100 feet (30 m) high. The geyser reliably repeats this process in five to ten minutes. Strokkur is the Danish word for "churn."

Akureyri, are heated during the winter to keep them snow- and ice-free. Some of the steam is used to spin turbines and generate electricity, although over 80 percent of the country's electricity is generated by running water (hydro power).

Tourism and energy generation go together at the Svartsengi Power Plant near Keflavik on the Reykjanes Peninsula, where excess hot water from the plant flows into a nearby lake, Blaa Lonith (Blue Lagoon). The water keeps the lagoon at a temperature of about 95°F to 104°F (35°C to 40°C) all year round, hence it has developed into a popular bathing resort and tourist attraction. The mineral-rich geothermal waters of the lagoon are a vivid blue due to the presence of blue-green algae and white silica mud. They are reputed to have therapeutic benefits for ailments including skin conditions such as psoriasis and eczema.

GREAT GEYSIR

The name geyser, as used to describe the geological phenomena of a jet of boiling water and steam periodically shooting upward from an opening in the ground, was derived from the original Great Geysir located in the Geysir thermal field in Haukadalur Valley, Iceland. The word means "to erupt" or "to gush."

RIGHT One of the rare occasions when Geysir is in full eruption. It can lie dormant for many years and its eruptions are highly unpredictable. First recorded in 1294, some eruptions have been violent, involving earthquakes.

The earliest accounts of the Great Geysir date back to the year 1294, when it is believed to have become active after large earthquakes shook that part of southern Iceland. Iceland's popularity as a tourist destination increased over the centuries as the fame of the Great Geysir grew. No other spectacle like it was known in Europe, and it was so unusual that it was long considered a supernatural phenomenon, with people flocking there to see it.

Many theories were proposed to explain the wonder, such as the presence of large steam-filled subsurface caverns beneath the area. It was not until German chemist, Robert Bunsen, came to Iceland in 1846 to inspect Geysir, that the presently accepted mechanism was understood. He was the first to realize that the eruptions were caused by water being heated up to boiling point deep within Geysir's pipe. As this water periodically flashed into steam, the sudden volume increase would force the water upward in the pipe, creating a sudden spout of water.

ABOVE The bright blue mineral pool Blesi is one the many hot pools in the Geysir region of Iceland. The vibrant blue is caused by minerals that are suspended in the water.

GREAT GEYSIR'S SPECTACULAR ERUPTIONS

Geysir erupted from a water-filled vent, 8 feet (2.4 m) wide located in the center of a shallow circular pool, 60 feet (18 m) across. The vent and pool sit atop a mound of sinter that had been built up from silica deposited by Geysir's mineral-rich waters over the centuries.

Geysir's famous eruptions would begin as a bubble of water over the vent, which burst open, shooting spouts of boiling water to heights of 60 to 100 feet (18 to 30 m) from the pool surface. This phase of the spectacle lasted for about 30 minutes before ending abruptly. Then the pool drained completely into the vent and all was quiet for a minute or two before the second phase of the eruption took place. A sudden narrow and powerful jet of water would burst from the dry cone, commonly reaching heights of 165 to 200 feet (50 to 60 m). On occasion, powerful eruptions approaching 300 feet (100 m) were recorded. As the water gradually ran out, a roaring column of high-pressure steam replaced it, which could continue for several hours.

EARTHQUAKES AFFECT GREAT GEYSIR

It makes sense that small changes to the plumbing of numerous underground joints and fractures will affect the performance of such a finely tuned natural system as a geyser. Slight ground movement associated with earthquakes will make geysers become active or dormant by opening and closing joints or changing the level of the water table. Some geysers have even been destroyed by people throwing rubbish into them, while others have stopped spouting when nearby geothermal power plants have begun to extract steam, thus upsetting the natural balance required to form jets of water.

A SHADOW OF ITS FORMER SELF

The Great Geysir stopped spouting in 1915, reawakening only briefly in 1935, before becoming dormant again. During this dormancy period, eruptions were sometimes induced for special occasions, such as Icelandic National Day, by adding soap to the geyser. The induced eruptions were significantly smaller than natural ones, and eventually it was considered that the process might be damaging to the geyser's natural plumbing system.

Geysir became active again after large earthquakes struck Iceland in June 2000. Initially it erupted about eight times a day, then this gradually lessened to about three times daily. Its timetable of eruptions is now very irregular. The geyser has never repeated its former display or regained its former heights. Nowadays it only erupts to heights of 25 to 30 feet (8 to 10 m). As the spouting water falls back onto the mound, it cools the water in the surrounding pool and the boiling stops.

Geysers and Hot Springs of North America

Thousands of thermal springs, from tiny seeps to voluminous outpourings of boiling water, are found in a broad belt running down the western part of the North American continent. They mostly occur within the Western Cordillera and are related to the areas of highest heat flow—where hot rocks lie close to the surface, due to tectonic interaction between the Pacific and North American plates.

RIGHT Liard River Hot Springs in British Columbia is on the Alaska Highway. With the water temperature around 104°F (40°C), it is a popular stopover in winter.

The belt of abundant hot-spring activity is at its widest in the United States and extends as far as the eastern edge of the Rockies. It continues south into Mexico and Central America to join the Andes thermal belt, and to the north through western Canada, Alaska, and the Aleutian Islands. A less intense belt of warm spring activity (with temperatures less than 122°F or 50°C) along the Atlantic seaboard corresponds to the Appalachian Mountains. Hot Springs National Park in the Ouachita Mountains of Arkansas appears as an unusual anomaly on the hotspot distribution map. It was also the first hot springs area to gain US federal protection, in 1911.

WELL-KNOWN HOT SPRING PARKS

Some of the best known and most-visited hot springs are protected as national parks to ensure their equitable use and appreciation. In the USA, these include Hot Springs National Park in Arkansas, Death Valley and Lassen Volcanic national parks in California; Yellowstone National Park in Wyoming; Big Bend National Park in Texas; and Olympic National Park in Washington.

In Canada, Banff Upper Hot Springs led to the establishment of Banff National Park, Canada's first, in 1885. Other Canadian hot springs include Radium Hot Springs in Kootenay National Park, Liard River Hot Springs Provincial Park, and Miette Hot Springs in Jasper National Park.

NORTH AMERICA'S GEYSERS

Geysers are also found in the thermal belt, but are far less common than hot springs because a specific set of geological requirements is necessary for them to occur. Of world renown are the geysers of Yellowstone National Park—the world's largest grouping of geysers—but there are several other lesser-known fields found in North America.

NEVADA'S LOST GEYSER FIELD

The now-extinct Beowawe Geyser Field was located in a small basin in central Nevada between the towns of Elko and Battle Mountain. The geyser field lay at the bottom of a hill slope, and could be seen from the highway. Its decline began in the

ABOVE An American bison (*Bison bison*) bull stands in the steam from a geyser to keep warm in Yellowstone National Park. This park is home to up to 3,500 bison and is the only place where a wild herd still remains. The American bison is also known as the American buffalo.

BELOW Some steam wells on private property in the Black Rock Desert in Nevada's Great Basin still produce geyser-like activity. These travertine domes constantly spray out boiling water.

1950s when, after missing out on being declared a national monument, drilling for geothermal power began. Steam extraction for the nearby dual-flash power plant, which commenced operation in 1985, finally upset the delicate natural balance of water supply and underground heat, and all geyser activity ceased. The sinter shield from geothermal activity is still visible, but now steam only issues at times from several steam wells at the top of the hill.

STEAMBOAT SPRINGS, NEVADA

The Steamboat Springs geyser field is located several miles south of Reno, along the highway to Carson City. According to the Geyser Observation and Study Association (GOSA), as many as 21 geysers were observed spouting from 1984 to 1987, ranging in height from a heavy outflow bubbling up a few inches to as tall as 50 feet (15 m). With this number of active geysers, Steamboat Springs was ranked the fourth or fifth largest geyser field in the world. After 1987, the unfortunate combination of a nearby geothermal power station along with a regional drought caused all geyser activity to cease. By 2003, the groundwater level had dropped down to 30 feet (10 m) below the surface.

UMNAK ISLAND'S REMOTE GEYSERS

The Geyser Bight geothermal area on Umnak Island contains Alaska's hottest and most extensive thermal springs. This remote site is the only one found in Alaska that boasts geysers as well as hot springs. Heat for the geothermal system is derived from magma from the Mt Recheshnoi volcano.

At least 12 geysers have been observed during the years since 1947. Geyser heights of up to 6½ feet (2 m) have been recorded, but in general they are much smaller. With the destruction of Nevada's Beowawe and Steamboat Springs, Alaska's Umnak Island now stands as North America's largest geyser field after Yellowstone.

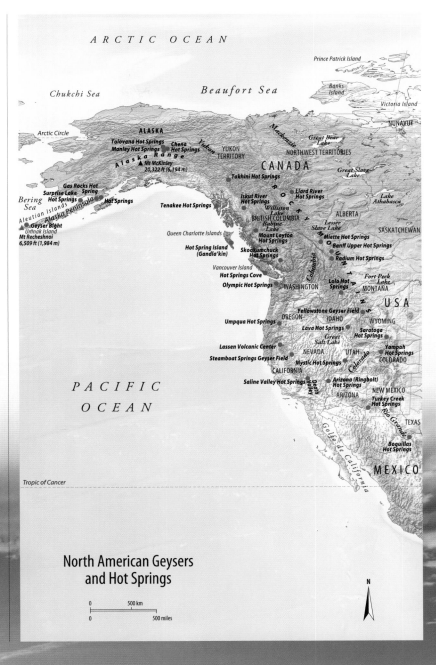

North American Geysers and Hot Springs

YELLOWSTONE NATIONAL PARK

Yellowstone National Park in northwestern Wyoming, USA, contains a phenomenal 500 geysers. That is approximately half of the world's total number, and these occur in nine geyser basins within the park. Old Faithful is best known, but Steamboat Geyser in the Norris Basin is currently the world's tallest. Major eruptions of Steamboat Geyser can reach 400 feet (120 m).

BELOW The geyser affectionately known as Old Faithful shoots up through a winter blanket of snow. No matter what the outside temperature is, the water temperature just before an eruption is always 204°F (95.6°C). Being a cone geyser, the water erupts with a jet-like action, which accounts for the great heights it can reach—up to 184 feet (56 m).

Visitors to Yellowstone can take raised boardwalks through the different basins to marvel at the various and often brightly colored thermal features, and glimpse geysers that often erupt to quite a precise schedule. Old Faithful was the first geyser to be named in the park in 1870 for this reason, being the most frequently erupting and reliable of the larger geysers. Currently, its eruptions can be predicted to within 10 minutes.

OLD FAITHFUL ERUPTS ON CUE
Eruptions of this geyser shoot between 3,700 and 8,400 gallons (14,000 to 32,000 L) of boiling water skyward to a height of about 100 to 200 feet

(30 to 60 m). The duration of an eruption varies from $1\frac{1}{2}$ to 5 minutes, and they occur about 65 to 92 minutes apart. Over 137,000 eruptions of Old Faithful have been studied, and most interestingly, a mathematical relationship exists between the size of an eruption and the interval between eruptions, with longer recharge times needed following larger eruptions. The reliability of Old Faithful is due to the fact that its subterranean plumbing system is not connected to the other thermal features of the Upper Geyser Basin.

Over the years, the interval between eruptions has increased, which may be the result of earthquakes affecting subterranean water levels.

LEFT From the air it is easy to see the bands of color that make up Grand Prismatic Spring, in the Midway Geyser Basin. The bright blue is its natural color. When the water exits the spring, it is close to boiling point. As the water cools in the pool—more so on shallower terraces—the various heat-loving bacteria tint the water with different colors in the surrounding rings. The orange ring is around 158°F (70°C).

Temperature and video probes have been lowered up to 70 feet (20 m) into its conduit, where water temperatures of 244°F (118°C) were measured. This has not changed since temperatures were first measured in 1942.

COLORFUL POOL OF GRAND PRISMATIC SPRING

With a diameter of 370 feet (113 m), Grand Prismatic is the largest hot spring found in Yellowstone. The pool sits upon a wide sinter mound over which water flows evenly on all sides, forming a series of small stair-step terraces. The spring discharges at an estimated 560 gallons (2,120 L) per minute.

The Hayden Expedition in 1871 named Grand Prismatic because of its exquisitely beautiful spectrum of different colors. A deep blue center is surrounded by pale blue, then in turn by a green shallow edge. Outside the scalloped rim, a band of yellow blends into orange, then finally red marks the outer border. White steam shrouding the spring often reflects the brilliant colors, which are produced by different species of thermophilic bacteria that live in narrow temperature ranges. The reddish bacteria at the outer edge survive in the coolest water, with the yellowish and greenish bacteria living in progressively hotter water. With a temperature of 147° to 188°F (64° to 87°C), the central blue area of the spring is too hot to support any of these bacterial species.

LEFT The exquisite silica beads in the beaded geyserite pools of Artemesia Geyser are precipitated around the vents of geysers. These pools are in the Upper Geyser Basin of Yellowstone National Park.

SITTING ON A SUPERVOLCANO

Yellowstone National Park resides in a giant caldera, which is perhaps one of the most potentially dangerous volcanoes in the United States. Heat from a large body of magma beneath the caldera fuels the park's more than 10,000 hot pools and springs, geysers, and bubbling mudpots. From ground level it doesn't look much like a supervolcano as its caldera is just so big. Nevertheless, it has been the site of three very significant eruptions in the last 2 million years. It is estimated that these would have had a volcanic explosivity index of 8, meaning that the ash and lava volumes erupted would have dwarfed those from the 1980 Mt St Helens eruption. The Yellowstone supervolcano is still active, as instruments placed in the caldera record a pattern of continued heaving and bulging that falls by as much as $^2/_5$ inch (1 cm) each year. Yellowstone is not a subduction volcano like those of the Cascades. It is located further inland, and sits atop of a rising mantle plume or hotspot, which is located beneath the continental crust.

Geysers and Hot Springs of the Andes

South America's Andean countries, like so many others positioned along the Pacific "Ring of Fire," experience extremely high levels of geothermal heat flow. The heat is generated by the region's abundant volcanic activity, which is caused by the Pacific Plate subducting beneath the South American Plate. Numerous heated groundwater springs are found throughout the Andes.

The hot springs range from those surrounded by elegant five-star health and relaxation resorts to remote and completely natural pools in a variety of contrasting settings. Geothermal water can be found bubbling out of the ground near sea level in the lush rainforest settings of Ecuador or above 13,000 feet (4,000 m) in the stark bare surrounds of the world's driest desert in Peru and Chile.

Perhaps most remarkable are the springs of Argentina and Chile's Tierra del Fuego region. Here against a backdrop of snow- and ice-covered mountains, steaming hot springs and thermal pools are found adjacent to near-freezing coldwater streams, providing a heart-stopping contrast for anyone willing to experience it.

ABOVE Chilean flamingoes live in the lakes of the Andes. Pink flamingoes start life a dull gray and become pink during adulthood as they consume algae in the lakes they feed in, along with brine shrimp. They are closely related to flamingoes in Africa, Asia and the Caribbean.

CHILE'S FAULT-CONTROLLED HOT SPRINGS

In Chile alone there are several hundred hot springs, often located along major faults that act as conduits along which the groundwater can travel. Near Liquine, a small Andean village in southern Chile, numerous geothermal springs occur along the Liquine–Ofqui Fault. This major fault runs along the Andes for over 600 miles (1,000 km) from Llaima Volcano to Hudson Volcano.

Near Liquine village, water bubbles up from the ground at about 176°F (80°C), and must be cooled before being fed into the bathing pools. The Liquine hot springs are rich in silica, calcium, lithium, iron, potassium, sodium, and sulfur. Puyehue National Park in Chile's Lake District is also well known for its hot springs, mud, and sulfur baths.

The Pocuro Fault in central Chile gives rise to some 35 springs, including the historic Cauquenes hot spring. A hotel built here in 1885 features baths made of European marble.

THE HIGH LAKES OF THE ANDES

A number of striking colored lakes can be found around Licancabur, a 19,455-foot (5,930 m) dormant volcano that straddles the border between Chile and Bolivia. The lakes have evocative names such as Laguna Blanca (White Lagoon), Laguna Verde (Green Lagoon), and Laguna Colorada (Red Lagoon). Different mineral salts color the lakes, except Laguna Colorada, which owes its intense red color to algae. The aquamarine Laguna Verde

is fed by volcanically heated springs. Long-legged pink flamingoes flock in the lakes, with the majestic Chilean flamingo (*Phoenicopterus chilensis*) reaching over 3 feet (1 m) tall.

GEYSERS AND HOT SPRINGS IN THE DESERT

El Tatio geyser field (Los Géiseres del Tatio) and nearby La Puritama hot springs are located in the Atacama Desert of northern Chile, some 13,800 feet (4,200 m) above sea level. The closest regional center is the small oasis town of San Pedro de Atacama. El Tatio Valley is an extraordinary place of geysers, colorful pools, terraces, and mud pots against a backdrop of active volcanoes. The surrounding desert landscape is harsh and treeless, apart from some greenery softening the edges of the thermal pools.

El Tatio contains at least 80 active geysers, which makes it the world's third largest geyser field—after Yellowstone in the USA and Dolina Geizerov in Russia. Although activity is vigorous, the geysers are not very tall, with the highest eruption observed having been around 20 feet (6 m). The average geyser eruption height at El Tatio is about 30 inches (80 cm). With the demise of many of New Zealand's geysers due to geothermal power extraction, El Tatio has become the largest geyser field in the southern hemisphere.

GEYSERS STOP SPOUTING

Gold and other metals are associated with subduction volcanism and, consequently, are richly concentrated in areas of high geothermal activity. This has made the Andean countries highly attractive to mining companies. Mineral exploration in the vicinity of South America's second-largest area of geyser activity, the Puchuldiza Geyser Field in Chile's Isluga Volcano National Park, caused the field's geysers to dry up in May 2003.

ABOVE At dawn, the El Tatio geyser field in the Atacama Desert paints a serene picture, yet beneath the ground lie channelways of boiling water. With 80 true geysers and 30 perpetual spouters documented, it is one of the world's largest fields.

BELOW The reddish pink hues of Laguna Colorada in southwestern Bolivia are caused by pigment in the algae that live in this lake. It is one of several colored lakes of a volcanic region that lies high in the Andes near the Sol de Mañana geyser basin.

Hot Springs and Spas of Europe

The Belgian town of Spa, nestled in a valley in the Ardennes Mountains, is well known across Europe as a site of therapeutic hot mineral springs. Travelers have frequented the springs since the fourteenth century, and since then, the word spa has become synonymous with any natural water source believed to possess special health-giving properties.

OPPOSITE Three naturally heated outdoor pools and several indoor pools are open for public bathing at Szechenyi Baths, in Budapest, Hungary. This is Europe's largest geothermal spa, first opened in 1913. The hot spring was discovered when drilling for a well.

It is believed the name Spa was derived either from the Latin word *espa*, meaning fountain, or the Latin phrase *salus per aqua* meaning "health by water." The earliest spas, or therapeutic baths, date back several thousand years to a number of civilizations, including those of Mesopotamia, Egypt, and Ancient Greece. The Greek historian Herodotus (484 to 410 BCE) described hot springs with curative properties and recommended spa therapy. Radiogenic hot water springs at the city of Therma on the Greek island of Ikaria have been used for these purposes since the fourth century BCE. Along Ikaria's coastline it is possible to swim in warm water at a number of places where the radiogenic hot springs flow into the sea.

THE RITUAL OF THE ROMAN BATH

During the time of the Roman Empire, the concept of public baths or spas began to evolve. Initially, the Romans used natural hot springs and thermal baths for the health and recuperation of their soldiers. Over time, however, thermal and mineral bathing evolved into elaborate rituals for socializing and relaxation, as well as continued medical treatment. Bathing became important as part of the daily regimen for men and women of all classes and elaborate public facilities, known as *balnea* or *thermae*, were developed.

Bathing was a communal activity, though separate times were reserved for men and women. Generally, men could bathe from 2:00 to 3:00 p.m. onward, after finishing their working day. Public baths spread rapidly throughout Europe. They were built over existing hot springs, such as those at Bath in England, and towns grew up around them, such as the German and Austrian spa (*bad*) towns, or Italian spa (*terme*) towns. Even in places where no hot springs existed, the Romans constructed special furnaces to heat the water and buildings.

ROMAN BATHING PROCEDURE

Entering via the *apodyterium* (changing room), patrons would undress, leaving their belongings in cubicles under the care of slaves. Exercise in the gymnasium, such as weight lifting, discus throwing, or ball games, was followed by a shower, before entering the main bathing areas. The main hall, known as the *tepidarium*, featured warm pools and was often richly decorated with marble and mosaics. The hottest room in the complex was the *caldarium*, where bathers would recline in a hot pool sunk into the floor, or perhaps alternate for short times in the *sudatorium* (steam room), and the *laconicum* (dry sauna). These

Europe's Hot Springs

were designed to open and cleanse the skin's pores by promoting sweating. In the *frigidarium*, an invigorating dip in the cold pool would complete the process by closing the pores.

EUROPEAN SPAS TODAY

In more modern times, spas still fill many of the same needs as they did thousands of years ago, including relaxation, health, and stress relief. They incorporate many of the same techniques developed by the ancients such as a variety of hydrotherapy treatments. Body scrubs and massage therapy are the most widely used spa treatments.

Budapest is a major spa center with over 100 thermal springs bubbling from the ground along a geological fault separating the Buda Hills from the Great Plain. Here, over 10 million gallons (40 ML)

of hot mineral water gush forth daily. After the Romans, the bath culture was further developed during Turkish occupation in the sixteenth and seventeenth centuries. Some of the finest spas still operate today, such as the Kiraly, Rudas, and Racz.

TURKISH BATHS OR *HAMMAM*

The Turkish *hammam* combine the functionality of the Greek and Roman baths, with the Turkish-Muslim tradition of bathing, ritual cleansing, and respect for water. The Arabs began building their own version of the baths after they conquered Alexandria. These quickly evolved into public institutions, with one of the finest examples being the Çemberlitas Hammam in Istanbul, built in 1584. It was designed by the great Ottoman architect Mimar Koca Sinan, and is still in operation today.

ABOVE This Roman stone ornament was uncovered at Bath in 1790 while excavating the pump room. Whose face is on it has become a riddle. It has been variously interpreted as being the emblem from the shield of the goddess Sulis-Minerva, a gorgon head (without the requisite snakes), or the head of Oceanus, the last European Titan. Despite its mystery, it still remains the iconic symbol of Bath.

ROCK FEATURE: GEYSERITE AND TUFA

Geyserite and tufa are the terms used for mineral deposits that have precipitated around the edges of hot springs and geysers. These minerals precipitate out of supersaturated boiling groundwater when it experiences a rapid pressure and temperature drop on reaching the surface. These deposits can build up over time into spectacular volcano-like cones around geysers or form beautiful fluted rims around pools. Geyserite is an opal-like hyrated silica, while tufa is composed of calcium carbonate. The Romans constructed many buildings and bridges from tufa.

Hot Springs of Asia

The two main geothermal belts running through Asia are controlled by tectonic activity along the edges of the Eurasian Plate: the Himalayan Geothermal Belt, generated by heat produced from the collision of the Indian–Australian Plate with the southern edge of the Eurasian Plate, and the circum-Pacific Geothermal Belt, caused by the frictional heating of subduction as the Pacific Plate slides beneath the eastern edge of the Eurasian Plate.

ABOVE A group of people soak in an outdoor *kurhaus* in Beitou, a popular hot spring spot near Taipei. The Japanese *onsen* bathing culture has greatly affected the development of the hot spring and spa industry.

RIGHT Tourists enjoy a therapeutic spa in a popular hot spring resort, home to "doctor fish," in China's Chongqing municipality. Doctor fish are freshwater fish traditionally used in areas of Turkey for treating skin diseases such as psoriasis. They eat the affected and dead areas of the skin, leaving the healthy skin to grow.

The Himalayan Geothermal Belt is the easterly continuation of the Mediterranean Belt. This belt passes through China's western Sichuan Province and southern Tibet before turning southward through western Yunnan Province and into Thailand. Here it is joined by the Indonesian Geothermal Belt. The circum-Pacific Belt of Asia traverses through Kamchatka, Japan, and the islands of the Philippines.

CHINA'S HOT SPRINGS AND GEOTHERMAL ENERGY

Hot springs have been used in China for irrigation and domestic purposes since the Eastern Zhou Dynasty (770 to 226 BCE). During the Ming Dynasty (1368 to 1644), Li Shi-zeng, a well-known medical doctor, recognized the spring water's curative properties and began using it to treat diseases. Although there are over 3,000 hot springs in China concentrated at 254 specific sites, the country boasts only 10 active geysers. Of these, there are four found at Chaluo, in Sichuan; two at Chapu, in Xizang; two at Guhu, in Xizang; one at Tagajia, also in Xizang; and one at Balazang, in Sichuan.

China is the world's most populous country and second largest energy consumer after the USA. It is also the largest producer of coal, which accounts for 70 percent of its energy needs. On the negative side, however, it is anticipated that by 2050 China will overtake the USA as the largest source of carbon dioxide emissions. As a matter of priority, China's geothermal fields are being studied as sites of potential power generation, and there are over 200 fields with reservoir temperatures over 300°F (150°C) that may be suitable.

Taiwan is home to a wide variety of hot springs. Beitou's hot springs near Taipei were first mentioned in a 1697 manuscript, but it was not until 1893 that a German businessman built a small spa there. A few years later, in March 1896, Hirado Gengo from Osaka, Japan, opened the province's first hot spring hotel, which was named Tenguan. Beitou became known as the hot spring village and marked a new era of hot spring culture based on the Japanese *onsen* tradition in which the mineral-rich hot spring waters are be-lieved to offer many health benefits. As well as raising energy levels, the radon and minerals in spring water are believed to help treat conditions such as chronic fatigue, eczema, and arthritis.

The island's four major hot spring districts, Beitou, Yangmingshan, Guanziling, and Sichongxi, were developed in this style during Japanese occupation. The Japanese occupied the island for some fifty years, from 1895 to 1945.

The Ruisui hot springs in Hualien county have a very high iron content, which oxidizes on contact with the air, giving the water a not-too-pleasant rusty color and taste. Its interesting reputation, however, attracts many newlywed couples as it is said that frequent bathing in its waters will increase the chances of a woman bearing a male child.

PILGRIMAGE AND WORSHIP IN INDIA

India's hot springs are often places of worship, such as those of Ganeshpuri in Maharashtra state where the enlightened spiritual healer, Nityananda, spent his latter years. Pilgrims coming to visit this revered master would bathe and reflect in these pools. Nityananda is also credited with restoring the hot springs at Akoli.

Manikaran in Himachal Pradesh is a spring revered by both Sikhs and Hindus and surrounded by temples of various religious persuasions. Here, boiling sulfurous water, often noted for its healing properties, bubbles to the surface near the edge of a mountain river. Some pilgrims come to cook rice and dhal in the boiling water, and can stay in the temple complex. The pyramidal Ram Temple was built by Raja Jagat Singh sometime during the seventeenth century.

By the side of the Seerehole River, Bendre Theertha is another hot spring near the town of Puttur in Karnataka state. A dip in these springs is said to help rid one of skin diseases, such as eczema, allergic rashes, and other ailments.

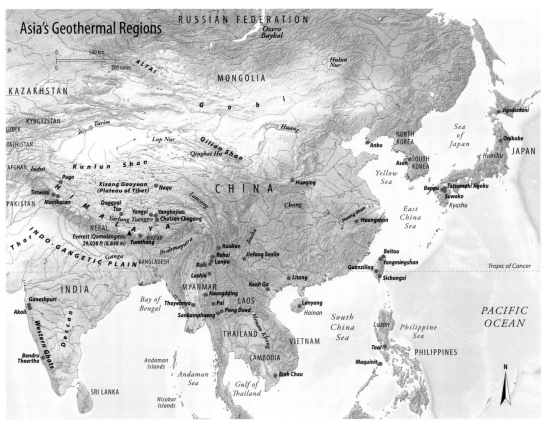

Asia's Geothermal Regions

ABOVE Over 1,200 species of orchids—a quarter of those found in Malaysia—flourish in Kinabalu National Park, which is home to Poring Hot Springs. Many rare endemic orchids are on display at the Poring Orchid Conservation Center in this park.

HOT SPRINGS AND *ONSEN* OF JAPAN

The volcanically active Japanese archipelago is presently still growing along the eastern edge of the Eurasian tectonic plate as it overrides the Pacific and Philippines plates. As such, Japan's islands form part of the circum-Pacific geothermal belt, and are home to many hot springs. Hot spring bathing has held a revered place in Japanese culture for millennia.

Natural hot springs (known as *onsen*) are highly popular throughout Japan and are an experience not to be missed by visitors. Every region of the country has its share of hot springs, and the numerous health and resort towns that have grown up around them. *Onsen* are not to be mistaken for those public bathhouses, or communal baths, which are filled with ordinary water heated in a boiler. The tradition of public bathing in Japan arose as a means by which many people, as they were unable to have bathrooms in their homes, could keep themselves clean.

THE *ONSEN* TRADITION IN JAPAN

Legend has it that ancient hunters first learned of the medicinal benefits of hot springs when they saw wounded game animals go into these bubbling sulfur-scented pools to relieve their pain.

When Buddhism reached Japan in 552 CE, it also popularized the *onsen*. The religion's teachings required bathing as a ritual of purification to cleanse oneself of sins and pray for benevolence. As a result, early Buddhist monks such as Kobo Daishi, Gyoki, and Ippen Shonin believed thermal spring waters were a gift from the Buddhist deities, and worked to develop many of the *onsen* centers.

Jesuit and Franciscan missionaries arrived during the sixteenth century and began to introduce Christianity into Japan. They had an aversion to human nakedness, and tried to prohibit public bathing. These initial attempts eventually failed as the Japanese populace saw the missionaries and their new supporters as dirty and unhygienic. Eventually the missionaries were expelled by Tokugawa Ieyasu in 1614, during the Tokugawa shogunate (1603 to 1868), and the Japanese continued to enjoy the pleasures of *onsen* bathing.

ONSEN CURATIVE PROPERTIES

In 1709, a doctor from Tokyo, Goto Konzan, noticed that hot spring bathing seemed to cure certain medical disorders and began the first medical study of hot springs. Since World War II (1939–45), over 50 national hot spring hospitals have been established, in which *onsen* are used in the treatment of chronic diseases such as rheumatism and hypertension. A particular *onsen* center may feature a number of specific curative baths, each with spring water of unique elemental composition dissolved during its long passage through different hot underground channelways.

BLOOD POND HELL AND SPOUTING GEYSERS

Beppu, located on the east coast of Kyushu Island, is one of Japan's main geothermal areas, and a favored holiday destination well

BELOW Visitors take in the pale milky blue waters in Shirahone Hot Springs. The water temperature here is 98° to 104° F (37° to 40°C). Set in a hanging valley in the mountains near Kamikochi, it lies in the Nagano Prefecture in the heart of Japan.

known for its hundreds of sacred *onsen*. As well as the bathing springs, there are other spectacular hot springs known as the "nine hells of Beppu." These include: Umi Jigoku (Sea Hell), a steaming pond of intense blue water; Oniishibozu Jigoku (Shaven Monk's Head Hell), a bubbling hot mud pool; Shiraike Jigoku (White Pond Hell), a pond of hot milky water; and Chinoike Jigoku (Blood Pond Hell), a pond of hot red water.

The Beppu geothermal area also boasts one of Japan's few geysers, Tatsumaki Jigoku (Geyser Hell), which spouts every 25 to 30 minutes to about 80 feet (25 m), and lasts about five minutes. Another of Japan's geysers at the Suwako Lake Geyser Center spouts water some 130 to 160 feet (40 to 50 m) into the air about once hourly. It is one of the world's tallest geysers. Visitors can view it while bathing in a large hot spring bath by the lake.

ABOVE The bright red water of Blood Pond Hell, one of the nine "hells" at Beppu, is aptly named: all their pools are far too hot for bathing. This resort produces the most geothermal water of any in Japan and in line with *onsen* traditions, offers many water, sand, mud, and steam baths.

JAPAN'S BATHING SNOW MONKEYS

An interesting behavioral trait of the Japanese macaque was observed in a hot spring in the Nagano Mountains on Honshu island. In 1963, a young female named Mukubili waded in to get some food that had been thrown in by the keepers. She appeared to enjoy the warmth of the 109°F (43°C) hot springs and soon other young monkeys joined her. Over several years the entire troop took up the behavior to escape the winter cold. Young monkeys have also learned how to roll snowballs, which seems to be a purely play activity.

Japanese macaques (*Macaca fuscata*) enjoy a warming bath. The world's most northerly monkeys, they are often called snow monkeys.

Geothermal Zone of Indonesia and New Guinea

A major geothermal zone runs through Indonesia and New Guinea, controlled by the northern edge of the Indo-Australian Plate. The volcanically active Indonesian archipelago is being built by magma from the melting Indo-Australian Plate as it pushes into the Java Trench and beneath the Eurasian Plate.

ABOVE Precipitated minerals from the constant spray of condensing hot water around a hot spring has caused a leaf to become lithified, or turned to stone. This is on Papua New Guinea's Fergusson Island.

Moving eastward through New Guinea, this geothermal zone joins the circum-Pacific geothermal belt. Here, the direction of subduction reverses and the Indo-Australian Plate pushes over the top of the Pacific Plate. The Pacific and a number of its sub-plates are being consumed and turned into the magma and heat that is fuelling the volcanic and geothermal activity of New Guinea and the Bismarck Archipelago. Volcanoes, earthquakes, and hot spring activity are commonplace in the lives of the people of this region. They have become accustomed to rebuilding after often-catastrophic events, such as the Indian Ocean earthquake and tsunami of December 2004.

GEYSERS OF INDONESIA AND NEW GUINEA

Indonesia has 16 documented geysers on the islands of Sumatra (Lampung-Semangko, Tapanuli, Kerinci, and Pasaman), Java (at Cisolok and Gunung Papandayan), the Celebes (Minahasa District), and the Maluku Group (Bacan), while Papua New Guinea boasts a total of 38 geysers. Most of these geysers are concentrated in three geothermal fields in Papua New Guinea. These are the Kasiloli field on New Britain Island (14 geysers), the Deidei field near Iamelele on Fergusson Island (12 geysers), and the geothermal fields on Lihir and Ambitle islands (9 geysers).

HAZARDOUS HOT SPRINGS IN EAST JAVA

Indonesia's high monsoonal rainfall feeds the region's abundant hot springs, but it sometimes also creates major hazards for the people living there. On December 11, 2002, several days of heavy rain triggered a mudslide that buried the Padusan Air Panas hot springs resort in Pacet village in East Java. Witnesses told of a massive wave of mud and water descending the mountain at 40 miles (70 km) per hour. At about 3:30 p.m., there was a thundering sound before the mud, mixed with rocks and wood, poured into the hot pools where over 50 people had been bathing. At least 30 people were killed, unable to get out of the gate in time, and hundreds of victims were treated in public hospitals and clinics in Pacet.

HOT SPRINGS HERALD UNTOLD GOLD AND MINERAL RICHES

Hot water circulating through the metal-rich magma chambers of the collapsed volcanic cones, or calderas, on Lihir and Ambitle islands is gradually building up world-class gold deposits that are currently being discovered and mined. The gold, which dissolves in the hydrothermal waters at depth, is being deposited in horizontal enriched bands near the surface where the temperature and pressure (and hence the gold solubility) are lower.

The Lihir gold mine, which commenced its operations in 1997, is located in the center of a caldera that was formed from the explosive destruction of the Luise volcanic cone. Miners work right in the heart of the active geothermal area where rock and groundwater temperatures range from 140° to 310°F (60° to 200°C), and

LEFT Vibrant yellow deposits of sulfur build up around fumaroles of volcanic craters in Indonesia that emit sulfur-rich steam. These miners are working in a sulfur mining field in Banyu-wangi, in eastern Java. Hundreds of people work here, often under dangerous conditions.

BELOW Two different hot pools on an island off New Guinea. The top photo shows a pool that lies dormant, while the one below is bubbling with boiling water. The escaping steam deposits its dissolved minerals around the rim, building up a cone.

continuous pumping is needed to lower the groundwater levels in order to keep the mining pit dry.

At Kasiloli, New Britain, metal sulfides, including pyrite and marcasite, have been found in the sediments surrounding the hot springs. Bacteria in the thermal waters cause the precipitation of these minerals. The bacteria attack and remove oxygen from the soluble sulfates, converting them into insoluble metal sulfides.

FERGUSSON ISLAND'S HOT SPRINGS

An active thermal field, with abundant hot springs, mud pools, fumaroles, and geysers spouting 65 feet (20 m) into the air, is found on Fergusson Island, located in the center of the D'Entrecasteaux Island group at the eastern tip of Papua New Guinea. The Iamalele-Fagululu area on the southwest part of the island is made up of several rhyolitic lava

ABOVE A captive Sumatran tiger (*Panthera tigris sumatrae*) takes a cooling dip in the water. The smallest of all subspecies of tigers, there are believed to be fewer than 500 individual animals left in the wilds of Sumatra.

domes and obsidian lava flows of Holocene Epoch age. Here, near Iamalele village, mineral-rich thermal waters from the Dei Dei hot spring have precipitated a flat concrete-like area of siliceous sinter. The thermal area is considered to be sacred by the Fergusson islanders. They are indigenous subsistence horticulturalists living in small traditional settlements, and use the mud pools and hot springs to cook bananas, clams, eggs, and other food. These items are usually wrapped in leaves and lowered into the boiling water using sticks. The islanders also bathe and wash their clothes in the nearby warm-water streams.

GUNUNG PAPANDAYAN VOLCANO

Nowhere is the relationship between hot spring activity and volcanically derived heat more clearly visible than at Papandayan Volcano in Java, Indonesia. The forest-covered stratovolcano erupted in 1772, triggering an unstable portion of its steep cone near the summit to collapse toward the northeast. This catastrophic debris avalanche wiped out 40 villages and killed nearly 3,000 inhabitants. Gunung Papandayan then lay dormant until an explosive eruption heralded its reawakening in November 2002.

The stratovolcano has four summit craters, the youngest of which is known as Kawah Mas, or the golden crater, as it is colored by sulfur. These vivid yellow deposits are building up around the mouths of the fumeroles or solfataras that noisily issue sulfur-rich steam from various places within the crater. Mud volcanoes also explode from the floor of the crater. These mud cones build up to about 4 feet (1.2 m) high before erupting violently and ejecting a spout of hot muddy water into the air. These eruptions are repeated about every 20 to 25 seconds.

BIRDS INCUBATE THEIR EGGS USING GEOTHERMAL HEAT

Spectacular geysers, fumaroles, hot springs, and boiling mud pools are found in the Pokili Wildlife Management Area in West New Britain. Scrubfowls (*Megapodius freycinet*) take advantage of the warm earth in this thermally active area to help hatch their eggs. The birds have covered an area of about 22 acres (9 ha), known as the Pokili breeding grounds, with tens of thousands of nest mounds. These birds are members of the megapode bird family, which are all mound-builders. Other megapodes also use the same incubation strategy where geothermal heat is available. Studies of *Macrocephalen maleo* nesting grounds of northern Sulawesi, Indonesia, show that they can determine the ground temperature, perhaps by picking up and testing the sand with their bills. They dig shallower nests in hotter ground closer to the center of the geothermal field and deeper nests further away where the ground is cooler. Local village people have been aware of these nesting grounds for centuries and come to collect megapode eggs, which are then eaten or traded. Battles have even been fought over customary egg collection rights on the nesting grounds. Once collected, the eggs can be conveniently cooked in the nearby hot springs.

Geothermal Zone of Indonesia and New Guinea

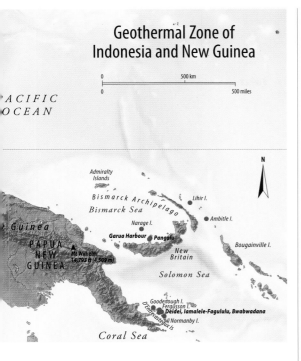

A river flows through the crater. It begins as a cool clear stream in the forest, but quickly becomes heated and cloudy with dissolved sulfuric acid and mud as it flows across the crater floor. Percolating through cracks in the hot rocks, the rain and river water is also recirculated to the surface to feed the sulfurous hot springs and steaming fumaroles.

INDONESIA'S GEOTHERMAL POWER POTENTIAL

Although geothermal power presently supplies only about one percent of Indonesia's total energy needs, the volcanic archipelago is one of the most attractive regions in the world for developing geothermal power. Planners, following the National Committee on Climate Change recommendations, are now moving ahead rapidly to develop renewable energy sources, such as geothermal and hydropower, in order to reduce the country's oil dependence. Indonesia's Ministry of Energy and Mineral Resources (MEMR) has identified 217 geothermal areas in the country, of which more than 70 locations show potential for geothermal power development.

One of these sites lies on the flanks of Mt Kendang, a "young" volcano in central Java. The Darajet Steamfield Project feasibility study in 1988 led to the design and commissioning of three reliable, highly efficient geothermal power plants utilizing the site's unusually dry high-pressure steam. In 2002, the Darajat project was recognized for its engineering excellence when the Institution of Professional Engineers New Zealand awarded it their prestigious Merit Award for Innovation.

ABOVE The water from these hot springs on the island of Bali in Indonesia is used to treat a variety of skin diseases. Bali has numerous hot springs that are visited by tourists to help sooth their ailments or to relax.

LEFT The edges of this hot spring in New Guinea are surrounded by flat terraces, each with a small raised rim that forms a shallow pool (rim pools). These terraces are mineral deposits left by the spring water as it cools.

New Zealand Geysers and Hot Springs

Geyser, an Icelandic word meaning 'to gush," is the name of a rare and fragile phenomenon that requires the uncommon convergence of hydro-geological features. Only 1,000 active geysers are in existence in the world today, and some of them are scattered over New Zealand's North Island.

BELOW Champagne Pool in Waiotapu Thermal Reserve on New Zealand's North Island is so-called because tiny bubbles rise up from its depths. The orange tint in the water is caused by its mineral content.

New Zealand's Taupo Volcanic Zone contains the world's fourth largest concentration of geysers, after Yellowstone in the USA, Dolina Geizerov in Russia, and El Tatio in Chile. At the end of the nineteenth century New Zealand had five major geyser fields—Rotomahana, Whakarewarewa, Orakeikorako, Wairakei, and Spa. Sadly, today fewer than 15 of over 130 original geysers remain, the rest destroyed by human activity such as extraction of hydrothermal energy. Today, most of New Zealand's geysers are found at Whakarewarewa.

THE WHAKAREWAREWA THERMAL RESERVE
The geysers of the Whakarewarewa Thermal Reserve are set amid 500 geothermal features including mud and chloride pools, silica terraces,

fumaroles, and other hot springs. Whakarewarewa is home to more than 65 geyser vents as well as seven active, inter-connected geysers. This includes Pohutu, which is New Zealand's largest geyser. The geysers are set on a sinter plateau known as "Geyser Flat" and they are aligned on a north–south axis above a common fissure.

Pohutu is the most active remaining geyser at Whakarewarewa and erupts regularly once or twice an hour to a height of 66 feet (20 m). The nearby Prince of Wales Feathers geyser became active after the 1886 eruption of Mt Tarawera, and has been erupting regularly since 1992.

WAIKATO: GEOTHERMAL CENTRAL
Almost 80 percent of New Zealand's geothermal features are found in the Waikato region of the North Island and these include geysers, hot springs, and crystalline structures such as sulphur crystals, sinter terraces, and mud pools.

Waikato also hosts rare chloride springs, deposits of boiling mineralized water heavy in alkaline, chloride, and silica. These are easily degraded, by the extraction of geothermal fluids, used for human consumption.

The number of sinter springs and active geysers in the region has remained relatively stable since the early 1960s, despite a large number being destroyed as a result of construction of the Wairakei power station in 1958. Many more were lost to flooding with the completion of the Waikato River Hydro Scheme in 1961 and the creation of Lake Ohakuri.

WAIKATO'S TAUPO VOLCANIC ZONE
The greatest concentration of geothermal activity in New Zealand can be found within Waikato's Taupo Volcanic Zone. Seventeen geothermal fields here abound in a region where the temperature of Earth's crust exceeds 660°F (350°C) at a depth of approximately 3 miles (5 km).

The world's newest hydrothermal system, the Waimangu Volcanic Valley, was formed here as a result of the eruption of Mt Tarawera in 1886, the only geothermal field created in recorded history. Prior to a landslide in 1904 which changed the water table, the area contained the world's largest known geyser, the Waimangu Geyser, which reached heights of 1,500 feet (500 m).

Within the Taupo zone's boundaries is the Orakeikorako geothermal area. This is a sinter-covered expanse of hot pools, springs, geysers, and mineral deposits with surface temperatures approaching 500°F (265°C). The Ketehahi Hot Springs, on the northern flank of Mt Tongariro, feature a 75-acre (30-ha) expanse of over 40 fumaroles, mudpools, and springs.

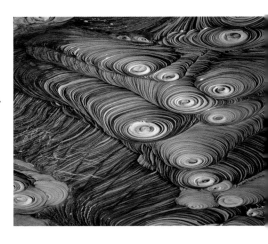

ABOVE These boiling mudpots are in the Whakarewarewa State Forest Park near Rotorua. There are over 500 hot springs here that vary from clear water to thick smelly mud. The mud is used in cosmetics and spas.

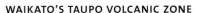

LIFE IN A HOT SPRING
Microscopic organisms were first discovered in the hot springs of Yellowstone National Park by Thomas Brock of the University of Wisconsin in 1966. These single-celled organisms called extremophiles are also present in the geothermal fields of New Zealand. Their genetic material is dispersed throughout the cell rather than enclosed within a nucleus like fungi, plants, and animals. This enables the extremophiles to thrive in conditions of high temperature, toxicity, salinity, and acidity.

There are three types of extremophiles: thermophiles (heat loving), acidophiles (acid loving), and thermoacidophiles (heat and acid-loving). They survive by creating energy from inorganic compounds, possessing temperature-resistant proteins, and thriving on a diet of ammonia, arsenic, hydrogen sulfide, and a variety of other dissolved chemicals.

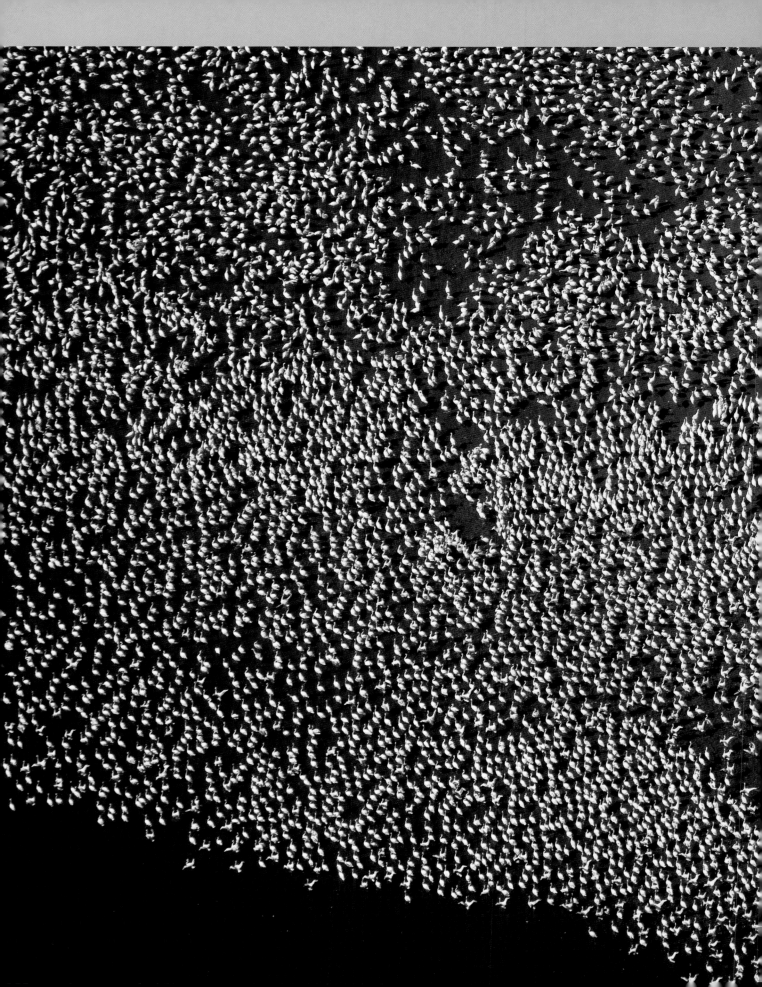

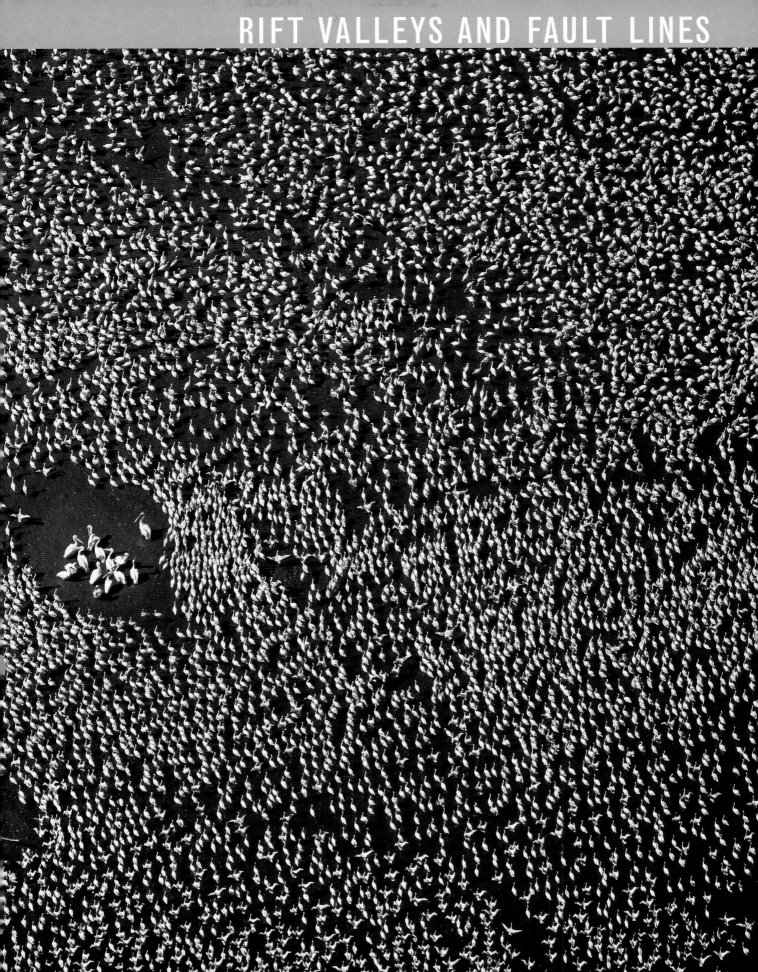

U nlike most valleys, rift valleys have not been cut by erosion. Instead, the valley bottoms have dropped away, sliding downward along lengthy parallel faults to create deep canyon-like slots that may be thousands of miles long. Rift valleys are immediately recognizable by their flat floors and very straight steep walls formed by the fault planes.

Rift valleys, also known as grabens, act as collection areas for sediment and water. They contain Earth's largest and deepest freshwater lakes, such as Lake Baikal in Siberia, Lake Tanganyika in Africa, and Lake Superior in North America. These lakes contain a rich selection of minerals, and oil and gas rise up out of the organic rocks that had been buried on the bottom of the lakes.

The East African Rift is probably the best known of the great rift valleys. The vast thicknesses of sediments that have built up in these valleys make them the world's best storehouses of fossils.

Other significant rift valleys and fault zones include Iceland's Mid-Atlantic Rift, the Red Sea–Gulf of Aden Rift, Turkey's North Anatolian Fault Zone, the Dead Sea–Jordan Valley Rift, faults and rift valleys of the North American west coast, and also the Spencer Gulf–Lake Torrens Rift in southern Australia.

ABOVE Lake Bogoria National Park lies within the eastern branch of the East African Rift. The lakes here have no outlet to the sea and therefore have a high concentration of minerals.

RIGHT This large volcanic crater is located in Kenya, within the East African Rift. The rift valley extends 3,700 miles (6,000 km), from Djibouti to Mozambique.

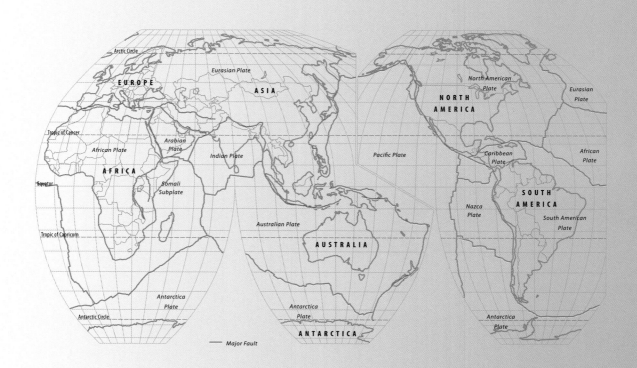

CONTINENTAL UPLIFT, STRETCHING AND FAULTING

Rift valleys form as a result of uplift and tension in Earth's crust. The initial upward doming is caused by hot convection currents that rise within Earth's mantle and push against the base of the crust. These currents also create the strong tensional forces that eventually tear the crust apart. Initially, at the point of maximum tension, three rift valleys develop, oriented at 120 degrees to one another, radiating in a star pattern. This is referred to as a triple junction. These tensional features usually begin to develop toward the middle of great continents, but eventually tear through to the edges and become flooded with seawater. Earth's textbook example of a triple junction can be seen where the three arms of the Great Rift meet at the Afar Triangle of Djibouti—the East African Rift, the Red Sea, and the Gulf of Aden.

Tension in the arms of the triple junction continues, and the walls of the rift valleys very slowly move apart. Eventually, the divergence or spreading becomes concentrated along two of the rift valley arms (known as the active arms), while the third one slows down and eventually stops opening (known as the failed arm or aulacogen). In the case of the Afar Triangle of Djibouti, the

active arms are the Red Sea and Gulf of Aden, while the failed arm is the East African Rift Valley.

CONTINUED RIFTING AND THE BIRTH OF NEW OCEANS

As spreading along the active rift arms continues, volcanic activity is commonplace. Basalt lava is emplaced as dikes, along the opening fractures between the diverging plate margins. The lava fills up the fractures as they grow apart. This process continues at only inches per year but leads to the growth of, at first, narrow seaways, and, finally,

ABOVE In Iceland's Thingvellir National Park, rifting between the Eurasian and the North American tectonic plates has formed awe-inspiring grabens and steep escarpments.

LEFT Lake Baikal is a long and narrow continental rift lake in southern Siberia, not far from the Mongolian border.

enormous oceans over hundreds of millions of years. Earth's most active rift valleys are all, therefore, found at the bottom of the oceans, and located along the crests of the mid-oceanic ridge systems as a result of seafloor spreading. Continental rift valleys are usually the result of a failed arm of a triple junction. Examples of continental rift valleys include the Mississippi Embayment and the Reelfoot Rift in North America, and the Cameroon Rift of Africa.

BASALT IN RIFT VALLEYS

Today, basalt is the most common rock found at Earth's surface. Erupted along rift valley faults, basalt forms the floors of all the world's oceans. It has also erupted from many of the continental volcanoes and lava fields. This tough, black to dark-gray, fine-grained rock cools from lavas that mostly have come directly from Earth's mantle and are barely affected by the processes of the crust. Often, erupting basalts bring up fragments of unmelted mantle rock that scientists study to learn about processes occurring in the mantle. Mid-oceanic ridge basalt cools relatively quickly, so its crystals

are too small to be seen with the naked eye. However, a microscope reveals that its main minerals are pyroxene, feldspar, and olivine.

MINERAL AND ENERGY
WEALTH OF THE RIFT VALLEYS

Initially rift valleys fill with shallow seas; however, the process is not uniform. Alternate periods of flooding and drying out create alternating seas, salt lakes, and saltpans. Over time, massive bodies of salt and other water-soluble minerals, known as evaporates, build up on the floor of the rift valley. The minerals precipitate from the water as it evaporates into the atmosphere. Some of the most common examples include halite (common salt), anhydrite, gypsum, calcite, and potassium, as well as magnesium salts such as sylvite, carnallite, kainite, and kieserite. Nitrate evaporate minerals are frequently mined for use in the production of fertilizer and explosives.

Eventually, when the warm shallow seawaters become more permanent in the rift valley, they house a biological soup of minute aquatic algae

BELOW The Njamusi people live at the south end of Lake Baringo, in the East African Rift Valley. This part of the valley has many shallow salt (soda) lakes, including Baringo, Bogoria, Nakuru, Natron, and Magadi lakes.

and bacteria. These microorganisms live and die, accumulating on the sea bottom in vast numbers, where thick piles of younger marine sediments eventually bury them. Temperatures rise as the depth of burial increases and these organic-rich rocks eventually "mature" into the source rocks of oil and gas. These rocks are slowly "cooked" in Earth's giant pressure cooker until the longer organic chains of carbon and hydrogen break down into smaller lighter molecules of oil and gas that rise upward out of the source rocks.

An important factor in the formation of economic hydrocarbon reserves is the presence of impermeable geological structures that can trap the oil and gas, and prevent them from reaching the surface. Fine-grained sedimentary rocks, such as mudstones, will not let oil and gas pass, and if these have been folded into anticlines or dome shapes then they make perfect traps. Rising domes (or diapirs) of low-density salt from the underlying evaporate beds will also push up the overlying sediments into closed structures, which can act as hydrocarbon traps. Out of all the world's 910 giant oilfields (those with more than 500 million barrels of ultimately recoverable oil or gas), more than two-thirds of them are associated with either continental spreading margins facing open ocean basins or continental rift valleys. The North Sea and West Siberian oilfields are important examples of hydrocarbon deposits that formed in a rift valley setting, while the Gulf of Mexico, Northwest Australian, and West African oilfields are classified as facing an open ocean basin.

CAMEROON RIFT, AFRICA

The Cameroon Rift of west and central Africa is an extremely long and straight continental rift valley that formed when South America and Africa began to split apart about 150 million years ago. A triple junction, centered on present-day Cameroon, was created as a result of uplifting of the crust by over a mile (2 km), followed by collapse of the arch crests along normal faults. The collapse formed the Cameroon Rift Valley, but it failed, and the two active rift arms formed the southern- and western-facing Atlantic coastlines. The faulting, however, caused substantial volcanic activity in the Cameroon Rift.

THE 1986 LAKE NYOS DISASTER

On August 21, 1986, a minor eruption or landslide disturbed the tranquil waters of Lake Nyos, nestled in a 400-year-old explosion crater in the rift zone of Cameroon's Northwest Province. This trigger caused the lake to suddenly overturn, belching out a dense cloud of carbon dioxide and other gases. A

rapid mixing of the gas saturated deep water with the upper layers of the lake. The accompanying pressure reduction allowed the stored gas to effervesce out of solution, like uncapping a soda bottle. The gas poured down two nearby valleys, displacing all the air and suffocating about 1,800 people who were living in the rural villages within about 13 miles (21 km) of the lake. Some 3,500 head of

ABOVE The San Andreas Fault runs through the US state of California. The Gulf of California is one of the best examples of a continental rift in the process of forming a young ocean basin.

DEGASSING DANGEROUS LAKES

As toxic outgassings similar to the 1986 Lake Nyos disaster could occur every few decades, scientists proposed that pipes should be extended into the deeper parts of the lake, to allow controlled and continuous release of the stored gases. The first pipe was installed in Lake Nyos by an international team in 2001 and degassing is continuing steadily. Following the Lake Nyos tragedy, other African lakes were examined to see if a similar phenomenon could happen elsewhere. Geologists found Lake Kivu in Rwanda to be supersaturated at depth, and also found evidence for outgassing events around the lake about every 1,000 years.

Toxic gases pour out of Lake Kivu following the January 2002 eruption of Mt Niragongo.

LAKE BAIKAL'S YOUTHFUL RIFT

Lake Baikal, also known as Dalai–Nor, or "Sacred Sea" in the Buryat and Mongol languages, lies in a classic young continental rift valley. The lake is surrounded by up-domed mountains, and stretches for 400 miles (630 km) across southern Siberia, but only attains a width of 50 miles (80 km). The water is an amazing 5,370 feet (1,637 m) deep, making Baikal the world's deepest lake, and also Earth's greatest natural freshwater reservoir. It contains over one-fifth of the world's liquid fresh water. The capacity of Lake Baikal is 5,500 cubic miles (23,000 km³) of water, which is as much as the contents of North America's entire Great Lakes system put together. The solid rock bottom of the Baikal rift lies more than 5 miles (8 km) below sea level, and the valley is filled with some 4 miles (6.5 km) of sediment. This makes the Baikal valley the deepest continental rift on Earth.

The rift graben is actively widening at approximately ⅘ of an inch (20 mm) per year. Every few years, sizable earthquakes are recorded in the region indicating that slippage is still occurring along the graben-bounding faults. On August 29, 1959, during an earthquake of magnitude 9, the Baikal lake bottom was displaced for perhaps as much as 65 feet (20 m). Thermal spring activity in the area also demonstrates that hot volcanic rocks lie quite close to the surface.

Programs that involve deep drilling of Lake Baikal's thick sediments are enabling scientists to map the history of the lake and rift as far back as 5 million years, including its climate and ecology. The long and unbroken record presents a unique opportunity to understand how continents begin to break apart, and give rise to new ocean basins.

A UNIQUE LAKE ECOSYSTEM

Some people refer to the spectacular Lake Baikal as the "Galapagos of Russia," owing to its rich and unusual freshwater fauna. The lake waters are well oxygenated throughout, despite their great depth, thus creating unique biological habitats unequalled elsewhere in the world. Lake Baikal's enormous biodiversity includes 1,085 species of plants and 1,550 species of animals, with over 60 percent of the animals being endemic. The Baikal seal (*Phoca sibirica*), which is the only mammal living in the lake, is found nowhere else. Also present in the lake are many endemic subspecies of fish, including the bottom-feeding sculpin and the omul fish. The omul fish (*Coregonus autumnalis migratorius*) is smoked and sold around the lake as a regional delicacy. Lake Baikal is protected within the boundaries of Zabaikalsky National Park, which is part of the Lake Baikal World Heritage Site.

livestock also perished. Around 4,000 inhabitants managed to flee the area, although they suffered from respiratory problems, burns, and paralysis as a result of the gases. Following the event, the normally blue waters of the lake had turned red (due to oxidization of the deep iron-rich lake water). The lake level had also dropped by about 3 feet (1 m), indicating the volume of gas released to have been about a quarter of a cubic mile (1 km³).

HOW RIFT VALLEYS ARE FORMED

Uplift and tension in Earth's crust can eventually lead to the formation of rift valleys or grabens. A. Doming and stretching creates a series of deep fractures in Earth's continental crust. Fire fountains of basaltic lava issue forth, and down-dropping of huge blocks begins the formation of a rift valley. B. Stretching continues, causing the rift valley to deepen to the point where the sea is able to enter. Volcanism continues while marine sediments and evaporates accumulate on the rift valley floor.

A   B

THE MISSISSIPPI EMBAYMENT
AND NEW MADRID SEISMIC ZONE

During the Tertiary and Cretaceous periods, the Mississippi River flowed southward along an ancient continental rift valley (the Reelfoot Rift) and into a large embayment in the shoreline of the Gulf of Mexico coast known as the Mississippi Embayment. The ancient shoreline would have extended as far inland as the confluence of the Ohio and Mississippi rivers at Cairo, Illinois. Since that time, the Mississippi River has filled the topographically low-lying basin with Cretaceous Period to recent sediments and built its present delta.

The Reelfoot Rift is an ancient failed continental rift valley that formed during the Precambrian breakup of the supercontinent Rodinia around 750 million years ago. The old rift was then partly reactivated during the relatively more recent opening of the Atlantic Ocean and Gulf of Mexico during the breakup of Pangea about 180 million years ago. The rift is still active today, as indicated by the intense seismic activity known as the New Madrid Seismic Zone. It was the site of the destructive New Madrid Earthquakes of 1811 to 1812.

The New Madrid Earthquake of February 7, 1812, was the second-largest earthquake ever recorded in the United States (the largest being the magnitude 9.2 Alaskan Good Friday Earthquake on March 27, 1964). This earthquake followed three other major quakes: two on December 16, 1811, and one on January 23, 1812. These earthquakes destroyed about half the town of New Madrid, Missouri, altered the course of the Mississippi, and changed the shorelines of lakes.

ABOVE The Afar Region of Djibouti and Ethiopia is a very important rift valley fossil site. It was here that Donald Johanson discovered the fossilized skeleton of a hominid (*Australopithecus afarensis*) in 1974. She was named "Lucy."

The Mid-Atlantic Rift, Iceland

Iceland is a significant destination for geologists seeking a better understanding of how oceans form. The island sits atop an enormous line of rift volcanism along which the Americas are separating from Africa and Europe, opening up the Atlantic Ocean, and it is the only place on Earth where the Mid-Atlantic Ridge can be seen on land.

RIGHT Clear waters fill some of the smaller rifts in Thingvellir National Park. The wishing pond in Nikulasargja, sometimes called Penny Canyon, is full of coins that have been thrown in. People toss coins hoping for some good luck.

The small North Atlantic island nation is in the process of being torn apart along giant north–south rift valleys running down its middle. Basalt lavas fill the fissures in the valley floors as the valley's walls move apart. Iceland is only 5 to 6 million years old, so geologic maps of the country show basalts of that age along the east and west edges of the island with the basalts becoming progressively younger toward the midline.

BIRTH OF THE ATLANTIC

About 190 to 220 million years ago, during the late Triassic Period to the early Jurassic Period, the Atlantic Ocean was born. Hot currents in Earth's mantle rising beneath the supercontinent Pangea started to stretch it apart. The crust split, initiating the development of a long rift valley much like the East African Rift. The east coast of North America began separating from the west coast of Africa and so the continents have been moving apart ever since at the rate of about $\frac{3}{4}$ inch (2 cm) per year; slower than the rate a fingernail grows.

A hotter and more active area of the mantle lies beneath the Icelandic section of the Mid-Atlantic Ridge and this caused it to rise higher and emerge above sea level.

THINGVELLIR NATIONAL PARK

Thingvellir is one of Iceland's stunning national parks where the rifting between the Eurasian and North American tectonic plates has created some spectacular grabens and steep escarpments. It is also highly significant to the nation's history being the site of the world's first parliament, "Althingi," founded in the year 930 CE. Icelandic Vikings began meeting here annually, gathering around a giant rock formation, Logberg, or Law Rock, to create new laws and change old ones. Criminals were also dealt with at these meetings in the Drekkingarhylur or drowning pool. The independence of Iceland was proclaimed on June 17, 1944, at this historic place on the shores of Iceland's biggest lake, Thingvallavatn. Owing to its cultural significance, Thingvellir was added to the World Heritage List in 2004.

Map labels

Denmark Strait
Arctic Circle
Greenland Sea
Grímsey
Ísafjardardjúp
Skagafjördur
Eyjafjördur
Skjál-fandi
Öxar-fjördur
Thistil-fjördur
Bakkaflói
Vopna-fjördur
Húnaflói
Krafla 2,133 ft (650 m)
Mývatn
Mid-Atlantic Ridge
Breidafjördur
ICELAND
Askja 4,974 ft (1,516 m)
Langjökull
Hofsjökull
Langjökull 4,462 ft (1,360 m)
Bárdarbunga 6,562 ft (2,000 m)
Grímsvötn 5,659 ft (1,725 m)
Vatnajökull
Mid-Atlantic Ridge
Faxaflói
Thórisvatn
Hvannadalshnúkur 6,952 ft (2,119 m)
Krísuvík 1,243 ft (379 m)
Reykjanes 755 ft (230 m)
Thingvallavatn
Hekla 4,882 ft (1,488 m)
Mid-Atlantic Ridge
ATLANTIC OCEAN
Katla 4,961 ft (1,512 m)
Eldfell 915 ft (279 m)
Heimaey
Vestmannaeyjar
Surtsey

The Mid-Atlantic Ridge - Iceland

0   150 km
0   150 miles

ABOVE It is common to come across the solidified remains of a lava flow in Iceland as the rift that runs through this island is constantly being filled with fresh outpourings of lava.

SURTSEY ISLAND—BIRTH OF A RIFT VOLCANO

Surtsey Island was born when a new rift opened in the ocean floor near the Vestmannaeyjar archipelago south of Iceland. Eruptions of lava onto the sea floor 425 feet (130 m) below sea level built an edifice that reached the ocean surface in late 1963. On the morning of November 14, a cook on the trawler *Isleifur II* spotted a dark column of smoke rising from the ocean. Thinking it might be a ship in distress, he ran to alert the crew. Captain Tomasson knew what was going on but to be sure he brought the ship closer. Through his binoculars he could see the ocean water boiling vigorously and plumes of black ash shooting upward.

Explosions were continuous, and after a few days the new scoria cone had reached a height of 150 feet (50 m) and over 1,600 feet (500 m) in length. Water coming into contact with the hot lava caused the initial eruptions to be very explosive, but by early 1964, the island had grown to a size that seawater could no longer reach the vents. Lava fountains and flows then became the main form of activity. These covered the loose volcanic pile with a hard cap of erosion-resistant basalt and

protected the island being washed away. The eruptions lasted until June 5, 1967, when the island reached its maximum size of about 1 square mile (2.6 km^2). Other smaller islands also reached the surface, such as Syrtlingur and Jolnir, but these were eroded away relatively quickly. The sea also eroded the softer parts of Surtsey. By 2005, the island was reduced to half its maximum size.

ABOVE Thingvellir translates roughly as "parliament plains" and is the site of the first government. This is perhaps the best place in the world to clearly see the actual rift of a rift valley, as the two sides of the rift are so close to the chasm in between.

THE ERUPTION OF HEIMAEY

On January 23, 1973, a fissure opened up on the outskirts of the Icelandic fishing port of Vestmannaeyjar on Heimaey Island, and began erupting a fiery curtain of lava. The alarm was quickly raised and within six hours most of the town's 5,300 residents were calmly evacuated to the Icelandic mainland by boat, airplane, and helicopter. Only volunteers and members of the Icelandic State Civil Defense Organization remained to coordinate the salvage operation.

Falling lava spatters built up into ramparts around the edges of the fissure. The volcano grew quickly and began engulfing outlying farmsteads. Lava bombs thrown from the vent smashed windows and set some homes alight. Volunteers nailed sheets of corrugated iron to windows and shoveled ash from house roofs to prevent them from collapsing. Houses that could not be saved were stripped of their furniture and fittings before they caught fire and succumbed to the advancing lava. Within two weeks, a 700-foot (210-m) volcano, named Eldfell,

stood at the edge of town disgorging lava and slowly demolishing block after block of houses. Lava was also flowing into the sea and threatening to block the harbor.

RESIDENTS CONQUER THE VOLCANO

Scientists proposed that the volcano could be conquered by spraying it with seawater to freeze the lava flows. It worked, substantially slowing the lava flows. Water pipes and massive pumps were moved onto the hardened lava itself, working uphill to freeze a larger area. Lava entering the harbor was hosed with water cannons mounted on ships. The fight would have eventually been lost if the eruption had not finally stopped on June 26, 1973, five months after it had started. The campaign's delaying efforts saved Vestmannaeyjar and its harbor. The lava had destroyed one-third of the town's 1,200 houses, one fishing plant, and stopped just 175 yards (160 m) short of closing the harbor.

Most of the residents have since returned to Vestmannaeyjar, and cleared the town of ash and lava. The fields are green again, and Eldfell lies in the background as a reminder of the people's heroic fight. Today the volcano is a tourist attraction and its heat is now being harnessed for the town's benefit. Steam generated from water pumped into holes drilled in the volcano's flanks is being used to heat the buildings.

NEW LIFE ON SURTSEY ISLAND

The new island Surtsey was named after Surtur, the fire god from Norse mythology, and its birth generated worldwide interest. Biologists used the barren volcanic island to study the process of life's gradual colonization. Insects were first detected in 1964, initially flying insects, believed to have reached the island under their own power or been blown across. Moss and lichen appeared as early as 1965 and now cover much of the island. Seals began breeding on the island's shores in 1983. Gulls also arrived, and a permanent colony has been present since 1986. The nesting birds improved the soil conditions with their guano, and helped spread seed allowing more advanced species of plants to colonize and survive. In 1998, the first bush, a willow (*Salix phylicifolia*), was found on the island. These plants can grow up to 13 feet (4 m) tall. A total of 30 plant species have become permanently established on Surtsey Island, and it is estimated that a further two to five new species arrive each year.

Higher forms of land life colonized Surtsey Island's soil. An earthworm was collected in 1993, perhaps carried over from the Icelandic mainland by a bird. Slugs were found in 1998, and spiders and beetles have also become established. As the vegetation improved, the island has become a more regular stopping-off point for migrating birds. Atlantic puffin nests were found on Surtsey Island in 2004.

ABOVE The Atlantic puffin (*Fratercula arctica*) is abundant throughout Iceland, feeding largely on fish, diving to depths of up to 200 feet (60 m). They have special spines on their tongues that allow them to hold several fish at one time.

BELOW Steam rises off the waters of Blue Lagoon, a natural spa in Iceland with an average water temperature of 104°F (40°C). The water comes from the geothermal power station behind the lagoon, where hot water is used to drive turbines.

Faults and Rift Valleys of the North American West Coast

The North American continent is like a huge whiteboard eraser. Pushed westward by the slow tectonic opening of the Atlantic Ocean, it is obliterating the Pacific Ocean along its western edge. The unique geologic setting of the American west coast began during the Tertiary Period, when the North American Plate actually began to override the still-active Pacific spreading ridge known as the East Pacific Rise.

ABOVE Liard hot springs are in a boreal spruce forest in the far north of British Columbia, Canada. The water temperature varies from 107°F to 126°F (42°C to 52°C). The area is frequented by moose and bears.

OPPOSITE, TOP Along the Juan de Fuca Trail on Canada's Vancouver Island, a waterfall cascades onto Mystic Beach. The Queen Charlotte–Fairweather fault line runs near this island and the area is subject to earthquakes and tremors.

BELOW The Gulf of California in Mexico is a flooded rift valley. Shallow magma melts beneath the fault zone are being tapped for their geothermal energy.

The long thin peninsula of Baja California is dramatic evidence of the effects of an active spreading ridge beneath continental crust. Tensional effects on the base of the continent caused the thin strip of land to begin tearing away from the Mexican mainland forming a rift valley, which was eventually flooded by the sea forming the Gulf of California.

THE GULF OPENS UP A YOUNG OCEAN

The Gulf of California is one of the best modern examples of a continental rift in the process of forming a young ocean basin. In places the basin floor and spreading ridge are overlain with thick layers of sediment that has been washed in from the edges, which have been intruded by basalt. The southern Gulf hosts numerous active hydrothermal mounds and chimneys composed of silicate, carbonate, sulfate, and sulfide minerals.

VAST VOLCANIC PLAINS

The East Pacific spreading ridge disappears beneath the North American continent under the Colorado River delta at the head of the Gulf of California. It is responsible for the uplift, crustal stretching, rifting, and volcanism of the US southwestern states, giving rise to the classic basin and range (graben and horst) topography. Upwelling hot currents within Earth's mantle probably destroy any residual oceanic crust, then begin to act directly on the base of the continental crust creating deep and pervasive tensional fractures. These are thought to sometimes act as conduits for the very fluid flood basalt, which can quickly spread over the land surface forming shield volcanoes and vast volcanic plains. The extensive 17-million-year-old Columbia River Plateau basalts of southern Washington, Oregon, and northern California are believed to have formed in this way.

THE RIDDLE OF THE ROCKY MOUNTAIN TRENCH

Other manifestations of sub-continental tension include the enigmatic Rocky Mountain Trench. This tectonic feature consists of a chain of valleys, stretching in a straight line for over 1,000 miles (1,600 km) from Flathead Lake, Montana, to the Liard Plain near the Yukon border in Canada, and is clearly visible from space.

The trench marks a line of weakness between the younger Rocky Mountains to the east and the older Columbia and Cassiar Mountains on its western side. The trench is 3 to 8 miles (5 to 13 km) wide with mountains on each side rising steeply 3,300 to 6,600 feet (1,000 to 2,000 m) above the flat-bottomed floor. Nine rivers, including the Fraser, Columbia, Peace, and Liard, drain the trench, most entering and leaving through canyons.

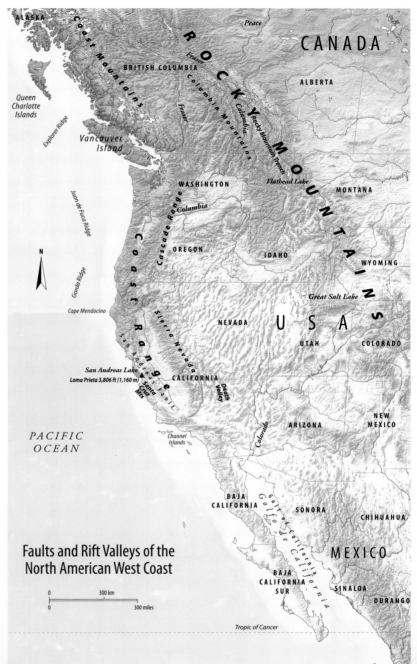

THE EARTHQUAKE CENTER OF THE WORLD

From Los Angeles to Cape Mendocino, north of San Francisco, movement between the Pacific and North American plates occurs along a transform fault boundary. This is well known as the San Andreas Fault, although rather than a single fault, it is a series of hundreds of sub-parallel and inter-woven smaller faults that together make up a wide zone of activity. Here, the plates slide horizontally alongside one another. They periodically lock up then break free, releasing the stored energy as earthquakes. The people of California have become accustomed to living under the constant threat of earthquakes. Lessons learned from the biggest one, the magnitude 8.3 earthquake that destroyed San Francisco in 1906, have meant California's building codes are among the toughest in the world.

THE QUEEN CHARLOTTE–FAIRWEATHER FAULT

The Pacific spreading ridge reappears briefly off-shore, north of Cape Mendocino as a series of short segments; the Gorda, Juan de Fuca, and Explorer ridges, then disappears again beneath the continent north of Vancouver Island. The boundary between the North American and Pacific plates continues north as the Queen Charlotte–Fairweather transform system. Earthquakes occur along this fault also, but there are no large population centers at risk.

Faults and Rift Valleys of the North American West Coast

SAN ANDREAS FAULT EARTHQUAKES

The San Andreas Fault is part of the Pacific "Ring of Fire," a massive seismic belt that rings the Pacific Ocean. It is responsible for about 80 percent of the world's earthquakes. California alone records thousands of shocks each year. Of these, 500 are large enough to be noticed by many people and the state averages about one earthquake of destructive magnitude every year.

RIGHT Picturesque Point Reyes National Seashore in California is situated west of the San Andreas Fault. Due to its location near the fault line, this peninsula is slowly moving northwards. During the earthquake in 1906 that destroyed the city of San Francisco, it shifted about 20 feet (6 m) north of the mainland.

The Pacific and North American tectonic plates continuously grind against one another at about $1\frac{1}{3}$ inches (3.3 cm) per year. Their jerking and shuddering is what causes the earthquakes that regularly devastate America's populous West Coast, including the major cities of San Francisco and Los Angeles, without warning. The biggest of these earthquakes include Southern California in 1857, Owens Valley in 1872, San Francisco in 1906, Santa Barbara in 1925, Long Beach in 1933, San Fernando in 1971, and Loma Prieta in 1989. The 1906 earthquake was the worst to strike a major city in the history of the USA. With West-Coast population numbers ever increasing, the situation is now at critical point—in San Francisco alone, the population has grown from 400,000 in 1906 to over six million in 2007. Notwithstanding preparations, the death toll and damage during the next event could be horrendous.

SAN FRANCISCO EARTHQUAKE 1906

On Friday, April 18, 1906, at 5:30 a.m., a 270-mile (430 km) segment of the San Andreas Fault suddenly gave way near San Francisco. It released, perhaps, over 150 years of pent-up energy. Any natural features and man-made structures crossing the fault were displaced by up to 20 feet (6 m) by a continuous rupture or tear in the ground surface that was visible for a distance of nearly 200 miles (320 km). The earthquake of Richter Scale magnitude 8.3 lasted only 45 to 60 seconds, during which time San Francisco and surrounding areas

were destroyed. The death toll was estimated at over 3,000, with 225,000 to 300,000 people left homeless. Overall damage was estimated to be US$400 million (1906 value), though the great fire that raged for days afterward was responsible for 80 percent of the total damage.

LOMA PRIETA 1989

On October 17, 1989, at 5:04 p.m. a Moment Scale magnitude 6.9 earthquake lasting 15 to 20 seconds occurred near Loma Prieta peak in the Santa Cruz Mountains, 10 miles (16 km) northeast of the city of Santa Cruz. Although the fault rupture did not break the ground surface, this was the largest earthquake to occur in the San Francisco Bay area since 1906. The earthquake caused damage throughout the San Francisco Bay area and in downtown Santa Cruz. The death toll was 62,

BELOW A panoramic view of San Francisco, a city in ruins after the earthquake and resulting fires of 1906. Residents reported the streets moving in waves and buildings collapsing into the streets. Fires were exacerbated by arson as some homes were not insured against earthquakes.

with about 3,757 people injured. The repair bill was US$6 to 8 billion (1989 value). Most deaths occurred when the resonant vibration collapsed 50 of the 124 spans of the Cypress Viaduct. This raised freeway with two traffic decks formed part of Oakland's Nimitz freeway. Its reinforced concrete span supports were built on unconsolidated soil and, unfortunately, the natural frequency of those spans coincided with the frequency of the earthquake ground motion. The viaduct structure thus amplified the ground motion. Cracks formed in the support frames, until finally, the upper road–way collapsed onto the lower road.

A span of the top deck of the San Francisco Bay Bridge also collapsed, falling at an angle onto the lower deck. One car drove off the edge of the gap, falling to the lower deck and killing the driver.

WHAT IS AN EARTHQUAKE?

Earthquake energy is released when stored stress in the crust overcomes frictional forces along a pre-existing fault, or ruptures unbroken rock. The point of rupture is known as the "focus" from which the energy radiates outward as compression (P-waves) or shear (S-waves) waves. The magnitude of the earthquake is measured from seismographs and zones of intensity can be delineated based on the reported damage that it causes. The modified Mercalli Scale ranges from an intensity of one (no movement felt) to 12 (complete destruction with the ground moving in waves).

East African Rift Valley

The East African Rift Valley is a classic geologic structure—a line of weakness along which the African continental plate is starting to tear apart. The rift zone is characterized by deep straight valleys bounded by sheer-walled fault escarpments. Voluminous outpourings of lava periodically issue forth from deep cracks opening along the rift.

ABOVE A lioness (*Panthera leo*) sets out on a hunt. Lions once inhabited vast areas of Eurasia, the Middle East and Africa but they are now found in the wild only in India and southern and eastern Africa, including East Africa's Rift Valley.

BELOW Flamingoes line the lakeshores of Lake Nakuru, a highly alkaline shallow lake in Kenya's Lake Nakuru National Park. One of the 450 species of birds known to inhabit this lake, the intense pink color is due to the absorption of a pigment from the abundant algae they feed on.

Hot rising convection currents within Earth's mantle are lifting the overlying continental crust in this part of Africa and stretching it apart. This stretching has led to the collapse of sections of the crust along long, straight, normal faults creating a tectonic feature known as a graben or rift valley. Such valleys act as a trap for water and sediment, making them excellent places in which to look for accumulations of fossils.

THE AFRICAN CONTINENT SPLITS APART

The structure stretches 3,700 miles (6,000 km)—halfway down the length of Africa—from Djibouti on the Gulf of Aden in the north to the Zambezi River delta in Mozambique. On its long journey south it passes through the countries of Ethiopia, Kenya, Uganda, Zaire, Rwanda, Burundi, Tanzania, Zambia, Malawi, and Mozambique; in some cases, it demarcates their political boundaries. The valley resembles a huge arc-shaped bite out of the east side of Africa along which "continental drift" may eventually carry an enormous chunk of Africa out into the Indian Ocean.

The spreading of the East African Rift has been much slower than that of the Red Sea and Gulf of Aden, where the shorelines are now 150 to 200 miles (250 to 300 km) apart and their valleys have been permanently flooded by the sea.

THE EAST AFRICAN RIFT CONTINUES TO SPREAD

Numerous earthquakes along the length of the rift indicate that spreading is still in progress. Recent global positioning satellite (GPS) measurements show that the East African Rift Valley is widening

ROCK FEATURE: CARBONATITE

Carbonatite is a rare igneous rock consisting mainly of calcium, magnesium, and iron carbonate minerals, principally calcite, dolomite, and ankerite. It occurs as lava flows or small intrusions, mostly found in continental rift tectonic settings such as the East African Rift. Oldoinyo Lengai is the only volcano where carbonatite lavas (pictured) have erupted in historic times. The black lava turns white within hours of eruption as its unstable minerals react with the atmosphere.

at about $\frac{1}{5}$ inch (4 mm) per year compared to a widening of $\frac{4}{5}$ inch (20 mm) per year for the Red Sea–Gulf of Aden Rift.

The lines of earthquakes and landforms indicate that the rift valley splits into two, with one branch running around the eastern side of Lake Victoria and the other around its western side. They rejoin on Lake Victoria's southern side. The incipient rift valley then divides again, branching in a southeasterly direction along Lake Malawi and southwest into Zambia and Botswana along the line of the Luangwa and Zambezi rivers.

In this way, the African continental tectonic plate is being fragmented into two or more smaller tectonic plates, with the Somalian Plate on the east and Nubian Plate on the west.

RIGHT The scattered deposits of mineral salts around Lake Magadi in Kenya, the rift valley's southernmost lake, starkly reveal the lake's high salinity. The lake is only partly liquid; at times up to 80 percent of its surface is made up of solid mineral salts called trona that lie up to 130 feet (40 m) thick on the lake bed.

RIFT VALLEY LAKES

The floor of the East African Rift Valley is marked by a necklace of lakes that lie in the deeper portions of the graben. The larger freshwater lakes occur in the western branch of the African rift, and the main rift to the Zambezi River delta. These are lakes Albert, Edward, Kivu, Tanganyika, and Malawi. Lake Albert and Lake Edward drain north into the White Nile, Lake Tanganyika and Lake Kivu empty west into the Congo River system, while Lake Malawi drains south via the Shire River into the Zambezi. Lake Tanganyika is the deepest lake, reaching a maximum depth of 4,820 feet (1,470 m) and holds the most water.

The lakes that occur in the eastern branch of the rift are smaller, shallower, and have no outlet to the sea. As a result, these lakes have a high mineral content. Continuous evaporation of their water in the hot equatorial climate leaves the dissolved salts behind in increasing concentrations, often as thick carpets of mineral salts.

These lakes include Lake Turkana, Lake Naivasha, Lake Magadi (which is almost pure soda—sodium carbonate), Lake Natron, Lake Ambroseli, Lake Manyara, Lake Eyasi, Lake Elmenteita, Lake Baringo, Lake Bogoria, and Lake Nakuru. All of these lakes are strongly alkaline, with Lake Turkana being the world's largest alkaline lake.

VICTORIA: AFRICA'S LARGEST FRESHWATER LAKE

Lake Victoria lies in an enormous basin, which is located between the curved eastern and western branches of the East African Rift. The lake started filling about 400,000 years ago, when a number of westerly-flowing tributaries of the Congo were cut off and dammed by the rising western branch of the African rift valley. It is the largest freshwater lake on the African continent, although it is relatively shallow and has dried out three times since it formed, during periods of low rainfall. The lake's waters were shown to be the source of the Nile River by American explorer Henry Stanley, who circumnavigated the lake in 1875 and discovered its outflow at Rippon Falls. The lake's waters flow over Rippon Falls then, via Lake Kyoga and Lake Albert, into the White Nile.

Lake Victoria was once home to an incredible diversity of freshwater fish that provided food for the people of the lake but many were wiped out by an introduced species, the Nile perch.

ABOVE Until recently, more than 500 different species of cichlid fish were found in Lake Victoria. The Nile perch was introduced in the 1950s and unfortunately this highly predatory fish has managed to wipe out over half of the lake's native cichlid species, affecting food supplies for people who live around the lake.

ABOVE In a remote corner of Tanzania stands an unusual volcano called Oldoinyo Lengai, where lava fountains harden in midair, then shatter like glass.

BELOW The magnificent glacier at the summit of Mt Kilimanjaro in Tanzania is shrinking rapidly. Ice that once thwarted climbers at its summit has shrunk by more than 80 percent and has almost completely disappeared. Current speculation is that it will be completely gone by 2015.

RIFT VALLEY VOLCANOES

As the East African Rift continues to spread apart, basalt lava from Earth's mantle continually outpours to fill the space between the separating plates. The lavas have built significant volcanic cones in several places along the length of the rift.

Mt Kilimanjaro, a dormant strato-volcano in northeast Tanzania, is the highest mountain in Africa with an elevation of 19,340 feet (5,895 m). Due to Kilimanjaro's near-equatorial location and its height, climbers begin their ascent in tropical rainforest and pass through every conceivable climate before reaching ice-covered alpine desert at the summit. The well-known Furtwängler Glacier and other patches of ice near the summit are small remnants of an enormous ice cap that once crowned Mt Kilimanjaro. During the twentieth century, this ice cap retreated dramatically, with 82 percent of the mountain's glacial ice disappearing between 1912 and 2000. Seasonal snowfall will still cover the mountaintop for several months of the year, but this meltwater will be insufficient to keep streams and springs running all year round.

Geologically, the white-topped Oldoinyo Lengai in northern Tanzania is one of the most interesting volcanoes. It is the only active volcano in the world that erupts a rare alkali-rich carbonatite lava. Carbonatite is an igneous rock that can contain economic concentrations of rare elements such as

RECENT SPREADING

Geologists from the University of London, University of Oxford, and Addis Ababa University in Ethiopia used satellite data to observe a crack widen to 26 feet (8 m) and grow to a length of 37 miles (60 km) along a section of the rift between the African and Arabian plates.

The cracking occurred along a fault in Ethiopia's Afar Desert. It began with a large earthquake on September 14, 2005, followed by a continuous series of smaller tremors. A week later, volcanic ash started to erupt from the crack as it continued to widen. The whole event, the first of its kind observed by scientists, took just three weeks. The event demonstrates that the spreading process does not take place smoothly, but in a series of sudden rupture events. It is expected that within several million years, the Red Sea will break through the highlands surrounding the Afar Depression and flood into the East African Rift Valley, eventually forming a new seaway along its entire length.

barium, fluorine, titanium, uranium, zirconium, and others. Small amounts of carbonatite lava can be seen continually erupting from the floor of the summit crater. These runny lavas have some of the lowest eruption temperatures measured, around 1,000°F (500°C). Periodic violent eruptions have occurred in 1917, 1926, 1940, 1958–67, 1983, and 1993. Other volcanoes along the rift include Mt Dallol, Erta Ale, Nyiragongo, Nyiamuragira, and various vents in Ethiopia's Danakil Depression.

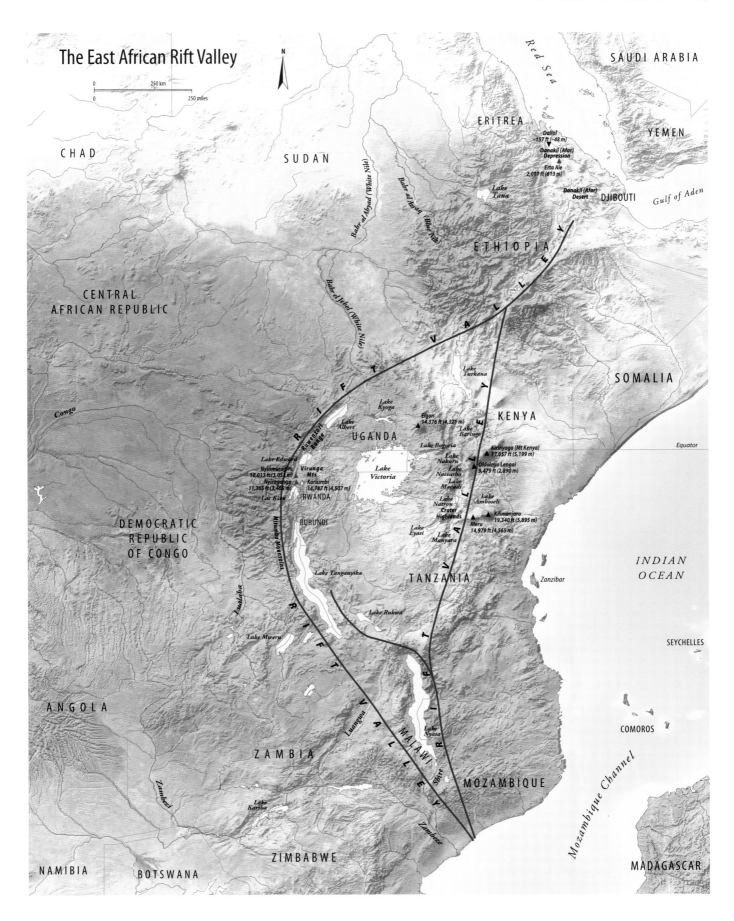

The East African Rift Valley

N

0 — 250 km
0 — 250 miles

SAUDI ARABIA

CHAD

SUDAN

ERITREA

YEMEN

Red Sea

Dallol
-157 ft (-48 m)

Danakil (Afar)
Depression

Erta Ale
2,011 ft (613 m)

Lake
Tana

Danakil (Afar)
Desert

DJIBOUTI

Gulf of Aden

ETHIOPIA

CENTRAL
AFRICAN REPUBLIC

Bahr al Abyad (White Nile)

Bahr al Azraq (Blue Nile)

Bahr el Jebel (White Nile)

SOMALIA

Lake
Turkana

KENYA

Congo

Lake
Kyoga

Elgon
14,376 ft (4,321 m)

Kirinyaga (Mt Kenya)
17,057 ft (5,199 m)

Equator

Lake
Albert

Lake
Baringo

Lake Bogoria

RUWENZORI RANGE

UGANDA

Lake
Nakuru

Lake
Naivasha

Oldoinyo Lengai
9,479 ft (2,890 m)

Lake Edward

Lake
Victoria

Lake
Magadi

Nyamuragira
10,013 ft (3,051 m)

Virunga
Mts

Nyiragongo
11,365 ft (3,465 m)

Karisimbi
14,787 ft (4,507 m)

Lac Kivu

RWANDA

Lake
Natron

Lake
Amboseli

Kilimanjaro
19,340 ft (5,895 m)

DEMOCRATIC
REPUBLIC
OF CONGO

BURUNDI

Crater
Highlands

Meru
14,979 ft (4,565 m)

Lake
Eyasi

Lake
Manyara

INDIAN
OCEAN

Mitumba Mountains

Lake Tanganyika

TANZANIA

Zanzibar

Lualaba

Lake Rukwa

Lake Mweru

SEYCHELLES

ANGOLA

Luangwa

Lake
Nyasa

COMOROS

Mozambique Channel

ZAMBIA

MALAWI

Shire

Zambezi

MOZAMBIQUE

Zambezi

Lake
Kariba

ZIMBABWE

MADAGASCAR

NAMIBIA

BOTSWANA

RIFT VALLEY

RIFT VALLEY

RIFT VALLEY

EARLY HUMANS

The growth and development of the East African Rift Valley is believed to have played an important role in pushing forth pre-human species on their evolutionary journey to become *Homo sapiens* and eventually modern humans. The changing environmental conditions created as the rift opened up presented new challenges or opportunities and saw the development of a more intelligent and adaptable creature: ourselves.

The 50-million-year journey from early primitive hominids, such as *Adaptis* and *Proconsul*, to modern *Homo sapiens* involved the development of bipedal movement, an increase in height, a change from a stooped to upright posture, loss of body hair, and a decrease in jaw and tooth size. A huge increase in brain size occurred, enabling speech and the development of sophisticated tool-making skills. It is interesting to explore what factors drove these rapid—and quite recent—evolutionary changes. The earliest fossils of *Homo sapiens*, or anatomically modern humans, date back only 130,000 years.

RIFTING CHANGES EAST AFRICA'S CLIMATE

Broad uplift and stretching of the crust of East Africa began about 10 million years ago. The crust tore apart along a series of faults, producing deep and wide rift valleys with the rift's shoulders uplifted on either side. These shoulder ranges formed vast highlands stretching from Ethiopia in the north to Mozambique in the south, and reached up to 2½ miles (4 km) in height. The effect of rifting on the climate was dramatic. The rising highlands impeded the passage of moist tropical air, creating a dry area or rainshadow over East Africa, and the original rainforest was eventually replaced by savanna grassland.

This change may have caused the evolution of bipedalism (the ability to walk upright) east of the rift about 6 million years ago, while west of the rift the other African apes continued to live in the trees. This savanna hypothesis proposes that our need to walk upright instead of living in the trees was driven by the replacement of rainforest by grasslands and the need to travel larger distances between food sources as efficiently as possible.

RIFT VALLEY LAKES

The large lakes lying along the rift valleys would have provided an ideal environment for early humans; plenty of fresh water and an abundance of food in the form of game living on the grasslands and coming to the lakeshores to drink. Sediment washing into the lakes from the surrounding highlands has since preserved a rich and continuous fossil record of life that once existed by the lakeshores. From this record comes evidence suggesting life may have not been easy. It appears that hominids did not evolve their larger brains just simply lounging around in an idyllic lakeside setting. Evidence from lake sediments at Baringo Basin in central Kenya suggests that these lakes may have appeared and disappeared rapidly on cycles of 20,000 years or less, perhaps within several generations. Alternating cycles of severe drought and abundance would favor the survival of those creatures adaptable and intelligent enough to survive.

SIGNIFICANT RIFT VALLEY FOSSIL SITES

Important discoveries of hominid remains have been made at Olduvai Gorge, in the rift valley between Lake Natron and Lake Eyasi in Tanzania. They began with the discovery of 1.8 million-year-old *Homo habilis* in the early 1960s by paleontologists Louis and Mary Leakey. In 1972, their son Richard found the skull of 2-million-year-old *Homo rudolfensis*. In 1984, at Koobi Fora, near Lake Turkana in Kenya, the nearly complete skeleton of a *Homo erectus*, Turkana Boy, was discovered by Kamoya Kimeu. In 1999, Richard's wife, Meave Leakey, discovered a 3.5-million-year-old skull there, believed to be a new branch of early hominid. She named the new genus *Kenyanthropus platyops*, which means "the flat-faced man of Kenya."

ABOVE This fossilized skull of an *Australopithecus afarensis* child, probably a female, was uncovered by paleontologists along with a nearly complete skeleton. The child is estimated to have died at the age of three at least 3 million years ago.

RIGHT The cinder cone of an extinct volcano rises up from Kenya's Lake Turkana, its volcanism defining the current zone of rifting. The region around this lake has yielded an extraordinary number of significant fossil finds of ancient humans.

RIGHT An anthropologist exam-
ines the famous fossil footprints
found by Dr Mary Leakey in
Laetoli, Tanzania, in 1975. Parallel
tracks of 70 prints were made
3.6 million years ago by two
Australopithecus afarensis
walking across the plains.

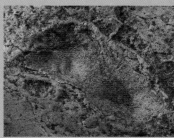

ABOVE The footsteps of our
ancestors: a closeup of one of
the footprints left in rain-wet
volcanic ash in Laetoli in close
view shows all toes are parallel
to the axis of the foot. This is an
indication the feet that made
them were human-like.

The Afar Region of Ethiopia, and Djibouti,
especially the Middle Awash, is also known as one
of the cradles of hominids. It was here that Lucy,
the first fossilized skeleton of *Australopithecus
afarensis*, was found by Donald Johanson in 1974.
The following year, one of his students found fossil
fragments of an entire *A. afarensis* family.

Jordan Valley–Dead Sea Rift

The Jordan Valley–Dead Sea Rift marks the north–south transform fault boundary that lies between the African Plate on the western side and the Arabian Plate on the eastern side. It runs over 600 miles (1,000 km) from the Taurus Mountains in Turkey to the Red Sea. The highly distinctive topography of this rod-straight active fault is most visible when viewed from space.

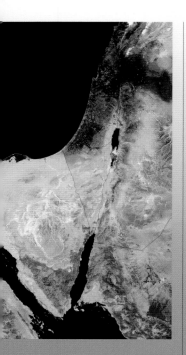

ABOVE The Jordan Valley–Dead Sea Rift is ruler-straight as it emerges from the Gulf of Aqaba (the right fork). It cuts through the Middle East between Jordan and Israel via the Dead Sea, then Lebanon and Syria in the north.

This transform-fault system has been caused by differential movement between the African and Arabian plates. Both plates are colliding with the Eurasian Plate to push up the Alps–Taurus–Zagros collision-mountains, but while they are both moving northward, they are doing so at different speeds. This faulting has created a spectacular deep flat-bottomed graben bounded by steep, straight escarpments.

A LONG LINE OF VALLEYS

The Dead Sea Fault is occupied by a line of lakes, rivers, dry valleys (wadis), and green spring-fed oases. It is known as the Beqaa Valley in Lebanon and the Hula Valley in Israel. Further south, the valley is occupied by the Jordan River, which flows into the Sea of Galilee and then continues south through the Jordan Valley into the Dead Sea. South of the Dead Sea the rift is known as the Wadi Arabah until the point where it becomes the Gulf of Aqaba and the Red Sea.

During the course of the last few million years, the Mediterranean Sea has flooded the Dead Sea Rift Valley several times, each time evaporating and leaving behind salt. Repeated inundations over time eventually built up an immense salt layer some 2 miles (3.2 km) thick. Finally, uplift of the Mediterranean coast cut off the valley from the sea permanently, leaving it as a hot dry basin.

THE DEAD SEA BASIN

The Dead Sea is 42 miles (68 km) long by 11 miles (18 km) wide and about 1,000 feet (300 m) deep. The basin occupies the lowest point on the Earth's surface at 1,371 feet (418 m) below sea level. Being the terminal point for the valley's drainage, and an area of intense evaporation, the Dead Sea has become one of the saltiest bodies of water on Earth. It has a salinity of up to 350 parts per thousand (35 percent) dissolved salts, which is an order of magnitude more than seawater. Anyone can easily float on the Dead Sea because of the high density of its water. It was called "dead" because no fish can live in it.

The Dead Sea salt is different from that of seawater, consisting of approximately 53 percent magnesium chloride, 37 percent potassium chloride, and 8 percent sodium chloride (common salt), whereas seawater salt is 97 percent sodium chloride.

Early in the twentieth century, it was realized that the Dead Sea was a natural deposit of potash and bromine. In 1929, a solar evaporation plant was built on the north shore of the Dead Sea at Kalia, to extract potash (potassium chloride) from the brine. Further plants were built as demand grew.

Today, potash, bromine, caustic soda, magnesium metal, and sodium chloride are extracted from the brines using extensive evaporation pans covering the entire southern end of the Dead Sea.

BELOW The Dead Sea is so saline a person will easily float on top of its surface. The level of buoyancy is extremely high, making it difficult to swim. Many people come here because they believe the water and mud pools have therapeutic properties.

ABOVE The village of Dana in Jordan overlooks steep cliffs and terrain that plunges into the rift valley and toward the Wadi Arabah in the desert lowlands below. Parts of this wadi have undergone extensive tectonic activity over the last 20 million years or so, which continues to pull this rift apart.

REFILLING THE DEAD SEA

Overdevelopment is damaging the Dead Sea. Particularly alarming is the falling water level due to diversion of water from the Jordan River for irrigation and other projects. Between 1930 and 1997 the Dead Sea has dropped 70 feet (21 m).

On May 9, 2005, a proposal known as the Two Seas Canal was agreed on by Jordan, Israel, and the Palestinian Authority. The plan would see water from the Gulf of Aqaba pumped and tunneled for 100 miles (160 km) along the Arabah section of the rift valley. From the top of the low dividing range separating the Red Sea and the Dead Sea, the water would flow downhill into the Dead Sea. As the majority of the water's journey is downhill, the plan is to use the excess energy to generate electricity and produce freshwater by desalination.

BELOW Vast fields of salt deposits dot the shallows of the Dead Sea, which is directly cut through by the rift valley. Evaporation ponds are used to produce vast quantities of salts for plastics manufacturing and other chemical industries.

North Anatolian Fault Zone, Turkey

The North Anatolian Fault Zone (NAFZ) is a boundary between two tectonic plates sliding against one another, very similar to the San Andreas Fault of California, USA. The Anatolian Plate, pushed by the African and the Arabian plates, is sliding westward at a rate of about ¾ to 1 inch (2 to 2.5 cm) per year against the Eurasian Plate.

An amazing earthquake predictability is possible along this creeping fault. Like an opening zipper, rupture and stress-release in the form of earthquakes are propagating systematically along the fault. Slippage during one earthquake appears to pass the stress westward along to the next segment of fault, which in turn is forced to rupture. These are known as remotely triggered earthquakes.

Earthquakes have been propagating from east to west along a 600-mile (1,000-km) length of this fault since the 1939 Erzincan earthquake. Those that followed occurred in 1942, 1943, 1944, 1951, 1957, 1967, and 1999. This sequence enables a prediction to be made of where the next earthquake is likely to take place.

THE PREDICTED ISTANBUL EARTHQUAKE

Attention is focused on Turkey's next earthquake, which is expected within a decade or so. It is now the turn of the locked-up section of the North Anatolian Fault to the west of the Izmit section to rupture. This will occur near Istanbul, and the outcome for the highly populous capital will be devastating. Many tens of thousands may die and the country's economy will be severely affected. Many of Istanbul's buildings are older, and the newer ones are not likely to meet their building code, so collapses will be significant. Fire will also be a serious problem in this earthquake because of Istanbul's older wood-frame buildings.

THE DIRECTION AND PATH OF THE FAULT

The North Anatolian Fault runs in a westerly direction through the towns of Erzincan, Susehri, Ilgaz, and Gerede. Its path is then marked by the narrow Gulf of Izmit which widens into the Marmara Sea. A southern branch of the fault system can also be seen controling the topography of Iznik Lake. Bathymetric surveys—research that explores the topography of the deep ocean floor—show that both faults extend beneath and are responsible for the shape and position of the Marmara Sea. The fault zone then continues west into the North Aegean Sea.

BELOW The great Hagia Sophia emerges through the fog at sunrise in the Sultanahmet district of Istanbul. This vast metropolis of over 12 million people is on the section of the fault line predicted to next experience a major earthquake, which could occur at any time.

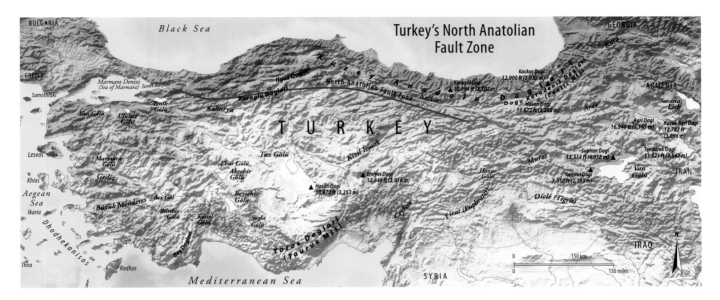

Turkey's North Anatolian Fault Zone

EARTHQUAKE HITS IZMIT IN 1999

On August 17, 1999, at 3:02 a.m., heavily industrialized western Turkey was struck by a magnitude 7.4 earthquake, right where it was anticipated to occur. The epicenter was on the NAFZ at the head of Izmit Bay. It was the largest event on record to have devastated a modern industrialized city since the 1906 San Francisco and 1923 Tokyo earthquakes.

The fault ruptured over a length of 70 miles (110 km) sliding up to 16 feet (5 m), with vertical displacements reaching 10 feet (3 m), with the north side of the fault dropping down. Unfortunately most people were at home sleeping and had no chance to escape from their multistory apartments. Poor construction and failure to comply with construction codes meant that many buildings collapsed completely. A total of 77,300 buildings were destroyed and 244,500 damaged over a wide region including Izmit, Adapazari, Gölcük, Karamürsel, Yalova, Düzce, and Istanbul. The official death toll was 17,225 (with other sources reporting over 30,000 deaths).

Slumping and subsidence caused extensive and permanent flooding of large areas along the Marmara Sea coast in the vicinity of Gölcük. Underwater slumping associated with the earthquake also generated a tsunami that struck the popular holiday coast of Izmit Bay. Minutes after the earthquake, the sea receded from the shore then returned as a 20 feet (6 m) tsunami laden with yachts and cruise ships. It surged across the promenade and slammed into seaside shops, hotels, and resorts. This massive tsunami was followed by numerous smaller ones caused by the wave energy bouncing backward and forward between the opposite shores of the bay.

ROCK FEATURE: MYLONITE

Mylonite is a fine-grained metamorphic rock often found along major fault lines. It is formed during prolonged grinding between the continually moving walls on either side of a fault plane. The rock material is fractured and refractured, and individual crystals break into smaller pieces, creating a microscopically fine-grained rock. It can have a crude foliation parallel to the direction of fault movement or stretched out "eyes" of lesser-deformed wall rock in thicker bands.

ABOVE An aerial picture taken on August 23, 1999, reveals the extent of the damage caused by the earthquake in a residential area near Izmit, a city in Turkey that lies along the fault line. The total repair bill was estimated at US$6 billion at the time and 44,000 people were injured.

Spencer Gulf–Lake Torrens Rift, South Australia

Spencer Gulf and the neighboring Gulf St Vincent are two large inlets near the city of Adelaide on the southern coast of Australia. They open into the Great Australian Bight and are separated by the boot-shaped Yorke Peninsula. Spencer Gulf, the larger of the two, is 200 miles (320 km) long and 80 miles (130 km) wide at its mouth.

ABOVE A sea lion (*Neophoca cinerea*) on its main breeding ground of Kangaroo Island, opposite the entrance to Gulf St Vincent. They feed in shallow waters near the shoreline.

ABOVE RIGHT Kangaroo Island probably formed during the initial tearing apart of Australia from Antarctica when basalt erupted through the newly forming rift as these two continents began separating.

These inlets are spectacular fault-bounded rift valleys that have been flooded by the sea. The region is Australia's most seismically active, with numerous small earthquakes revealing that movement is still taking place along these faults. Traveling north along the rift valley, inland of Port Augusta at the head of Spencer Gulf, the climate dries rapidly to desert. Here, the valley floor is occupied by Lake Torrens. This unique, internally draining, saline rift lake is 150 miles (240 km) long with an area of 2,200 square miles (5,700 km^2), and is protected within the Lake Torrens National Park. The lake is usually dry and has only been filled with water once in the past 150 years. Other dry salt lakes lie further north, and Lake Eyre is the largest of these.

AUSTRALIA AND ANTARCTICA SEPARATE

About 80 to 100 million years ago, during the early Cretaceous Period, Australia began rifting from Antarctica with the development of young ocean floor between the two continents. The Southeast Indian Ocean and Pacific–Antarctic mid-oceanic spreading ridges formed the wide inverted V-shape of the Great Australian Bight.

The Spencer Gulf Rift Valley began to develop at the apex of the inverted V-shape at this time, but failed to mature—much in the same way as the well-known East African Rift Valley has failed to mature into a young ocean as quickly as the Red Sea–Gulf of Aden rifts. The Gippsland, Bass, and Otway basins in Victoria and the offshore basins of South Australia and Western Australia also began to form and fill with sediments at this time. These basins host significant oil and gas reserves.

THE WORLD'S LARGEST SALTPAN

The Lake Eyre Basin is the world's largest saltpan, with an area of approximately 450,000 square miles (1,170,000 km²), which is about the size of France, Germany, and Italy combined. The basin covers about one-sixth of the Australian continent.

All the riverbeds in this vast, mostly flat, area lead inland toward Lake Eyre, which lies at the basin's lowest point, at 52 feet (16 m) below sea level. These include those of the Georgina, Diamantina, Thomson, and Barcoo rivers, and Cooper Creek from central and western Queensland, and the Finke, Todd, and Hugh Rivers from central Australia.

Water from the wetter edges of the basin reaches a vast area of multiple braided channels and ephemeral waterholes. Most years, it is absorbed by the earth or simply evaporates on its journey toward Lake Eyre. Only during rare flood years do the rivers reach the lake. For short periods, long-dormant aquatic creatures multiply in the lake and large flocks of waterfowl arrive to feed and raise their young before the waters evaporate once more.

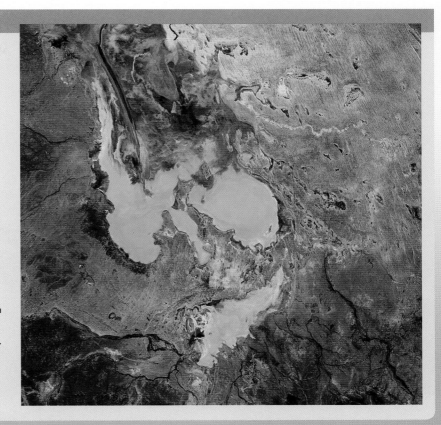

A false color thematic satellite image of Lake Eyre reveals the extent of flooding that occurred in 1999. During the twentieth century, it completely filled only twice: in 1950 and 1974.

LAKE EYRE BASIN SINKS

The Lake Eyre Basin began to sink about 60 million years ago, during the early Tertiary Period, forming a vast internally drained depression covering a large part of central Australia. Rivers began to deposit sediment into the shallow basin, which is still slowly sinking today. Back then, rivers supplied the Lake Eyre Basin with abundant fresh water and it was extensively forested. About 20 million years ago, large shallow lakes grew and covered much of the central basin. These lasted about 10 million years, but as Australia drifted further north, the climate gradually became more arid. The lakes and floodplains started to dry, becoming today's deserts.

In contrast to the dry surface rivers, a vast volume of water flows underground through sandstone aquifers known as the Great Artesian Basin. Tapped by numerous bores, this water resource is vital to the region's graziers.

BELOW Floodwaters flow into Lake Eyre during flood season in South Australia's channel country. This massive salt lake is in a vast depression located 430 miles (700 km) north of Adelaide, South Australia's capital city.

Earth's mountain ranges are more than just places of grandeur and great beauty. They are storehouses of water, and sources of sediment, and they play a major role in influencing climate and vegetation patterns. Mountain ranges have long controlled the movements and settlement of native cultures, and they still form geopolitical boundaries today. Mountains are also keys that help us to understand the internal geologic workings of our planet—both past and present.

ABOVE Mt Everest was formed around 60 million years ago, and is part of the Himalayan range of collision mountains. Currently listed as 29,028 ft (8,848 m) high, geological forces are gradually increasing Everest's height.

RIGHT The MacDonnell Ranges spread out across hundreds of miles either side of the town of Alice Springs in Australia's Northern Territory. These fold mountains appear as parallel ridges and contain many gorges.

The major mountain belts of our planet are still forming today, all along the boundaries of Earth's actively moving tectonic plates. Mountain building is most aggressive where the plates are converging and colliding with one another, and zones of intense and often hazardous earthquake activity mark their upward growth. The vertical rise of the land is matched by the forces of erosion, which work to reduce all mountains to sea level.

There are two quite distinct classes of mountain range that form at convergent plate margins: collision and subduction ranges. The most spectacular collision mountain belt is the Himalayas. Subduction mountain ranges encircle the Pacific Ocean basin. A third kind of mountain range is totally submerged beneath the world's oceans.

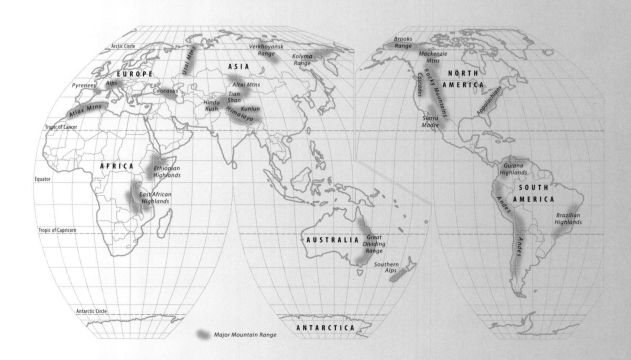

Major Mountain Range

the shadows, unlike the snow on the sunny slopes facing the equator. Alpine icefields also release water continuously during the warmer months.

ALTITUDE-DEPENDENT CLIMATIC ZONES

One of the most enjoyable aspects of hiking up into the mountains is seeing the changing vegetation types that occur in altitude-dependent belts on the mountainsides. The vegetation types are mostly influenced by the decline in air temperature with increasing altitude, known as the adiabatic lapse rate. Lapse rates are generally between 1.8°F to 3.6°F per 1,000 feet (1°C to 2°C per 300 m), but vary according to the moisture content of the air. Wet air will cool more slowly. Thus, plants at the higher elevations must endure lower temperatures and shorter growing periods than plants at the mountain base. The changes seen include a general reduction in plant size due to slower growth rates, and communities composed of fewer, but more cold-tolerant, species.

Sometimes a walk of several hours from a mountain's rainforested base to its alpine peak will cover a span of climatic zones that would in other circumstances require thousands of miles of travel from equatorial to polar regions.

Mountainous regions remain as some of the most sparsely populated in the world due to the difficulties of living and farming there. Indigenous communities have developed ingenious terracing and irrigation systems to cultivate the lower valleys. They also venture onto the higher alpine slopes with their livestock to take advantage of the summer pastureland. But due to difficulties of access and cultivation, mountains, especially those more remote, still contain some of the last remaining refuges of endangered animal and plant communities, and are places to be preserved and studied.

BELOW This traditional building (kasbah) in the village of Ait Arbi blends in with the red rock of the Dades Gorge, Morocco's "Grand Canyon." The gorge is in the High Atlas Mountain Range in central Morocco.

North American Cordillera

The North American Cordillera forms a wide mountainous backbone running through western Canada, the United States, and Mexico. Its sub-parallel, overlapping, linear ranges tell a complex but interesting geological story. The cordillera comprises two main high belts that bound an intermontane system of plateaus and basins.

OPPOSITE, TOP The sun sets over the snow-covered Mt Olympus in the Olympic Ranges, Washington, USA. This lofty peak—part of the Coast Ranges—reaches a height of 7,965 feet (2,429 m).

BELOW The Rocky Mountains in Montana's Glacier National Park clearly show the shale layers that were originally laid down on an ancient seafloor. These mountains were sculpted by the erosional forces of glaciation.

In the USA and Canada, the Coast Ranges, Cascades, and Sierra Nevadas make up the western or Pacific coastal belt, with the imposing Rocky Mountains forming the eastern or inland range. The cordillera continues south into Mexico to become the Eastern and Western Sierra Madre ranges and the mountainous spine of Baja California. The North American Cordillera is continuous with the Aleutian Range to the north, and with the Central and South American Cordillera, which includes the Andes, to the south.

Collectively these ranges form a massive mountain belt known as the American Cordillera that almost spans the globe from north to south—

Earth's longest subaerial mountain range, and part of the tectonically active Pacific "Ring of Fire," a zone marked by earthquakes and active volcanoes.

RISE OF THE NORTH AMERICAN CORDILLERA

The cordillera started to grow as the American Plate began its northwesterly push over the Pacific Ocean floor, swallowing or subducting it in a deep submarine trench. The process was driven by the newly opening Atlantic Ocean. Two major mountain-building events, known as orogenies, shaped the geology. During these dramatic tectonic events, the North American continent collided with island archipelagoes—equivalent in size to Japan—along with other strips of continental crust that were earlier torn from the edge of the continent. These outliers were all crushed back against the Pacific coastline—being heavily folded and metamorphosed in the process.

About 175 million years ago, the Columbia Orogeny caused huge masses of rock to crack and slide over one another. The intense compressive forces formed the Columbia Mountains and, in turn, the Rocky Mountains. The Laramide Orogeny occurred about 85 million years ago, initiating a whole new compressive cycle that uplifted the modern Rocky Mountains as tilted crustal blocks bounded by thrust and reverse faults. Then, during the mid Tertiary Period, there was a change from compression to crustal extension, with volcanism and uplift of the entire southern Rocky Mountains and Colorado Plateau regions. Rivers entrenched deep canyons, and severe erosion of the landscape took place, forming the plateaus, buttes, peaks, and ridges of today's landscape.

THE PACIFIC OCEAN'S DISAPPEARING MID-OCEANIC RIDGE

The dramatic tectonic changes of the Tertiary Period came about when the North American Plate actually began to override the still-spreading Pacific oceanic ridge, known as the East Pacific Rise. Today, where the East Pacific Rise meets the North American continent, it has torn away the Baja California peninsula from the Mexican mainland forming the Gulf of California, a

drowned rift valley. The ridge disappears beneath
the continent under the Colorado River delta
at the head of the gulf and is responsible for the
uplift, crustal stretching, rifting, and volcanism
of the US southwestern states—the classic basin
and range topography.

From Los Angeles to Cape Mendocino, north
of San Francisco, the boundary between the
Pacific and North American plates manifests
itself as a transform boundary, the well-known
San Andreas Fault. Here, the plates slide horizontally alongside one another, periodically locking
up then breaking free and releasing the stored
energy as earthquakes in the process. The people
of California have become accustomed to living
under the constant threat of violent earthquakes.
Lessons learned from the biggest one, the Richter
Scale magnitude 8.3 earthquake that destroyed San
Francisco in 1906, leaving some 300,000 people
homeless, have meant the state's building codes
are among the toughest in the world.

The spreading ridge reappears briefly offshore,
north of Cape Mendocino. It is responsible for the
parallel chain of active volcanoes, including Mt St
Helens in Washington state, in that section of the
cordillera, before it disappears again beneath the
continent to the north of Vancouver Island.

AMERICA'S MOUNTAIN-FED GIANT RIVERS

The North American Cordillera shapes the climate
of the entire North American continent, as do the
Andes in South America. Air masses journeying
westward across the continent collide with the
eastern foothills of the Rockies and are forced
to rise, cool, and release their moisture as orographic rainfall. This rainfall and snowmelt feeds
the numerous tributaries of the mighty rivers such
as the Missouri and the Arkansas that flow eastward across the continent's Great Plains and then
southward to the Gulf of Mexico.

ABOVE The Brooks Range in
northern Alaska, along with the
Alaska Range, together form
the northernmost end of the
North American Cordillera.
The Brooks Range creates a
geological boundary that separates Arctic Alaska from the rest
of the state. This glaciated landscape runs nearly as far as the
Chukchi Sea, which is located
on the west coast.

ROCK FEATURE: SCHIST

Schist is formed by the metamorphism of clay-rich rocks. It is a highly foliated rock due to a preferred alignment of its constituent minerals, particularly muscovite and biotite mica. These platy crystals align themselves perpendicular to the principal direction of regional stress. They form part of a sequence of increasing metamorphic grade, starting from shale, then slate, then phyllite, and then schist. At higher metamorphic grades, "knotted" schists can develop as rounded garnet crystals begin to grow within the rock fabric. Schists are common in collision mountain belts.

Depleted of its moisture, the air continues its journey down the cordillera's western flanks, heating rapidly as it descends toward the Pacific Ocean. This drying air is the reason why North America's major deserts are found west of the great range, between the latitudes of 25°N and 40°N. These include the Great Salt Lake Desert, Mojave Desert, Gran Desert, and Viscaino Desert.

VAST MINERAL RESOURCES

The cordillera is a vital storehouse of mineral and energy resources. Vigorous volcanism, metamorphism and re-melting associated with the formation of the cordillera caused molten granitic bodies to rise into the crust, bringing with them significant concentrations of metals, including copper, gold, lead, molybdenum, silver, tungsten, and zinc.

Meanwhile, the intermontane basins and giant lakes lying between the high ranges have acted as giant traps for huge thicknesses of eroded sediment that host reserves of coal, natural gas, oil shale, and petroleum. Of particular significance

are the La Brea Tar Pits of the Los Angeles Basin, where the unique conditions led to the preservation of a wide variety of Pleistocene Epoch wildlife, including saber-toothed cats, mammoths, and bison. Some 40,000 years ago these animals, coming to drink at the pools, became the unsuspecting victims of a sticky tar hidden beneath carpets of leaves. Predators, moving in for a feast on the struggling animals, also quickly became victims.

PEOPLE OF THE AMERICAN CORDILLERA

In a series of epic Ice-Age migrations, people first reached the Americas via Siberia and Alaska about 9,000 to 15,000 years ago. These hardy nomadic hunter-gatherers moved eastward, via land bridges or short island-hops, in pursuit of mastodons, prehistoric bear, bison, and caribou. Continuing south into Mexico, Central, and South America, they eventually populated the entire Americas. These cordilleran people organized their lives around the seasons—migrating between the plains in winter, where they hunted bison, and the mountains in spring and summer, where they lived on fish, deer, elk, roots, and berries. These "paleo-indians" would later become the Native American tribes—the Apache, Arapaho, Bannock, Blackfoot, Cheyenne, Crow, Flathead, Shoshoni, Sioux, Ute, and others.

Arrival of Europeans to the shores of the Americas altered the native cultural lifestyle irreversibly. The mountains were the last areas to be affected, but finally the "mountain men" colonists came—first the Spanish, hungry for gold, then the French and British in search of animal fur and mineral wealth. Later came the permanent settlers. The discovery of gold in California in 1848 would

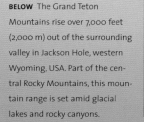

BELOW The Grand Teton Mountains rise over 7,000 feet (2,000 m) out of the surrounding valley in Jackson Hole, western Wyoming, USA. Part of the central Rocky Mountains, this mountain range is set amid glacial lakes and rocky canyons.

herald the beginning of the gold rushes. Thousands of fortune seekers flooded into the region, forever changing the way of life in the cordillera. Today, the economic activity of the cordillera is centered on mining, forestry, agriculture, and, increasingly, on recreation and tourism, with the populations of the region's cities continuing to grow strongly.

LIFE ADAPTS TO THE CORDILLERA

Plants have adapted to life in zones of increasing altitude on the cordillera. A hiker climbing the mountains will traverse deciduous forest, grassland, and taiga before reaching the cold alpine region just below the snowline. Animals of the cordillera have also evolved in various ways: hoofed species,

such as bighorn sheep and mountain goats, are very surefooted on the rugged slopes; moose, caribou, and elk migrate to lower and warmer elevations during the winter months. Others remaining in the winter snow moult their brown fur and replace it with white fur to blend in, such as the snowshoe hare, which, as the name suggests, has also developed especially large feet.

Some animals, such as bears, marmots, and squirrels, will drop their metabolic rate during the winter period of hibernation. Today, national parks of the cordillera provide refuges for many species, such as the gray wolf and black-footed ferret, which are endangered, or have vanished from other parts of the continent.

ABOVE A Rocky Mountains goat (*Oreamnos americanus*) grazes in Glacier National Park, Montana, USA. These goats live throughout the Rockies and Cascades. Their surefootedness is an evolutionary adaptation to the rugged landscape, as it allows them to escape predators by hiding in the rocky landscape.

The Andes

The Andean or South American Cordillera is the longest mountain range in the world; it is only the Himalayas that are higher. The Andes run some 5,000 miles (8,000 km) from the Caribbean Sea in the north to Tierra del Fuego, at the southernmost tip of South America.

OPPOSITE PAGE, TOP Black shale caps the jagged peaks of granite in the Cuernos del Paine in the Andes in Chile's Torres del Paine National Park. The metamorphic rocks here were baked harder by the intense heat of a granite intrusion 12 million years ago. The peaks were sculpted by glacial action after an ice sheet covered the mountains around 100,000 years ago.

BELOW The ancient city of Machu Picchu in Peru was one of the few Incan cities not found by the Spanish during their sixteenth-century conquest of South America. It remained unknown until archeologist Hiram Bingham, of Yale University, discovered it in 1911.

The tallest peak is Mt Aconcagua, which stands 22,831 feet (6,959 m) above sea level. The Andes shape the entire physical geography and climate of South America. From their eastern flanks flow some of the most significant rivers, irrigating the continent on their journey to the Atlantic Ocean, while on the western side lies the world's driest desert. Historically, the Andes have controlled the human settlement of the region and played a vital role in setting the present political borders.

MOUNTAINS CONTINUE THEIR DRAMATIC RISE

The Andes are still forming today, marking the boundary of two colliding tectonic plates. The Pacific oceanic plate is being thrust beneath the more-buoyant South American continental plate, which in turn is rising and riding over the top. The driving force of the American Plate's westward push is the opening of the Atlantic Ocean, which began about 130 million years ago. Each year the Pacific Ocean shrinks by a few inches as the Atlantic Ocean grows, so the Andes continue to rise but, as they are being eroded at similar rate, there is no net increase in elevation.

Rocks along the South American Pacific coast are being pushed up and folded into long high

RICHES OF THE ANDES

The continuous cycle of subduction, melting, and volcano-building along the South American Pacific coast leads to a concentration of elements, such as gold, silver, copper, tin, and other metals that are otherwise rare in the crust. Exploration in the Andes has uncovered some of the world's largest economic metal deposits and rarest gemstones, such as lapis lazuli. Chile produces about 30 percent of the world's copper from massive open-pit mines such as Chuquicamata and Escondida. These are the world's largest mines, with an annual production of 680,000 tons and 800,000 tons of copper metal respectively. The name Chuquicamata comes from the Aymara language. It refers to the Chuqui Native South American group, who first worked the copper deposits to make tools and weapons. Today, enormous trucks carry loads of low-grade copper-bearing rocks out of these colossal pits. In the weathered zones of these mines, many rare and unusual green copper minerals are commonplace (such as antlerite and atacamite). Many are highly soluble minerals that can only survive in the region's dry deserts.

ranges, known as cordillera, running parallel to the coastline. In places, the mountain ranges separate leaving vast basins in between. In these basins lie large inland lakes—Lake Maracaibo in Venezuela is the largest lake in South America, and Lake Titicaca, which sits between Peru and Bolivia, is the highest commercially navigable lake at an elevation of 12,500 feet (3,800 m) above sea level. The lakes are fed by rainfall and melt-water from the surrounding mountains. In Peru, Bolivia, and Chile, the high-altitude valleys between the cordillera are known as the *altiplano*. Here, living in some of the world's highest permanent settlements, are people who eke out a meager existence tending to their crops and livestock.

A VOLATILE EARTHQUAKE-RIDDEN REGION

As the crust of the Pacific Ocean is consumed, it melts and the molten magma moves upward, erupting explosively to build massive volcanic edifices along the length of the mountain chain. The best-known active volcanoes of the cordillera

are Sabancaya (Peru), Cotopaxi (Ecuador), Nevado del Ruiz (Colombia), and Lascar (Chile). The subduction process is accompanied by numerous earthquakes. Their position and depth, when plotted on a map, reveal the location of the eastward-dipping thrust-fault plane that lies between the under- and over-thrusting plates. The earthquakes range in size from small tremors, felt on a near-daily basis, to the largest on record, a magnitude 9.5 earthquake which struck Valdivia, Chile, on March 27, 1960.

On that fateful afternoon, the sea floor tore apart like a giant zipper opening along a length of about 680 miles (1,100 km) parallel to the southern Chilean coastline as the Pacific Plate suddenly thrust 66 feet (20 m) eastward beneath the South American Plate. It sent out a destructive Pacific-wide tsunami and triggered numerous rockfalls and landslides in the mountains, killing 2,300 people in Chile alone. Such natural hazards have become a way of life for the people of the Andes.

FROM THE EQUATOR TO THE ANTARCTIC

The Andes stretch from the Caribbean, through the equatorial tropics, to practically touch the Antarctic continent, where they are spectacularly carved by permanent ice sheets and glaciers. Along their length it is possible to experience all imaginable climates. Moisture-laden air traveling westward across the flat South American continent is pushed upward on reaching the eastern foothills of the Andes. The massive mountain barrier forces the air upward, where it cools and is forced to drop its precious cargo of moisture as rain. This process, known as orographic uplift, is responsible for the vital water feeding the tributaries of the giant river systems that flow to the Atlantic, such as the Amazon, Orinoco, and Parana. With all of its moisture spent, air descending the Andes' western slopes is hot and dry, and parches the lands over which it travels. The Atacama Desert, flanking the Pacific shore, is one of the world's driest places.

LIFE ADAPTS TO THE ALTITUDE OF THE ANDES

Unique life evolved to survive in the often harsh, dry, and cold high-altitude conditions of the Andes. Best known are the camel-like animals, known as guanaco. Their long necks and surefootedness enable them to spot and avoid predators such as pumas and leopards, while their stomach structure and gastric secretions permit them to survive on the sparse and nutrient-poor grasses. Guanaco have

ABOVE The Top of Andes lies between the cities of Mendoza in Argentina and Santiago on the Chilean coast. During his expedition on the *Beagle*, Charles Darwin first sighted the Top of Andes from Patagonia in 1834.

ABOVE RIGHT *Polylepis* trees form a brightly colored forest that is said to be the world's highest. These trees grow at high altitudes—up to 15,000 feet (4,500 m). Native to South America, they are endemic to the Andes and are most common in Ecuador, Peru, and Bolivia.

BELOW A herd of guanaco (*Lama guanicoe*) grazes in front of the Andes in Torres del Paine National Park, Chile. Native to South America, these camel-like animals have adapted to the cold with a double-layered coat: a soft undercoat that retains warmth close to the skin, and a more rugged outer coat.

been used by the native Indians for some 5,000 years, and domesticated into the alpaca and llama. These animals, renowned for their fine warm wool, are able to carry loads, and their dung is used as a cooking fuel.

The world's largest bromeliad, *Puya raimondii*, also evolved in the harsh Andean conditions. Found only around 13,000 feet (4,000 m) in the Andes of Peru and Bolivia, *Puya*, which is Mapuche for "point," takes about 80 to 100 years to reach maturity and flower, producing a flower spike 30 to 35 feet (9 to 10 m) tall. Its radiating leaf structure directs rainwater or condensation to the base of the plant. The Mapuche Indians used the *Puya*'s young leaves to make a salad. Sadly, the plant is now threatened with extinction.

LOST PEOPLES OF THE ANDES

The Inca Empire rose from its heartland in Cuzco, in the Andes of present-day Peru, during the 1400s, dominating the independent smaller native groups to become the largest nation. At its peak, Incan society had over 6 million people in an empire stretching almost the length of the Andes—an area equal in extent to that of the Roman Empire. The Incas maintained their empire through wise and meticulous government.

Cobblestone roads, often still in use today, and irrigation canals were built through the mountains. They developed Quechua as a universal language, and communication throughout the empire was by *chasqui* or runners. Each of these young men would run a short distance to the next village or outpost and faithfully relay the message.

Irrigation permitted agriculture, even in remote and drier regions. To utilize the steeper slopes, stone walls were built to create raised level fields in a series of step-like terraces along the sides of hills. These terraces created more usable land and kept the topsoil from washing away in heavy rains.

Unfortunately the Incan empire was short-lived. Its great wealth led to its ultimate downfall, for the Spaniards, on reaching the New World, learned of the abundance of gold in Inca society and set out to plunder it. Through deceit and trickery, King Atahualpa was killed and the Inca Empire fell to a small army of 180 men led by Pizarro in 1532. Rampant disease, introduced by the conquerers, hastened the empire's downfall.

After the Spanish empire in turn lost its hold during the various battles for independence, the Andean region fractured into the independent countries we know today—Venezuela, Colombia, Ecuador, Peru, Bolivia, Chile, and Argentina.

The Andes

| 0 | 400 km |
| 0 | 400 miles |

Caribbean Sea

Golfo de Venezuela

EL SALVADOR

NICARAGUA

COSTA RICA

PANAMA

Golfo de Panama

Pico Cristobal Colon 18,947 ft (5,775 m)

Lago de Maracaibo

Pico Bolívar 16,427 ft (5,007 m)

VENEZUELA

Sierra Nevada del Cocuy 18,022 ft (5,493 m)

Llanos

Orinoco

GUYANA

SURINAME

FRENCH GUIANA

Nevado del Ruiz 17,713 ft (5,399 m)

Cerro el Nevado 14,961 ft (4,560 m)

Guiana Highlands

Nevado de Huila 18,865 ft (5,750 m)

COLOMBIA

Volcán de Purace 15,243 ft (4,646 m)

Negro

Amazon Basin

Nevado de Cumbal 15,630 ft (4,764 m)

Volcán Cayambe 18,996 ft (5,790 m)

Volcán Cotapaxi 19,347 ft (5,897 m)

Equator

ECUADOR

Chimborazo 20,702 ft (6,310 m)

Amazonas

Islas Galapagos

Selvas

BRAZIL

PERU

Nevado Huascaran 22,205 ft (6,768 m)

Cerro Yerupaja 21,765 ft (6,634 m)

Cerro Pumasillo 20,492 ft (6,246 m)

Nevada Auzangate 20,945 ft (6,384 m)

Nevado Illimani 21,194 ft (6,460 m)

Planalto do Mato Grosso

Brazilian Highlands

Nevado Coropuna 21,079 ft (6,425 m)

Lago Titicaca

BOLIVIA

Paraguay

Nevado Sajama 21,391 ft (6,520 m)

Lago Poopo

Salar de Coipasa

Salar de Uyuni

Tropic of Capricorn

Volcán Licancabur 19,455 ft (5,930 m)

PARAGUAY

Salar de Atacama

Gran Chaco

Volcán Llullaillaco 22,057 ft (6,723 m)

CHILE

Desierto de Atacama

Volcán Antofalla 20,013 ft (6,100 m)

Parana

Nevados Ojos del Salado 22,572 ft (6,880 m)

Cerro Bonete 22,546 ft (6,872 m)

Salinas Grandes

Laguna Mar Chiquita

Lagoa dos Patos

PACIFIC OCEAN

Cerro de Las Tortolas 20,774 ft (6,332 m)

Cerro de Olivares 20,610 ft (6,282 m)

Cerro Champaqui 9,449 ft (2,880 m)

Lagoa Mirim

Cerro Aconcagua 22,835 ft (6,960 m)

Volcan Tinguirrica 14,108 ft (4,300 m)

Volcán Maipo 17,356 ft (5,290 m)

Pampas

URUGUAY

Volcan Peteroa 13,419 ft (4,090 m)

Cerro Nevado 12,500 ft (3,810 m)

Rio de la Plata

Volcán Domuyo 15,449 ft (4,709 m)

ARGENTINA

Volcán Llaima 10,249 ft (3,124 m)

Volcán Lanin 12,388 ft (3,776 m)

Negro

Cerro Tronador 11,660 ft (3,554 m)

Lago Nahuel Huapi

Patagonia

ATLANTIC OCEAN

Volcan Corcovado 7,546 ft (2,300 m)

Minchinmavida 8,104 ft (2,470 m)

Monte Maca 9,711 ft (2,960 m)

Monte Melimoyu 7,874 ft (2,400 m)

Monte San Valentin 13,314 ft (4,058 m)

Lago Colhue Huapi

Lago Buenos Aires

Lago General Carrera

Monte San Lorenzo 12,139 ft (3,700 m)

Volcan Lautaro 11,089 ft (3,380 m)

Lago Viedma

Cerro Murallon 11,811 ft (3,600 m)

Lago Argentina

Falkland Islands (Islas Malvinas)

Monte Burney 5,741 ft (1,750 m)

Estrecho de Magallanes

Tierra del Fuego

South Georgia

Monte Darwin 8,163 ft (2,488 m)

Cabo de Hornos

CORDILLERA OCCIDENTAL

CORDILLERA CENTRAL

CORDILLERA ORIENTAL

CORDILLERA CENTRAL

CORDILLERA OCCIDENTAL

CORDILLERA ORIENTAL

A N D E S

N

The Alps

The Alps are a massive collision mountain system stretching 750 miles (1,200 km) through southern Europe, with a number of its peaks rising above 10,000 feet (3,000 m). The ice-covered summit of Mt Blanc on the French–Italian border is the highest of these, with an elevation of 15,771 feet (4,807 m).

ABOVE The Aiguilles de Chamonix are in the French Alps, near the ski resort of Chamonix. Also within the region is Mt Blanc, the Alps' and France's highest peak at 15,771 feet (4,807 m).

The Alps run in a giant arc from the Riviera on the Mediterranean coast, through southwestern France, northern Italy, Switzerland, southern Germany, and Austria.

At their eastern end, this mountain system branches into two, encircling the Great Hungarian Plain, with the Carpathian Mountains to the north and the Dinaric Alps to the south. At their western end, the Maritime Alps curve back then continue around the headwaters of the Po River, known as the Lombardy Plains, and head down the boot of Italy to become the Apennine Mountains.

MOUNTAINS RISE FROM AN ANCIENT SEA

The Alps were born during the Oligocene and Miocene epochs as a result of the ancient Tethys Sea being slowly crushed and closed by the northward movement of the African continent.

As Africa approached the Eurasian continent, beds of sandstone, shale, limestone, and dolomite that had been deposited in the shallow Tethys Sea during the Triassic and Jurassic periods were uplifted. They were pushed into giant recumbent folds that piled up on top of one another like a huge rumpled carpet. Continued pressure tore the

ROCK FEATURE: SLATE

Slate is formed through the low-grade metamorphism of clay-rich rocks. During this process, minute platy crystals of mica and chlorite begin to grow at right angles to the direction of maximum pressure, giving the rock a perfect slaty cleavage. This means that the rock can be easily split into large, strong but thin sheets, and as such is ideal for a roofing material. Many houses in Britain and Europe have slate roofs.

As metamorphic grade increases, slates may become "spotted" as minerals begin to grow larger, finally changing to phyllite and ultimately schist.

folds from the basement and forced massive slabs of rock to slide over the top of one another along long, low-angle, thrust-fault planes.

Today, abundant marine fossils and even entire fossil reefs found in limestone beds high above sea level in the Alps serve as a silent reminder of the region's turbulent past.

RIGHT At 12,457 feet (3,799 m), Grossglockner is Austria's tallest mountain and one of the highest peaks in the Alps. It is in the Hohe Tauern Mountains, a range of the Eastern Alps. The Pasterze Glacier at its foot is the largest of the 925 glaciers found in Austria, all of which are rapidly shrinking.

THE QUATERNARY ICE AGE

During the Pleistocene, vast ice masses moved through the valleys of the Alps, transforming them into deep U-shaped troughs with steep walls. The glaciers carved and shaped the Alps, bulldozing the piles of debris outward in the form of moraines. As these glaciers retreated, this rubble dammed the streams and rivers producing the region's lakes. Lake Geneva is one of the largest moraine-dammed lakes. The results of the glacial erosion are the classical landforms that make the Alps a popular tourist destination.

Geologists first studied glacial landforms in the Alps, so today similar snow- and ice-affected areas around the world are known as alpine. Many terms used to describe glacial landforms were derived from the names of those features found in the Alps. For example, a steep-pointed peak carved by the heads of several glaciers mowing outward from that point is known as a horn, of which the Matterhorn is the most classic example. Arêtes are sharp jagged ridgelines separating two adjacent glacial valleys. Cirques are the circular-shaped depressions, frequently filled by a picturesque lake, found at the heads of U-shaped glacial valleys.

Although the Pleistocene Ice Age has long passed, leaving most of the formerly glaciated valleys empty, some parts of the Alps remain above the permanent snowline and are home to a number of significant glaciers. Europe's largest glacier, the Aletsch in the Bernese Alps of southern Switzerland, is 17 miles (27 km) in length. These areas of permanent ice have been shrinking and it is forecast that the European Alps could become nearly ice-free by 2100.

BELOW A bank of clouds fills a valley in the Swiss Alps, which form part of the Alps. This mountain range makes up more than 60 percent of Switzerland's geography. Dufourspitze on the Swiss–Italian border is the highest peak in Switzerland, reaching 15,203 feet (4,637 m) feet.

The Apennines

The Apennines, as a geographic entity, stretch the length of the Italian peninsula. They cover a distance of some 840 miles (1,350 km) and are about 70 to 80 miles (110 to 130 km) wide. Geologically, the Apennines range is a continuation of the Western or Maritime Alps. It runs down the Italian peninsula then swings around into the toe of Italy. The mountain chain then crosses the Strait of Messina into Sicily after which it continues across the Mediterranean Sea to join with the Atlas Mountains of North Africa.

ABOVE A meadow full of spring crocus (*Crocus vernus*), a common meadow flower in Italy during spring. It is native to the Apennines and the Alps and is related to the valuable saffron crocus, *Crocus sativus*.

The highest peak in the Apennines is Corno Grande. It is 9,554 feet (2,914 m) and hosts Europe's southernmost glacier, Ghiacciaio del Calderone, on its northern flank. Several important rivers are born in the ranges—the Ofanto in southeastern Italy, and the Arno, Tiber, Volturno, and Garigliano rivers, which drain its western slopes. The once-forested slopes of the Apennines, teeming with deer, ibex, chamois, and wolves, have been heavily cleared over the centuries to create extensive olive groves, vineyards, orchards, nut plantations, and pastureland for sheep and goat grazing. Most people here live in the mountain valleys and fertile basins.

RISING OF THE APENNINES

The Apennines consists mainly of Triassic to Tertiary period sea-floor sediments that were deposited in the shrinking basin between the closing African and Eurasian continents. During the Eocene Epoch, the sea began to grow shallow and the deposits of limestone became progressively more sandy and pebbly. The squeezing of these sedimentary beds between the two continents uplifted, folded, and metamorphosed them, giving rise to the existing Apennine chain. This compression continued to the end of the Miocene Epoch.

The Apennine chain contains several active volcanoes. These include Mt Vesuvius near Naples, Mt Etna near Catania in Sicily, and the Aeolian Islands of Stromboli and Vulcano, north of Sicily. This volcanic unrest lies above the subduction zone where the African Plate is pushing northward beneath the Eurasian Plate. The creaking and groaning descending plate is melting to release viscous gas-charged explosive andesitic magma, which rises to feed the dangerous volcanoes. Naples, a city of over 3 million located on the flanks of Mt Vesuvius, is highly at risk from eruptions. The tectonic unrest is also marked by numerous and

THE PURE WHITE MARBLE OF CARRARA

One of the most famous marble quarries is that of Carrara in the northwestern Apennines. It was used in the Pantheon and highly favored for many of the famous Renaissance sculptures, such as Michelangelo's statue of *David*. Marble Arch in London and the medieval cathedral, Duomo di Siena in Italy, are also made from this well-regarded stone. Blasting with explosives in the stone quarries is generally avoided because of the possibility of fracturing and damaging the rock. Great care was taken in the past to mine the marble using traditional labor-intensive sawing methods. Today, large diamond-impregnated cable saws are used to cut out enormous blocks for export. The Carrara area is also popular with mineral collectors. Here, rare but well-formed and spectacular crystals of minerals such as quartz, feldspar, sphalerite, gypsum, fluorite, pyrite, and sulfur can be found in the white matrix of the Carrara marble.

The 17-foot (5-m) tall statue of *David* was sculpted from a block of Carrara marble by Michelangelo between 1501 and 1504.

sometimes devastating earthquakes, such as the Messina Strait earthquake of 1908, which killed about 70,000, and the 1980 earthquake near Campania that left 4,800 people dead.

MINES AND STONE QUARRIES

Valuable deposits of borax, copper, iron ore, lignite, mercury, and tin are mined in the central and northern Apennines, while power is generated in the many hydroelectric stations located along the mountain's major rivers.

The world's best-known serpentines and marbles are also quarried from the Apennines, with some quarries dating back to Roman times. These beautiful rocks are highly sought after for building facades, statues, columns, tiles, and table-tops. Serpentine from Polcevera Quarry, known as *verte des Alpes* (Alps green), is a beautiful dark green, mottled with waves of light green and white. As well as pure white marble, gray, pink, and black marble is also quarried in Italy—including veined and mottled varieties.

ABOVE LEFT Ponte Vecchio in Florence lies on the Arno River. The Arno originates from Mt Falterona, one of several rivers sourced from the Apennines.

BELOW One of many mountain lakes in the Apennines. The southern part of the range also features mineral springs, crater lakes, and fumaroles.

Ural Mountains

Long considered the traditional boundary between Europe and Asia, the Urals are an ancient mountain range separating the Northern European Plain from the West Siberian Lowlands. The mountains run north–south for about 1,500 miles (2,400 km), from the flat grassland steppes that lie to the north of the Caspian Sea, to the coastline of the Arctic Ocean. The sickle-shaped mountainous island of Novaya Zemlya, separated from the mainland by the Kara Strait, represents a continuation of the Urals into the Arctic Ocean.

ABOVE The Arctic fox (*Alopex lagopus*) lives largely to the north of the Arctic Circle and inhabits the Ural Mountains. Like most Arctic mammals, its thick white fur is an adaptation to the cold, as each shaft is hollow to provide optimum warmth as well as camouflage in the snow.

OPPOSITE PAGE, TOP Remote Novaya Zemlya is one island in an archipelago in the Arctic Ocean. The northern tip of the Ural Mountains, the islands —two large and several small— are mountainous with many glaciers on the main island.

RIGHT A soft blanket of snow coats the Ural Mountains in early winter. These low-lying mountains are somewhat less dramatic than other European ranges such as the Alps, but they contain a treasure trove of hidden mineral wealth.

The mountain chain was thrust upward in a rock-crushing collision between two major continental landmasses during the late Carboniferous Period, which marked the final stages of the formation of the supercontinent Pangea. At this time, the Siberian continent rammed into the giant continent already containing most of Earth's land— the combination of Laurasia (Europe and North America) and Gondwana (Antarctica, Australia, South America, Africa, and India).

A VERY ANCIENT MOUNTAIN RANGE

Although Pangea has long-since broken up into today's continental configuration, Europe and Siberia have remained firmly joined along the line of the Urals. The Urals are possibly the world's oldest mountain range still standing as a prominent topographic feature. They are arbitrarily divided into four sections: the Southern (between latitude 51°N and 55°N), Central (between latitude 55°N and 61°N), Northern (between latitude 61°N and 64°N), and the Subarctic and Arctic Urals (north of 64°N). The highest peak is Mt Narodnaya at 6,217 feet (1,895 m). Traveling northward along the range, taiga or coniferous forests slowly give way to tundra grasslands in the northern section until Arctic conditions finally prevail.

A RICH MINERAL INDUSTRY

Russia's major mineral resources are found in the Urals, exposed through deep erosion of this ancient mountain chain. These include rich deposits of aluminum, coal, copper, gold, iron ore, manganese, and potash. There are oilfields and refineries along the Kama and Belaya rivers in the western Urals. Gemstones such as amethyst, chrysoberyl, emerald, and topaz are also mined. Perhaps the most rare and beautiful of the minerals is alexandrite, a variety of the mineral chrysoberyl. Named for Russian Czar Alexander II, it displays a remarkable color change under different lighting: red under artificial light and green in daylight.

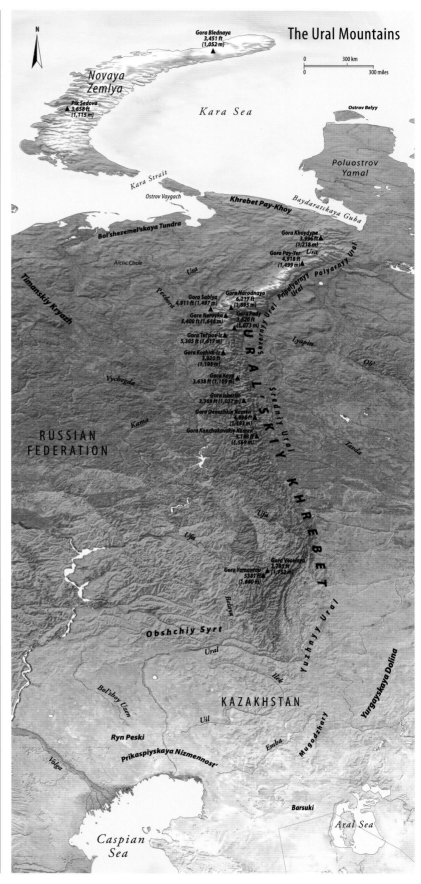

The metallurgical region, in the central and southern Urals and the adjacent lowlands, is fueled by hydroelectric and coal power. Huge Soviet-style industrial centers have grown up at Berezniki, Chelyabinsk, Magnitogorsk, Nizhni, Orenburg, Orsk, Perm, Tagil, Ufa, Yekaterinburg, and Zlatoust. Unfortunately all of this industrial concentration has come at a price—the severe environmental degradation of many of the region's mountain habitats.

WORLD-HERITAGE KOMI FORESTS

One of the more positive aspects regarding the Urals is that an area of 12,600 square miles (32,600 km²) was declared a World Heritage Site in 1995. It is located in Russia's Komi Republic, and contains one of the largest areas of virgin coniferous forest in Europe. This unspoiled ecosystem of Siberian spruce, larch, and fir forests, supporting mammal populations of reindeer, mink, sable, and hare, was saved from being logged and mined as a result of this international action.

NUCLEAR TESTING ON NOVAYA ZEMLYA

During the Cold War, the USSR decided that the remote unpopulated island of Novaya Zemlya would make the ideal site for testing of its nuclear weapons. The North Test Site facilities were completed in 1955 and were used until the last Soviet Union test in 1990, after which the site was decommissioned. The largest atmospheric test was the enormous 50-megaton "Tzar Bomba," exploded in 1961. After the Limited Test Ban Treaty was signed in 1963, atmospheric explosions ceased in favor of underground nuclear testing. The largest underground test on Novaya Zemlya was a 4.2-megaton blast involving four nuclear bombs exploded simultaneously on September 12, 1973. The energy released in this test was equal to that released by a natural earthquake with a magnitude of 7 on the Richter Scale and the explosion triggered a major landslide that dammed two glacial streams, creating a 1⅕-mile (2-km) long lake.

Atlas Mountains

The Atlas Mountains form a chain stretching some 1,200 miles (2,000 km) across Tunisia, Algeria, and Morocco in northwestern Africa—a high divide separating the moist climate of the Atlantic and Mediterranean coasts from the dry Sahara Desert on the southern side of the range.

The highest peak, Jbel Toubkal, with an elevation of 13,665 feet (4,168 m), lies 50 miles (80 km) south of the ancient Moroccan city of Marrakech. The High Atlas of Morocco split into two ranges surrounding the Moroccan and Algerian High Plateau, becoming the Tell Atlas range to the north, and the Saharan Atlas range to the south. Joining again in eastern Algeria and Tunisia, they become the Aures Range and continue east across the Mediterranean Sea to join the Apennines of Sicily and the Italian peninsula.

A HISTORY RECORDED IN STONE

The story of two significant geological events can be read in the rocks of the Atlas Mountains. The first happened about 300 million years ago, during the Carboniferous Period, when the two former continents of Euramerica (containing Europe and North America) and Gondwana (containing Africa) collided with one another to form the supercontinent Pangea. The collision event, known as the Appalachian Orogeny, was so intense that it pushed up a massive north–south trending mountain range to heights exceeding those of the present-day Himalayas. The Anti-Atlas Mountains of Morocco and the Appalachians of North America are the diminutive eroded remains of this formidable collision event.

The second more-recent event occurred in the Tertiary Period when the African continental landmass collided with Eurasia at the southern end of the Iberian Peninsula. As the landmasses approached one another, they crushed the marine sediments lying in the intervening basin, thrusting up the Atlas Mountains. This same collision uplifted the Alps and the Pyrenees. As is typical of collision ranges, the Atlas Mountains are rich in resources, including antimony, copper, gold, iron, manganese, lead, marble, mercury, phosphate, rock salt, silver, and zinc, as well as coal and natural gas.

REGULAR DEADLY EARTHQUAKES

Abundant earthquakes show that the Atlas Mountains are still rising. The larger ones regularly devastate the poorly constructed traditional mud-brick dwellings of the local Berber people. In 1716, 20,000 people were killed in a strong earthquake that completely destroyed the Algerian city of Blida in the Tell Atlas. In 1980, this destruction was repeated in Cheliff when it was flattened by a magnitude 7.7 earthquake, with 4,500 lives lost. The most recent earthquake in 2003 killed 2,250 people in Boumerdes, and has prompted attempts to introduce safer building codes.

THE DEVELOPMENT AND DESERTIFICATION OF THE ATLAS

The fertile Mediterranean slopes of the Atlas Mountains are well watered by moisture-laden air masses forced upward over the ranges. The once-common verdant forests of Atlas cedar (*Cedrus atlanticas*) carpeting the northern slopes have been cut down over the last few thousand years and replaced by irrigated farmland. The drier slopes facing south are covered with shrubs and grasses, and salt lakes

BELOW The ruins of a kasbah at Ait Arbi lie among the eroded carmine volcanic rocks of Dades Valley in Morocco's High Atlas Mountains. The arid landscape yields a surprising number of oases and is dotted with hundreds of kasbahs—ancient fortresses built to protect against marauding invaders.

and salt flats occur on the Sahara Desert's northern edge. Sheep grazing is an important activity here.

The movement of humans into the Atlas area has led to the eradication of many native species adapted to living here, mostly through intensive hunting and habitat loss—some only very recently. The last northern bubal hartebeest was reportedly killed in 1923. The Barbary or Atlas lion, famous as Rome's fierce Colosseum beast, was finally eradicated from its Atlas home in 1922. The region's most significant environmental problem is desertification—the loss of fertile grasslands as the desert advances northward on the Atlas. It is due to a combination of population pressure, over-cultivation, soil erosion, a falling water table, and climate change.

AMMONITES OF THE ATLAS

The Atlas Mountains are popular with fossil collectors, and are particularly renowned for the well-preserved spectacular ammonite fossils (pictured) found here. Ammonites were ancient cephalopods, squid-like sea creatures that became extinct along with the dinosaurs at the end of the Cretaceous Period. Their remains were preserved in the sea-floor sediments that were later thrust upward to become the Atlas Mountain Range.

ABOVE A false-color thematic satellite mosaic shows the Anti-Atlas Mountains, which form part of the Atlas Mountain Range. These contain the highest mountains in Africa—several peaks in the High Atlas rise over 13,000 feet (4,000 m). The region contains vast mineral resources.

Zagros Mountains

The Zagros Mountains stretch for about 1,100 miles (1,500 km) from eastern Turkey, through Iran, along the northern shores of the Persian Gulf, to the Pakistan border. They are part of the massive mountain belt that extends from the European Alps to the Himalayas.

ABOVE Date palms (*Phoenix dactylifera*) are intensively cultivated for their sweet fruit in the southern oases of the Zagros Mountains. Iran produces over half a million tons of dates annually for export.

The mountain ranges of the region surround a vast dry basin in Iran. This central area is known as Dasht-e Kavir, a great sandy desert, an almost alien landscape home to numerous dry salt lakes that are intermittently fed by rivers draining inward from the surrounding ranges. Tributaries draining the western flanks of the Zagros feed the Tigris River, which flows into the Persian Gulf.

The Zagros Mountains are a collision belt made up of a number of parallel ranges, which from space look like a vast rumpled carpet of anticlinal and synclinal folds. The ranges were thrust upward and wrapped around the leading edge of the Arabian continental landmass when it rammed into the southern edge of the Eurasian landmass, about 50 million years ago.

THE ZAGROS RISE FROM THE SEA

Sediments of different types, destined to become the Zagros, were deposited in a shrinking ancient sea that lay between the approaching continents—including silt, sand, gravel, and pebbles. Very thick beds of salt also built up as the ancient sea began to evaporate. When the Tethys Sea finally closed, compressive forces folded and pushed these deep marine sediments upward, with the highest peak, Zardeh Kuh, reaching an elevation of 14,921 feet (4,551 m). Though hard to believe, these mountains were once sediments deposited in a deep ocean. Rock outcrops containing spectacular ammonite fossils, weighing up to 5 pounds (2 kg) each, found at altitudes of 10,000 feet (3,000 m) leave no doubt that this was indeed the case.

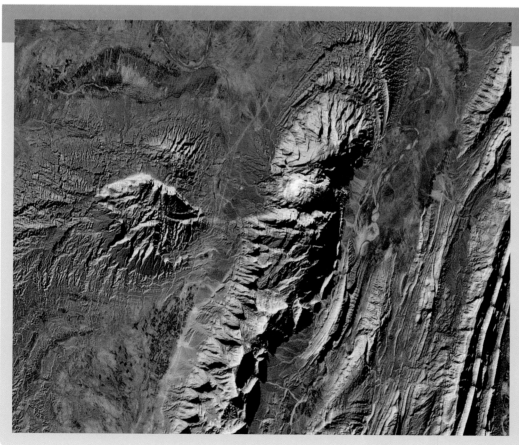

SALT VOLCANOES AND THE OILFIELDS

Huge salt domes in the topography of the Zagros Mountains are readily visible as circular structures. Salt, being less dense than other rocks, will slowly rise in vertical columns called diapirs. When these reach the surface, the salt extrudes and flows, like lava, from a salt volcano.

Salt diapirs "dome up" the surrounding and overlying sedimentary layers, which can then act as traps for huge amounts of oil and gas. Iran's great oilfields are among the most productive in the world.

Zagros salt domes are enticing exploration targets, particularly those with telltale clues, such as natural bitumen seeps and burning naphtha springs, naptha being a kerosene-like chemical.

This image of the Kuh-e-Namak salt dome in the southwestern Zagros Mountains was taken by an astronaut while orbiting Earth.

RIGHT When viewed from space, the rippling folds of the Zagros Mountains reveal the ultimate result of two tectonic plates colliding: pushing up a vast mountain range. The waterways are the Persian Gulf and Gulf of Oman, narrowed by the Strait of Hormuz, which runs between Iran (seen on the left) and Qatar, the United Arab Emirates, and the Oman (visible on the right).

Intense earthquake activity along the boundary between the Arabian Plate and the Eurasian Plate shows that mountain building is still in progress. The destruction of the mud-brick city of Bam, and its historic citadel in 2003 by a magnitude 6.5 earthquake was a clear reminder of the dangers of living in such an tectonically active region. The loss of over 26,000 lives in Iran, when an earthquake of similar magnitude killed only a handful in the US, highlights the urgent need for improved building codes.

THE ZAGROS: THE CRADLE OF HUMAN SETTLEMENT

The Zagros Mountains were one of the first areas where crops were planted and animals domesticated some 10,000 years ago by ancestors of the present-day Kurds, Lurs, Bakhtiaris, and Kashkais. The northern Zagros are rugged, forested, and snowcapped mountains, dotted with volcanic cones and basin lakes such as Lake Urmia. Most people live and farm in the temperate high-level fertile valleys in the northern region. Tribal herders seasonally move sheep, goats, and cattle between high-altitude summer and lower-altitude winter pastures. The southeastern Zagros are mainly bare rock and shifting dunes, with the occasional green desert oasis supporting intensive cultivations of food crops such as dates and cereals.

Millennia of intensive human use, however, have destroyed many of the region's deciduous forests of native oak and juniper, and a number of protected areas have been established to combat the tide of habitat degradation. These serve as refuges for a variety of animal life, including the Persian fallow deer, brown bear, eagle, wolf, and leopard, which have long made the Zagros their home.

LEFT The citadel at Bam before it was leveled by the 2003 earthquake in the cold, dark, early hours of December 26. Very few buildings in this 2,000-year-old walled city survived the intensive shaking as they were largely constructed of mud bricks.

The Himalayas

The Himalayas are Earth's most vigorously growing mountain range. The name Himalaya means "land of snow" and is derived from the Sanskrit words, *hima* for snow, and *alaya* for abode. This near-vertical wall of rock contains all 14 of the world's highest peaks (those over 26,000 feet or over 8,000 m), including the tallest, Mt Everest, which stands at 29,028 feet (8,848 m).

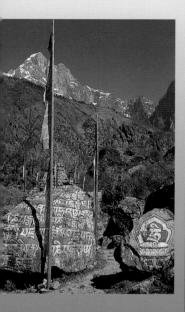

ABOVE Prayer stones line the roads between villages in the Nepalese Himalayas. They are usually inscribed with a Buddhist mantra that is recited by devotees as they pass by, touching the stone; or repeated during meditation.

By comparison, the highest peak in the Andes, Mt Aconcagua, is only 22,835 feet (6,960 m) tall. The Himalayas stretch across Bhutan, China, India, Nepal, and Pakistan and form part of a continuous mountain range running right across central Asia and Europe. These fold mountains include the Hindu Kush, Makran Range, Zagros Mountains, and the Alps, all of which are being pushed skyward by the powerful collision of enormous continental landmasses.

A UNIQUE GEOLOGIC SETTING

The Himalayan story begins about 130 million years ago, during the early Cretaceous Period, when India tore away from its Gondwanan connection with Australia and Antarctica. It began its long journey northward, traveling at the geologically phenomenal speed of 3½ inches (9 cm) per year. As the gap between the Indian and Asian continents diminished, the ocean in between, known as the Tethys Sea, became progressively shallower and warmer. The Tethys Sea floor would have been consumed in a subduction trench along the edge of the Eurasian continent, with the molten magma from its destruction forming a line of active volcanoes—in much the same way as is happening today along the coast of Japan or Chile.

About 40 to 50 million years ago, when the continental shelves of India and Asia touched, the situation changed dramatically with neither continent able to slide beneath the other. Instead, the two continental edges crumpled and folded like two cars meeting head on. Sedimentary rocks, such as sandstone, shale, and limestone borne below sea level, were tightly folded and metamorphosed. They became pushed miles into the air, and today trekkers in Nepal can pick up marine shells and coral high in the mountains they are climbing.

ENDLESS LINE OF PARALLEL RANGES

Tall ranges of folded rocks and deep mountain valleys now stretch as far as the eye can see, parallel to the join line between the colliding continents. Collectively, the Himalayas are comprised of three well-recognized geological units. These are, from south to north: the Siwalik or sub-Himalayan foothills that abut the plains of India; the Lesser Himalaya, consisting mainly of folded and overthrust sedimentary rocks of the former Tethys Sea; and the Great Himalaya, composed of a granite and gneiss crystalline core with some sedimentary remnants on the summits. The Great Himalaya is the tallest of the three ranges, with elevations rarely dropping below 18,000 feet (5,500 m).

ROCK FEATURE: GNEISS

Gneiss is a coarse-grained, high-grade metamorphic rock made up of irregular light-colored (quartz- and feldspar-rich) bands and dark-colored (biotite-mica rich) bands. It is formed from quartz-rich sedimentary or igneous rocks that were subjected to extremely high temperatures and pressures. Gneiss is form-

ing in present-day collision mountain belts such as the Himalayas and is also an indicator of ancient mountain belts that have long since been eroded away. When polished, gneiss is a very attractive decorative stone for tiles, benches, and tabletops.

SAVAGE EARTHQUAKES

The collision, folding, and uplift of the Himalayan ranges continues today as attested by the powerful earthquakes that regularly rock the region. One of these devastated the region of Kashmir in 2005, claiming over 90,000 lives, and destroying some 570,000 poorly constructed homes. Such tremendous forces of uplift, however, are equally matched by the ferocity of the forces of erosion in the area, which are simultaneously working to flatten the mountains to sea level.

POWERFUL RIVERS, GIGANTIC DELTAS

Snowmelt and rainfall on the Himalayan ranges feed some of the world's most powerful rivers. The Ganges River drains the southern wall of the Himalayas, its sediment-laden water flowing westward into the Bay of Bengal. Here many tons of deposited gravels, sands, and silts are continuously creating the giant Ganges delta, on which the villages and cities of Bangladesh have been built. The Indus and the Brahmaputra rivers capture water and sediment from the northern slopes of the Himalayas. The Indus River carries its load to the northwest and then to the south, onto the Indus delta, while the Brahmaputra River loops east, south, and then west to join the Ganges River.

The sheer weight of these massive deltas on the sea floor of the Bay of Bengal and the Arabian Sea is deforming them into enormous basin-like depressions. Due to their annual top-dressing of sediment, these deltas provide some of the richest and most fertile agricultural land in the world. Millions of people toil to support their families on the deltas, under the ever-present threat of devastating annual floods.

ABOVE A vista of the Himalayas as seen on the pass to Mt Everest. The rocks that form this mountain range were originally laid down on the floor of an ancient sea, the Tethys.

BELOW About 130 million years ago, after Pangea broke apart, India began moving north, closing the ancient Tethys Sea. It rammed into Asia about 40 to 50 million years ago, the collision setting off a rapid uplift that pushed up the Himalayas.

ABOVE Xegar, also known as Shekar Dzong or Tingri Shekar, is a small village at 14,891 feet (4,542 m) altitude in the Chinese Himalayas. It has been used as the departure point for numerous expeditions to Mt Everest.

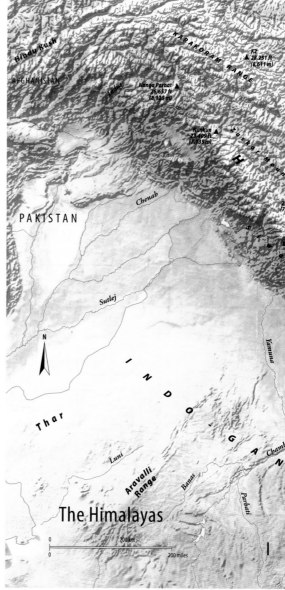

A FRAGILE WATER SOURCE

Hundreds of millions of people live in the enormous basins surrounding the Himalayas. They are highly dependent on the rivers that rise in the mountains as a vital water source for agriculture and industry. Over thousands of years, abundant snowfall on the mountains has built up impressive storehouses of glacial ice, such as the Siachen Glacier, one of the world's largest. Presently, seasonal monsoonal rainfall, combined with snowmelt from the mountains, ensure that the region's rivers flow all year round. However, shrinking and retreat of glacial ice in the Himalayas, due to the effects of global warming, is concerning many scientists. It will result in declining permanent water supplies for those living in the surrounding regions in decades to come.

OPPOSITE PAGE, FAR RIGHT The endangered snow leopard (*Uncia uncia*) has a patchy distribution in alpine areas of the Himalayas and high plateaus and deserts of Central Asia. It has been hunted for its fur and bones, which have been used in Chinese medicine.

LIFE MOVES HIGHER

Historically, the Himalayas formed an impenetrable barrier to the people of the region, resulting in entirely different language and cultural groups developing in the surrounding lowland regions. The great ranges remained under a cover of snow and glaciers until only about 10,000 years ago, when the Ice Age at that time began to lose its grip. The rising snowline allowed flora and fauna to move up to increasingly higher elevations.

Humans first reached the Himalayan pasturelands in about 1500 BCE. Struggle for survival at altitude was intense, with animals and humans both being naturally selected for powerful lungs and blood with a higher-than-normal capacity to carry oxygen. Animals, such as the snow leopard,

developed large forepaws and superb balance for moving rapidly in rugged and often snowy terrain, as well as long hair with dense woolly underfur providing warmth. Life was hard and highly attuned to the seasons, with the semi-nomadic herders moving their livestock between the winter lowland pastures and the summer uplands that lie at elevations as high as 10,000 to 13,000 feet (3,000 to 4,000 m). Permanent agriculture is only possible below about 5,500 feet (1,700 m), where picturesque tiny flat fields are terraced into the valley hillsides. These are irrigated for rice paddies, or planted with maize, millet, potatoes, and barley.

Shifting cultivation or *jhum* is still practiced in parts of the Eastern Himalaya, where vegetation is more luxuriant. It is the easiest form of land use

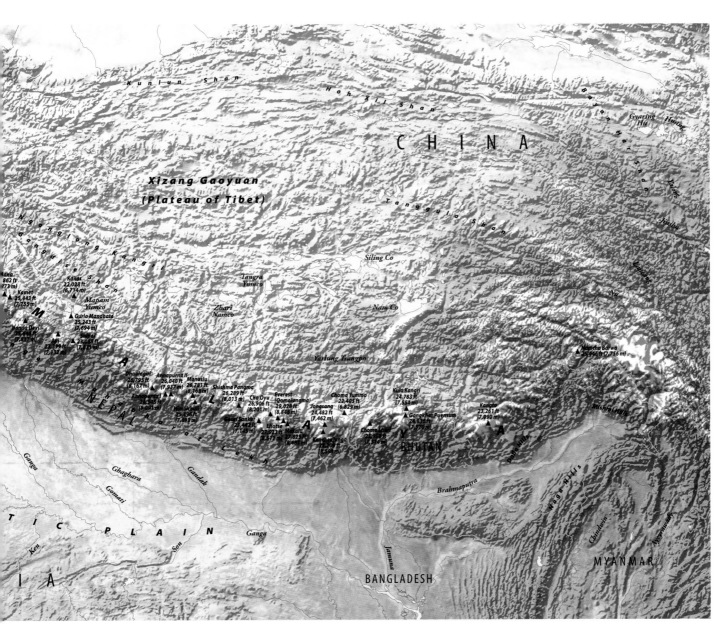

Kunlun Shan

Hoh Xil Shan

CHINA

Gyaring Hu

Bayan Har Shan

Xizang Gaoyuan (Plateau of Tibet)

Tanggula Shan

Gangdise Shan

Kailas
22,028 ft
(6,714 m)

Mapam Yumco

Gurla Mandhata
25,243 ft
(7,694 m)

Zhari Namco

Tangra Yumco

Siling Co

Nam Co

Nz

Nainchu Barwa
25,446 ft (7,756 m)

Kamet
25,443 ft
(7,755 m)

Nanda Devi
25,637 ft

Api
23,399 ft
(2,132 m)

Saipal
23,068 ft
(7,031 m)

M

A

NEPAL

Yarlung Tsangpo

Dhaulagiri
26,795 ft
(8,167 m)

Annapurna II
26,040 ft
(7,937 m)

Manaslu
26,781 ft
(8,163 m)

Shishma Pangma
26,289 ft
(8,013 m)

Cho Oyu
26,906 ft
(8,201 m)

Everest
(Qomolangma)
29,026 ft
(8,848 m)

Choma Yummo
22,405 ft
(6,829 m)

Kula Kangri
24,783 ft
(7,554 m)

Kangto
23,261 ft
(7,090 m)

Mishra

Dhaia hills

Khangchul
26,503 ft
(8,001 m)

Nimalchuli
25,896 ft
(7,893 m)

Gauri Sankar
23,442 ft
(7,145 m)

Jongsang
24,482 ft
(7,462 m)

Lhotse
27,923 ft
(8,571 m)

Makalu
27,825 ft
(8,463 m)

Gangkhar Puensum
24,836 ft
(7,570 m)

Kanchenjunga
(8,598 m)

Choma Lhari
23,996 ft

BHUTAN

Naga Hills

Ganga

Ghaghara

Gomati

Gandak

Son

Ganga

Brahmaputra

Tamura

Jamuna

Chindwin

Ayeyarwadi

TIC PLAIN

IA

Ken

BANGLADESH

MYANMAR

---

because suitable land is simply cleared of vegetation and burned prior to seed planting. No terracing or plowing is needed, but once the land's natural fertility is gone, after about two to three years, the patch is abandoned to rest and another area is taken up and cleared.

LOSS OF A UNIQUE ECOSYSTEM

A rapidly growing population and its quest for survival is causing severe environmental problems. Excessive land clearing is leading to a loss of soil fertility and increased soil erosion. Severe flooding of the valley floors and deltas, and landslides on the steeper slopes, are all becoming tragically commonplace. Much of the region's unique flora and fauna is now threatened or endangered.

Traditional practices, supported by illegal trade, may push many species over the edge into extinction: tigers and snow leopards are highly sought after, not only for their beautiful fur, but also for their bones, which are used in Asian traditional medicines. Male musk deer are hunted for the musk gland in their abdomen. The musk, believed to have aphrodisiac properties, is highly valued for perfumes and Chinese medicines. The Himalayan black bear is taken, largely in order to extract bile from its gall bladder, one of the most coveted of Oriental medicines. The list of animals rapidly in decline is a long one and includes the one-horned rhinoceros, swamp deer, pygmy hog, wild elephant, hispid hare, wild yak, red panda, and the gharial crocodile.

MT EVEREST

Mt Everest, the world's highest peak, stands in the Himalayan Mountain Range at 29,028 feet (8,848 m) above sea level. Edmund Hillary and Tenzing Norgay conquered its rarified summit in 1953, and though followed by many others, it is certainly not a climb for the fainthearted.

Out of some 4,000 climbers who have made an attempt to reach the summit, only 660 have made it, and 142 people have died trying. When asked the reason for all of this fervent climbing activity, mountaineer George Mallory famously answered, "because it is there." He is one of the 120 people whose bodies still lie high up on the mountain; his is just below the summit. Whether or not he made the peak may never be known.

THE EVEREST ENIGMA

The top of Everest is made of limestone formed in an ancient seabed, so perhaps one of the most curious questions we must ask is: how did it get up there? Furthermore, why is Everest so high? And why is the mountain still there, considering the ferocity of the agents of erosion, ice, wind, and water working constantly to reduce it to sea level?

The answer is that this massive mountain is located in one of Earth's most tectonically active regions, at the focal point of folding and upthrusting in a massive collision zone between two continental landmasses, the Indian Plate and the Eurasian Plate. The collision began about 50 million years ago, and the Himalayas themselves started rising about 25 to 30 million years ago. Today, India is still ramming northward and Everest's summit is continuing to rise by about $\frac{1}{5}$ to $\frac{2}{5}$ inches (0.5 to 1 cm) per year. So it is now a footstep higher than when Hillary and Norgay famously first reached its summit.

British surveyor Sir George Everest declared Peak 15 to be the highest in the world in 1852, and it was later named for him. Global positioning instruments that are now located on the peak continuously monitor its changing height.

Mt Everest is shaped like a three-sided pyramid. Three flat faces form the sides and join along steep ridges. Both the North Face and the East Face rise above China, while the Southwest Face rises above Nepal. The southeastern and northeastern ridges are the main climbing routes. The upper slopes

and summit are so high in Earth's atmosphere that the amount of breathable oxygen there is only a third of what it is at sea level. Apart from fungi and lichen, the development of plant or animal life is prevented by the lack of oxygen, powerful high-level winds, and the extremely low temperatures.

Everest is made up of three distinct rock units. The bottom layer is highly folded and metamorphosed schist and gneiss, although most of this part of the mountain is permanently covered with snow and ice. The middle layer is metamorphosed shale and sandstone intruded by granite, characterized by rugged dark cliffs and pinnacles. The top section of the mountain is gray limestone with a distinctly visible dipping yellow limestone bed, known as the yellow band, exposed in Everest's sheer pyramidal faces.

THE DEATH ZONE

An attempt on the summit of Mt Everest requires entry into the death zone. This is the region above about 26,000 feet (8,000 m) where no human can stay for prolonged periods, even with extensive training and acclimatization. Many climbers carry oxygen into this zone, but even so, the body's functions begin to slow and shut down. The mind becomes hazy, extremities go numb, and digestion of food to produce energy ceases. With prolonged exposure to that altitude, climbers risk fluid on the lungs and brain (edema), and ultimately death.

Babu Chiri Sherpa is one of Everest's legendary figures, having reached the summit without oxygen 10 times, and holds two records—longest time of 21½ hours on the summit, sleeping overnight in a specially made tent, and for the fastest southern ascent, taking just 16 hours and 56 minutes. Unfortunately he too became one of Everest's victims when he fell into a crevice while leading an expedition on April 29, 2001.

ABOVE Pack animals such as donkeys and yaks are used to transport climbers' equipment and supplies up the mountains to the Everest base camp from Namche village in Khumbu Province, Nepal. Trekkers often spend a day or two here to acclimatize to the high altitude.

ABOVE Expeditioners heading toward Mt Everest receive their first glimpse of the mountain along the trail not far past Namche village. The summit separates Nepal from China.

LEFT The trail heading toward Mt Everest hugs the sides of the valley, winding its way through the villages of Thyangboche, Deboche, Dingboche, Pangboche, Dugla, and Lobuje to the base camp at Gorak Shep. During the spring months, the valley blazes with the brightly colored blooms of rhododendron shrubs.

Scotia Arc and Antarctic Peninsula

At the windswept southernmost tip of South America, a most unusual tectonic-topographic feature can be seen. Here the south-trending Andes range sweeps sharply to the east, creating the characteristic "tail" of Tierra del Fuego. As the glacially carved mountains descend into the sea, they become hundreds of fjord-dissected islands, with the easternmost point being Cabo San Juan.

BELOW Precariously tilted layers of rock strata in the mountains of South Georgia, one of many subantarctic islands in the Scotia Arc, attest to the vigorous geological forces that shifted the rocks in these islands.

BELOW Prince Gustav Channel cuts through the Antarctic Peninsula in the Weddell Sea. This waterway is near the northernmost tip of the peninsula, which forms the southern boundary of the Scotia Arc.

A prominent underwater ridge continues east through the island of South Georgia, then wraps around to the south in an elegant arc of active volcanoes known as the South Sandwich Islands. Heading west, the ridge forms the Orkney Islands and the South Shetland Islands before joining the northern tip of the Antarctic Peninsula.

AN UNDERWATER MOUNTAIN RANGE

The Scotia Plate, a small D-shaped oceanic tectonic plate that sits between the southern end of the Andes Mountains and the northern tip of the Antarctic Peninsula, is responsible for the chain of island stepping-stones. This largely submarine mountain range, joining the two continents, and marking the Scotia Plate's eastern perimeter, is known as the Scotia Arc. The tectonic setting is similar to that of the Lesser Antilles islands forming along the eastern edge of the Caribbean Plate.

THE SOUTH SANDWICH ISLANDS

The Scotia Plate, complete with its own small segment of spreading ridge, has pushed eastward into the Atlantic Ocean, dragging with it the tips of South America and the Antarctic Peninsula. It is overriding the Atlantic oceanic crust (the South American and Antarctic plates), which is being forced on a downward journey into the South Sandwich Trench. Rising magma—from the destruction of the Atlantic—is constantly building the South Sandwich Islands. This chain of 11 principal volcanic islands bends in an arc 250 miles (400 km) long and is less than 5 million years old. The islands, first visited by seal hunters in 1818, are uninhabited due to extreme weather and are mostly covered by glaciers.

THE ANTARCTIC PENINSULA

The majority of Antarctica lies within the Antarctic Circle, or the 66.5° line of latitude south, where extended darkness is experienced for the whole or part of winter. The Antarctic Peninsula extends outside this circle and its tip is Antarctica's furthest point from the South Pole. As the weather here is the mildest on the continent, a few flowering plants are found in addition to Antarctica's sparse vegetation of mosses, lichens, and algae. The peninsula is a long glaciated mountain range running through Graham Land and Palmer Land, paralleled on the west by an arc of volcanic islands (South Shetland Islands, Palmer

RIGHT Grytviken on South Georgia is home to Antarctica's first whaling factory, which was operational from 1904 until 1960. Over 175,000 whales were caught off South Georgia during this time. The rusted remains of buildings lie where whales were once gutted, largely for their blubber.

Archipelago, Alexander Island, and Biscoe Islands). On the eastern side of the peninsula, glaciers flow from the mountains into the extensive Ronne and Larsen ice shelves. The Larsen Ice Shelf experienced significant breakup during 2002 and studies carried out by the British Antarctic Survey and the US Geological Survey show the peninsula's glaciers are also in rapid retreat due to rises in air and sea temperatures. Many research stations and outposts have been established on the northern part of the Antarctic Peninsula and outlying islands, making it one of the busiest places on the continent. The population on the peninsula varies between 3,000 at the height of the summer season (not counting tourists), and less than 1,000 during the dark winters.

THE SOUTH GEORGIA GROUP

South Georgia (also known as San Pedro) is a windy forbidding island surrounded by a number of smaller islets. Steep cliffs plunge into the sea and glacially carved fjords form numerous large bays and inlets, providing a natural habitat for wildlife and excellent deep-water anchorages. Over half of the mountainous island is covered with permanent ice, such as the Fortuna Glacier, leaving the remainder sparsely vegetated by grass, moss, and lichen. Geologically, South Georgia is distinct from the South Sandwich Islands. Being made up of metamorphic gneiss and schist, it is more akin to the rocks found in the Andes Mountains.

South Georgia is a vital wildlife breeding ground, with massive colonies of southern fur seals, southern elephant seals, king penguins, albatrosses, and other sea birds, which cover the slopes with their nests.

ABOVE A southern elephant seal (*Mirounga leonina*) mother and her pup have come ashore on South Georgia. These seals breed on subantarctic islands.

ABOVE The giant water lily *Victoria amazonica* is native to the Amazon River. Several species of water lily are native to wetland areas, but some are becoming endangered because their habitat has been drained.

RIGHT The Ganges River begins on the southern slopes of the Himalayas. It meets the Gandak River at the Indian towns of Patna and Sonpur (pictured), whose communities thrive on the fertile alluvial floodplains.

An Earth without rivers would have no valleys, no waterfalls, no floodplains, and very little, if any, life on land. Gravity causes water to flow from higher to lower elevations, first in the higher altitudes as small trickles that merge to form streams and then, at the lower altitudes, form rivers. The drainage pattern that results resembles the branches of a tree joining at the trunk.

But how does water manage to reach the higher elevations? It can't flow uphill against gravity. Water falls on land as rain, or perhaps as snow, from clouds of condensed water evaporated by solar energy from the oceans. Although about 75 percent of Earth's surface is covered by water, 98 percent of it is the salt water of the oceans. During evaporation, the salt is left behind so that the vapor is the pure water that condenses to give us fresh rainwater, the purest form of natural water.

Water falling as rain or snow in mountainous regions flows to lower levels as surface water. However, not all the water flows over the surface; some percolates through the soil and rock to form a layer of groundwater. At times, gravity causes this to seep back out into the river system as springs through porous and fractured rocks. When groundwater is confined at lower levels by surrounding rock, it forms an aquifer.

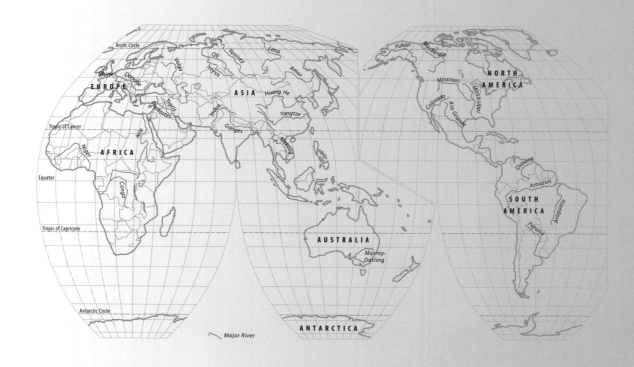

Missouri and Mississippi Rivers, North America

The Missouri and Mississippi rivers share the largest watershed in North America—the world's third largest watershed after those of the Amazon in South America and the Congo in Africa. The combined length of these two North American rivers—over 4,000 miles (6,400 km)—makes them the world's fourth longest river system.

BELOW The Mississippi River begins broadening into the low-lying backwaters and marshes of the Upper Mississippi River National Wildlife Refuge at Black River Delta. This park was created as a refuge for local waterbirds to breed, and is a prime habitat for many migratory birds.

The lengths of the Missouri and Mississippi rivers are remarkably similar, and incredibly difficult to determine exactly because of the engineering of channels to make the rivers more accessible to navigation and because of the natural processes, such as flooding, that change channel courses. However, it can be argued that the Missouri when measured at 2,341 miles (3,767 km) is longer than the Mississippi at 2,320 miles (3,705 km).

Mississippi is derived from the old Native American Ojibwe word, *misi-ziibi,* meaning "great river," while *Missouri* is a Missouri Indian name meaning "town of large canoes."

THE VAST MISSISSIPPI WATERSHED

The Mississippi Basin or watershed drains a funnel-shaped area between the highlands of the Rocky Mountains to the west and the Appalachian Mountains to the east; an area well over 1,200,000 square miles (3 million km²). This is over 40 percent of the area of continental USA, and takes in 31 states and two Canadian provinces. The source waters of the Mississippi come from a number of small streams that flow into Lake Itasca—a lake of glacial origin—some 1,475 feet (450 m) above sea level in Clearwater County, northwestern Minnesota. It drops to 725 feet (220 m) at St Anthony Falls in Minneapolis— the only waterfall along its length.

From its source in Montana, the Missouri River flows into the Mississippi River near St Louis in Missouri. The lower river is characterized by extensive meandering channels that twist and turn through the broad alluvial plain, which varies in width from 25 to 125 miles (40 to 200 km). Many loops of the meanders have been cut to form oxbow lakes and these, together with

TAMING A RIVER SYSTEM

Although there have been many tremendous natural changes to the course and nature of the Missouri–Mississippi River system in the recent geologic past, the most significant changes have been those made by humans in those relatively few years since early settlement. Extensive engineering projects have kept the rivers open for commercial navigation and kept floods under control.

Human influence has impacted greatly on the river system, and it is said that only 1 percent of the length of the Missouri remains uncontrolled by human activity. Currently, there are efforts being made to reclaim some of the altered habitats, such as wetlands, and return them to their former natural state.

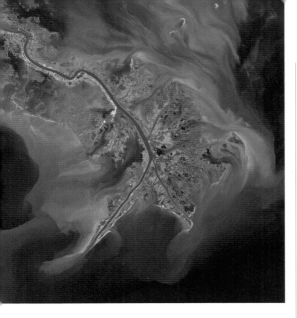

extensive marshes, indicate remnants of the river's former channels. The Mississippi River then continues its journey south to Louisiana on the Gulf of Mexico, where it forms one of the largest bird's-foot deltas in the world.

FROM PRECAMBRIAN RIFTING TO ICE AGE

Geologically, the Mississippi Basin is a vast syncline that funnels down toward the Gulf of Mexico. It was formed through rifting during extensive Precambrian Era tectonic activity some 750 million years ago and is the underlying structure of the Mississippi embayment, which filled when sea levels rose during Cretaceous Period tectonic activity (around 100 million years ago). During this period, a subsequent fall in sea levels caused the ocean to retreat, leaving the embayment as the valley of the Mississippi. Large amounts of sediment were deposited by the river as it flowed toward the Gulf of Mexico.

The many rapid sea level changes during the Pleistocene Epoch glaciations—between 1.6 million and 10,000 years ago—together with abundant meltwater from the glaciers—resulted in the Mississippi forming deep valleys and channels when the sea levels were at their lowest and river terraces at their highest.

From the time of the initial rifting, the syncline continued to subside and fill with massive amounts of sediments to a present basin depth of several miles. These are covered with a thick sequence of Quaternary Period deposits that blanket the many ancient channels and other valley structures.

Investigations show there have been many changes in river channel numbers and courses since the time of its inception. Some changes in channel direction have been caused by earthquakes along the New Madrid Fault Zone, which is associated with the original rifting. Present-day seismic

activity along these faults has caused some of the largest earthquakes ever recorded in North America, such as those occurring in the early 1800s.

On the lower Mississippi Delta, channel switching causes changes to course direction in cycles that occur roughly every 1,000 years. As the sediments are moved further toward the sea, the delta advances into the Gulf of Mexico, moving the coastline seaward from 15 to 50 miles (24 to 80 km) over the past 5,000 years.

A DIVERSITY OF RIVER LIFE

The Mississippi River has a vast range of habitats and an extensive biodiversity. Extraordinary numbers of fish, amphibians, reptiles, and invertebrates are found in the many unique habitats associated with the wetlands, open water, and floodplains. Some species are relicts of the glacial environments of the Pleistocene Epoch.

The biota of the river ecosystem is greatly threatened by farming activities, including land clearing, intense cultivation, and pollution through nutrients and pesticides. Wastewater and sediment from urban development pose further threats. These are apart from the engineering of the major hydrological modifications that have destroyed habitats from the time of white settlement.

LEFT The Mississippi River Delta drains into the Gulf of Mexico. This satellite image shows the vast quantities of sediment—topsoil, sand, and vegetation—that were deposited along the shores of Louisiana and into the gulf following a big storm.

BELOW A group of midland painted turtles (*Chrysemys picta marginata*) bask on a log with an American alligator (*Alligator mississippiensis*) in the lower reaches of the Mississippi River.

Missouri and Mississippi Rivers, North America

Niagara Falls, North America

Three separate waterfalls—the Horseshoe Falls in Canada, the American Falls, and the smaller Bridal Veil Falls in the USA—collectively make up Niagara Falls, formed by the Niagara River plummeting over a 167-foot (51 m) precipice.

RIGHT The vast volume of water in Horseshoe Falls erodes the underlying rock, causing the falls to continually retreat. The rate of erosion has been measured since 1842. Between that year and 1905, these falls eroded at a rate of 3¾ feet (1.1 m) per year. Following the diversion of river waters after 1905, the rate of erosion slowed to 2⅓ feet (0.7 m) per year.

BELOW American Falls close up appear powerful and constant. Yet flow of water varies depending on the timing and demands of hydroelectric power generation. During the night and also outside the peak tourist seasons, the falls are slowed by roughly half in order to generate power.

North America's Niagara Falls is very young when it comes to the geological time scale. It was only around 12,500 years ago—after the end of the last Ice Age—that meltwater from the vast ice fields of the Great Lakes Basin began to flow toward the Atlantic Ocean along what became the Niagara River. The Niagara River and Falls form the boundary between the United States and Canada, and the name "Niagara" is derived from a Native American word for "at the neck."

Horseshoe Falls are 2,200 feet (670 m) wide, while American Falls total 1,075 feet (328 m) in width. The Niagara River carries water northward from Lake Erie down a drop of about 300 feet (100 m) to Lake Ontario. Rapids and the falls account for most of the drop in elevation. The 634,000 gallons (2,400,000 L) per second that pass over Canada's Horseshoe Falls, make it one of the largest falls by volume on Earth. The force of the water causes erosion that is responsible for pushing the rock face back upstream by as much as 7 feet (2 m) per year.

GEOLOGY OF THE GREAT LAKES REGION

Precambrian Era rocks dating to about 3 billion years ago form the very ancient foundation to the Great Lakes region of North America. From that time, the geological history includes periods of intense volcanic activity, mountain building, and the formation of sedimentary rocks beneath seas that repeatedly inundated and retreated over the hundreds of millions of years of history up to the Quaternary Period with its Ice Age. This started with the advance of the first glacier more than 1 million years ago and, little by little, the glaciers built up into ice sheets as much as 1¼ to 1¾ miles (2 to 3 km) thick. Pressure from the weight of the ice depressed the underlying land. These massive sheets of ice behaved like giant bulldozers as they moved south, scouring the land, carving valleys, leveling hills, and depositing vast amounts of glacial sediments.

The basins that became the Great Lakes were formed when river systems that existed before the Ice Age were scoured into deep large valleys that filled with water from melting ice when the climate began to warm. As the glaciers retreated and pressure on the underlying land was released, the land rebounded isostatically, with the resulting uplift dramatically changing the early drainage patterns. The most recent drainage is through the Niagara River and over the Niagara Escarpment—the cliff face formed by the differential erosion of layers of sedimentary rocks of varying hardness. Over a period of about 5,500 years, the escarpment has retreated several miles southward at a rate of from 2 to 10 feet (0.6 to 3 m) per year. Recent engineering intervention has slowed

erosion, but it is possible that erosion could cause the escarpment to retreat right back to Lake Erie, which could eventually empty because its bottom is higher than the bottom of the Niagara River.

NIAGARA'S FRAGILE ECOSYSTEMS

The ecosystems of Niagara Falls and its river have been severely disturbed by human practices that have changed the natural drainage patterns, causing the loss of wildlife habitats and destruction of wetlands. The river has been affected by pollution from homes, agriculture, and industries to the point where wildlife population and biodiversity have declined. The natural vegetation around the falls once included a cedar swamp.

In 1990, UNESCO declared Ontario's Niagara Escarpment a World Biosphere Reserve. The reserve stretches 450 miles (725 km) from Lake Ontario near Niagara Falls to the tip of the Bruce Peninsula. The biosphere reserve includes the provincial Niagara Escarpment Plan Area and two national parks with two major forest types, wetland complexes, cliff faces, slopes, and aquatic systems.

OUTLOOK FOR THE GREAT LAKES

Today the four upper lakes of the Great Lakes Basin contain about 20 percent of the world's fresh water, which is in effect "fossil water" remaining from the ice melt all those thousands of years ago. There is comparatively little water entering the system today—rain and other run-off account for less than 1 percent input annually. There has been some recent evidence to suggest the levels of the lakes are falling and, as climate change sets in, they are likely to continue to fall.

BELOW Chinook salmon are among the 94 fish species that regularly tumble down the falls at Niagara. Nearly all fish survive the plunge as they are adapted to the fast-moving, high-pressure water. A tourist was once hit by a salmon while strolling along the boardwalk at the falls.

Amazon River, South America

There are a number of ways to determine just what it takes to be the greatest river on Earth. Length is usually chosen as the ultimate measure—but it is not always that simple. When it comes to the Amazon River, at 3,968 miles (6,385 km), it is just short of the Nile, which has a length of 4,160 miles (6,693 km). However, there are other factors that put the Amazon way ahead of any other river.

BELOW This satellite image shows a small part of the Amazon River, which carries 20 percent of all river water discharged into Earth's oceans. Surrounding the river is the Amazonian rainforest, one of the most ecologically diverse regions on the planet, much of which has been deforested.

The Amazon is the widest of all rivers, releasing up to 32 million gallons (121,000,000 L) of water per second into the Atlantic Ocean in the rainy season—a greater volume of flow than the next six largest rivers combined. This volume of freshwater is about 16 percent of Earth's total freshwater runoff. Not surprisingly, the Amazon Basin is also the largest of all drainage areas, or watersheds, covering some 40 percent of the total area of South America.

ROCK FEATURE: SHALE

Shale, sometimes called mudstone, is a soft gray- to black sedimentary rock. It is made up of fine clay particles that were transported in water and deposited under extremely quiet conditions such as lagoons, floodplains, or marine environments far from the coast. Shale is finely layered as the flat clay flakes align themselves horizontally as they settle. The rock is easily split along those layers often revealing a treasure trove of fossils. Shale is mined for clay, which is used for brick and ceramic manufacture.

AN EXTRAORDINARILY VAST RIVER

The Amazon Basin encompasses extensive plains, bounded by a series of mountain ranges—the Andes to the west, the Brazilian Highlands to the south and east, and the Guiana Highlands to the north. As the largest on the planet, the watershed incorporates Brazil, Bolivia, Peru, Ecuador, Colombia, and Venezuela, draining an area of over 2,700,000 square miles (7,000,000 km²).

Much of the watershed is fed by the heavy rains that occur when moisture-laden clouds—driven from the east by trade winds—are blocked by the Andes Mountains. Because this is the longest mountain range in the world, huge volumes of water feed the many tributaries over a vast distance—from just above the equator down to 20°S. The most remote tributaries collect water from the east Andean slopes, not far from Peru's western coast where the range rises steeply from the Pacific Ocean. The thousands of tributaries—of which 17 are more than 1,000 miles (1,600 km) long—come together to flow east through Peru and

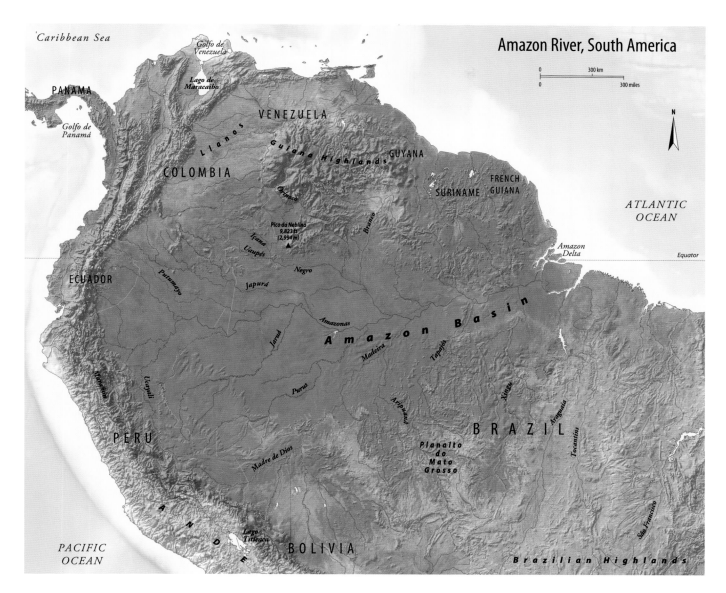

Amazon River, South America

Caribbean Sea

PANAMA

Golfo de
Venezuela

Lago de
Maracaibo

Golfo de
Panamá

VENEZUELA

Llanos

Guiana Highlands

GUYANA

COLOMBIA

Orinoco

SURINAME

FRENCH
GUIANA

ATLANTIC
OCEAN

ECUADOR

Pico da Neblina
9,823 ft
(2,994 m)

Içana

Uaupés

Branco

Negro

Japurá

Amazon
Delta

Equator

Putumayo

Amazonas

Amazon Basin

Juruá

Madeira

Tapajós

PERU

Marañon

Ucayali

Purus

Arinã

Xingu

Araguaia

Tocantins

BRAZIL

Madre de Dios

Planalto
do
Mato
Grosso

ANDE

Lago
Titicaca

BOLIVIA

Brazilian Highlands

Sta Francisco

PACIFIC
OCEAN

Brazil, forming the approximately 4,000-mile (6,400 km) long trunk of the Amazon River, which is about 7 miles (11 km) at its widest. It enters the Atlantic Ocean at the equator where the mouth is around 200 miles (300 km) wide. However, these dimensions change dramatically during the wet season, when flooding triples the area covered by water, and channel widths reach 25 miles (40 km) in some regions.

Because of its immensity, the Amazon is sometimes called The River Sea and its waters remain drinkable well out into the ocean. It was because of this that the first European found the Amazon in 1500. Vicente Yáñez Pinzón, in command of a Spanish expedition, observed that although he was out of sight of land, he was sailing in freshwater some 200 miles (300 km) out to sea. This observation led him to trace its source to the mouth of the Amazon River.

As most saltwater organisms cannot survive for long in freshwater, ships sometimes anchor in the Amazon's outflow, taking advantage of the fresh to brackish water to remove marine life attached to their hulls.

THE MAKING OF THE AMAZON

The Amazon River has not always followed the drainage pattern it has today. From the time of the Paleozoic Era more than 300 million years ago, many tributaries flowed into a huge marine inlet that opened to the Pacific Ocean. The westward flow was reversed when the uplift of the Andes Mountains reached a height of almost 23,000 feet (7,000 m) by the time of the Miocene Epoch, some 20 million years ago. Geologically, the uplift was the result of the tectonic processes that caused the subduction of the

ABOVE Roughly half of Ecuador is home to rivers that give rise to the Amazon. The Ecuadorian Amazon is known as El Oriente and the country is often referred to as the headwaters of the Amazon. These rivers start high in the eastern Andes and feed into the Amazon Basin.

Nazca Plate under the South American Plate along the western margin of South America. Following this initial uplift, the Amazon drained toward the Caribbean, but with continuing rise throughout the Miocene Epoch, the river changed its course and flowed toward the Atlantic along the valley formed between the gently dipping Precambrian Era shields—the Guianan to the north and the Brazilian to the south.

THE ENDANGERED AMAZON

Deforestation and overfishing have placed the Amazon River region under ever-increasing threat, and many species of animals and plants have been lost as their habitat is destroyed. Commercial and agricultural interests, along with increasing human settlement along the river—and the associated infrastructure this brings with it—have impacted on the once-pristine rainforest region and its fragile ecological balance.

Economic factors have seen developments in land use that are changing the landscape of the region, as logging activities, along with the introduction of cattle farming, make their presence felt. Overfishing of the waters, coupled with mercury pollution generated by gold mining in the region, has seen its fish populations dwindle. The drive for clean renewable energy has seen dams built, with more on the planning agenda—and the impact of such developments is felt across the basin as water flow is affected and long-established migratory routes are cut. Species and environments that rely on the status quo will come under further threat as future changes take effect.

An aerial view of the lower Amazon River floodplain in Brazil, where cattle are farmed.

Over the most recent of all geological periods, the Quaternary, which began 2 million years ago, there were a number of climate changes that caused water levels within the Amazon Basin to rise and fall in response to sea level changes. Enormous lakes formed during sea-level rises and into these were deposited the vast amounts of sediments that characterize the modern Amazon Basin. With the lowering of sea level, the lakes were drained by rivers that cut down into the sediments to form the system of the widely separated, broad, flat river valleys of today.

Because of the low elevation of most of the Amazon Basin—about 40 percent is between zero and 300 feet (0 and 100 m) and another 29 percent at less than 600 feet (200 m)—the river channels meander across the valley flats, transporting and depositing sediments mostly derived from erosion of the Andes. These sediments cause turbidity that makes the water appear milky so that the tributaries are called white water or *varzea*. There are also the black (*igapó*) and clear water rivers that have their catchments within the Amazonian forests. The Rio Negro, the largest of the black-water tributaries, derives its color (more like the color of strong tea) from dissolved organic matter or humus, while the clear-water tributaries have more dissolved minerals than humic acids. Where the black water meets the white water of the main trunk, it takes time before the waters mix, so that a distinct margin between the dark and the light waters is evident for some distance downstream.

During cycles of annual flooding, vast layers of sediments are deposited over the floodplains of the basin. This, however, does not account for all the sediment transported by the Amazon. Some billions of tons per year of finer sediment are discharged into the Atlantic, forming deposits on the continental shelf. Unlike other rivers, these sediments do not form a delta because of the rate of flow and also because of extreme tidal currents. Variations in the current allow the deposition of some sediment to form a belt of small, alluvial, mangrove-covered islands that in turn are eroded and carried out to sea when the rate of flow increases.

THE AMAZON'S ABUNDANT WILDERNESS

With the Amazon having such an enormous watershed involving so many different habitats, it follows that it supports the most abundant wilderness on Earth with the greatest diversity of living organisms and the most extensive of all rainforests. The basin has climatic zones ranging from the vast tropical equatorial region of the floodplains to the dramatic variability of the Andes, where climate changes with altitude and rainfall. There, temperatures range from those of the permanent snow-covered peaks, through those of the temperate cloud forests, to the steamy tropical heat of the lower slopes of the mountains and the seasonal savannas along the lowlands.

Although there is a range of habitats, often with their own unique biota, it is the tropical rainforests that are most associated with the Amazon. There are more species in these rainforests and aquatic environments than in any others of the world. Many have yet to be discovered, let alone identified.

A remarkable feature of these rainforest ecoregions is the flooded forests. These occur during each wet season, when flooding causes the river to rise over 30 feet (10 m) and invade extensive areas of forest, forming the world's largest habitat of this type. Only the canopies of the trees can be seen above the floodwaters, and for months the trunks and lower branches are totally submerged. Leaves and fruits fall from the trees to provide food for the many fish, reptiles, and other aquatic animals that migrate annually into the flooded forests. The trees are specially adapted to overcome oxygen shortage through submergence and they quickly regenerate their leaves when the waters recede.

UNIQUE TO THE AMAZON

The Amazon Basin is a unique environment that supports a vast range of species—in fact, it is home to some 30 percent of the world's species of flora and fauna. The entire Amazon Basin contains the largest diversity of freshwater fish in the world—estimated at more than 3,000 species. White-blotched river stingrays, which grow to about 16 inches (40 cm) across, are one of a number of freshwater stingrays found in the waterway; piranhas swim in shallow water, and many species of catfish populate the river. Among the more unusual species found in the Amazon's waters are several species of river dolphin, one of the world's few surviving species of lungfish, the manatee, and the black caiman.

Above the water, the region boasts poison arrow frogs, scarlet macaws, and uakaris. Palms, orchids, and bromeliads grow beneath the canopy of taller trees—such as brazil nut, kapok, and cocoa trees.

An Amazon poison arrow frog (top), a pink river dolphin (*Inia geoffrensis*; center), and a piranha (bottom).

Iguazú Falls, South America

The Swiss botanist Robert Chodat aptly described Iguazú: "The waters of the deluge falling abruptly into the heart of the world, by divine command, in a landscape of memorable beauty, amidst an exuberant, almost tropical vegetation, the fronds of great ferns, the shafts of bamboos, the graceful trunks of palm trees, and a thousand species of trees, their crowns bending over the gulf adorned with mosses, pink begonias, golden orchids, brilliant bromeliads and lianas with trumpet flowers; all this added to the dizzying and deafening roar of waters that can be heard even at a great distance, makes an indelible impression, moving beyond words."

ABOVE A toco toucan (*Ramphastos toco*) perches on a limb in Iguazú National Park, Argentina. The largest of the 37 species of toucan, it is found throughout eastern South America and lives in the rainforest canopy.

RIGHT While most of the falls lie within Argentina, the panoramic Garganta del Diablo Falls, or Devil's Throat, straddles the border between Argentina and Brazil and is best viewed from the Brazilian side. The falls measure 500 feet (150 m) wide and run for 2,300 feet (700 m).

This truly remarkable series of waterfalls form where the borders of the three South American countries, Argentina, Paraguay, and Brazil, connect. Higher than Niagara Falls and wider than Victoria Falls, the Iguazú Falls are often said to make up the largest waterfall in the world.

The source of the Iguazú River is the Serra do Mar near the Atlantic Ocean in Brazil. The river flows west for more than 800 miles (1,300 km) before cascading spectacularly over drops of up to 270 feet (80 m) high and 8,800 feet (2,700 m) wide. The flow rate, which varies seasonally, causes the water to split into as many as 275 falls and rapids, the most impressive being the Garganta del Diablo, or Devil's Throat. Here, the water flows over a 500-foot (153 m) wide U-shaped cliff—it is this part of the river that forms the border between Argentina and Brazil.

HISTORY REVEALED IN LAYERS OF LAVA

The falls are formed by the Iguazú River as it cascades over a series of massive flat-lying to undulating basalt flows. These are a section of

extensive layers of flood basalts that cover parts of southern Brazil, Paraguay, Argentina, and Uruguay. The Paraná basalts formed when the supercontinents Gondwana and Laurasia broke up, causing basalt lavas to surface through tectonic faults and cracks without forming volcanic cones. The basalt formed deep layers, and the falls comprise 19 major drops over these layers. Below the falls, the river flows through a

narrow 260-foot (80 m) deep canyon formed
by erosion, which in turn has caused the falls to
recede 17 miles (27 km) upstream to their present
location. It is possible that some 20,000 years ago
the falls were located where the Iguazú River
joins the Paraná River.

DIVERSE INHABITANTS, LUSH LANDSCAPE

Surrounding the Iguazú Falls are Argentina's
Iguazú National Park, with an area of 136,000
acres (55,000 ha), and the Iguaçu National Park
in Brazil, with an area of roughly 450,000 acres
(182,000 ha). They were designated World
Heritage Sites in 1984 and 1986 respectively.

In a region where much of the land has
been deforested to make way for agriculture,
these parks form a haven of wilderness. Humidity
caused by the clouds of spray from the falls produces lush vegetation—subtropical wet forest
rich in lianas and epiphytes, including bromeliads
and more than 60 species of orchids. In total, there
are some 2,000 plant species living in the parks,
comprising many large trees more than 100 feet

(30 m) in height, intermediate-size trees, shrubs,
and a variety of herbaceous ground flora.

Within this lush environment, a diverse range of
wildlife—some endangered, some under threat—
can be found. Vibrant birds, such as the spectacular
toucan, take to the skies overhead, along with the
great dusky swifts that nest behind the falls, while
butterflies flutter among the vegetation. Below,
giant otters, giant anteaters, howler monkeys,
tapirs, jaguars, ocelots, spectacled caiman, and a
variety of snakes can be found in the area.

ABOVE The massive overall drop
in the height of the Iguazú River
after it hits the falls can be most
clearly seen from above. The falls
drop dramatically into a narrow
chasm with an average depth of
230 feet (70 m).

LEFT Caimans in the Pantanal
wetlands of Brazil wait to catch
fish as they come downstream.
Part of the alligator family, caimans live exclusively in Central
and South America. The black caiman (*Melanosuchus niger*) of the
Amazon is the largest and can
reach up to 20 feet (6 m) long.

Danube River, Europe

Europe's Danube River could easily be called the "Romantic River." Writers, musicians, and artists from the Danube River Basin are icons of our classical and romantic heritage. The Danube might not be the longest river in the world, but its basin—which encompasses 19 countries and 81 million people—does claim the title of the world's most international river basin.

ABOVE By night, the lights on Budapest's parliamentary buildings and the Chain Bridge are reflected in the calm waters of the Danube, which flows through Hungary and forms part of its border with Slovakia.

At 1,770 miles (2,848 km), it is second to the Volga—the longest river in Europe at 2,290 miles (3,688 km). The watershed of the Danube River Basin has a drainage area of about 310,000 square miles (800,000 km²) from 19 countries—Albania, Austria, Bosnia and Herzegovina, Bulgaria, Croatia, Czech Republic, Germany, Hungary, Italy, Macedonia, Moldova, Montenegro, Poland, Romania, Serbia, Slovak Republic, Slovenia, Switzerland, and Ukraine. The river itself flows through, or forms the borders of, 10 countries. The river flows east from its westernmost source in the Black Forest region of Germany through to the Black Sea via its extensive delta shared by Romania and Ukraine.

THE MAKING OF THE DANUBE

The Danube's original river system, the Urdonau, was much larger than today's river. Before the Pleistocene Epoch Ice Age, the Urdonau carried the bulk of the waters from the Alps, and the younger Rhine started in the southwestern Black Forest and flowed to the east. There are exposures of the ancient Urdonau channels in the waterless canyons of the Swabian Alb, a mostly porous Jurassic Period limestone responsible for the most extensive karst region in Germany. Karst is formed when water percolates through soluble limestone, forming underground channels and caverns, sinkholes, and numerous other geological features.

Today's Danube headwaters are much smaller, with significantly more of its original source now going to the Rhine. Erosion of the Upper Rhine Valley changed the direction of the flow of water from the Alps, causing it to feed the Rhine instead. With the water level of the Rhine being lower than that of the Danube, the subsurface stream channels—known as the Danube Sink—carry a significant volume of water from the Danube to the Rhine. The enormous volume of subterranean

A LONG HISTORY OF SETTLEMENT

The Danube River has been home to centuries of human settlements—from the Stone Age Neolithic times with sites containing some of the earliest human culture from about 6000 BCE, through the Roman occupation of the third century CE, to the 81 million people now living within the Danube River Basin. The river is of immense economic importance. It provides drinking water, supports agriculture, industry, and fishing, and, being navigable for much of its length, it is essential to commerce.

Sadly, as human population has increased, so too have the problems of water quality and reduced biodiversity through pollution and other human disturbances.

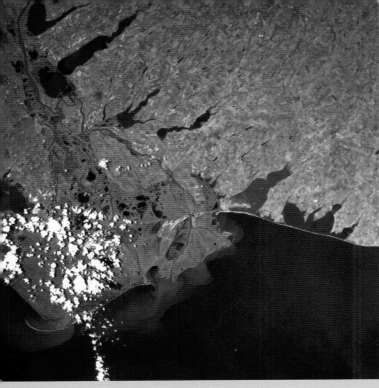

(40 m) per year as 67 million tons (74 million tonnes) of alluvial sediments are deposited from the river. Geologists have shown that the delta is composed of a number of lobes, starting with the earliest that would have formed about 6,000 years ago.

THE DANUBE'S DIVERSITY OF HABITATS AND LIFE

The climate and topography of the Danube River Basin is very diverse. Some 30 percent is mountainous, with the remainder made up of hills and plains, including the delta. The climate varies with an average precipitation ranging from 20 inches (510 mm) per year on the plains to 80 inches (2,000 mm) per year in the higher regions. Forest ecosystems dominate the basin, and a number of reafforestation projects are seeing an increase in forest cover with ranges of 20 percent for Hungary, and from 20 to 50 percent and more in Slovenia, and Bosnia and Herzegovina. Sadly, as the forests increase they are affected by pollution, insect attacks, and extreme weather events.

The karst itself has a number of unique ecosystems containing many endemic species that are under threat because of the extreme sensitivity of limestone to pollution.

Wetland areas—and particularly those of the delta, which at times are seasonally submerged—contain habitats for a diversity of biota and a home for rare and threatened species.

LEFT The delta of the Danube River is a large low-lying wetland region in northeastern Romania. A biosphere reservation, this unique ecological area borders the northwestern corner of the Black Sea, and covers roughly 2,200 square miles (5,700 km²).

BELOW A fox swims through the floodwaters of the Danube River at Gemenc Forest of Danube-Drava National Park in April 2006. Central Europe battled swollen rivers that drove thousands from their homes and caused a dozen deaths.

BELOW The Danube River flows through Vienna, the capital city of Austria. Over 2 million people live within its metropolitan area and the heart of the old city is contained by the Danube Canal.

water continually erodes the limestone to the point where it could be possible for the upper course of the Danube to flow entirely into the Rhine, a process known as "stream capturing."

The Rhine Valley now separates the Black Forest from the Vosges mountain range in France, with the Rhine flowing north to the North Sea. The divide in southern Germany is sometimes known as the European Watershed, however with water flowing through the sinks in seasonally varying volumes and distances that cross the divide, its line becomes a little fuzzy.

Of great geological significance is the delta of the Danube River. It is a typical triangular shape with sides about 45 miles (70 km) long that covers an area of about 2,300 square miles (6,000 km²). It is growing outward at the rate of about 130 feet

Nile River, Africa

The Nile competes with the Amazon River of South America for the title of "longest river on Earth." With a length of 4,160 miles (6,693 km) from its headwaters in Burundi to its mouth at the Mediterranean Sea, and the Amazon at 3,968 miles (6,385 km), the Nile is almost 200 miles (300 km) longer. However, some will argue that the Amazon is up to 4,195 miles (6,712 km) long.

OPPOSITE, TOP A waterfall on the Blue Nile near Lake Tana, Gondar, in Ethiopia. Lake Tana is perched high on the Ethiopian Plateau, and once the Blue Nile's waters leave, they rapidly drop from around 6,000 feet (1,800 m), flowing through a steep gorge. Consequently, there are myriad rapids and falls along its length.

There are two great tributaries of the Nile—the White Nile and the Blue Nile. The Blue Nile is the major source of the Nile's water, while the White Nile is the longest. Another tributary is the Atbara River, which is usually dry, flowing only when there is rain in Ethiopia.

The actual drainage basin of the Nile—over 1,200,000 square miles (3,250,000 km²)—is about 10 percent of the area of Africa. In the northern section, the river flows almost entirely through desert from Sudan to Egypt and empties into the Mediterranean Sea through one of the world's largest deltas. The Nile Delta covers around 150 miles (250 km) of coastline, spreading from Alexandria in the west to Port Said in the east. From north to south, the delta is approximately 100 miles (160 km) in length.

RIVERS WHITE AND BLUE

Although the source of the Nile is considered to be Lake Victoria in equatorial East Africa, the lake itself is fed by a number of large rivers, with the most distant rising in the Nyungwe Forest in Rwanda. This river is joined by the Rukarara, Mwogo, Nyabarongo, and Kagera rivers, before flowing into Lake Victoria in Tanzania, near the town of Bukoba. After the Nile leaves Lake Victoria at the Ripon Falls in Uganda, it is known as the Victoria Nile, which then flows for a further 300 miles (500 km) through Lake Kyoga to Lake Albert. Here it becomes the Albert Nile, which then flows into Sudan, becoming the Bahr el Jebel then joining other rivers at Lake No to become the Bahr al Abyad, or the White Nile, because it is colored by clay particles suspended in its waters.

From there, the river flows to Khartoum, Sudan's capital.

Compared with its cloudy counterpart, the waters of the Blue Nile appear clearer—hence the river's name. Originating in the Ethiopian Highlands at Lake Tana, the course of the Blue Nile flows in a northwesterly direction for about 850 miles (1,400 km) before joining the White Nile at Khartoum, from which point on, this famous waterway is known as the Nile River.

The contribution to total Nile water volume is shared between the two tributaries. For much of the year, the rivers draining Ethiopia flow weakly,

LEFT An aerial view shows the two rivers, the White Nile and the Blue Nile, where they converge near Khartoum, the capital of Sudan. An Arabic poet once tenderly described the confluence of these two rivers as "the longest kiss in history."

and some dry up completely, but during summer, when rains drench the Ethiopian Plateau, the Blue Nile is responsible for almost the entire volume of water and sediments to the Nile. During the dry season—from January to June—the White Nile becomes the major contributor, discharging between 70 to 90 percent of the total flow.

The Atbara River originates north of Lake Tana in Ethiopia and travels around 500 miles (800 km) before it joins the Nile approximately 185 miles (300 km) north of Khartoum. This confluence marks a rough midway point along the course of the Nile, and from this point on, the Nile diminishes in flow due to evaporation. The Atbara River contributes to the Nile River system during the rainy season only, as throughout much of the year, its flow is significantly diminished.

THE CHANGING DELTA

An aerial view of the Nile Delta reveals an arc-shaped (arcuate) form that resembles a lotus flower—in Ancient Egyptian times this was the emblem of Lower Egypt. However, the famed waterway of the Nile and Nile Delta has been influenced by both natural and artificial developments. Erosion along the seaboard coastline of the delta has seen salinity levels rise in the region, as the salt waters of the Mediterranean Sea permeate the area. The construction of the Aswan Dam brought to an end the flow of nutrient-rich waters and sediment, leaving the once-rich floodplains in need of fertilizers to maintain their fertility.

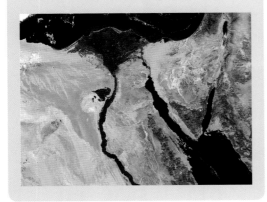

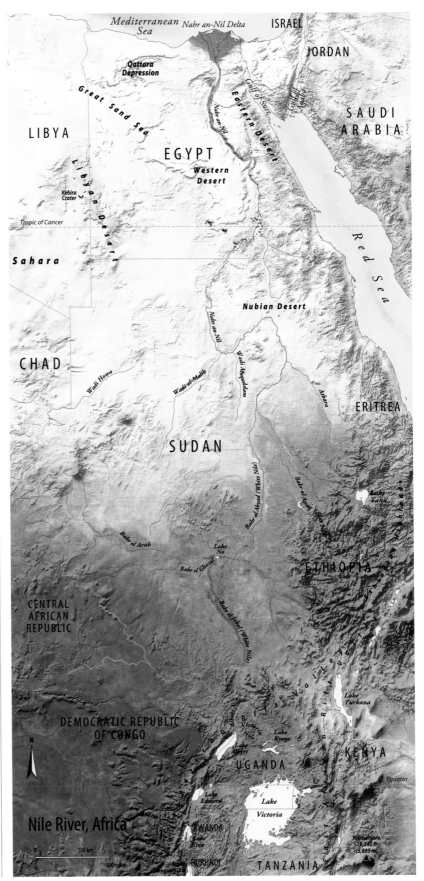

Nile River, Africa

The oldest of these "paleorivers," named Eonile, was identified by satellite imagery that showed its dry watercourses in the desert to the west of the present Nile. Sediments were carried to the Mediterranean, which was a closed basin during the late Miocene Epoch. Its level was approximately 5,000 feet (1,500 m) below that of world sea level and so the Eonile cut down to base level forming a huge canyon as it carried its sediments into the Mediterranean Sea. The canyon has since been filled with sediments, some as deep as 4,600 feet (1,400 m). Gas fields have been found in some of these sediments.

Following the Eonile, three other river systems have been identified up to the more recent Pleistocene Epoch, when as little as 130,000 years ago, the Nile system of the time captured the Blue Nile, forming the Nile River as we know it today.

The area is still highly active tectonically, with seismic and volcanic activity continuing along the East African Rift Valley in Ethiopia. This is the region of Africa where hominid species are believed to have evolved, eventually, into the modern human being, *Homo sapiens*—the only hominid species still in existence.

FIVE NILE RIVERS

The Nile River system has been greatly influenced by late Tertiary Period tectonic activity. This resulted in the fracturing and faulting of the continent, uplift, and changes in relative sea level, forming the Red Sea, Gulf of Aden, and East African Rift Valley. Uplift formed highlands along the Rift Valley and at least five different Nile river systems have flowed north from the western flank since the late Miocene Period (around 10 million years ago) to the present.

THE RIVER THROUGH TIME

The Nile River valley is steeped in human history from ancient times. The Nile has been the lifeline of Egyptian culture from the Stone Age through to modern times. Most Egyptian cities lie along

A THRIVING WATERWAY

Many of the world's great and ancient cities occur along the Nile including Khartoum in Sudan and in Egypt, Aswan, the ancient city of Swan; Luxor, the site of Thebes; and Cairo, formerly Giza. The first of six cataracts is at Aswan to the north of the Aswan Dam and is the closest to the mouth of the river, and the sixth is at Sabaloka, north of Khartoum. Below the cataracts, cruise ships and traditional wooden sailing boats, known as feluccas, are part of the regular tourist route to and from the north. Sadly, security concerns have shut down the northernmost portion for many years.

the valley and nearly all of Ancient Egypt's cultural and historical sites are found along the riverbanks.

Possibly as long ago as 8,000 years, the pastoral lands of Egypt became very arid, forming the Sahara Desert. This may have been the result of climate change, overgrazing, or both, so that those working the land were forced to migrate toward the river, where they developed an agricultural economy in a centralized society. The regular flooding of the Nile River made the valley highly fertile. The cultivation of wheat and other crops provided food for the general population. Water buffalo and camels were used for food, plowing, and other tasks, as well as transportation. The Nile itself also provided a convenient and efficient way to transport people and goods.

THE NATURAL HISTORY OF THE NILE

Being so fertile a river valley, the Nile has been home to an abundance of fauna and flora. Papyrus sedge, *Cyperus papyrus,* grows in the Upper Nile less abundantly now than in the past. It was the papyrus plant that the early Egyptians processed for writing material as a written history emerged, and is the source of the English word "paper."

The delta is the wintering site for hundreds of thousands of migratory birds such as the world's largest concentrations of little gulls and whiskered terns. Gray herons, Kentish plovers, shovelers, and cormorants also nest in the delta. Other birds common to the Nile are the falcon, kite, goose, crane, egret, heron, plover, pigeon, ibis, vulture, and owl.

Many animals living around the Nile were considered sacred by the Egyptians. Sacred birds included the falcon (or hawk), which was thought to be the guardian of the ruler and depicted with its wings spread protectively behind the head of the pharaoh, and the ibis, which appeared as the head of the god, Thoth. The ibis has become far less abundant since the nineteenth century.

Fish found in the river include carp, perch, and catfish, while striped mullet and sole populate the delta. Various fish were also considered sacred in Ancient Egypt. Other animals found in the delta include frogs, turtles, tortoises, mongooses, and the Nile monitor. Although Nile crocodiles are no longer found in the delta, they occur in other parts of the Nile River and are not endangered.

ABOVE Two nile crocodiles (*Crocodylus niloticus*) chew on a bone. One of three species of crocodile in Africa, they range from the Nile River as far south as South Africa, and from the east coast to the west coast.

BELOW The Nile meanders past the temples of ancient Egypt at Luxor. These were mortuary temples, built by successive kings as preparation for eternal life.

Congo River, Africa

The Congo or Zaire River of central Africa draws its water from a rainforest-clad watershed—a region made famous by British explorers, Stanley and Livingstone, and which author Joseph Conrad called the "heart of darkness." Livingstone was famously found after "vanishing" while exploring the Congo.

ABOVE An adult male okapi (*Okapia johnstoni*), a rare relative of the giraffe, is found only in rainforests of the Congo. Some of the better-known species of the Congo region include the lowland gorilla, the chimpanzee, the bonobo, the Congo peacock, and the forest elephant. The river itself is home to many species of fish and crocodiles.

BELOW Two capital cities on opposite banks of the Congo River are viewed from the International Space Station. Brazzaville is on the northern side of the river, while Kinshasa is on the south. They lie at the end of a widening in the river known as Pool Malebo.

With a length of around 2,700 miles (4,300 km), the Congo is the second longest river in Africa after the Nile and the ninth longest river in the world. The width of the river can reach up to 10 miles (16 km) in places.

When it comes to volume, the Congo, with a flow of 1,500,000 cubic feet (43,000 m^3) per second, is rivaled only by the Amazon River of South America for world honors. The relatively stable flow and high volume of water is a product of the constant rains that feed the watershed. The river's course crosses the equator twice, so that it collects rain from the tropics of both hemispheres during their respective rainy seasons.

THE MASSIVE CONGO BASIN

The watershed, or Congo Basin, is the second largest in the world, draining an area of 1.5 million square miles (3,900,000 km^2). This is an area that embraces the countries of the Democratic Republic of Congo (DRC), Republic of the Congo, Angola, northern Zambia, western Tanzania, and southern Central African Republic. The source of the Congo is considered to be streams that enter from the savannas into the southern end of Lake Tanganyika, from which the only river to leave is the Lukuga River that flows westward to join the Lualaba River.

Of the many tributaries, some come together in the Lualaba and Luvua rivers, which form a wider and more turbulent river that enters a 75-mile (120-km) long canyon of impassable rapids, known as the "Gates of Hell."

THE MAKING OF THE CONGO BASIN

Geologically, Africa has been the center of extensive tectonic activity for millions of years, particularly since the breakup of Gondwana during the early Mesozoic Era (about 220 million years ago). Rifting associated with the opening of the Atlantic Ocean in the Jurassic Period is one of central Africa's most impressive characteristics. Earth's crust continues to be pulled apart by convection currents in the mantle, deep below the surface. Consequently, continental Africa features long and deep rift valleys bounded by faults and fractures in the crust.

Lake Tanganyika, one of the deepest and longest freshwater lakes in the world, is a feature of the Western Rift of the Great Rift Valley and once formed part of the course of the Congo River. The lake is bounded on either side by fault blocks, which rise to 6,500 feet (2,000 m) above the lake's surface on the western side. Its only discharge now is to the west through the Lukuga River, which eventually empties into the Congo.

The basin's structure is a result of geological processes associated with rifting. Large blocks of underlying Precambrian Era rocks were pushed downward to form a basin shape that filled with huge amounts of sediments throughout the Jurassic and Triassic periods. These are now exposed over much of the basin and are the source for the deposits carried by the river westward to the coast a million or so years ago. Tectonic forces during the Tertiary Period altered a classic delta into present day structures that have been found to contain significant economic deposits of oil and gas.

LIFE IN THE CONGO BASIN

The basin is covered by a blanket of dense rainforests, particularly in the valleys. This cover makes it the second largest tropical rainforest region in the world. The southern highlands are cooler and drier, so that there is a considerable diversity of habitats and ecoregions within the basin.

OPPOSITE The Odzala Forest lies deep in the heart of the Congo River Basin in the Democratic Republic of Congo. The rainforest is occasionally interrupted by a patchwork of small clearings, or *bais*.

Victoria Falls, Africa

Mosi-oa-Tunya—or "the smoke that thunders"—is the indigenous name for Victoria Falls, which form part of the border between southern Zambia and northwestern Zimbabwe in southern Africa. In 1855, David Livingstone—the first European to sight the falls—named them in honor of Queen Victoria.

BELOW Tourists in Zimbabwe viewing Victoria Falls from the top of the gorge are continually showered by mist from the spray that rises up from the vast volume of water that passes over the falls every minute.

Victoria Falls is among the most spectacular falls in the world, and is the world's greatest sheet of falling water. The Zambezi River is roughly 1 mile (1.6 km) wide at the point where it plunges from a height of 300 feet (90 m) deafeningly down a series of basalt gorges, raising a shimmering mist that can be seen more than 20 miles (30 km) away.

Rainbows add to the splendor when they materialize in the mist. The falls are made all the more spectacular by the channelling of over 300,000 cubic feet (9,000 m³) per second of water into the narrow 100-foot (30 m) wide Batoka Gorge. It is possible to view the falls from the other side of the narrow gorge directly opposite.

NATIONAL PARKS AND WILDLIFE

The falls are part of two national parks on either side of the Zambezi River—Mosi-oa-Tunya National Park in Zambia and Victoria Falls National Park in Zimbabwe. Both are relatively small, covering areas of 25 square miles (65 km²) and 9 square miles (23 km²) respectively and both were placed on the World Heritage List in 1989.

A riverine "rainforest" lies within the waterfall splash zone. Growing on sandy alluvium, this discontinuous forest relies on the plume of spray arising from the falls to supply moisture and create the high humidity necessary to maintain this fragile ecosystem. Tree species within this forest are many and include acacias, ebony, and the ivory palm and date palms.

In the parks around the falls are sizable populations of elephants, buffalo, and giraffes and a general abundance of other animals, including a large population of hippos. There are some 400 species of birds in the Victoria Falls region, and the gorges below the falls are a breeding site for the Taita falcon (*Falco fasciinucha*), Verreaux's eagle (*Aquila verreauxi*), and peregrine falcon (*Falco peregrinus*).

CULTURAL HERITAGE

In the area near Victoria Falls, various stone artifacts have been found. These have been identified as belonging to the hominid species, *Homo habilis*, from 3 million years ago. Stone tools indicate that there was prolonged occupation of the area in the Middle Stone Age 50,000 years ago, and weapons, adornments, and digging tools dated between 10,000 and 2,000 years ago point toward the presence of Late Stone Age hunter-gatherers. Farming communities more recently displaced these early inhabitants.

Victoria Falls is the physical boundary that distinguishes the regions of the upper and middle Zambezi River—with each river region populated by distinct fish species.

FRACTURES AND FISSURES

The Zambezi River flows for a considerable distance over a level sheet of basalt through a valley bounded by low and distant sandstone hills. Numerous tree-clad islands dot the course of the river, increasing in number in the approach to Victoria Falls, which marks a rough midway-point along the river's course.

Uplifting of a region known as the Makgadik-gadi Pan some two million years ago caused the Zambezi River to cut through the basalt plateau along fractures that created east–west trending fissures. Over some many thousands of years, the falls have been receding upstream and relics of seven past waterfalls now form a series of sharply zigzagging gorges downstream from the falls. The Devil's Cataract in Zimbabwe is presently the start of a part that is now being cut back to form an

eighth waterfall that will eventually leave the present crest high above the river in the canyon below.

At the end of the first gorge, the river has hollowed out a deep pool—called the Boiling Pot—that measures about 500 feet (150 m) across. The surface water of the Boiling Pot is smooth when the flow is low, but, as the flow over the falls increases, slow enormous eddies form in the churning water. From the Boiling Pot, the river channel turns sharply westward and enters the next of the zigzagging gorges that have walls around 400 feet (120 m) high.

ABOVE The flow of the Zambezi River into the falls can vary considerably depending on the climatic conditions at the time. During the middle of 2006, a severe drought caused the river to dry considerably and reduced its flow so significantly that the river braided into channels.

ABOVE LEFT A bloat of hippopotamuses (*Hippopotamus amphibius*) gathers in the Zambezi River. They have adapted to the harsh African sun by secreting a reddish orange pigment through their skin that has sunscreening properties. Curious evidence from 47-million-year-old whale fossils supports a theory that whales are related to hippopotamuses.

Ganges River, Asia

The Ganges, or Ganga (feminine of Ganges) in most Indian languages, is the sacred river of India. It has been revered as a goddess throughout Indian history. Bathing in the river helps Hindus attain salvation and the forgiveness of sin. They believe that to make life complete, it is essential to bathe in its waters at least once.

With a length of about 1,700 miles (2,700 km), the Ganges is by no means one of the longest rivers in the world, but its basin claims the highest human population density on Earth—more than 500 million at an average of 155 people per square mile (400 people per km²). The total population of this area (392,328 square miles or 1,016,130 km²) represents about one in every 12 people on Earth (8.5 percent of world population).

GATHERING THE WATERS OF THE GANGES

The source of the Ganges is the massive Gangotri Glacier at 14,000 feet (4,300 m) in the foothills of the Himalayas—the range that is the northern boundary of the Ganga Basin. Ice-melt from the glacier feeds the Bhagirathi River, which is joined by the Alaknanda River to form the Ganges at the picturesque town of Devprayag, buried deep in a craggy canyon within the Himalayas.

The Ganga Basin—part of a larger complex, the Ganga–Brahmaputra–Meghna Basin—draws its water from Bangladesh, China, India, and Nepal. To the west, the basin is bounded by the Aravalli Range. The Vindhyas and the Chotanagpur Plateau form the southern boundary, while the Brahmaputra Ridge marks the eastern boundary.

Snow-melt from September to March causes the swollen Ganges to thunder through the valleys of the Himalayas on to the Gangetic Plains, where enthusiasts enjoy white-water rafting. The river is joined by numerous large tributaries bringing further water to the rushing Ganges as it flows past some of India's most populous cities, where the population depends on its waters for sustenance and economic viability.

The river forms a number of distributaries in the immense delta; the first is the very large Hooghly River, on which lies the huge city of Kolkata (formerly Calcutta) and its population of 14 million people. The main channel of the Ganges continues into Bangladesh, where it becomes the Padma River and this, in turn, becomes the Meghna River after its confluence with the very large Brahmaputra River. The Meghna River forms the main channel of the Ganges Delta, cutting through the region on its way to the Bay of Bengal.

A YOUTHFUL TECTONIC HISTORY

The tectonic history of India explains just why the Ganges and Brahmaputra are such mighty rivers. The uplift of the Himalayas is a relatively "young" event geologically. When Pangea broke up about

ABOVE The area around the delta is known as the Sundarbans, meaning "beautiful forests," named for the predominant trees that form the largest mangrove ecoregion in the world. It is home to the endangered Bengal tiger (*Panthera tigris tigris*; above), Indian python (*Python molurus molurus*), Asian elephant (*Elephas maximus*), clouded leopard (*Neofelis nebulosa*), and crocodiles.

200 million years ago, the Indian tectonic plate began its rapid move northward to collide with the Eurasian tectonic plate about 50 million years ago. The impact of their collision eventually caused both plates to crumple, forcing them upward into the highly contorted and jagged peaks of the Himalayas. The uplift continues today and accounts for the disastrous earthquakes at times experienced in this part of the world.

The steepness of the mountains, together with extreme precipitation, causes extensive erosion that results in the heavy sediment load carried by the Ganges and its tributaries. The sediments are deposited as alluvium in the downwarped "basin" of the Indian Plate, forming the Gangetic Plains.

The Ganges Delta is an arcuate type of delta that covers an area of well over 40,000 square miles (100,000 km²), making it the largest in the world. Its alluvial plains are cut by a maze of interwoven channels, swamps, and lakes. Sediments carried beyond the delta form a fan in the Bay of Bengal, which is the largest submarine accumulation of sediment on Earth. Important gas reserves have been discovered in the delta and exploration is underway in the fan. Success could play a major role in the future of the region, and hopefully help the people out of their poverty.

ECOSYSTEMS AND LIFE OF THE GANGES

With the highest mountain range in the world forming one of its boundaries, and the Ganges emptying through the largest delta, it is no wonder that the basin has very diversified ecosystems supporting some remarkable biota. The dense forests of the mountainous regions, sparsely forested hill districts and the fertile Gangetic Plains are all home to a variety of fauna and flora. The Ganges River itself is home to the extremely rare true freshwater river dolphin, the Ganges dolphin (*Platanista gangetica*), while the Irrawaddy dolphin (*Orcaella brevirostris*)—which tolerates freshwater— enters the delta from the Bay of Bengal.

ABOVE LEFT From the heights of the Himalayas, numerous rivers flow into the Brahmaputra River Valley in Assam, India. The river turns southward at the border of Bangladesh and is joined by the Ganges, flowing in from the left. It then splits into channels as it runs down to the Bay of Bengal.

BELOW Long considered a sacred river, the Ganges is frequently the site of religious festivals. During Allahabad, a huge crowd of Hindu pilgrims gathers by the river for the Kumbh Mela, or Pitcher Festival, which attracts more people than any other religious gathering on Earth.

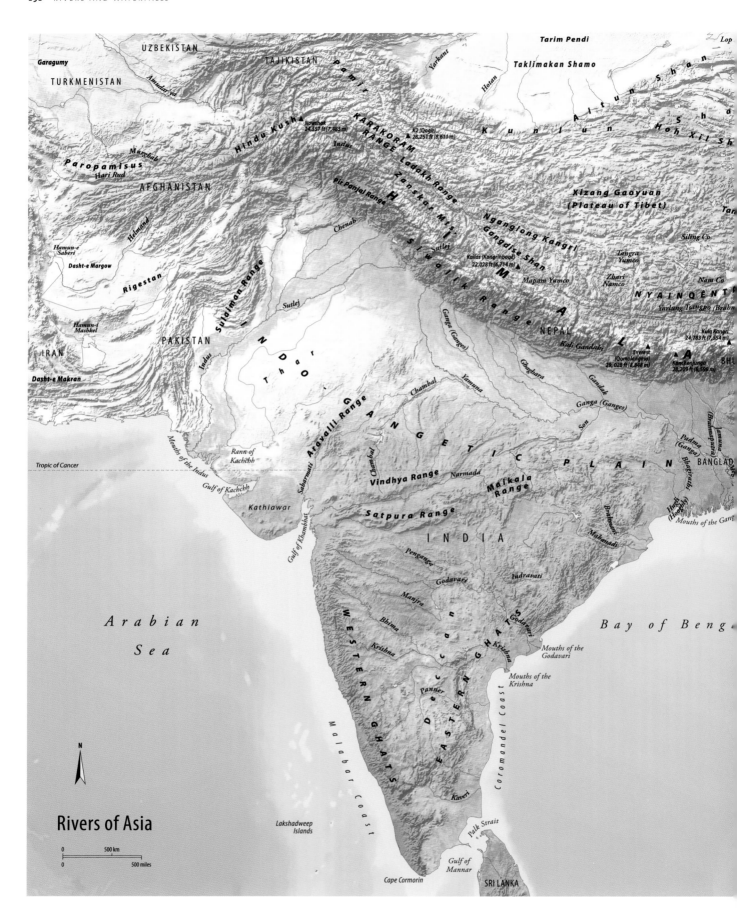

UZBEKISTAN

Garagumy

TURKMENISTAN

TAJIKISTAN

Pamir

Tarim Pendi

Yarkant

Taklimakan Shamo

Hotan

Lop

Amudar'ya

Kun l u n S h a n

Altun Shan

Hoh Xil Sh

Hindu Kush

Murghab

KARAKORAM

Novshak
24,557 ft (7,485 m)

Indus

K2 (Qogir)
28,251 ft (8,611 m)

Paropamisus

Hari Rud

RANGE *Ladakh Range*

**Xizang Gaoyuan
(Plateau of Tibet)**

Tan

AFGHANISTAN

Zanskat Mts

Nganglong Kangri

Siling Co

Hamun-e
Saberi

Helmand

Pir Panjal Range

Chenab

Gangdise Shan

Kailas (Kangrinboqê)
22,028 ft (6,714 m)

Mapam Yumco

*Tangra
Yumco*

*Zhari
Namco*

NYAINQÊNT

Nam Co

Dasht-e Margow

Rigestan

Sutlej

Sutlej

S i w a l i k R a n g e

Yarlang Tsangpo (Brahm

Kula Kangri
24,783 ft (7,554 m)

Hamun-i
Mashkel

PAKISTAN

Sutlej

Indus

T h a r

Ganga (Ganges)

NEPAL

Kali Gandaki

Ghaghara

Everest
(Qomolangma)
29,028 ft (8,848 m)

Kanchenjunga
28,205 ft (6,598 m)

BH

IRAN

I N D O

D e s e r t

Indus

Gandak

L

A

Dasht-e Makran

A R A V A L L I   R A N G E

*Rann of
Kachchh*

Tropic of Cancer

Mouths of the Indus

Gulf of Kachchh

Kathiawar

Sabarmati

Gulf of Khambhat

Chambal

Chambal

Yamuna

Vindhya Range

Satpura Range

Narmada

*Malkala
Range*

G A N G E T I C

P L A I N

Son

Ganga (Ganges)

*Padma
(Ganga)*

Bhagirathi

*Jamuna
(Brahmaputra)*

BANGLAD

*Hugli
(Hooghly)*

Mouths of the Gang

I N D I A

Penganga

Manjira

Godavari

Indravati

Brahmani

Mahanadi

*Mouths of the
Godavari*

B a y   o f   B e n g

*A r a b i a n
S e a*

W E S T E R N   G H A T S

Bhima

Krishna

D e c c a n

Godavari

Krishna

*Mouths of the
Krishna*

E A S T E R N   G H A T S

Penner

C o r o m a n d e l   C o a s t

N

M a l a b a r   C o a s t

Kaveri

Rivers of Asia

Lakshadweep
Islands

Palk Strait

0   500 km

0   500 miles

*Gulf of
Mannar*

Cape Cormorin

SRI LANKA

QILIAN SHAN

Mu Us Shamo
(Ordos)

Tengger
Shamo

Liaodong
Bandao

Bo Hai

Mouths of the
Huang (Yellow)

Korea
Bay

dam Pendi

Qinghai
Hu

Huang (Yellow)

Huang (Yellow)

Lüliang Shan

Fen

Shandong
Bandao

Bayan Har Shan

Gyaring
Hu

Huang (Yellow)

Taihang Shan

Great Plain of China

Huang (Yellow)

Yellow
Sea

Ma (Yellow)

CHINA

Hongze
Hu

(la) Shan

Ningjing Shan

Dri (Yangtze)

Yalong

Wei

QIN LING

Han

Daba Shan

Huai

Chao Hu

Chang (Mouth of
the Yangtze)

Chiama Ngu (Salween)

Dza (Mekong)

Jialing

Three
Gorges

Han

Chang (Yangtze)

Tai Hu

Hangzhou Wan

SHAN

Litang

Dadu

Sichuan Pendi

Hengduan Shan

Mishmi
Hills

Min

Hua

Dongting
Hu

Poyang Ha

East
China
Sea

Hkakabo Razi
19,295 ft (5,881 m)

Jinsha (Yangtze)

Yalong

Jinsha
(Yangtze)

Dalou Shan

Wu

a Hills

outra

Naga Hills

Jiuling Shan

Kumon
Range

Xiang

Gan

WUYI SHAN

Chindwin

NAN LING

Min

Nu (Salween)

Nanpan

Hongshui

Fuchun

East
China
Sea

Yuan

Wuliang Shan

Xun

Xi

Xin

Ayeyarwady (Irrawaddy)

Lancang (Mekong)

Thanlwin (Salween)

Fan Si Pan
10,312 ft (3,143 m)

Taiwan Strait

Keokradong
735 ft (1,230 m)

MYANMAR

Song
(Red)

Pratas Island

Batan
Islands

Tanen
Taunggyi

Doi Inthanon
8,497 ft (2,590 m)

Phou Bia
9,245 ft (2,818 m)

Mouths of the
Sông (Red)

Leizhou
Bandao

Luzon Strait
Babuyan
Islands

Arakan Yoma

Pegu Yoma

Irrawaddy

Menam Khong (Mekong)

Chaine Annamitique

Hainan

Gulf of
Tongking

Luba

Dawna
Range

Menam Khong (Mekong)

LAOS

Cape Bojeador

Mouth of
the Thanlwin
(Salween)

THAILAND

South

Mouths of the
rwady (Irrawaddy)

Gulf of
Martaban

Bilauktaung Range

Chao Phraya

VIETNAM

China

Sea

PHILIPPINES

Andaman
Islands

Andaman
Sea

CAMBODIA

Mekongè (Mekong)

Mergui
Archipelago

Tonlé
Sap

Chuor Phnum
Krâvanh

Phnum Aôral
5,810 ft (1,771 m)

Nicobar Islands

Gulf of
Thailand

Cuu Long
(Mekong)

Mouths of the
Cuu Long (Mekong)

Yangtze River, Asia

The Chinese name for the Yangtze River is "Chang Jiang," meaning "long river," and at around 3,900 miles (6,300 km) long, it is the longest river in Asia and the third longest river in the world. Tang Dynasty poet Li Bai described it as "a river flowing on the borders of heaven."

The Yangtze River is one of six major rivers, the waters of which stem from the Tibetan Plateau—the highest and largest plateau on Earth. The river is fed by a glacier in the Dangla Mountains on the eastern side of the Tibetan Plateau and enters the plains of the Sichuan Basin, where water volume is greatly increased by the confluence of many very large tributaries. At this stage, the river drops in elevation from above 16,000 feet (5,000 m) to under 3,000 feet (1,000 m) flowing southward through a steep-sided valley toward Yunnan. The valley is one of three very deep parallel gorges of Three Parallel Rivers of Yunnan Protected Areas, declared a World Heritage Site through the United Nations Environment Program. The Three Gorges, as it is famously known, is renowned for its spectacular scenic wonders.

The Protected Areas, extending over some 4,200,000 acres (1,700,000 ha), encompass the watersheds of three great rivers—the Yangtze, the Mekong, and the Salween—which flow from north to south through steep parallel gorges that are around 6,000 feet (2,000 m) deep in some places.

From the Three Gorges, the river enters the Yangtze Plain, where it continues to increase in volume from waters of numerous lakes and tributaries as it flows eastward to Shanghai and the river delta on the East China Sea. The delta has a population of over 18.5 million people, making it one of the most densely populated regions of the world. Shanghai itself has a population of more than 20 million people. Not surprisingly, it is also

ABOVE In China's Sichuan Province, terraces are carved into the hillsides all along the Three Gorges of the Yangtze River to allow the rugged landscape to be cultivated. Known as the Red Basin owing to the intense color of the soil, originally deposited by an ancient inland sea, it is intensively farmed, with the main crop being rice.

THE GREAT DAM OF CHINA

In the past, regular severe flooding of the Yangtze River has been a threat to the millions of people living in its catchment. In an effort to control flooding, generate power, and provide irrigation, the Three Gorges Dam was built as the largest such project in the world. As a result, there has been the submergence of many towns and massive changes to the ecology. Ancient cultural sites were also inundated. Evidence of human remains dating back to 2 million years have been found in the Three Gorges region putting into question the origin of the Chinese people. It is generally agreed that humans evolved in Africa and migrated much later than 2 million years ago.

Consequently, there were many groups and individuals who opposed the Three Gorges Dam and critics claim that it will have little or no effect on flooding and in fact, it could even make some floods worse.

one of the most polluted regions of the world, and the Yangtze Delta is said to be the major cause of Pacific Ocean pollution.

CARVING OUT THE LANDSCAPE

The uplift of the Tibetan Plateau started about 50 million years ago, at the time of the collision between the Indian and Eurasian plates, which was also the cause of the dramatic uplift of the Himalayan Mountains. The pressure caused by the Indian Plate pushing under the Asian Plate set off vast faulting and shearing of the overlying crust, creating high jagged mountains with their spectacular vertical relief. The zone of the Three Gorges is controlled by a complicated system of faults, along some of which the rivers flowed, cutting deep valleys as the uplift progressed over tens of millions of years.

When the last Ice Age commenced about four million years ago, the entire Tibetan Plateau was covered with sheets of ice, including glaciers that advanced and retreated a number of times until about 10,000 years ago. The erosive power of the glaciers further enhanced the major topographical features of serrated peaks and ridges and filled the extensive valleys with huge volumes of glacial sediments comprising silts, sands, and gravels. Because of the low gradients involved, the major rivers of the Tibetan Plateau flow as braided streams over the layers of sediment. By the time they reach the edge of the plateau, their volume has increased through confluence with other rivers, and the greatly increased erosive power results in the rivers cutting deep narrow canyons and gorges into the underlying bedrock. Some massive rapids and dramatic waterfalls are a feature of this process.

PRESERVING A DIVERSE ENVIRONMENT

A river the size of the Yangtze, together with a considerable diversity of ecoregions, becomes home to a greatly varying biota. Although the river itself suffers at least two critically endangered species, the Chinese alligator and paddlefish, and was home to the now extinct Baiji river dolphin, the region of the Three Parallel Rivers of Yunnan Protected Areas almost makes up for it. As one of the world's least disturbed ecoregions, the Protected Areas National Park supports a very rich diversity of flora and fauna. The park includes a number of vastly differing climatic zones and a huge range in altitude. There is still the need for extensive scientific investigation into biodiversity, but it is said that the region possibly supports over 25 percent of the world's animal species.

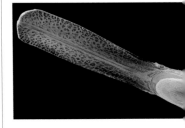

BELOW Paddlefish (*Polyodon* species) have a long paddle they use to sense electrical fields to detect prey as they skim along the muddy river bottom. The Chinese paddlefish lives in the Yangtze River and the American paddlefish lives in the Mississippi River.

BELOW A view along the Yangtze River of the Three Gorges in the People's Republic of China. Once the dam is fully operational, anticipated to be in 2009, much of this scenery will no longer exist. Up to 2 million local residents could be displaced.

Murray–Darling River System, Australia

The two million residents of Australia's Murray–Darling Basin live with the unpredictable and extensive drought conditions caused by the effects of the El Nino–Southern Oscillation (ENSO). As a result, river flows have dropped alarmingly and salinity is increasing dramatically—a disaster for the region.

The dilemma that faces Australia is that over 80 percent of the basin's flow is removed for irrigation, industrial use, and domestic supply. The area, known as Australia's "food basket," provides some 40 percent of the nation's food, and is a major contributor to Australia's economy. With the effects of climate change, the total system inflow over the five years to 2006 had been the lowest on record.

The Murray–Darling Basin, which encompasses the catchment areas of the Murray and Darling Rivers and their many tributaries, covers an area of over 405,000 square miles (1 million km²) or about 14 percent of Australia, making it one of the largest drainage systems in the world. When measured as a continuous river system, the Australian government states an official length of 2,095 miles (3,370 km) —the seventeenth longest in the world.

The River Murray itself rises in the New South Wales section of the Australian Alps and flows for 1,566 miles (2,520 km) before reaching the Southern Ocean at Encounter Bay in South Australia. Most of the water in the river comes from the southeastern section of the basin, either as rainfall or as snow-melt from the Australian Alps. The river forms the border between the states of New South Wales and Victoria.

THE MURRAY BASIN BACK IN TIME

The Murray Basin is a 311-mile (500-km) wide saucer-shaped structure with a basement of rocks dating back to the Cambrian and Ordovician periods. The "saucer" is filled to a depth of approximately 2,000 feet (600 m) by sediments that have been deposited over the past 65 million years, when the area started to subside and the eastern Alps rose due to the tectonic processes that caused Australia and Antarctica to separate.

Over that time, the climate was wetter so that rainforests, lakes, and wetlands covered the basin. About 32 million years ago, the area was inundated by shallow seas following a number of sea-level changes. Sediments eroded from the Alps were deposited into the basin by large rivers—the early Murray system.

After the sea had advanced and retreated a number of times—each time depositing more and more sediment layers in the basin—some two million years ago, a massive freshwater lake, Lake Bungunnia, was formed through an uplift that

ABOVE A stately river red gum (*Eucalyptus camaldulensis*) bows over the Murray River at Albury. These trees, vital in maintaining riverbank stability, have been considerably stressed by both a lengthy drought and the way the river system has been managed since the area was settled.

dammed the Murray. Again, a great deal more sediments were deposited into the subsiding basin.

The lake emptied after it breached some 700,000 years ago and for the past 500,000 years or so, the region has remained fairly stable, apart from increasing seasonal aridity. The basin is relatively flat and lies at only about 650 feet (200 m) above sea level. Its landforms relate to the deposits of sand and fine clay, climatic variation, and the interaction between fluctuating saline ground waters and surface processes.

The Darling Basin has a completely different geological history to that of the Murray. This basin was formed by crustal extension and faulting in the underlying earliest Devonian basement (around 400 million years old). Many of these structures are associated with widespread mineralization of considerable economic value.

LANDSCAPE AND INHABITANTS

Being so extensive, the Murray–Darling Basin has a vast range of climatic and environmental conditions. There are cool and humid rainforests, temperate mallee regions, inland subtropical areas, and up to 80 percent of the area comprises hot, dry, semi-arid, and arid lands. In some of the drier areas, tree-lined watercourses meander slowly through mulga or mallee scrub, grasslands, or chenopod shrublands comprising salt-tolerant drought-resistant plants, such as saltbush.

Not surprisingly, because of the range of environments, the basin is very important for its biodiversity. At the time of European settlement, about 28 percent of Australia's mammal species, about 48 percent of its birds, and some 19 percent of its reptiles were found there. Of these species,

THE DARLING RIVER SYSTEM

For a river system that has a potential evaporation rate twice that of the precipitation rate, and an overall environment that is naturally saline because of its soils and geology, it has always been a difficult task to manage water use.

About 90 percent of the water is used for agricultural irrigation—for crops such as rice (pictured below), tobacco, and cotton. Scientists have endeavored to show that such a dry region cannot support plants requiring so high a water consumption.

Management of the Murray–Darling River System has for years been the responsibility of governments of the states through which the rivers flow. Over-allocation of water to those upstream has always been a controversial issue. Only now, on the brink of total disaster, is it being considered to appoint a single body to manage basin-wide policies.

20 mammals are now extinct, and 16 mammals and 35 birds are endangered across the country.

There has been widespread degradation with loss of flora and fauna habitats through the clearing of over 50 percent of the original vegetation, soil erosion, river siltation, interference with groundwater aquifers and subsequent discharge of saline groundwaters to rivers, dryland salinity, and invasion of pest plants and animals.

BELOW River red gums line the banks of the slow-moving waters of the Murray River at Bruce's Bend in Yarrawonga Regional Park, in Victoria. This is a popular site for bush camping.

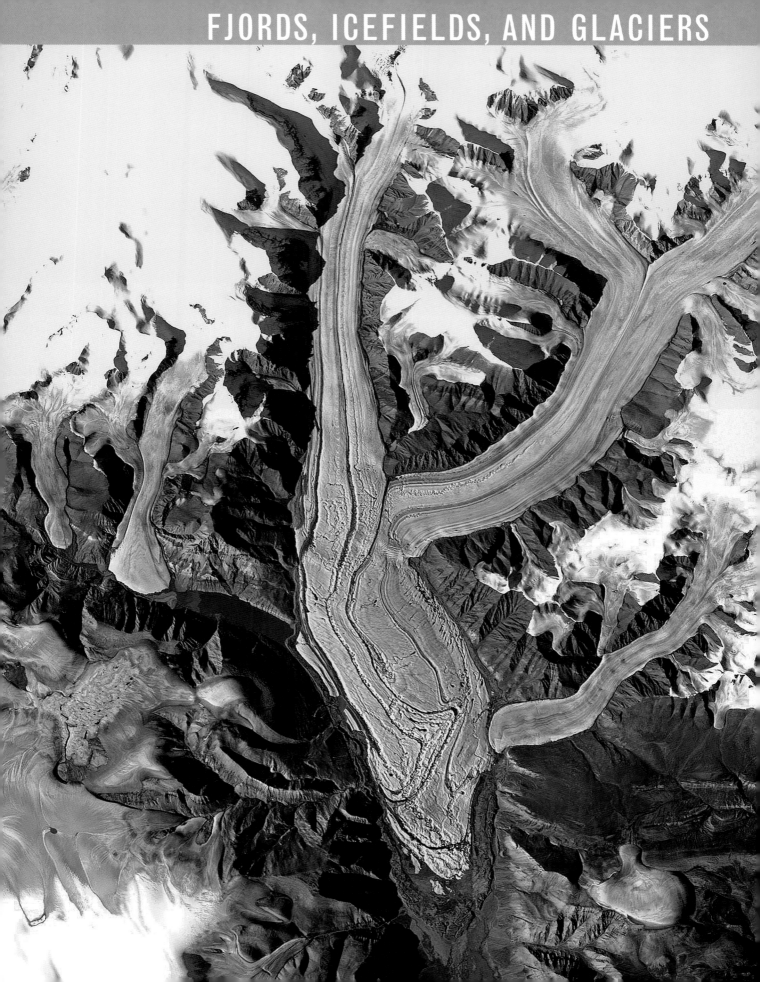

F or most of its history, Earth has been ice free and relatively warm, even at the poles. However, the delicate balance between solid ice and liquid water on the planet's surface is in a constant state of change. When ice builds up on the polar landmasses, average global temperatures drop, sea levels fall, and the planet experiences a glacial epoch.

Since Earth formed some 4.6 to 5 billion years ago, there have only been a few major glacial epochs. These are the Huronian from 2.4 to 2.1 billion years ago, the Cryogenian (Sturtian–Marinoan) from 850 to 635 million years ago, the Andean–Saharan (Ordovician Period) from 450 to 420 million years ago, the Karoo (Carboniferous–Permian periods) from 360 to 260 million years ago, and the Late Cenozoic Era ice epoch that began about 30 million years ago—the epoch that we are currently experiencing.

ABOVE Perito Merino is one of 47 large glaciers in Argentina's Los Glaciares National Park, which is a World Heritage Site. Located in the southern part of the park, along with Spegazzini and Upsala glaciers, it flows into Lake Argentino.

RIGHT Humpback whales (*Megaptera novaeangliae*) spend the summer in the southern coastal waters of Alaska, where fjords are commonplace.

For Earth's entire existence, all the glacial epochs put together have only occupied around 10 percent of geological time. During major glacial epochs, great ice sheets form and thicken in the high latitudes and push equator-ward to cover as much as 40 percent of Earth's land surface. During glacial epochs, the temperature may drop as much as 25°F (14°C) in the mid-latitudes.

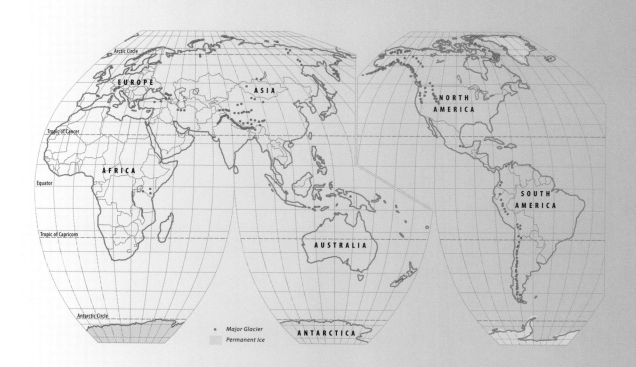

Arctic Circle

EUROPE   ASIA   NORTH AMERICA

Tropic of Cancer

AFRICA

Equator

SOUTH AMERICA

Tropic of Capricorn

Antarctic Circle

• Major Glacier
   Permanent Ice

AUSTRALIA

ANTARCTICA

MAJOR ICE EPOCHS CONTROLLED BY EARTH'S DRIFTING CONTINENTS

It is the position of the continents in relation to Earth's poles that determines whether or not the planet will experience a major ice epoch or not. Normally, the ocean currents distribute heat from Earth's equatorial regions, ensuring reasonably uniform global temperatures; however, if a landmass positions itself over the pole this balance is upset. At present, with Antarctica positioned over the South Pole, and with the northern hemisphere continents crowding around the Arctic Ocean, ice caps have been able to build up in both the north and south polar regions. It is a unique situation.

The previous ice epoch, the Karoo, was initiated when the Laurasian (northern) and Gondwanan (southern) continents assembled into a supercontinent called Pangea. The landmass disrupted ocean circulation and also stretched into the Antarctic region. Ice soon began to build up within the Antarctic Circle and advance northward, and also outward from the higher altitude alpine regions.

GLACIAL AND INTERGLACIAL PERIODS WITHIN ICE EPOCHS

Within each major ice epoch there are Ice Ages, which alternate with shorter, warmer periods

known as interglacials. At the present time, Earth is experiencing an interglacial period that has followed the last Ice Age. The climate began to warm around 18,000 years ago, then 15,000 years ago glacial advance halted and sea levels began to rise. Ice Age megafauna (mammals, birds, and reptiles) became extinct some 10,000 years ago, and then approximately 8,000 years ago the Bering Strait land bridge was submerged, cutting off the human and animal migration routes into North America. The Holocene Epoch interglacial warm period peaked around 5,000 to 6,000 years ago.

ABOVE New Zealand's Milford Sound is a fjord that formed when New Zealand was part of the supercontinent Gondwana. Ice has gouged sheer walls, creating a unique deepwater environment inhabited by sea stars, fish, sponges, and a rare coral.

LEFT A satellite view shows why the West Fjords make up over half Iceland's coastline, despite the region covering less than an eighth of the country's land area. The Tertiary Period basalts here are among Iceland's oldest rocks. Many waterfalls line the fjords. In summer, birds, particularly puffins, populate the seacliffs.

It is now accepted that barely noticeable eccentricities in Earth's orbit may affect these glacial and interglacial cycles approximately every 100,000 years. These eccentricities include variations in Earth's elliptical orbit, tilt of its rotational axis, and a wobble or precession in Earth's axis. These are named Milankovitch cycles, after the twentieth-century Serbian civil engineer and geophysicist

who first suggested that these fluctuations could account for the Ice Age–interglacial pattern.

ABRUPT MINI ICE AGES AND WARM SPELLS

Even shorter cycles of colder weather and warmer weather have been recognized within the current interglacial period, which started approximately 15,000 years ago. Some 5,000 to 6,000 years ago,

HOW FJORDS ARE FORMED

Fjords are constructed by glacial ice build-up during an Ice Age. A. Steep river valleys have been incised in the landscape, probably controlled by joints or planes of weakness in the rocks. Such V-shaped valley profiles are typical of youthful energetic rivers trying to cut down to sea level. B. During an Ice Age, glaciers widen the valley into a broad U-shape. The cutaway at the base of the glacier shows how the moraine, or rocks imbedded in the ice, act as tools to carve, scour, and polish the walls and base of the valley. C. When the glaciers retreat, the empty U-shaped valleys are flooded by the rising sea to form fjords. They are spectacularly beautiful, sheer-walled deepwater harbors with waterfalls often cascading from hanging tributary valleys.

A   B   C

Holocene Epoch interglacial temperatures reached an average of about 3½°F (2°C) warmer than present-day temperatures. This period was followed by a much colder spell during the Iron Age, around 1000 to 500 BCE. This cooling sparked re-glaciation of the coastal mountain ranges. Temperatures then rose and peaked between 1000 and 1350 CE, known as the Medieval Warm Period. They then fell again from 1400 to 1860 CE during the "Little Ice Age" and have been rising since 1860 to the present day, known as the "Industrial Age" global warming cycle. The driving force behind these shorter cycles is probably dependent on a combination of factors including ocean circulation, volcanic eruptions, meteorite impacts, solar variations, and greenhouse gas concentration in Earth's atmosphere.

It was long believed that these cycles came and went slowly on time scales measured in centuries. Current studies of ice cores taken from drill holes in Earth's icefields, however, reveal the potential for far more rapid temperature changes. These are known as "abrupt transitions," in which major shifts in Earth's climate occur on very short timescales. Abrupt transitions are explained by the recent discovery that large-scale ocean circulation in the North Atlantic can occur in one of two patterns, or stable states, that can switch suddenly. In one state, the warm Gulf Stream flows along the eastern coast of the United States and continues northward, past the British Isles to the Norwegian Sea, thereby keeping the climate of northwestern Europe relatively warm. In the second possible state, the northward extension of the Gulf Stream is "turned off," resulting in a much colder climate for all who live downstream including the populations of Europe.

Evidence in ice cores from Greenland shows that the switch between warm and cold states has taken place in the past within a span of five to ten years. The weakening of the current is caused by a reduction in the salinity of surface waters in the high latitude North Atlantic Ocean. This can perhaps be linked to an increase in the release of icebergs from glaciers and icefields, which on melting contribute large volumes of freshwater into the ocean, which in turn reduces the local salinity. Less salt means that the seawater is not as dense, therefore it will not sink to the ocean bottom and circulate away to the south during normal wintertime cooling. This lack of deepwater southward flow means that the warm water of the Gulf Stream cannot flow north into the region.

Whatever the exact cause of these fluctuations, we now know that in the North Atlantic, at least, the climate system can change very rapidly, and is most likely to do so again in the future.

V-SHAPED AND U-SHAPED VALLEYS

Rivers drain the landscape on their journey to the sea. They wind their way downward, seeking weak joints, fractures, and cracks to pass through, avoiding anything that offers resistance. As a result, water slowly and ceaselessly carves down through the rocks forming valleys. Valley widening accompanies valley deepening, creating the familiar V-shape. A glacier, on the other hand, is a river of ice making its way out from above the permanent snowline. Though glacial ice moves only a few dozen yards a year, it is relentless in its advance, bulldozing everything standing in its path. Spurs or bars of resistant rock in the valley are truncated by the moving ice, and ground into small rocks and glacial flour. All this debris, or moraine, is continually carried out of the valley and deposited at the front of the glacier. The original V-shaped valley is gouged and scoured into a smooth, broad U-shaped valley.

Wokkpash Lake sits in a U-shaped valley in Muskwa Range, British Columbia, Canada.

GLACIAL MECHANICS

Icefields form in Earth's cooler regions where more snow has fallen over the winter months than will melt during the following summer. In terms of tectonics, both high mountainous regions—such as those forming at colliding plate boundaries—and larger continents that have drifted into polar regions can act as zones of ice accumulation. The yearly snow buildup slowly compacts to form ice, which begins to move outward or downslope under its own immense weight—and it is in this way that glaciers are born.

The glaciers act as huge conveyor belts, moving ice and enormous amounts of rocky debris from the zone of accumulation above the snowline to the zone of ablation (melting) at lower elevations or latitudes, below the permanent snowline. Glaciers may move slowly but they transform landscapes, crushing and pulverizing the rocks as they pass over. Bands of debris form starkly contrasting black stripes against the brilliant blue-white of the glacial ice, either lying within the glacier itself (medial moraine), along the edges of the glacier (lateral moraine), or as mountainous piles in front of it (terminal moraine). Glaciers are responsible for the thick, rich soils of North America and northern Europe, which in turn

ABOVE A glacial waterfall cascades down a valley in Beagle Channel, Tierra del Fuego, at the tip of South America. The cold wet summers experienced in this archipelago just off southern Chile help to preserve the ancient glaciers there.

ABOVE Lake Louise, in Canada's Banff National Park, is a classic glacial tarn. Suspended silt particles give it a lovely blue color.

OPPOSITE Norway owes much of its beautiful glacial and fjordland scenery to the Eurasian ice sheet, which once covered the whole of northern Europe.

BELOW The Wrangell–St Elias National Park and Preserve in Alaska was largely sculpted by glaciers. Today, glaciers cover around 20 percent of the 13 million acres (5.2 million ha) within the park and preserve.

allowed for the development of numerous strong agricultural-based societies in these regions.

Vast numbers of meteorites around the edge of Antarctica had geologists puzzled. It was only recently understood that they were carried and deposited there by the outward-moving glacial ice.

FJORDS AND GLACIAL RETREATS

Although glacial ice is continually moving down its U-shaped valley, the front edge or terminus of the glacier may either advance or retreat. The terminus moves forward or backward depending on how fast the ice melts away compared to how fast ice is brought to the front by the glacier. During warmer interglacial periods, glaciers will shrink and retreat within their broad U-shaped valleys. All the extra meltwater from the ice causes global sea levels to rise, leading to the flooding of low-lying coastlines and coastal valleys.

The term fjord is used for a flooded U-shaped glacial valley. Fjords are characterized by smooth, near-vertical walls that rise majestically from deep-water estuaries. Their protected waters are often filled with abundant marine life. Waterfalls often cascade out spectacularly from the hanging valleys. Fjords are commonplace along the coastlines of once heavily glaciated lands such as Norway, New Zealand, Chile, and Alaska. New Zealand's famous Milford Sound is a deepwater fjord that formed when New Zealand was part of the Gondwana supercontinent. The extraordinary fjordland scenery and wildlife make these coastlines extremely popular cruise-ship destinations.

GLACIATED LANDSCAPES

As icefields and glaciers shrink, they leave behind a distinctively carved landscape. Pyramidal horns are found at high points from where several glaciers once radiated. At the base of each side of these horns is usually found a circular-shaped depression or cirque. They are carved by the immense weight of ice building up at the glacier's head. Cirques are often filled with tarns, or tranquil glacial lakes, which are often a beautiful cornflower blue due to the fine suspended silt particles known as "rock flour." Knife-edge jagged ridges or arêtes usually separate the empty glacial valleys.

The bare bedrock surfaces of these U-shaped valleys are often covered with glacial striations, gouged by rocks that were dragged along with the glacial ice. Geologists have been able to re-create the original flow directions in ancient ice sheets based on these glacial striae.

EARTH'S PRESENT GLACIAL CONFIGURATION

At present Earth is in an Ice-Age configuration, with a large landmass lying over its South Pole, Antarctica, and another, Greenland, close to the North Pole. This means that even a small drop in global temperature can cause abundant water to become locked up as ice on these polar landmasses. When this occurs, sea levels will fall and glaciers will surge outward from the polar and mountain regions as they have done during numerous glacial periods in the distant past.

Presently, glacial ice covers about 10 percent of Earth's surface—that is some 6 million square miles (15.5 million km²). The majority of ice is locked up in Antarctica (5.2 million square miles; 13.5 million km²) and Greenland (650,000 square miles; 1.7 million km²). All the rest of Earth's glaciers and ice sheets, including those found in the world's mountainous areas, account for only a very small percentage of the total.

All of Earth's glaciers are in a state of constant change, and what happens next is the subject of much debate. The last great ice age stretches back two million years. Holocene Epoch warming, which began approximately 18,000 years ago, and its associated glacial retreat, allowed mass migrations from Asia to the Americas and across northern Europe. Only 3,000 years ago, the lower global temperatures sparked re-glaciation of the coastal mountain ranges. Recent dates from numerous glaciated areas around the world indicate Earth is in a stage of accelerated glacial melting due to increased atmospheric greenhouse gases. Whatever the climatic future may hold, human settlement and land-use patterns will ultimately have to make significant adjustments to the changes.

Fjords and Icefields of Greenland

Greenland is experiencing a recurrence of what happened to its climate in the lead-up to the Medieval Warm Period (MWP). During this period of unusually warm weather, lasting from the tenth to fourteenth centuries, conditions there became more hospitable. Glacial ice retreated, leaving mounds of debris, exposed fjords, and wide valley floors, and these areas rapidly became lushly grassed, encouraged by the longer milder summers.

BELOW The glacier at Ilulissat (Jakobshavn Isbræ) is Greenland's largest outlet glacier. These massive walls of ice are actually the leading edge of the glacier calving into the bay.

During this four-century-long warm spell, seafaring Vikings took advantage of the ice-free conditions to colonize the fertile southern coast. These small Scandinavian communities thrived on farming, hunting, and trading for nearly 500 years before they mysteriously vanished. It is now thought that crop failures, brought about by a return of the colder conditions, caused their demise. This "Little Ice Age" lasted from the sixteenth to the mid-nineteenth century, and was followed by the current period of global warming.

GREENLAND'S NATIVE PEOPLES

The Inuit arrived in Greenland over 4,500 years ago in the first of several waves of migrations from North America that crossed the narrow Robeson Channel. Today Greenland, or Kalaallit Nunaat as it is known locally, supports a small population (over 56,000) living mostly on the ice-free western coast. The native Inuit make up about 85 percent of the population, and many pursue their traditional activities of hunting and fishing. Their traditional kayaks are disappearing, being replaced by

motorized fishing boats from a growing local shipyard industry. The country earns about 95 percent of its export income from shrimp fishing, and Greenland's handicraft and tourist industries are becoming increasingly important.

A MASSIVE ARCTIC ICE SHEET

A massive ice sheet with an area of 678,000 square miles (1.76 million km^2) covers about 80 percent of the country. It formed relatively quickly when smaller ice caps began to grow and coalesce in the late Pliocene or early Pleistocene epoch glacial period some 100,000 years ago. The ice sheet is second in size only to that of Antarctica.

The weight of ice, which is over 2 miles (3 km) thick at its center, is enormous. Resting on Greenland's continental crust, the ice has depressed the central part of the landmass into a huge basin with a depth of more than 1,000 feet (300 m) below sea level. The immense pressure is also squeezing ice, like toothpaste from a tube, radially outward as glaciers. Known as outlet

glaciers, these have carved out U-shaped valleys through the mountainous edges of the continent, and pushed their way down to the coast, creating some stunning fjordland scenery.

These outlet glaciers are like valves or taps that let Greenland's ice sheet run into the sea. Whether the ice sheet is growing or shrinking in any given year is dependent on the delicate balance between snowfall accumulating on the ice sheet and the volume of ice being discharged into the sea (by calving or melting) via its outlet glaciers. For many years, scientists have assumed this balance to be quite static, requiring thousands of years for changes to occur, but this is clearly not the case. Greenland's largest outlet glacier, Ilulissat Glacier (Jakobshavn Isbræ), took just 6 years (between 1997 and 2003) to nearly double its ice discharge.

CLIMATE AND SEA LEVEL
CONTINUALLY ON THE MOVE

Ice cores from holes drilled into the center of Greenland's ice sheet have enabled European and American researchers to create a record of climate

ABOVE Looking toward the sea from the fjord at Ilulissat, these icebergs floating in the bay are remnants of the calved glacier. The amount of ice that is discharged by the glacier can vary over a very short period of time.

GREENLAND'S GLACIERS SPEEDING UP ALARMINGLY

Glaciers draining the Greenland ice sheet have been retreating dramatically due to locally increasing sea and air temperatures. Glaciers are also traveling down their valleys faster and, as a result, are thinning, which, in turn, is speeding up the entire retreat process. From the 1970s to 2001, the front of Helheim Glacier remained in the same place. But during 2001, it began retreating rapidly in response to a rise in Greenland's temperature of more than 5°F (3°C) over the previous decade. Between 2001 and 2005, it moved back 4.5 miles (7.3 km), while ice flowage rates sped up to nearly 110 feet (35 m) a day, causing thinning by more than 130 feet (40 m). Glaciers in fjords can only thin until they reach a critical point—when they become light enough to float on seawater—after which they will rapidly crack up and disperse as icebergs. The Helheim's speedup is propagating rapidly up the glacier and, as the center of Greenland ice sheet is only 150 miles (240 km) inland, scientists are concerned this will lead to a thinning of the entire ice sheet, and its disappearance within a few hundred years.

ARCTIC OCEAN

Svalbard

Lincoln Sea

Kap Morris Jesup

Frederick E Hyde Fj.

J P Koch Fj.

Independence Fj.

Victoria Fj.

Hagen Fj.

Darimark Fj.

PEARY LAND

Ellesmere Island

Brikkerne Glacier

Flade Isblink

Robeson Channel

Ingolf Fj.

Stensby Glacier

Ryder Glacier

Petermann Glacier

Nioghalvfjerdsfjorden

Greenland Sea

Nares Strait

Humboldt Glacier

Jøkel-bugten

Tracy Glacier

Robertson Fj.

Heilprin Glacier

Storstrommen Glacier

Dove Bugt

Olrik Fj.

Inglefield

Bessel Fj.

Wolstenholme Fj.

Gade Glacier

Docker Smith Glacier

Ardencaple Fj.

Helland Glacier

Dedødes Fj.

Dietrichson Glacier

Gael Hamkes Bugt

Nansen Glacier

Waltershausen Glacier

Steenstrup Glacier

Baffin Bay

Kejser Franz Joseph Fj.

GREENLAND
(Kalaallit Nunaat)

Kong Oscar Fj.

Carlsberg Fj.

Nordvestfjord

Øfjord

Scoresby Sund (Kangertittivaq Fj.)

Baffin Island

Daugaard-Jensen Glacier

Fønfjord

Gåsefjord

Ukkusissat Fj.

Karrats

Uummannaq

Ilulissat Glacier

Christian IV Glacier

Disko

Gunnbjørn Field
12,139 ft (3,700 m) ▲

Disko Bugt

Kangerlussuaq Glacier

Davis Strait

Nansen Fj.

Miki Fj.

Kangerlussuaq Fj.

Nordre Strømfjord

Nordre Isortoq

Arctic Circle

Denmark Strait

Søndre Strømfjord

Nigertuluk

Kangertittivatsiaq

ICELAND

Søndre Isorta

Helheim Glacier

Niaqunngunaq

Sermilik

Godthåbsfjord

Ivertivaq

Ameralik Fj.

Gyldenloves Fj.

Groedefjord

ATLANTIC OCEAN

Frederikshab Isblink

Kvanefjord

Greenland's Fjords and Icefields

Sermiligaarsuk

Arsuk Fj.

Qoroq Glacier

Kangerluluk Fj.

Labrador Sea

Danell Fj.

Bredefjord

Lindenow Fj.

Prins Christian Sund

Tasermiut Fj.

Kap Farvel

N

0 ____ 200 km

0 ____ 200 miles

change in the northern hemisphere going back about 100,000 years. These show that Earth's climate is rarely stable for long periods of time.

Given its massive size and bulk, the melting Greenland ice sheet is now contributing significantly to global sea level rise, and will eventually raise global sea levels by about 20 feet (6 m). Although total melting is unlikely this century, even small amounts of melting will inundate any low-lying coastal cities and change ocean-circulation patterns in the North Atlantic.

ILULISSAT: THE TOWN OF ICEBERGS

Declared a World Heritage Site in 2004, Ilulissat Glacier, along with its spectacular fjord, is one of Greenland's best-known tourist destinations, frequented by cruise ships. The biggest glacier in the northern hemisphere, it calves majestic icebergs that drift right past the town on their way toward open sea. The region provides an opportunity to experience Inuit culture and view Arctic wildlife.

As one of this planet's great natural wonders, Ilulissat's glacier is perhaps also one of the most alarming examples of a rapid response to global warming in the Arctic. The lower extremity of the glacier has shrunk by more than 6 miles (10 km) in just a few years, after having remained relatively stable since the 1960s.

The Arctic Climate Impact Assessment Report, which was written by more than 250 scientists and published in November 2004, has sounded serious climate-change alarm bells for the region.

The report warned that, in less than a century, the Arctic sea ice could melt completely during summer, threatening the indigenous Inuit lifestyle and many Arctic species. Greenland's cold-adapted mammals include those that live on land, such as the Arctic fox, Arctic hare, ermine, lemming, musk ox, and wolf, plus those living on the ice and in the sea, such as the bearded seal, hooded seal, bow whale, narwhal, polar bear, ringed seal, and walrus.

ABOVE Quaint Itilleq village lies on a small island at the mouth of Itilleq Fjord. It is on the west coast of Greenland, a largely under-explored area with a vast system of fjords. The region is being drilled for potential oil and gas reserves, particularly offshore in the Nuussuaq Basin.

THE POLAR BEAR: PERFECT ARCTIC ADAPTATION

The polar bear has evolved to become one of the animals best-adapted to life on the Arctic ice and in its frigid waters. They are the largest land carnivore and can weigh up to 1,800 pounds (800 kg). Polar bear claws are adapted for running and climbing slippery ice, being smaller, more curved, and sharper than those of bears living outside the Arctic. To insulate their bodies, they have thick layers of fat and a dense coat of long, hollow, water-proof hairs. Polar bear hair is translucent and transmits the sun's heat to the base of the hair, and into the bear's black skin. White fur also serves as camouflage for stalking prey on the pack ice or under water. Their principal prey is ringed seal, but they also eat bearded seal, walrus, and occasionally beluga whale. An important adaptation to the difficulties in obtaining regular food is the polar bear's ability to slow its metabolic rate in order to conserve energy. Slowing begins about a week after not eating and lasts until the bear can find food again.

SCORESBY SUND

Scoresby Sund (Kangertittivaq) on the east coast of Greenland is the world's largest fjord system, stretching a distance of 220 miles (350 km) inland from the Greenland Sea. Its waters are up to 5,000 feet (1,500 m) deep. Traveling inland, the fjord branches into a labyrinth of smaller fjords, each with glaciers that reach back into Greenland's ice sheet.

Until just 18,000 years ago, this massive fjord system was full of ice, with glaciers still actively carving it out. After that, rising temperatures signaled the end of the last glacial period, and the polar ice sheets, including the Greenland ice sheet, began to shrink. The small town of Ittoqqortoormiit (with a population of just 537) lies at the mouth of Kangertittivaq Fjord. It is the main settlement and administrative center for the region. It is one of Greenland's most remote towns, only accessible by helicopter or by boat for a few months each year.

LIFE IN ITTOQQORTOORMIIT

Ittoqqortoormiit, which means "big house," was founded in 1925 by Ejnar Mikkelsen, who arrived with about 70 settlers on the ship *Gustav Holm*. They prospered in the new area, hunting the abundant populations of arctic foxes, narwhals, polar bears, seals, and walruses. The small community gained an income through trading meat and other animal products depending on their seasonal availability.

Interestingly, while out hunting, the settlers found tent rings and other ruins indicating the area had previously supported a dense Inuit population. Recently tourism has been growing in importance to the region.

Like much of Greenland, Ittoqqortoormiit lies above the Arctic Circle (latitude 66.5°N), where it enjoys midnight sun (day-round sunlight) during summer, but must also endure polar nights (day-round darkness) in winter. In the northernmost regions of Greenland, these winters can last for four months. One consolation, however, is the opportunity to see the aurora borealis or northern lights. This surreal display of undulating curtains of different colored lights occurs when atmospheric particles are ionized by solar and cosmic radiation. These charged particles are trapped in Earth's magnetic field and drawn into the polar regions.

GATEWAY TO THE WORLD'S GREATEST NATIONAL PARK

The Northeast Greenland National Park, the largest in the world, was established in 1974 by the Danish parliament, and extended westward in 1988. Today it protects 375,000 square miles (972,000 km^2) of ice sheet, glacier, and fjord landscape to the north of Scoresby Sund.

The park includes Peary Land in north Greenland's high Arctic, which interestingly is not covered by sheet ice. It is a bare cold desert or tundra because the air is too dry to produce snow. At 80°N, Peary Land is the northernmost

THE MUSK OX: AN ICE AGE RELIC

The Northeast Greenland National Park is home to 40 percent of the world's population of musk ox (*Ovibos moschatus*). These animals are the last surviving members of an extinct group of Ice Age oxen, and as such they are often referred to as Ice Age relics. The musk ox does not need to find shelter from Arctic blizzards as it has evolved a thick winter coat for protection. It will use its horns and feet to clear away large round patches of winter snow to reach edible vegetation. These food craters are fiercely defended against other musk ox. They live in social groups of about 10 to 20 animals. When threatened, adults of the herd encircle the young, with their heads facing outward, toward the predators.

ice-free landmass. The region is named for the American Naval Commander Robert Peary who, in 1892, discovered that Greenland is actually an island and that the North Pole, his ultimate goal, lay beyond its northern shores.

Scoresby Sund is named for the English Arctic explorer and scientist, William Scoresby (1789–1857), who made his first whaling voyage to Greenland with his father at the age of 11, then later returned as captain on other ships. During these voyages, he drew charts and collected valuable data on Greenland's deep-sea temperature, flora, and fauna. He later wrote several important books on Arctic geography.

Glaciers and Fjords of Alaska

The icefields of Alaska straddle the mountain ranges and the southeastern coast. They are part of a belt of glaciers that descends from the high mountain ranges ringing the Pacific Ocean. The fjords dotting the southeastern coast formed when coastal glacial valleys became flooded after the sea level rose.

These coastal ranges in Alaska are largely made up of volcanic peaks that are continually growing as the Pacific Ocean floor is being overridden and consumed by the neighboring tectonic plates.

The subduction process here punches up more volcanoes along the plate edges, creating mountain peaks. The icy conditions that occur in the northernmost latitudes are prime territory for the formation of ice shields and glaciers. Consequently the area covered by ice in this region is immense: glaciers in Alaska and neighboring Canada cover 35,000 square miles (90,600 km²) or roughly 13 percent of Earth's mountain glacier areas.

Maritime conditions, particularly along the coastline, are perfect for glacier formation. Here, moisture-laden onshore winds cool as they are pushed up and over the Coast Mountains. As the air cools, its ability to hold moisture becomes reduced and the freezing water falls as snow. Yearly snowfalls on the Juneau Icefield exceed 100 feet (30 m), and the short summers ensure that at high elevations winter snow accumulation is greater than summer snowmelt. Alaskan glaciers are found in a number of different settings—in U-shaped valleys, in hanging valleys, and in cirques, but some of most spectacular are the tidewater glaciers.

TIDEWATER GLACIERS

Tidewater glaciers are glaciers that reach the sea. In Alaska, they often occur at the heads of fjords—there are 16 in Glacier Bay alone. As a glacier advances into the sea, it bulldozes a wall of rocks and sediment—terminal moraine—ahead of its terminus, which protects it from the tidal effects of the seawater. If the glacier begins to retreat from its terminal moraine walls and the barrier is breached, it loses its protection. Being less dense than salt water, glacial ice can float on seawater, which generates a great strain within the glacier's terminus. Consequently, huge slabs of ice peel away from the front of the glacier, a process called calving. These fall into the sea below, creating an unstable vertical wall. Retreat by calving is rapid until the terminus reaches a stable position from where it can recommence building its terminal moraine.

Glaciers, Fjords and National Parks of Alaska

0 ___ 300 km
0 ___ 300 miles

Key to Glaciers
1. Harding Icefield
2. Portage Glacier
3. Matanuska Glacier
4. Columbia Glacier
5. Valdez Glacier
6. Wortmans Glacier
7. Schwan Glacier
8. Tickel Glacier
9. Tonsina Glacier
10. Nabesna Glacier
11. Bagley Icefield
12. Malaspina Glacier
13. Hubbard Glacier
14. Mendenhall Glacier
15. South Sawyer Glacier
16. Dawes Glacier
17. Stikine Glacier
18. Le Conte Glacier

Key to Fjords
1. College Fjord
2. Endicott Arm Fjord

Key to National Parks and Forests
1. Kenai Fjords National Park
2. Russell Ford Wilderness
3. Glacier Bay National Park
4. Tracy Arm-Fords Terror Wilderness
5. Tongass National Forest
6. Misty Fjords National Monument

GLACIAL FLORA AND FAUNA

Many plants and animals have adapted to live around Alaska's glaciers and fjords. Seals and sea lions often sleep on calved icebergs. Salmon spawn in the lakes and streams left behind by the retreating glaciers. Bears are drawn here by the plentiful food in spring. Eagles soar above the peaks.

Most fascinating to understand and observe is the succession of plant and animal recolonization that follows the temperate rainforest reclaiming the bare ground when it becomes exposed as a glacier retreats. This process commences with lichens and moss, which coat the bare rocks, followed by alder, which introduces nitrogen through its roots into the nutrient-poor soil. Later, cottonwood and spruce overtake the alder, which decay, adding further nutrients to the soil. Finally, tall hemlocks overshadow and dominate the vegetation, completing the development of old growth or climax forest. The entire process takes at least 350 years.

LAND OF 100,000 GLACIERS

With five percent of Alaska's surface covered with ice, this state is a favored tourist destination for viewing icefields, glaciers, and fjords first-hand, along with their associated landscapes and wildlife.

Alaska has roughly 100,000 individual glaciers, but among the better known are the impressive Columbia and Hubbard glaciers, two of Alaska's largest tidewater glaciers. Malaspina Glacier is the state's largest glacier, with an area of 1,500 square miles (3,800 km²), extending 50 miles (80 km) from Mt Elias toward the Gulf of Alaska. Glacier Bay National Park has one glacier that receded over 60 miles (100 km) between 1794 and 1916. Around Anchorage, Portage Glacier is one of many hanging-valley glaciers that surround the town of Girdwood, while a two-hour drive takes in 35-mile-long (48-km) Matanuska Glacier. Other well-known Alaskan glaciers include Mendenhall, which sits behind Juneau, the state's capital, Nabesna, Tonsina, Tiekel, Valdez, and Wortmans.

ALASKA'S RETREATING GLACIERS

Scientists have found that Alaskan glaciers are retreating much faster than they originally thought. They have been accurately measuring the surfaces of the glaciers from aircraft using laser altimetry systems. The data they collected has been used to calculate the reduction in volume of 67 Alaskan glaciers from the mid-1950s to the mid-1990s. The glaciers have thinned substantially during this period, but over the last five to ten years, there has been a marked acceleration in melting. This is possibly a result of human activities, such as burning fossil fuels, which are causing an unnatural temperature rise globally. The melting Alaskan glaciers have been responsible for at least nine percent of the global sea-level rise that has occurred during the last century.

ABOVE The Hubbard is Alaska's largest tidewater glacier. It has been advancing toward the sea since it was first mapped in 1895. Its calving drives the mechanics of its advance-and-retreat cycle; it is unlike most of Alaska's glaciers, which are retreating due to global climate change.

ABOVE The dark line running down the center of Shakes Glacier is the moraine, created by rocks ground in its path. Crevasses are also visible here.

TRACY ARM FJORD

Deep into the rugged terrain of Tracy Arm–Fords Terror Wilderness, cruise ships often make their way along the twists and turns of Tracy Arm Fjord, a challenging 30-mile (48-km) journey. Slow-moving glacial ice was the massive force originally responsible for carving out the sheer vertical walls of this fjord.

RIGHT An aerial view down Tracy Arm Fjord shows the terminus of South Sawyer Glacier in the foreground. The narrow passage typical of a fjord is fully revealed from the air. The torrents that create treacherously unstable currents in this fjord can be seen near the glacier's terminus.

The changing tides in the fjord bring strong, unpredictable currents, eddies, and whirlpools. Venturing up the treacherous waters of this steep narrow fjord, ships must navigate around drifting ice. Nearing the head, the water becomes dotted with larger pieces of ice. Around the last bend are the two Sawyer glaciers, surrounded by 7,000-foot-high (2,100-m) snowcapped mountains.

THE TWIN GLACIERS

As recently as the 1880s, these two tidewater glaciers were joined. Now Sawyer has retreated almost a mile (1.6 km) from the original junction, while South Sawyer is almost 3 miles (4.8 km) back. Blocks of ice the size of three-story buildings often break away or calve from the sheer face of the glacier's terminus. Each block emits a thunderous retort like a cannon blast as it cracks away, exposing a fresh face of sparkling blue ice. These descending calved blocks often generate significant waves that can swamp the smaller boats used to ferry passengers for a closer look at the icebergs.

The fjord and surrounding park is a haven for wildlife. Harbor seals and their pups often rest on the icebergs, and whales and sea lions may also be spotted in the water. Bald eagles soar above towering cliffs. A spruce and hemlock rainforest grows upward to 1,500 feet (450 m) giving way to steep, glacially cut rock faces. Roaming the forest are brown and black bears, black-tailed deer, and other smaller mammals, while higher up, mountain goats may be seen scaling the cliffs and ledges.

A TERRIFYING HISTORY

In 1899, US Navy Crewman Ford paddled up this narrow fjord and became trapped for over six hours by the treacherous changing tidal currents and massive crashing icebergs. Hence the name Tracy Arm-Fords Terror Wilderness came to be used when the US Congress designated the area in

ABOVE Glacial blue icebergs float in the waters of Tracy Arm Fjord. Though the water in this fjord appears to fill a V-shaped valley, the glacially carved bottom of the fjord is actually U-shaped.

TLINGIT CULTURE

The first settlers in the Juneau area, the Tlingit arose from ancestors who crossed into Alaska from Asia via the Aleutian Islands during the last glacial retreat, at least 10,000 years ago. Isolated by the jagged coastline and mountains, they got around in log canoes. They harvested fish using traps, and hunted seals and sea lions. Clams, seaweed, berries, and roots were also eaten.

Their stratified society was divided into two groups or moieties, "Eagles" and "Ravens," then into clans and family units, each with its own crest and family names. Each home's center post had a totem carved with their clan and family figures. Highly skilled traders, when the fur traders arrived here during the 1700s, they profited from trade between Russian and English ships and inland tribes. However, the situation broke down when Russian fur traders began kidnapping Tlingit women and children. Tlingit society was devastated by a smallpox epidemic during the 1800s and Klukwan is the only village still active in its traditions.

1980. The park covers 653,179 acres (264,537 ha) and is located 45 miles (72 km) south of Juneau; its eastern boundary extends back to the Canadian border. One-fifth of the park is covered by permanent ice. There are two fjords within the park— Tracy Arm Fjord and Endicott Arm Fjord, which has its own tidewater glacier. The glaciers of both fjords are fed from the adjacent Juneau Icefield, which receives 100 feet (30 m) of snow each winter, dumped by the rising moisture-laden air masses as they are pushed upward over the mountains by the winds buffeting the west coast.

Immense icy forces punched the deep U-shaped channel through the Coastal Ranges. The fjord's steep walls, polished smooth by ancient glacial ice, offer a vista of west-coast geology, recording in its zigzag-folded metamorphic rocks the collision of a series of younger crustal terrains. Between the Permian and Cretaceous periods, these crustal fragments were rafted-in on the subducting ocean floor, rammed against, and fused to, the ancient coastline that, at the time, was located back around the present Alaskan–British Columbian border.

Ice Caps and Glaciers of Canada

The neatly circular Hudson Bay in northeastern Canada, superficially resembling the crater of a massive meteor strike, is actually an enormous basin-shaped depression in Canada's continental crust. Together with Foxe Basin, it marks the former center of the extensive North American (Laurentide) ice sheet that only disappeared from Hudson Bay as little as 8,000 years ago.

During the Pleistocene Epoch, Canada and Alaska were repeatedly covered by ice sheets. Massive glaciers pushed southward into the northern United States, east toward the Labrador Sea and surged downward out of the mountains of the cordillera. Much of the fertile soil of the Canadian Prairies and the American Midwest is glacial in origin. Even the position of the Great Lakes and the drainage patterns of the Ohio and Missouri rivers were influenced by the advancing and retreating ice sheet. In North America, the most recent glacial event is the Wisconsin Glacial, which began about 80,000 years ago and ended around 10,000 years ago.

As a result of the massive crustal unloading that occurred as the ice melted, Hudson Bay became one of the fastest rising areas in the world. This process is known as isostatic rebound.

CURIOUS EVIDENCE OF UPLIFT

Evidence for the uplift or isostatic rebound of the Hudson Bay area takes the form of a spectacular staircase of former shorelines, sea cliffs, and beach ridges that now climbs several hundreds of feet above sea level. Carbon-14 dating on marine shells and driftwood from the raised beaches show that they are less than 14,000 years old. The greatest measured rates occur in the Richmond Gulf area of southeastern Hudson Bay, where the ice must have originally been the thickest. Here, Hudson Bay's shoreline is currently rising about $4\frac{1}{3}$ feet (1.3 m) per century. The rates of uplift decrease radially outward from Hudson Bay, reaching zero by about the Great Lakes.

HUDSON BAY AND ITS SURROUNDS

Hudson Bay is a relatively shallow extensive water body covering 473,000 square miles (1.23 million km²) that collects the runoff from most of Canada's land area. It was named for Henry Hudson, who explored the bay in 1610 on his ship the *Discovery*. Only a limited number of settlements exist along the inhospitable coast of Hudson Bay, some founded as trading posts in the seventeenth and eighteenth centuries by the Hudson Bay Company. Residents of these villages—such as Puvirnituq, Churchill, and Rankin Inlet—are mostly Cree and Inuit people.

The broad expanse of flat country to the west and south of Hudson Bay is known as the Hudson Bay lowlands, which, in summer, thaws into an insect-ridden maze of thousands of interconnected

Canada's Ice Caps and Glaciers

lakes, ponds, and muskeg marshes. This landscape was exposed as glaciers retreated and through gradual shrinkage of Hudson Bay.

CANADA'S RESIDUAL ICE CAPS AND GLACIERS

The remains of the once-extensive Laurentide ice sheet can still be found as ice caps on the islands of the Canadian Arctic, such as the Penny and Barnes Ice Cap on Baffin Island, Bylot Ice Cap on Bylot Island, Devon Ice Cap on Devon Island, and the Simmon Ice Cap on Ellesmere Island. All of these ice caps have been thinning and receding slowly. For example, since 1959 the Simmons Ice Cap has lost 47 percent of its surface area and, if the current climatic conditions continue, it will melt completely by about 2050.

Canada's ice caps, though relatively small, still represent most of Earth's polar land ice outside of Antarctica and Greenland and, as such, have the potential to contribute to sea-level rise. During the next 100 years, melting glaciers and ice caps outside of Greenland and Antarctica are expected to raise global sea levels between 8 and 16 inches (20 to 40 cm).

The only remaining fragments of the former Laurentide ice sheet found outside the Arctic Circle are the icefields and glaciers of the high cordillera of Canada and Alaska. These remnants, such as Canada's Columbia Icefield, have been preserved in the colder high altitudes.

ABOVE The shoreline near Pond Inlet (Mittimatalik) is surrounded by fjords and glaciers. It is on the tip of Baffin Island and features a raised beach that was uplifted by isostatic rebound, a geological process that occurs after the heavy pressure of ancient glacial ice has been lifted and the land eventually "springs" up.

ABOVE Berg Glacier is in the northern part of Mt Robson Provincial Park, British Columbia. One of Canada's few advancing glaciers, it calves into the turquoise waters of Berg Lake.

ABOVE A caribou (*Rangifer tarandus caribou*) on Baffin Island, in Hudson Bay, Canada. A type of wild reindeer, they lower their metabolism to cope with cold.

COLUMBIA ICEFIELD

The Columbia Icefield is the largest surviving remnant of the thick ice sheet that once mantled the western Canadian Cordillera. It lies on a wide elevated plateau in Alberta's Banff and Jasper national parks. The icefield is replenished by up to 23 feet (7 m) of snowfall per year from moisture-laden winds blowing across from the Pacific Ocean.

BELOW Athabasca Glacier spills out from the Columbia Icefield in the Rocky Mountains. It has been retreating ever since it reached its maximum extent in 1750, 1⅓ miles (2 km) ahead of its current terminus.

Accumulating ice continually flows radially outward from the icefield, squeezing through gaps in the surrounding high peaks of the Canadian Rockies, as great tongues of glacial ice. The Columbia Icefield feeds eight major outlet glaciers. These include the Athabasca, Castleguard, Dome, Columbia, Stutfield, and Saskatchewan glaciers. The entire field covers an area of 125 square miles (325 km^2) and is the largest glacial area in the northern hemisphere outside of the Arctic Circle. It lies on the continental divide of North America with its meltwater feeding rivers that ultimately flow north into the Arctic Ocean, east into Hudson Bay, and west into the Pacific Ocean.

Mountaineers J. Norman Collie and Hermann Woolley first reported the icefield in 1898 after they completed the first ascent of Mt Athabasca.

CANADA'S MOST ACCESSIBLE GLACIER

The Athabasca Glacier is located near the scenic Icefields Parkway between the towns of Banff and Jasper. The slow-moving tongue of ice is 3.7 miles (6 km) long by ½ mile (1 km) wide. It descends from the Columbia Icefield over a series of giant bedrock steps flowing slowly down its U-shaped valley. The broad front of its terminal snout is within easy walking distance from the road. As such, it has become the most-visited glacier on the North American continent. The adjacent Columbia Icefield Center is the starting point for trips onto the glacier during the summer months. Visitors travel onto the glacier in large

comfortable snowcoaches designed to negotiate the steep ice and snow, and are given an opportunity to walk on the ice during the tour.

A UNIQUE WORLD HERITAGE SITE

The combined national and provincial parks of the Canadian Rocky Mountains were declared a World Heritage Site in 1984, to protect the unique Columbia Icefield and its surrounding landscape of mountain peaks, glaciers, lakes, and waterfalls. The largest park in the Canadian Rockies is Jasper National Park, named after Jasper Hawes, who once operated a trading post in the region for the North West Company. It has an area of 4,200 square miles (10,900 km²). The park supports and protects an abundant range of wildlife adapted to the harsh seasonal conditions of the mountains. These include the caribou, elk, moose, mountain goat, bighorn sheep, grizzly bear, black bear, beaver, pika, and marmot.

THE COLUMBIA ICEFIELD SHRINKS

Because of a warming climate, the Athabasca Glacier has retreated some 4,900 feet (1,500 m) since the late-nineteenth century when its terminal snout flowed over the position of the present highway. Although its ice is continuously creeping forward at the rate of several inches per day, carrying with it massive quantities of crushed and ground rock, the glacier as a whole has thinned and lost over half of its volume since then. As it has moved back up its valley, the shrinking glacier has left behind a debris-strewn wasteland of ground-up rubbly moraine. Athabasca Glacier's rate of retreat has increased since 1980, following a period of slower retreat between 1950 and 1980.

Saskatchewan Glacier, also accessible from the Icefields Parkway, is approximately 8 miles (13 km) long and covers an area of 11½ square miles (30 km²). Between 1893 and 1953, it retreated a distance of 4,470 feet (1,360 m), with the rate of retreat between 1948 and 1953 averaging 180 feet (55 m) per year.

Disappearance of glacial ice is more than just cosmetic as much of western Canada relies on glacier-stored water being released into its river systems during the dry summer months, when it is most needed. As the glaciers shrink, the drop in late-summer water levels of glacier-fed streams will become more pronounced.

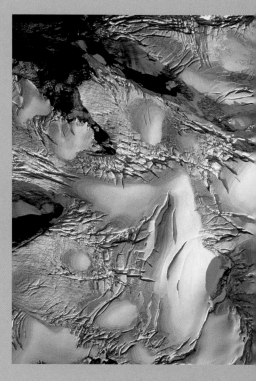

BELOW A bird's-eye view of the Columbia Icefield shows a portion of the overall extent of the ice. Tongues of ice spilling outward are glaciers flowing into the surrounding valleys. Their cracks and crevasses are clearly visible from the air.

BELOW The hoary marmot (*Marmota caligata*) is a type of ground squirrel that lives in the Rocky Mountains above the treeline at altitudes between 6,800 and 8,000 feet (2,100 and 2,400 m). They adapt to the cold by hibernating for up to eight months of the year.

Fjords, Ice Caps, and Glaciers of the Andes

Along the axis of the Andes Cordillera lie the remnants of a much larger icefield that existed during the Last Glacial Maximum (LGM) around 17,500 to 18,000 years ago. It was an elongated narrow ice sheet that would have extended from southern Chile and Argentina (the Patagonian Ice Sheet) to the Andes of Colombia and Venezuela in the north. The Patagonian ice sheet surged out of the high peaks of the Andes.

ABOVE The world's most southerly lizard, *Liolaemus magellanicus,* suns itself on a rock in Torres del Paine. Its low body temperature is an adaptation to the cold.

It pushed outward and downward onto the lower plains on either side. On the Chilean side, ice would have covered much of the country south of present-day Puerto Montt. Fed by snowfall from moisture-laden air masses blowing in from the Pacific Ocean, these glaciers would have reached the sea, carving the spectacular valleys and fjordland scenery enjoyed today, including the well-known sculpted peaks of Torres del Paine. On the eastern side of the Andes, glaciers pushed downward onto the Argentinian plains, only just barely reaching the Atlantic because of the dry climate. That side of the Andes was a desert, lying in the mountain's rain-shadow and probably even drier than it is today.

ANDEAN TOWNS ARE RUNNING DRY

From small villages to major cities, many Andean population centers in arid areas are dependent on glacial meltwater as their only source of fresh water during the dry summer months. Recently, with the glaciers melting at an unprecedented rate, there has been a deluge of excess water. In some areas of Peru, this water is flooding and damaging the network of old irrigation canals that supply the farms and mills. The big problem is that the glaciers are melting faster than they can replenish themselves, which means that by 2030, these towns will not have a water supply.

REMNANT OF A MASSIVE ICEFIELD

At its maximum, it is estimated that the Patagonian Ice Sheet covered an area of about 185,000 square miles (480,000 km^2). Today, only about 4 percent of that ice remains in two separate patches known as the Northern and Southern Patagonian Ice-fields. They are still, however, the largest masses of non-Antarctic ice in the southern hemisphere.

Retreating in a series of oscillatory stages since the Pleistocene Epoch (1.8 million to 10,000 years ago), the gradual melting of this field has contributed about 4 feet (1.2 m) to global sea levels.

From 1995 to 2000, the rate of ice loss from the Patagonian Icefields has more than doubled, to a rate that makes them the fastest-retreating area of glacial ice on Earth. Increased air temperatures and decreased precipitation are blamed, as well as the fact that the Patagonian glaciers are mainly calving glaciers. Calving glaciers shed icebergs into lakes or the sea from their terminal edge, and are far more sensitive to climatic change than those that terminate on land and retreat by melting.

The Southern Patagonian ice mass feeds numerous glaciers, some of the best known of which are the Upsala, Viedma, and Perito Moreno

LEFT A boatful of tourists watch the icebergs calving off Grey Glacier on Lago Grey at the northern tip of Chile's Torres del Paine National Park.

in Argentina's Los Glaciares National Park, and the Bruggen (Pío XI), O'Higgins, Grey, and Tyndall glaciers in Chile's Bernardo O'Higgins and Torres del Paine national parks.

TROPICAL GLACIERS OF THE ANDES

The northern Andes contain most of Earth's tropical high-mountain glaciers (those found within 23.5° of the equator). These include the Quelccaya Ice Cap, Peru's largest, Chacaltaya Glacier in Bolivia, and the Antizana Glacier in Ecuador. However, more than 80 percent of this tropical ice is found on the highest peaks in numerous tiny isolated glaciers of about one-half to one-third of a square mile (under 1 km²) in size.

All of these glaciers are currently shrinking due to the increasing temperatures in the high Andes. Chacaltaya Glacier is presently only 10 percent

ROCK FEATURE: TILLITE

Tillite is a sedimentary rock composed of unsorted glacial sediment (till) that has been cemented. These mixtures of clay, sand, gravel, and boulders may have once been a glacier's terminal, lateral, or medial moraine abandoned as they retreated. It differs from conglomerate as its component rock fragments are angular. Tillites are important as they mark the aerial extent of ice sheets during glacial periods. Secondary tillites are those that have undergone some fluvial reworking, such as in a glacial outwash plain.

of the size it was when it was first examined back in 1940 and, at the current rate of shrinkage, by 2010 to 2015, this glacier will most likely no longer exist. By 2030, it is expected that most of the large ice caps on the Andes will be gone.

LEFT The glacier-covered rocky peaks of the Grupo La Paz Mountains tower above a fjord. The Gamboa Mountains in the Cordillera Sarmiento in the foreground were first seen by sailors, their soaring peaks averaging 6,500 feet (2,000 m) in height.

Fjords and Glacial Scenery of Scandinavia

Northern Europe, particularly Norway, owes much of its spectacular glacial and fjordland scenery to the massive continental ice sheet that once covered the entire region. Known as the Eurasian ice sheet, it reached its zenith between 18,000 and 20,000 years ago during the Last Glacial Maximum (LGM), extending from the Arctic into Siberia and all of the countries surrounding the Baltic and North seas.

ABOVE A herd of reindeer (*Rangifer tarandus*) move across a snowfield in Norway. Reindeer are native to Scandinavian countries and some wild herds still exist in Norway. They also occur in other Arctic and subarctic regions.

Sea ice linked the Eurasian ice sheet with the Greenland ice sheet and that of North America, the Laurentide ice sheet. The center of the Scandinavian portion of the ice sheet was in the region currently known as the Gulf of Bothnia, a shallow lobe of the Baltic Sea separating Sweden from Finland. Here the ice reached a maximum thickness of about 2 miles (3 km). Pushing outward from this region, glacial ice scored huge U-shaped valleys through Scandinavia's mountainous spine and out into the Norwegian Sea. The enormous picturesque fjords of the Norwegian coastline, such as Sognefjorden, Trondheimsfjorden, and Hardangerfjorden attest to the erosional power of this mobile ice. Europe's major northerly flowing rivers were blocked from entering the sea by the growing ice sheet, and extensive lakes began to flood the landscape along the southern edge of this ice dam.

THE GULF OF BOTHNIA GLACIAL BASIN

At its center, the immense weight of ice depressed Earth's crust by about 2,600 to 3,300 feet (800 to 1,000 m) before retreating from the region as the climate began to warm from around 18,000 years ago. As the ice disappeared, the crust also began to rebound toward its original level but at a slower rate, leaving the region still depressed and largely below sea level, as is reflected in the shape of the present-day Gulf of Bothnia. It is estimated that the land has a further 300 to 400 feet (100 to 125 m) to rise before equilibrium is reached.

A RISING LANDSCAPE: SCANDINAVIA'S CONTINUING ISOSTATIC REBOUND

Repeated precise measurements of tidal gauges show that the northern Baltic Sea is rising at up to about 40 inches (100 cm) per century. This

recovery rate will progressively slow as isostatic equilibrium is approached such that the lithosphere —Earth's outer shell—may not completely reach its equilibrium before the next glacial episode.

Some of the best scenery illustrating the region's rapid uplift is found along Sweden's Gulf of Bothnia coastline, appropriately-named the High Coast. With its unusual tall cliff formations and raised beaches, the High Coast was recognized by the World Heritage Committee as the "type area" for research on isostacy and noted as the area where the phenomenon was first recognized and studied. At 935 feet (285 m), it holds the world's record for the highest isostatic land uplift following the last glaciation. For this reason, the High Coast achieved its inscription on the World Heritage List in the year 2000. The only other comparable area is Hudson Bay in Canada, where the equivalent uplift is 892 feet (272 m).

Continued uplift of the region will eventually create a land bridge between Finland and Sweden across the shallow Kvarken Strait, and the Bay of Bothnia will become Lake Bothnia.

PRESENT-DAY ICE CAPS AND GLACIERS

As the Scandinavian ice sheet retreated, it left isolated ice caps and glaciers in areas where colder conditions have allowed them to survive. These are principally along the mountain range separating Norway and Sweden, and on the islands within the Arctic Circle, including Spitsbergen, Franz Josef Land, North East Land, and Novaya Zemlya.

Studies in Sweden and Norway have shown that most of the Scandinavian glaciers have been retreating, as have glaciers in other parts of the world, since the end of the Little Ice Age. Interestingly, some Norwegian glaciers did have

brief periods of advance around 1910, 1925, and the 1990s. During the 1990s, a study of 25 Norwegian glaciers showed that 11 of these advanced due to above-average snowfall for several winters in a row. Since 2000, however, most of these glaciers had shrunk significantly as a result of several consecutive years of low winter snowfall, and the record warm summers of 2002 and 2003. By 2005, only one of the 25 glaciers was advancing. Engabreen had retreated 590 feet (180 m) during the period from 1999 to 2005, while Brenndalsbreen retreated 380 feet (116 m), and Rembesdalsskåka 675 feet (206 m) between 2000 and 2005. The Briksdalsbreen glacier retreated a record 315 feet (96 m) in 2004, which is its largest annual retreat measured since monitoring started in 1900.

ABOVE Ny Alesund is an international Arctic research settlement on Spitsbergen, in the Svalbard Archipelago. It is situated on the southern side of Kongsfjorden, the "king's fjord," a deep fjord yielding many calving glaciers at its head.

BELOW Efjord is one of many fjords that lie south of Narvik in far northwestern Norway between Tysfjord and Oftofjord. It is situated in Troms county, which has a highly indented coastline of far-reaching fjords and inlets, and many glaciers.

THE LITTLE ICE AGE

During the Little Ice Age, from the sixteenth to the mid-nineteenth centuries, average global temperatures dropped 2 to 3°F (1 to 1.5°C) below what they are today. This drop is attributed to a combination of decreased solar activity and increased volcanic activity. Many areas around the world are likely to have been affected, but it is the historical records of Europe and North America that best document the hardships that this imposed on the people's daily life. Falling crop yields and livestock losses led to famine in areas of Europe. It was reported by anglers that southern Hudson Bay remained frozen for about three weeks longer each spring and large amounts of sea ice blocked the North Atlantic.

Worldwide, alpine glaciers grew larger, and, in several cases, it was even reported that the glacial ice threatened mountain villages. Tree growth was also stunted due to shorter growing seasons during that period as attested by tree-ring data.

THE VAST LAKES DISTRICTS

In its retreat, the ice sheet left behind a debris-strewn hummocky landscape. Trapped behind ridges of moraine, the rivers could not drain

ABOVE A village along a fjord in Norway is blanketed by snow during wintertime. The communities along the myriad fjords of Norway are well serviced throughout the year by regular ferries that ply the fjords' waters.

properly and created an extensive labyrinth of forested islands and interconnected lakes, bays, and channels stretching across Sweden, Finland, and western Russia.

The best known of these is Finland's famous Lakes District, the place of many a Finn's memorable summer holidays spent boating, fishing, canoeing, or hiking. This "Land of a Thousand Lakes" actually has at least 55,000 lakes that are larger than 650 feet (200 m) across. Lake Saimaa is the largest with an area of more than 1,700 square miles (4,400 km²), and is the fifth-largest lake in Europe. These glacial lakes are generally very shallow, with their average depth being 23 feet (7 m).

Lake Saimaa, surrounded by Linnansaari and Kolovesi national parks, is home to the Saimaa ringed seal (*Phoca hispida saimensis*), an endangered lake seal found nowhere else in the world. Its ancestors appear to have become trapped in the lake when Saimaa was cut off from the Baltic Sea at the end of last Ice Age.

Also protected within the Finnish national park system are the habitats of various native animals, including the black grouse, hazel hen, wood grouse, willow grouse, bean geese, elk, hare, beaver, wild reindeer, and brown bear.

THE FAMOUS FJORDS OF NORWAY

Norway is renowned for its glacial scenery, such as that found in Jostedal Glacier National Park. This park surrounds one of mainland Europe's largest plateau glaciers, the Jostedalsbreen, and constitutes one of the largest wilderness areas in southern Norway. Spectacular scenic contrasts within a short distance vary from luxuriant fjords to U-shaped valleys in ice-covered mountains. Lowland farms in traditional agricultural landscapes give way to peaks such as Lodalskåpa, 6,834 feet (2,084 m) above sea level. Innumerable streams, rivers, and waterfalls lace the mountainsides during summer, fed by the ice-cap and snow melt.

Sognefjord is Norway's longest fjord, stretching more than 120 miles (200 km) inland from its outer coastal islands to its inner precipitous cliffs. The region where this deep fjord meets up with the glaciers and Norway's highest mountains is considered one of the world's most beautiful travel destinations. Another fjord, Hardangerfjord, is 110 miles (179 km) long and lined by a number of picturesque villages. Tourists visit this fjord to enjoy the Hardanger spring blossoming, when the district's fruit orchards flower against the backdrop of the snow-covered peaks.

ABOVE The leading edge of the Austfonna Ice Cap has a glacial front measuring 125 miles (200 km) in length. It covers the island of Nordaustlandet in the Svalbard Archipelago, which lies to the north of mainland Norway. With an area of over 3,000 square miles (7,800 km²), it is the second largest Arctic ice cap after that of Greenland.

Fjords and Glaciers of New Zealand

Fjords and glaciers are a striking feature of New Zealand's landscape. Fiordland National Park in the southwestern corner of New Zealand's South Island was created in 1952. It is the largest expanse of wilderness in the country, and one of the largest national parks in the world.

OPPOSITE Unlike most glaciers in the world, Fox Glacier in Westland National Park has been advancing slightly since the 1990s, after 100 years of retreat. This is largely due to colder temperatures and higher snowfall, attributed to the El Niño weather pattern present at this time, which will soon reverse.

BELOW A view along Lake Adelaide against the backdrop of the Darran Mountains in Fiordland National Park, shows the U-shaped valley that was originally carved by Pleistocene Epoch glaciers during the last Ice Age.

The 3 million acres (121 ha) of Fiordland National Park represent 10 percent of the total land area of New Zealand, and the 14 fjords located within the park's boundaries showcase some of the most spectacular transition points between land and sea to be found anywhere on the planet.

THE TOPOGRAPHY OF FIORDLAND

Gouged and shaped by glaciers during the Pleistocene Epoch, Fiordland is a complex mix of elongated high-altitude lakes and flooded U-shaped valleys. Fiordland's elevations range from the bed of Doubtful Sound, which reaches a depth of 421 feet (128 m), to mountain summits that rise up as high as 6,000 feet (1,800 m) above the vast glaciated expanses of plutonic and metamorphic rocks radically deformed by movements along the South Island's Alpine Fault.

The steep topography of Fiordland's mountains has been made possible by an amalgam of hard igneous and metamorphic rocks composed of diorite, granite, and gneiss. Their extreme resistance to erosion has preserved the ancient contours, inclines, and near-vertical rock faces of Fiordland that can reach as high as 5,000 feet (1,500 m).

FRESH WATER FJORDS

With an annual precipitation rate of around 300 inches (8,000 mm), Fiordland is one of the wettest places on Earth. Abundant waterfalls cascade down the region's near-vertical topography, bringing with them leaf litter, sediments, and tannins. These transform the upper layers of the fjords into a yellow-brown brackishness, floating above the denser saltier ocean waters below. These sediments inhibit the penetration of light, making photosynthesis difficult and reducing the limit of marine life to a depth of just 130 feet (40 m).

Rock debris at the mouth of these fjords, deposited during the retreat of glaciers during the last Ice Age, has formed barriers to ocean waters that today restrict the infiltration of seawater and give New Zealand's fjords extraordinarily meager levels of salinity—as low as 20 percent.

Despite the challenges presented by heavy sedimentation, poor light, and low salinity, the waters of Fiordland are home to subtropical fish, sponges, brachiopods, and black coral trees. Other permanent residents include the Fiordland crested penguin and New Zealand fur seals, as well as pods of bottlenose and dusky dolphins that are found in the waters of Milford and Doubtful sounds.

GLACIERS AND ANCIENT ICE SHEETS

A hundred thousand years ago, during the Waimaungan glaciation, New Zealand was dominated by massive ice sheets that extended from Fiordland to the west of Nelson at the northern extremity of the South Island and even into the North Island's Tararua Range.

All traces of glaciation from this period have long since been erased by erosion. The glaciers that dominate the landscape today are remnants of the much later Otiran glaciation that ended approximately 10,000 years ago.

These glaciers carved out the hollows now occupied by lakes such as Te Anau and Wakatipu. They formed and sharpened the great peaks of the Southern Alps and gouged out the glacial valleys of Fiordland that were later drowned by the rising oceans in the wake of the last Ice Age. Otiran glaciers also built the Canterbury Plains, lowlands

composed of fragmented rocks carried down from the alps in glacial moraines and later eroded by streams and water courses to eventually be deposited as broad swathes of sand and gravel.

GLACIAL MAPPING: SKETCHES TO SATELLITES

New Zealand's glaciers were initially recorded in the form of sketches in 1859, with the first precise measurements of movements and terminal positions taken by T. N. Brodrick in the Mt Cook region in 1889. It wasn't until the 1950s that researchers began to record and measure individual glacial volumes, water balance, and flow rates.

In recent years, satellite imaging has detected over 3,000 glaciers in the Southern Alps alone, with an approximate mass exceeding 12 cubic miles (50 km^3) and a combined area of almost 460 square miles (1,200 km^2).

New Zealand Fjords and Glaciers

MILFORD SOUND

The northernmost fjord to be found along the crenellated southwestern coast of New Zealand's South Island, Milford Sound contains some of the most awe-inspiring and dramatic fjord scenery found in the southern hemisphere.

ABOVE The kakapo *(Strigops habroptilus)* is an endangered flightless night parrot. Fewer than 50 birds existed in 1995, and a recovery program has seen its numbers rise to 86. One of the few places it lives is in a hanging valley in Milford Sound.

One of New Zealand's most-visited attractions, Milford Sound, along with the string of 6,000-foot (2,000-m) mountains that hem it in, were once described by the author Rudyard Kipling as the "eighth wonder of the world."

The Maori people first began to explore the region of Fiordland approximately 800 years ago and named Milford Sound "Piopiotahi," meaning Place of the Singing Thrush.

From the open ocean, Milford Sound is well hidden from view. It was twice missed by the great navigator Captain James Cook, once in 1770 on the *Endeavour*, and again in 1773—despite taking his ship, the *Resolution,* into nearby Dusky Sound for three months of repairs.

A WATER WONDERLAND

It wasn't until 1823 that John Grono, a Welsh whaler and fur sealer, located the sound after his vessel was forced into what appeared to be a bay that gradually opened up to reveal its hidden splendors. Grono named it Milford Haven, after his birthplace in Pembrokeshire. Milford Haven's name was later changed to Milford Sound by another Welshman, Captain John Stockes of the Royal Navy ship HMS *Acheron*, during a survey mission of the west coast in 1851.

Milford Sound begins its irregular 9-mile (14.5-km) long course to the waters of the Tasman Sea from its head at Lake Ada in the Arthur

Valley, picking up along its way the waters of the Harrison, Bowen, Stirling, and Sinbad rivers.

The flooded glaciated valleys of Milford Sound represent the true definition of a fjord, where rising ocean waters have entered deeply incised glacial troughs cut into the surrounding mountains. The peaks surrounding Milford receive over 250 inches (6,500 mm) of rain annually, which fall on an average of 182 days a year, producing the streams that enter the principal valley of Milford Sound as waterfalls. These include the Bowen Falls (520 feet; 160 m) and the permanently flowing Stirling Falls (480 feet; 145 m).

ANCIENT ROCKS AND GLACIAL RETREAT

Milford Sound's ancient rocks of diorite and gneiss have experienced 12 distinct glacial phases over the past 2 million years. The last retreat, during the Otiran glaciation, approximately 10,000 years ago, left a 1,000-foot (300-m) deep terminal moraine at its entrance that reaches to within

THE MILFORD TRACK

The Milford Track began as a Maori trail used to access deposits of greenstone, a valuable, translucent, metamorphic rock, also called New Zealand jade, found near the entrance to Milford Sound. Greenstone was used by the Maori people to make tools and ornaments.

In the 1880s, it was decided to commission the building of a track that would follow the old Maori trail from Lake Te Anau along the Clinton River and over the soon-to-be-named 3,507 foot (1,069 m) MacKinnon Pass. Continuing through the Arthur Valley, it would then join an earlier trail cut by the region's first European settler, Donald Sutherland, and bring to an end Milford Sound's isolation from the rest of the South Island.

Today, the Milford Track is universally recognized as one of the world's great walks. Its moderately demanding 33 miles (54 km) can be negotiated over five days and four nights. The number of walkers is strictly limited so intending trekkers must book well in advance.

90 feet (27 m) of the surface, preventing ocean swells from entering the sound and giving it an abnormally low rate of salinity.

Like many fjords, the depth of Milford Sound increases the further inland it extends. Halfway along its length, Mitre Peak rises to a height of 5,560 feet (1,694 m) and is surrounded by hanging valleys such as Sinbad Gully, one of the few remaining habitats of the kakapo flightless parrot.

MILFORD DEEP UNDERWATER OBSERVATORY

Located on the shoreline of Harrison's Cove, with Pembroke Glacier and Mitre Peak looming above it, the Milford Deep Underwater Observatory provides a unique perspective into the surrounding marine environment.

Milford Sound's tannin-rich waters have resulted in an unusual deepwater environment that consists of tubeworms, sponges, black coral, shellfish, and brachiopods. These animals are normally only found at far greater depths in the open ocean.

ABOVE A clear day on Milford Sound is a rarity—this part of New Zealand has one of the world's highest rainfalls, and a sunny day here is a luxury.

LEFT A rainbow prism strikes the mist coming off Stirling Falls in Milford Sound. This is one of the many waterfalls that plunge into the depths of its chilly waters. Here, the fjord reaches its greatest depth—1,280 feet (390 m) below sea level.

Ice Sheets and Glaciers of Antarctica

Antarctica, the coldest, driest, and windiest continent, is 98 percent ice-covered. The Antarctic ice sheet is the largest body of ice on Earth. Its thickness averages 1½ miles (2.5 km) and it contains 90 percent of Earth's ice. Melting of all of this ice would release enough water to raise the global sea level by about 200 feet (60 m).

Snow and frost accumulating on the continent has formed a continuous sheet containing some 7 million cubic miles (30 million km³) of ice. The Antarctic ice sheet is between 120,000 and 500,000 years old and is over 13,800 feet (4,200 m) thick in places. It is a weight so immense that it has depressed portions of the continent's surface to more than 8,200 feet (2,500 m) below sea level. Under its own weight, the ice sheet pushes outward from the center of the continent toward the sea, with some areas of ice moving much faster than others.

PENINSULA GLACIERS IN MAJOR RETREAT

In 2005, the British Antarctic Survey and the US Geological Survey jointly published a study of Antarctic Peninsula glaciers. Comparing thousands of satellite images and aerial photographs dating back to 1940, the study found that 87 percent of the region's glaciers are retreating at an average rate of about 165 feet (50 m) per year. It is a recent phenomenon as, up until the 1950s, most of these glaciers were advancing. The retreat of the glaciers, which lie mostly on the western side of the peninsula, has coincided with the collapse of ice shelves on the eastern side. Researchers believe this is due to human-influenced climate change.

PECULIAR FLOWING ICE STREAMS

Anomalously fast-flowing areas are known as ice streams, and these are separated by slower-moving ice ridges. These rivers of ice, up to 30 miles (50 km) wide, travel at speeds of 3,000 feet (1,000 m) per year, while the surrounding ice travels much slower. The reason for this phenomenon is that the ice stream is slipping on a water-saturated soft-sediment base, while the inter-stream ridges are frozen to hard bedrock. Ice streams are quite common in Antarctica, making up about one-tenth of the ice volume. They account for most of the ice leaving the ice sheet and entering the ice shelf.

FLOATING ICE PLATFORMS ON THE OCEAN

Ice streams reaching the coast push out onto the ocean bottom until they eventually lift off to form a thick floating platform of ice, known as an ice shelf. Antarctica's two largest ice shelves are the Ross Ice Shelf in the Ross Sea and the Filchner-Ronne Ice Shelf in the Weddell Sea. The line marking where the grounded ice ends and the floating ice shelf begins is known as the grounding

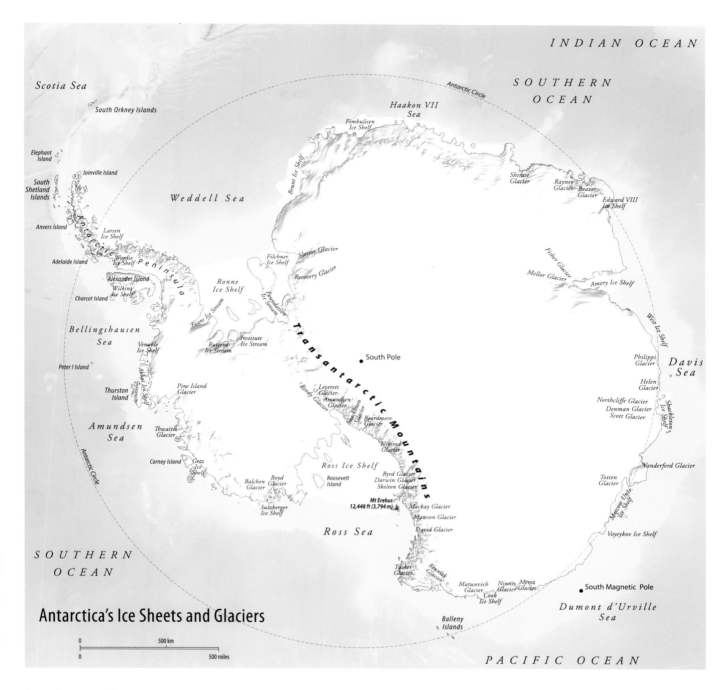

Antarctica's Ice Sheets and Glaciers

INDIAN OCEAN

SOUTHERN OCEAN

Scotia Sea

South Orkney Islands

Antarctic Circle

Haakon VII Sea

Fimbulisen Ice Shelf

Elephant Island

Joinville Island

South Shetland Islands

Shirase Glacier

Rayner Glacier

Beaver Glacier

Edward VIII Ice Shelf

Brunt Ice Shelf

Weddell Sea

Anvers Island

Larsen Ice Shelf

Antarctic Peninsula

Wordie Ice Shelf

Filchner Ice Shelf

Slessor Glacier

Fisher Glacier

Adelaide Island

Recovery Glacier

Mellor Glacier

Amery Ice Shelf

Alexander Island

Wilkins Ice Shelf

Ronne Ice Shelf

Foundation Ice Stream

Charcot Island

West Ice Shelf

Bellingshausen Sea

Evans Ice Stream

Rutford Ice Stream

Institute Ice Stream

Transantarctic Mountains

Philippi Glacier

Davis Sea

Venable Ice Shelf

Peter I Island

South Pole

Helen Glacier

Thurston Island

Abbot Ice Shelf

Pine Island Glacier

Leverett Glacier

Reedy Glacier

Amundsen Glacier

Northcliffe Glacier

Denman Glacier

Scott Glacier

Amundsen Sea

Thwaites Glacier

Shackleton Ice Shelf

Beardmore Glacier

Vanderford Glacier

Carney Island

Getz Ice Shelf

Ross Ice Shelf

Nimrod Glacier

Totten Glacier

Balchen Glacier

Boyd Glacier

Roosevelt Island

Byrd Glacier

Darwin Glacier

Skelton Glacier

Sulzberger Ice Shelf

Mt Erebus 12,448 ft (3,794 m)

Mackay Glacier

Mawson Glacier

Merow Upya Ice Shelf

David Glacier

Ross Sea

Voyeykov Ice Shelf

SOUTHERN OCEAN

Tucker Glacier

Rennick Glacier

Matusevich Glacier

Cook Ice Shelf

Ninnis Glacier

Mertz Glacier

South Magnetic Pole

Dumont d'Urville Sea

Balleny Islands

PACIFIC OCEAN

0   500 km
0   500 miles

---

line. The ice shelf continues to grow outward to its equilibrium point where the ice breaks off or calves from the seaward edge of the shelf. Usually, the shelf is about 300 to 3,000 feet (100 to 1,000 m) thick and its seaward edge will extend forward for years or even decades between periodic large calving events.

THE ICE SHELVES START COLLAPSING

Since 1995, several major sections of Antarctic shelf ice have broken away. Beginning that year, the Larsen A Ice Shelf, with an area of 600 square miles (1,600 km²), became detached from the

continent. In 1998, the 420-square-mile (1,100-km²) Wilkins Ice Shelf gave way. The largest was in March 2002, when a 1,250-square-mile (3,250-km²) segment of the Larsen B Ice Shelf collapsed. These collapses are most likely being triggered by higher air temperatures and warmer waters circulating beneath the ice shelves.

ANTARCTICA'S GLACIERS SPEED UP

Measurements made between 2000 and 2003 using satellite radar show that glaciers have been flowing into the ice shelves much faster than previously, perhaps because they are no longer being

OPPOSITE A Russian icebreaker, the *Kapitan Khlebnikov*, cuts through the sea ice in Antarctica near Ross Island. Passengers are able to disembark onto the sea ice, which is many hundreds of feet thick in some places. The Ross Ice Shelf is Antarctica's largest, and covers an area roughly the size of Spain.

ABOVE A massive glacier wall spills into Yankee Harbor on Greenwich Island. Situated in the Antarctic Peninsula, it is one of the South Shetland islands, many of which are covered by ice caps that yield many glaciers.

restrained by the surrounding ice shelves. For example, several glaciers flowing into the collapsed Larsen B Ice Shelf are traveling two to six times faster than before, and have thinned considerably.

LAKES DISCOVERED BENEATH ANTARCTIC ICE
An arterial network of hundreds of lakes has been discovered thousands of feet beneath the surface of the Antarctic continental ice sheet. They appear to fill before bursting suddenly and sending their waters cascading along the base of the ice sheet. Lake Vostok, discovered beneath Russia's Vostok Station in 1996, is the largest of these subglacial lakes. Seismic soundings indicate that the lake is comparable in size to North America's Great Lakes, with a water depth of 1,600 feet (500 m), and an extent of 30 miles (50 km) by 140 miles (225 km). Drilling is yet to reveal if Lake Vostok's waters, which may be as old as 35 million years, contain microbial life. The lake may be similar to those thought to exist below the frozen surface of Jupiter's moon, Europa.

AN ABUNDANCE OF LIFE IN ANTARCTICA
There are no vertebrates that live permanently on land in Antarctica. Invertebrate life includes microscopic mites, lice, nematodes, tardigrades, rotifers, and springtails. At just one quarter of an inch (6 mm) in size, the wingless midge (*Belgica antarctica*) is the largest land-based animal in Antarctica. However, Antarctica's sea bird and marine life is abundant. Tremendous blooms of marine phytoplankton (diatoms), thriving in the endless summer sunlight, form the base of the marine food chain.

VARVED SHALE
Varved shale is formed in glacial lakes. Each "varve" is a pair of light-colored silt and dark-colored clay bands, and represents one year of deposition on the floor of the lake. During summer, inflow of glacial meltwater into the lake is stronger and can carry and deposit coarser glacial silt or even sand. During winter, inflow greatly reduces or even ceases and the suspended fine-grained clay settles to the bottom, forming a thin dark layer. The varves in this sample (shown at right) have been cut by numerous tiny faults.

Above the phytoplankton are the vital Antarctic krill (*Euphausia superba*) and, in turn above them, sit a variety of krill eaters—penguins, seals, and baleen whales. The very top of the food web belongs to the orca or killer whale (*Orca orcinus*) —the most common predator of the large krill eaters—and the leopard seal (*Hydrurga leptonyx*), which feeds on penguins.

Seabirds are another important element in the Antarctic food web, with most feeding on krill and the larger ones on squid and fish. Besides penguins, the majority of Antarctic seabirds are skuas, gulls, terns, albatrosses, and petrels. The albatrosses and petrels are known as tubinares (tube noses) because of the tube-like nostrils on top of their bills. These contain glands that act like a desalination plant filtering excess salt from the bird's blood enabling them to drink seawater. Most of these seabirds nest during the Antarctic summer on sea-cliffs of the continent and the surrounding islands. They leave during the Antarctic winter when there is little food. The penguins of Antarctica and the subantarctic islands include the emperor, which breeds in Antarctica during the winter, the adelie, which breeds further south than any other penguin, the rockhopper, with distinctive eyelash-like feathers around the eyes, the king, chinstrap, and gentoo penguins.

THE SUBANTARCTIC ISLANDS

These are the islands of the Southern Ocean that mostly lie south of Tasmania, New Zealand's south island, South Africa's Cape of Good Hope, and South America's Cape Horn. The islands are claimed as territories of various nations for historic reasons, such as first discovery, or because they lie in the arc subscribed by that country from the South Pole. They include the Falklands,

AN ISLAND CARVED BY GLACIERS

South Georgia is a forbidding, cliff-bound, and glacially carved island. The highest peak is Mt Paget at 9,594 feet (2,935 m) high, and over half of the mountainous island is covered by ice that has been in retreat for about 17,000 years. The Fortuna Glacier in Cumberland Sound is the largest of the island's tidewater glaciers. Raised beaches occur up to 30 feet (10 m) above sea level, inland from the existing beaches. They appear as distinct tussock-covered flat terraces underlain by smooth beach cobbles and shingle. The beaches are rising in response to the unloading of the vast weight of ice as the glaciers retreat.

South Georgia lies in the subantarctic, west of the southern tip of South America; its rugged landscape was carved by glacial activity. The island has 11 mountains over 6,500 feet (2,000 m) in height.

Gough, Tristan da Cunha, South Georgia, South Sandwich, and South Orkney, all claimed by the United Kingdom; Norway's Peter Island and Bouvetøya; New Zealand's Campbell, Chatham, Bounty, Auckland, and Snares islands; Australia's Heard and Macquarie islands; France's Kerguelen and Crozet islands; and South Africa's Marion Island. San Pedro Island (South Georgia) and Las Malvinas (the Falklands) are also claimed by Argentina, which resulted in the war between the United Kingdom and Argentina in 1982.

Most of these inhospitable, windswept, and ice-covered islands are World Heritage Sites for various reasons; they are home to over half of the world's seabird population, some species of which are not found elsewhere. The islands support rare plants and animals that differ from their relatives found in more temperate climates and, despite human occupation on many, ecological restoration efforts have been highly successful.

Most of the islands are volcanic in origin, with many still active. On October 10, 1961, Tristan da Cunha's several hundred residents were forced to evacuate their island when a new volcanic cone began to erupt not far from the main settlement area. After being relocated to live in England for two years, the islanders lobbied to return, and were repatriated on November 10, 1963.

ABOVE A pair of black-browed albatrosses (*Diomedea melanophris*) nesting. These seabirds breed on subantarctic islands around the Southern Ocean. Known to live for up to 29 years, they have no more than one chick per year.

LEFT A satellite image of Antarctica shows the glacier on Mt Erebus, an active stratovolcano, flowing outward and over the Ross Ice Shelf. This glacial tongue is over 30 feet (10 m) high and 7 miles (11 km) long.

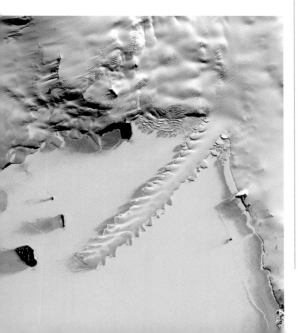

Deep canyons and gorges carved by powerful rivers tell us that we are indeed living on an active planet. These steep-sided valleys only exist wherever the land is rising quickly, giving water a vast amount of potential energy to erode rocks, carry sediment, and cut huge slots into the landscape.

Canyon is the term generally used in the United States, while the word gorge is more common in Europe and Oceania. Canyons and gorges are different from a normal river valley in that they have steep to vertical sides, as opposed to more gently sloping V-shaped or U-shaped valley walls. Canyons are more common in arid areas because weathering of their surrounding rock walls takes place more slowly than it does in humid tropical environments.

ABOVE This textbook example of an entrenched hairpin meander has been carved by the Colorado River in Canyonlands National Park, Utah, USA.

RIGHT About 1,500 years ago, the Anasazi people built a high-density city in the canyon walls of the Colorado Plateau, USA. The place is now in ruins.

Slot canyon is the term used for those canyons that are very narrow. It is often possible to touch both sides of a slot canyon at the same time, and their smooth polished walls may be so close together that it is impossible for a human adult even to squeeze between them.

Most of us know about the famous canyons such as Grand Canyon in the United States and Africa's Blue Nile Gorge. However, there are so many other spectacular canyons and gorges in the world for us to discover and explore.

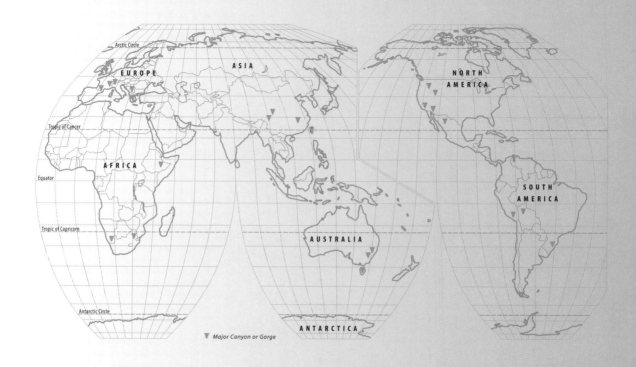

▼ *Major Canyon or Gorge*

CANYONS POINT TO AREAS
OF TECTONIC INSTABILITY

Canyons and gorges can be found everywhere wherever tectonic activity is creating active mountain ranges and volcanoes. These zones principally occur around colliding plate margins, where continents crush against one another, or are pushing over the top of the sea floor. The rising, volcanically active mountain ranges around what is known as the Pacific "Ring of Fire" mark the line along which the oceanic plates of the Pacific basin are being swallowed up by the surrounding plates. The Grand Canyon in Arizona, USA, and Copper Canyon in Chihuahua, Mexico, are just two examples of the very dramatic gorges that cut into these active mountain belts.

The great collision mountain belts that stretch from the Himalayas of Asia to the European Alps host some of the world's deepest and most impressive canyons, such as the Yarlung Tsangpo Gorge in Tibetan China. Canyons also exist where Earth's crust is uplifted and torn open along tectonic plate boundaries, known as spreading margins. Here, plates move away from one another and create a gap. These fault-bounded gorges, such as Tanzania's Olduvai Gorge in Serengeti National Park, are more correctly called rift valleys.

EARTH'S MOST IMPRESSIVE CANYONS

The Grand Canyon, in Arizona, USA, is probably the world's best-known and most visited canyon. The Colorado River has cut this spectacular desert canyon through sedimentary layers, exposing an enormous span of geologic time; the 1.7 billion-year-old Vishnu schist is exposed at the very bottom. The Grand Canyon is often quoted as being Earth's largest canyon, but it is actually the Yarlung Tsangpo Canyon that takes the prize. Here the Brahmaputra River has carved a mighty gorge in the Himalayas to a depth of $3\frac{1}{3}$ miles (5.3 km),

which makes the Grand Canyon seem somewhat small in comparison, with depths of approximately 1½ miles (2.4 km). Other giant gorges in the area include the Kali Gandaki Gorge in Nepal, and the Polung Tsangpo Gorge in China. It is the phenomenal rate of uplift in this area, brought about by the ongoing collision between the Indian subcontinent and Asia, which is leading to the formation of these incredibly deep canyons. The land here is rising at more than 1 inch (25 mm) per year.

All of Earth's canyons pale into insignificance in comparison to some of the mighty structures found on Mars. The enormous canyon system near the Martian equator known as Valles Marineris is 125 miles (200 km) wide and as much as 4⅘ miles (7.7 km) deep. The canyon was named after the *Mariner 9* Mars orbiter, which transmitted photographs of it back to Earth in 1972. Valles Marineris is believed to be a Martian rift system.

ENTRENCHED RIVERS ETCH THEIR PATHS
Canyons and gorges are both actually a form of entrenched river, or a river that is now no longer able to alter its course. A river will only etch its drainage pattern permanently onto the landscape when its ability to wear down through the rocks matches or exceeds the rate of uplift in that area. In some cases, even the slow-flowing meanders, typical of an old river flowing across its floodplain, can become heavily entrenched.

Classic examples of entrenched hairpin meanders (meanders that loop back on themselves) have been carved by the Snake River in Snake River Canyon, Idaho, and also by the Colorado River in Canyonlands National Park, Utah, USA. Another bizarre example can be seen on the Emu Plains, west of Sydney, Australia. Here, a meander of the Nepean River heads into the Blue Mountains, flowing around in a broad loop through a narrow canyon, before emerging again on the plains.

EXQUISITE CANYON PATTERNS
The several different patterns that canyon systems can carve into the landscape depend very strongly on the underlying geology. If there is no structural control, such as in areas of thick uniform sediment or soft rocks, a dendritic pattern of river drainage will develop. From the air, these look rather like giant tree roots with tributaries coming together to form larger and larger streams at acute angles until they have all joined into the main trunk stream. Typically, however, there is a superimposed or accordant pattern to the drainage, as most hard rocks, such as granite, are crosscut by at least two directions of jointing or faulting. These planes of weakness offer easier channel-ways for flowing water, and eventually the joints are cut deeper and

ABOVE The world's oldest river, China's Yarlung River, drops from about 10,000 feet (3,000 m) to 1,000 feet (300 m) over the length of Yarlung Tsangpo Gorge.

OPPOSITE Matkatamiba Canyon in Grand Canyon National Park, Arizona, USA, is a slot canyon. The sandstone was eroded over a long period of time to create this natural wonder.

HOW A SLOT CANYON IS FORMED
A. A low energy river or mature river flows sluggishly on its coastal floodplain, forming a series of broad hairpin meanders. These change with the seasonal floods, forming meander cutoffs and oxbow lakes. B. Rapid uplift of the land rejuvenates the river system, giving it energy to begin rapidly cutting into the rocks. The meanders entrench themselves to form a slot canyon and are no longer able to move freely. C. Once the river has finished downcutting, the valley slowly begins to widen as the processes of weathering and erosion attack the exposed sedimentary layers. The meanders remain fixed in place.

A   B   C

Canyoning is the adventure sport of exploring slot canyons of varying difficulty. Sparkling pools lit by shafts of sunlight, waterfalls shooting from between boulders, and wet, glistening, moss-covered, shady walls with dripping ferns are some of the sights that lure people into canyons. Some canyons are easily accessible on foot; others require more technical skills to be able to explore them including climbing, jumping, abseiling, swimming, and navigational skills. Some water-filled canyons with long stretches of wall-to-wall pools, such as those of the Blue Mountains in Australia, are negotiated using inflatable air mattresses. Wetsuits are often a necessity to keep warm while exploring the chilly water in canyons. Canyoning carries with it the risk of sudden flooding. Thunderstorms anywhere in a canyon's catchment area can lead to dramatic and sometimes unexpected rises in water level causing its calm still pools to become raging torrents within minutes—even though the skies may be clear overhead. This is known as flash flooding, which makes a deadly combination with the steep sides of a canyon. It is often impossible to leave the canyon before finishing the planned route. Flash flooding fatalities have occurred, such as the group of tourists who drowned in Saxetenbach Gorge, Switzerland, while on a commercial canyoning trip in 1999.

Many people are thrilled by the adventure—and the risks—involved in canyoning.

deeper. Most typically this creates a distinct rectangular drainage pattern with tributaries joining one another at right angles.

In areas of dipping or folded sedimentary rock sequences, the differential hardness of the strata will control the pattern of drainage. Long straight channels will develop along the softer strata in a parallel pattern, only rarely cutting across the more resistant strata to form larger channels. Smaller tributaries are very short, flowing from the adjacent ridges of hard strata on each side of the main channel and creating a trellis pattern.

Rapidly rising domes or growing volcanoes will force water to flow away from the highest point, carving gullies and canyons in a radial pattern. The reverse of this pattern occurs if the center of the dome is a soft sedimentary rock, such as salt. A circular depression or basin will form, encouraging the water to flow inward toward the center. This pattern is known as centripedal drainage.

SUBMARINE CANYONS

Submarine canyons are very steep to vertical sided gorges that cut into the sea floor of the continental slope. They are more closely spaced on the steeper continental slopes, and they cut into all rock types as well as loose sediment. They have only been discovered and explored by submersibles relatively recently. Some of the largest submarine canyons occur as offshore extensions of Earth's major rivers. The largest of these is Congo Canyon, which extends from the mouth of the Congo River. It is 500 miles (800 km) long, and 4,000 feet (1,200 m) deep. Other examples include the Amazon Canyon, the Ganga (Ganges) Gorge, the Hudson Canyon, and the Indus Gorge, which extend from the mouths of their respective rivers.

There is another group of submarine canyons and channels that have no connection with major rivers, including the Monterey Canyon, off the coast of central California. One of the longest of these is the Northwest Atlantic Mid-Ocean Canyon off the east coast of Canada. As many of these canyons are found cutting well into the flat or very gently sloping area known as the abyssal plain, it has been difficult to explain what has caused them. They are well below the maximum depth that has ever been exposed by sea level falls during glacial times, so rivers did not cut them.

THE MYSTERY OF SUBMARINE CANYONS REVEALED

It is now believed that submarine canyons have been cut by dense, sediment-laden fluid flows that move through the canyons like underwater rivers of sand. The canyons are continuously supplied by sediment that is brought to the deltas of the major rivers, or by sand that is carried along the beaches by long-shore drift and then fed into them. The movement of sediment through the canyons may also be sudden and sporadic, which some believe could be from collapsing masses of sediment triggered by earthquakes. These flows are known as turbidity currents or underwater landslides, and enormous sediment fans have been found on the abyssal plain at the mouths of submarine canyons as proof of this mechanism.

CANYON PEOPLE

The Anasazi, or Ancient Pueblo civilization constructed thousands of high-density, apartment-style dwellings as well as other structures in the canyon walls of the Colorado Plateau in the southwestern United States. These stone structures include underground pits and buildings for grain storage, circular stone watchtowers, and elaborate ceremonial structures called kivas. The settlement dates back to about 500 CE when the people became full-time farmers growing corn, beans, and squash, and no longer needed to move from place to place. The Anasazi learned to master the changing seasons by storing their produce over winter in enclosed granaries. They wove baskets out of plant fiber and made clay pots to carry and store food and other items of daily life. They also domesticated the turkey, and supplemented their diet by hunting, while living in harmony with their land. This abandoned stone city was only rediscovered in 1888 by two cowboys, Richard Wetherill and Charles Mason, and has since provided an amazingly rich source of archeological evidence about the way of life of this prehistoric people.

Archeologists now comb the many dwellings and caves for clues regarding the natural, social, or religious crisis that may have led to the collapse of this magnificent civilization during the thirteenth century. Some archeologists studying human bones

that had been found in the ruins concluded that the Anasazi began to suffer from malnutrition, shorter life spans, and increased infant mortality at around that time. Tens of thousands of Anasazi abandoned their cliff homes at Mesa Verde, and dispersed to the Hopi mesas in northeastern Arizona as well as to the Zuni lands in western New Mexico. They also constructed adobe villages in the Rio Grande catchment. Due to a lack of written records, it is still not fully understood what caused this to occur. Why were the people of this region starving and forced to move?

The most obvious theory was that a prolonged dry spell in the late 1200s, known as the Great Drought, drove these people from their homeland; however, it now appears that the situation was more complex than this. Tree-ring and pollen studies suggest that Anasazi farmers were also affected by erratic seasonal patterns caused by the worldwide cooling trend known as the Little Age Ice. This meant that the Anasazi were "climatically squeezed" from two directions. Lower elevations became too dry for farming, but farmers were unable to move to higher ground because it was too cold. Another theory proposes that the drier weather and food shortages drove the Anasazi to form hostile groups that would raid one another, and that this led to conflict and warfare. However, this theory fails to explain why the winners of the conflict didn't remain in the region.

ABOVE Archeologists have taken a keen interest in the rock dwellings carved by members of the ancient Anasazi civilization at Marble Canyon in Arizona, USA.

ABOVE LEFT These vertical canyons of red rock in Tanzania are located in the East African Rift Valley, a region of great archeological significance.

Canyons and Gorges of the United States of America

The climate of the western United States is extreme: summer temperatures exceed 100°F (38°C) and winter temperatures drop well below freezing. There can be severe droughts and floods, especially in a terrain dominated by desert and mighty rivers such as the Colorado and the Gunnison. In these harsh conditions and environments, fabulous canyons wait to be explored.

When the Pacific tectonic plate forced its way beneath the North American tectonic plate, both plates buckled. This resulted in huge disruptions about 70 million years ago, which began in California and gradually moved eastward through Utah, Arizona, Nevada, Idaho, and Oregon to Colorado. Over time, these disruptions caused mountain ranges such as the Rocky Mountains and Sierra Nevada to form, and the mountains channeled water into rivers and ice into glaciers. Plateaus such as the Colorado Plateau rose up above sea level.

The wild rivers and glaciers eroded their way down through the plateaus, creating stunning canyons that numerous species of plants, animals, and birds call home.

THE MULTI-HUED ROCKS OF BLACK CANYON

Black Canyon in Colorado is a 2,000-foot (610 m) deep rainbow. Nearly 2 billion years ago, coarse black metamorphic gneiss and micaceous schist formed in the planet's crust. As they cracked, pink pegmatite (igneous rock, containing shiny muscovite mica and pink potassium feldspar) squeezed into the fissures. An inland sea that had covered the area retreated 70 to 285 million years ago, leaving layers of red sandstone and mud that became Mancos shale (a slippery mudstone that still causes landslides today). Between 18 and 30 million years ago, volcanoes erupted. Ash mixed with rock and mud to form gray sediments.

When the Colorado Plateau rose, the Gunnison River cut through it at the rate of the width of a human hair each year. The river exposed the many rock layers, and the canyon is now a swirl of black, pink, red, and gray rock, complementing the emeralds and olives of the riverbank vegetation and woodland.

Sauropod dinosaurs with long tails and necks, theropod dinosaurs (like *Tyrannosaurus rex*), and *Stegosaurus* once lived in the area. Now horned owls, elk, golden-mantled ground squirrels, and

coyotes roam through the woodlands of aspen, singleleaf ash, and scrub oak, while the peregrine falcon, the fastest bird in the world, drops at over 200 miles (320 km) per hour in an aerial dive.

HOODOOS, LIVING SOIL, AND POTHOLES

Bryce Canyon in Utah is a series of amphitheaters, the largest being Bryce Amphitheater, which is 12 miles (19 km) long, 3 miles (5 km) wide, and 800 feet (250 m) deep. This spectacular canyon is

ABOVE Wavelike patterns occur in the Navajo sandstone of Monument Valley in Arizona. The intense red banding is caused by iron staining the water that seeped through originally white sandstone.

RIGHT A fresh blanket of snow settles on Bryce Canyon in Bryce Canyon National Park, Utah, as seen from Inspiration Point. This amphitheater of the canyon features an array of impressive formations including the many hoodoos, which were sculpted by the erosional activity of rivers.

named for Ebeneezer Bryce, a pioneer and cattle farmer, who described the canyon as "a helluva place to lose a cow."

Between 40 and 63 million years ago, shallow lakes deposited limestone, mud, and sandstone on top of the Dakota sandstone and Tropic shale (mud and silt containing fossilized ammonites) that had been laid down by an inland sea in the late Cretaceous Period.

When the Colorado Plateau uplifted, it rose unevenly, creating tension that formed vertical joints. Over the past 10 million years, these joints were eroded by wind and water, creating rock spires, called "hoodoos," up to 200 feet (60 m) high. The hoodoos are pink, orange, brown, red, yellow, and white, depending on the amounts and types of sediment they contain, and have names such as Poodle and Wall Street.

Also in Utah is Canyonlands National Park, a series of canyons over 2,000 feet (600 m) deep, crossed by the Colorado River and its tributaries. These canyons contain a black soil crust made up of cyanobacteria, lichens, mosses, green algae, microfungi, and other bacteria. Cyanobacteria,

or blue-green algae, are one of Earth's oldest life forms, and today work with the other organisms in the crust to create a haven for trees such as water birch, tamarisk, and box elder. These trees provide a habitat for birds such as yellow-breasted chats, blue grosbeaks, and spotted towhees, and are inspected by bald and golden eagles soaring overhead.

Naturally occurring depressions in the sandstone in the canyons, called potholes, collect rainwater and sediment, forming tiny ecosystems. These potholes range from fractions of inches to a few yards in depth, and vary greatly in temperature, depending on the climate. As water evaporates, animals move to larger pools or bury themselves beneath the mud to withstand dehydration. Creatures found living in such potholes include fairy and clam shrimp, spadefoot toads, and snails.

A series of attractive rocky spires, known as The Needles, can also be seen in the park. They are made of Cedar Mesa sandstone, which is around 250 million years old and was deposited when an inland sea covered much of western Utah. The white bands in the sandstone are sand, the red bands are sediment from mountain streams.

ABOVE A mountain lion or cougar (*Felis concolor*) leaps easily across rocks in Canyonlands National Park, Utah. Highly adapted to rocky terrain, they can leap cliffs up to 18 feet (5 m) high. Their large paws and long tails help them balance.

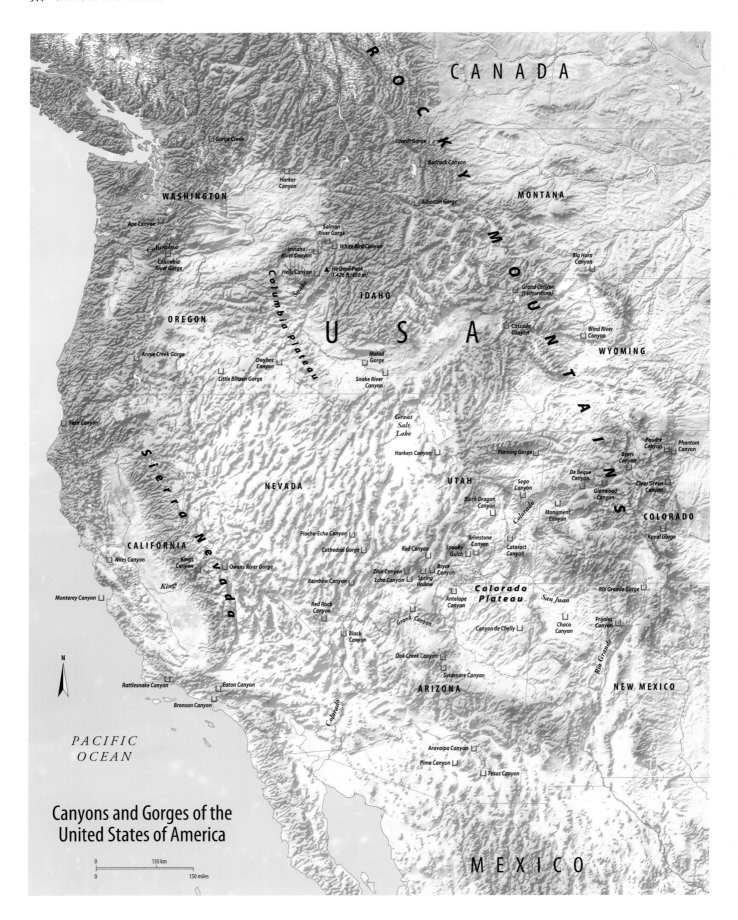

CANADA

ROCKY MOUNTAINS

USA

Gorge Creek

Sunrift Gorge

Badrock Canyon

Harker
Canyon

WASHINGTON

Alberton Gorge

MONTANA

Ape Canyon

Salmon
River Gorge

White Bird Canyon

Big Horn
Canyon

Columbia

Columbia
River Gorge

Imnaha
River Canyon

Hells Canyon

He Devil Peak
1,476 ft (450 m)

Grand Canyon
(Yellowstone)

Snake

IDAHO

Cascade
Canyon

OREGON

Columbia Plateau

Wind River
Canyon

WYOMING

Annie Creek Gorge

Owyhee
Canyon

Malad
Gorge

Little Blitzen Gorge

Snake River
Canyon

Great
Salt
Lake

Poudre
Canyon

Phantom
Canyon

Fern Canyon

Harkers Canyon

Flaming Gorge

Byers
Canyon

De Beque
Canyon

Clear Green
Canyon

Sierra Nevada

NEVADA

UTAH

Sego
Canyon

Glenwood
Canyon

Black Dragon
Canyon

Colorado

Monument
Canyon

COLORADO

CALIFORNIA

Pioche-Echo Canyon

Red Canyon

Brimstone
Canyon

Spooky
Gulch

Cataract
Canyon

Royal Gorge

Niles Canyon

Kings
Canyon

Owens River Gorge

Cathedral Gorge

King

Zion Canyon
Echo Canyon

Bryce
Canyon

Spring
Hollow

Colorado
Plateau

San Juan

Rio Grande Gorge

Monterey Canyon

Rainbow Canyon

Antelope
Canyon

Red Rock
Canyon

Grand Canyon

Canyon de Chelly

Chaco
Canyon

Frijoles
Canyon

Black
Canyon

Rio Grande

N

Oak Creek Canyon

Rattlesnake Canyon

Eaton Canyon

Colorado

Sycamore Canyon

NEW MEXICO

Bronson Canyon

ARIZONA

PACIFIC
OCEAN

Aravaipa Canyon

Pima Canyon

Texas Canyon

Canyons and Gorges of the
United States of America

MEXICO

0   150 km

0   150 miles

LEFT Cataract Canyon is in the heart of Canyonlands National Park in Utah at the confluence of the Green and Colorado rivers. This narrow gorge runs over 50 miles (80 km) and drops by 400 feet (120 m).

ABOVE Sockeye salmon (*Oncorhynchus nerka*), swim in the Snake River, which cuts through Hells Canyon. Other fish found here are lamprey and white sturgeon, and birds include the three-toed woodpecker and pygmy nuthatch. Mammals such as gray wolves, California wolverines, and mountain lions prowl amid native bunchgrasses, wildflowers, and mariposa lilies.

Utah's canyons are home to 50 mammal species. To escape the harsh winters, mule deer migrate to lower elevations. To cope with hot summers, kangaroo rats get water from the plants they eat, and sleep in a cool underground burrow. Beavers live in lodges on the riverbanks as the waterways are too large to dam. Indeed, when the Colorado River enters Cataract Canyon, there are 14 miles (23 km) of rapids that, at peak flows, may result in waves over 20 feet (6 m) high.

DEEP, DRAMATIC, AND DYNAMIC

Between the mountains of the Sierra Nevada in California lies Kings Canyon, North America's deepest canyon where Spanish Peak towers 10,051 feet (3,064 m) above the Kings River. The predominant rocks in Kings Canyon are granite, diorite, and monzonite, and were formed when molten rock—a by-product of the plates colliding—cooled far beneath Earth's surface.

The Sierra Nevada is a young mountain range, probably not more than 10 million years old. At least four glaciers have formed in the mountains, coating their slopes in ice and moving through them like a giant dragging knives and shovels cutting into the ground. The glaciers carved deep valleys and craggy peaks, among them Kings Canyon.

Hells Canyon, on the border between Idaho and Oregon, varies from 9,393 feet (2,866 m) high where He Devil Peak rises above the Snake River to just 1,480 feet (450 m) high at Granite Creek.

Most of the dark rocks along the canyon's lower walls come from volcanoes that erupted on islands in the Pacific Ocean around 300 million years ago. After volcanic activity stopped, the region was covered by ocean and limestone sediment was deposited. Next, molten granitic rocks rose up through Earth's crust into the older sediments

and lava, where they slowly crystallized. Lava flowed over the region, creating a nearly level plateau 15 to 17 million years ago. About 6 million years ago, the Snake River began to cut its way through the plateau. Its activity intensified during the last 2 million years when melting glaciers, intense rainfall, and flooding from large lakes added to its power. The river still continues to cut through the plateau today.

The canyon has arid deserts, woodlands, and high alpine areas, providing a home for an astounding diversity of plants and animals.

SLOT JACKPOT

Many amazing canyons less than 3 feet (1 m) wide and over 100 feet (30 m) deep can be found in North America's southwest. These slot canyons have been formed over 100 to 200 million years by violent flash floods and windstorms cutting through sandstone. Many found on the Colorado Plateau have been carved out of Navajo sandstone in glowing colors—crimson, vermilion, orange, salmon, peach, pink, gold, yellow, and white. The color depends on the mix and amounts of hematite, goethite, and limonite in the sandstone. Antelope Canyon in Arizona is a beautiful and much visited slot canyon. Others are Brimstone Canyon, Spooky Gulch, Echo Canyon, and Spring Hollow, which contains a waterfall and petroglyphs.

GRAND CANYON

No one visiting this place fails to be impressed by the Grand Canyon. It is one of Earth's greatest natural wonders, and it burns its image into the memory forever. Theodore Roosevelt, visiting in 1903, said: "You cannot improve upon it ... [it is] the one great sight which every American should see."

A World Heritage Site, the Grand Canyon encompasses 1,900 square miles (4,900 km²). It varies in height from 100 feet (30 m) above sea level in the west to 9,165 feet (2,795 m) in the north. Its south rim is, on average, 7,000 feet (2,100 m) above sea level while the higher, less visited, and wetter north rim is 8,000 feet (2,400 m) above sea level. Cutting into the Colorado Plateau, the canyon is about 277 miles (446 km) long and 15 miles (24 km) at its widest point.

ANCIENT GEOLOGICAL HISTORY
The Grand Canyon contains some very old rock at its base. The Vishnu schist is the remains of a mountain range 2 billion years old, which was eventually worn down by erosion.

Shallow seas covered the area during the Precambrian Era. The remains of algae caused layers of limestone to form. The seas retreated, leaving sandstone and mudstone. Over a span of 400 to 500 million years, erosion by wind and water carved away many of these deposits.

About 600 million years ago, during the Ediacaran Period, seas flooded the area again. Unlike earlier oceans, these contained animals similar to oysters or clams. Their remains formed enormous limestone layers. When the seas retreated, erosion wore away some of these layers along with layers of sandstone and shale. In certain areas, where erosion wore away so much sediment, the "youngest" rocks are 250 million years old.

The cycle of oceans flooding, then waning, leaving the area to be worn away by erosion, continued until about 70 million years ago, when accumulated layers began to rise above sea level due to volcanic activity, which created the San Francisco Peaks, 70 miles (110 km) away, and the Rocky Mountains to the east.

The Colorado River was formed by water draining from those mountains, and it began to cut its way through the rock. Due to compression in the crust caused by friction between the North American and Pacific tectonic plates, the Colorado Plateau lifted up about 17 million years ago. As the plateau rose, the river continued to cut its way through it. The plateau stopped rising 5 million years ago and the river settled into its current course, continuing to erode and widen the canyon.

ABOVE Looking along the Colorado River at sunrise from the Toroweep lookout in the Grand Canyon at 3,000 feet (915 m) gives an overall impression of the depth of this canyon.

RIGHT The Grand Canyon as seen from the southern rim. In 1540, a party of Spaniards exploring the area was led to the canyon by Hopi guides. These Native Americans believed people were created by the earth and they first emerged through an opening known as the *sipapu*, with the Grand Canyon being a typical example. The park receives 5 million visitors annually.

A VARIETY OF FLORA AND FAUNA

Home to a dozen endemic plants and many animals, the Grand Canyon yields much evidence of ancient life. Its rocks house the fossils of brachiopods, mollusks, sponges, and shark's teeth.

The region contains three broad habitat areas—the Colorado River corridor and vegetated banks, canyon desert scrub with its succulent plants and pink and orange rock, and coniferous forests. At elevations between 6,500 and 8,200 feet (2,000 and 2,500 m) above sea level, there are ponderosa pines with cinnamon-colored bark that smells like vanilla (with which the tree shares a chemical compound). At elevations higher than 8,200 feet (2,500 m) above sea level, the landscape features Douglas fir trees, blue spruces, lupines, and asters.

Around 1,500 plants (12 of which are endemic), 355 birds, 89 mammals, 47 reptiles, 9 frogs, and 17 types of fish live in the canyon.

The California condor, an endangered species with an impressive wingspan of 10 feet (3 m), soars overhead, sharing the skies with birds such as ravens and pinon jays.

Bighorn sheep thrive on the canyon's rocky slopes, which keep them safe from coyotes and mountain lions. Their horns continue to grow throughout their lives, those of the male sheep growing into a full curl.

The gila monster, the largest and only venomous lizard in the United States, shelters under overhanging rocks—this shy animal has a painful but not deadly bite.

ABOVE North America's gila monster (*Heloderma suspectum*), is one of only two poisonous species of lizard in the world. They eat small rodents, ants, and the eggs of other animals.

Canyons and Gorges of Mexico, Central America, and South America

Mexico, Central America, and South America encompass a region dominated by extremes in its topography and geography. Its rising mountain ranges are dissected by canyons carved by powerful sediment-laden rivers and glacial activity.

ABOVE The capybara (*Hydrochaeris hydrochaeris*) is a large rodent that lives in Iguazu Falls gorge and other parts of South America where thick vegetation lines riverbanks, lakes, marshes, or swamps. A highly efficient digestive system means they can extract more nutrients from plants than other mammals.

OPPOSITE, TOP RIGHT Yareta (*Azorella compacta*) at Cerani Pass in the Colca Canyon, South America. These cold-resistant, compact cushion plants blossom below 16,000 feet (4,880 m). They grow close to the ground to take advantage of the warmth generated by the loose soils.

RIGHT The Copper Canyon in Chihuahua, Mexico, was formed by the Rio Urique. The canyon contains a variety of microclimatic zones, ranging from its forested cliff tops to the vegetation on its desert floor.

In Central and South America, water has dominated and carved the landscape to create long deep gorges into which waterfalls crash or hanging glaciers swoop. Iguazu Falls, some of the largest in the world, fall into a deep canyon 1¾ miles (2.7 km) long on the border between Brazil and Argentina. The canyon is at its deepest at Garganta del Diablo, or Devil's Throat, where water the color of caffe latte cascades 2,300 feet (700 m) into the Iguazu River. Swifts roost in the canyon walls while fox-like coatis and capybara—the world's largest rodent—roam the banks.

In Chile, turbulent white waters boil through the many canyons and evergreen forests of Parque Nacional Queulat. Ventisquero Colgante, a beautiful turquoise hanging glacier, laces the cliff face and drops into a deep gorge.

MEXICO'S INTERCONNECTED CANYONS

The Copper Canyon, or Barranca del Cobre, is a 25,000-square-mile (64,750-km^2) system of six interconnected canyons in the Sierra Tarahumara in Chihuahua, Mexico. The canyons are narrow and steep, with sheltered valley floors that have much warmer temperatures than the rims. As a result, the tops of the canyons are covered in pine forests while their valley floors display desert vegetation. At 6,136 feet (1,871 m) deep, Urique Canyon (Barranca del Urique), is the deepest canyon in the system, while Copper Canyon—which gives the system its name—is 5,770 feet (1,760 m) deep.

About 40 million years ago, the Pacific Plate was subducted beneath the North American Plate, creating massive volcanic eruptions that spewed out rhyolite, a thick lava rich in silica. The walls of the canyon are composed of layers of rhyolite volcanic ash and mud.

At the same time, the area where Copper Canyon is located was also uplifted and tilted and subsequently eroded by a number of rivers, including Rio Verde, Rio Batopilas, Rio Urique, and Rio Fuerte—which began carving the canyons about 29.5 million years ago.

The canyons are a haven for birds such as the zone-tailed hawk, hairy woodpecker, yellow-eyed junco, Mexican chickadee, and painted redstart.

ANDEAN VOLCANIC CANYONS

The canyons of the central Andes in Peru demonstrate the power of water erosion and volcanic explosion. Mighty rivers have carved canyons that bear their name—Rio Cotahuasi has carved the Cotahuasi Canyon and Rio Colca has eroded into rock to form the Colca Canyon. The walls of both of these canyons are largely composed of explosive volcanic ignimbrites. As part of the "Ring of Fire"—a fault line stretching round the Pacific Rim—the area where these canyons are based is still subject to disturbances. Volcanoes still steam, earthquakes still threaten, and rivers still flood.

Cotahuasi Canyon—arguably the deepest canyon in the Americas—is 11,000 feet (3,350 m) deep and is becoming deeper every year. Studies have found that the river eroded between 2 to 3 cm of rock every hundred years between 3.8 and 15 million years ago—when the canyon's creation began—and has slowed down to eroding 0.5 cm of rock per year since that time.

Colca Canyon is another superb geological feature. At its deepest point, it is 10,725 feet (3,270 m) deep. The turbulent green Rio Colca, beloved by whitewater rafters, buffets through its 37-mile (60-km) long stretch of multi-colored ignimbrites, sandstone, and limestone. The predominant colors are pink and gray.

The canyon is also home to the Andean condor (*Vultur gryphus*), a large black-and-white bird with a huge wingspan. The birds find the canyon an ideal habitat as they prefer to roost in high places from where they can launch without much effort. The canyon's cliffs provide them with the thermals they need to help them rise effortlessly into the air, where they hover in search of prey.

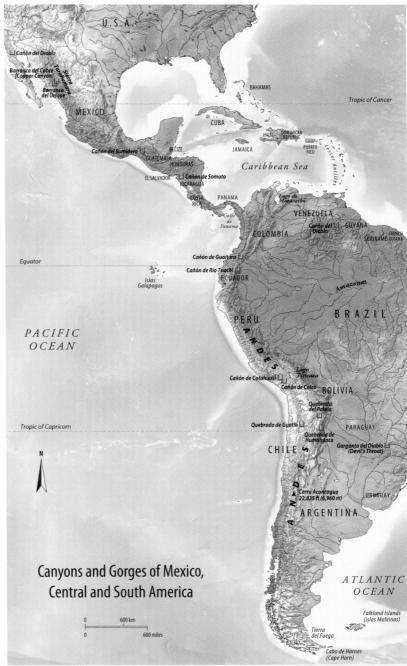

Canyons and Gorges of Mexico, Central and South America

Canyons and Gorges of Europe

The geologically variable continent of Europe has a rich diversity of gorges. Some are composed of limestone and contain turquoise rivers or glacial lakes; some are long winding corridors of granite; yet others are made of sandstone, oolite, chalk, and gypsum.

BELOW The Gorges du Verdon in Provence, France, were formed when the Alps were uplifted. Cutting its way through the limestone is the emerald-hued Verdon River. These gorges are very popular with climbers.

Many gorges—including three of Europe's deepest—have a limestone karst geological formation. Limestone is a sedimentary rock mainly composed of calcium carbonate. Karstification occurs when rainwater absorbs carbon dioxide, becomes acidic, and dissolves the limestone, so cracks and fissures widen and eventually become canyons.

FRAGRANT HERBS AND FORESTS

Europe's deepest canyons have a lot in common. All have beautiful turquoise or emerald rivers flowing between walls of red and gray limestone, are situated in national parks, and shelter many species of plants and animals.

Gorges du Verdon in Provence, France, is 2,300 feet (700 m) deep and 13 miles (21 km) long. It has cone-shaped peaks, gigantic cliffs, and valleys where rosemary, mint, and thyme perfume the air. It was also a favorite place for people to commit suicide until barriers were put up in 1980. In November 2006, officials removed the hulks of 10 cars from the site, some of which dated back to the 1930s.

Vikos Gorge lies in the North Pindus Mountains in Epirus, Greece. At its deepest point, the gorge reaches a depth of 5,840 feet (1,780 m), is 7.5 miles (12 km) long, and about 3,600 feet (1,100 m) wide. Its deciduous forests contain orchids, beech and maple trees, and mammals such as bears, foxes, and deer.

Tara River Canyon, 4,265 feet (1,300 m) deep, 50 miles (80 km) long, narrow in places and at other points over 3 miles (5 km) wide, is located in Durmitor National Park, a World Heritage Site in Montenegro. It was formed by a combination of river erosion and tectonic forces that caused the mountains to rise as the river cut down. The canyon is composed mainly of limestone from the mid- and late Triassic, the late Jurassic, and the late Cretaceous periods.

One of Europe's last black pine forests clings to the walls of the canyon, and some of the pine trees are 165 feet (50 m) high and 400 years old. Gray wolves, wild cats, weasels, and stoats roam the area to the music of the yellow-bellied toad's howl, the long-tailed tit's whistle, and the thrush's song.

LIMESTONE CAVES IN CHEDDAR GORGE

Cheddar Gorge is the United Kingdom's largest gorge and one of its most visited. J. R. R. Tolkien included it in his honeymoon tour of 1916, and the gorge is said to have been the inspiration for Helm's Deep in his book, *The Lord of the Rings*. Britain's oldest human skeleton, the 9,000-year-old

Canyons and Gorges of Europe

Cheddar Man, was found here. Situated in the Mendip Hills in Somerset, England, the gorge is up to 370 feet (110 m) deep.

Its life began 425 million years ago, when the continents of Baltica and Laurentia collided, joining the northern and southern halves of Britain. Subsequent volcanic activity, uplift, and erosion, between 370 to 408 million years ago, formed Old Red sandstone (sedimentary rock up to 36,000 feet [10,980 m] thick). Shelled marine creatures in advancing and retreating waters between 299 and 359 million years ago were the basis of Carboniferous Period limestone that formed on top of the sandstone. During the Ice

LEFT Carboniferous Period limestone cliffs of Cheddar Gorge, in Somerset, England, also contain several limestone caves. Fossils abound here and the 9,000-year-old skeleton of Cheddar Man was found in this gorge.

located southeast of Stuttgart in Germany. Almost 200 million years ago, during the Jurassic Period, the area was covered by a sea, which deposited dark limestone, clays, marl, and oil-shale. Shells and carbonate skeletons of coral and sponges created a layer of white limestone. After the sea retreated, volcanic activity raised the land. During the mid-Jurassic Period, between 180 and 159 million years ago, fine-grained clays were deposited. The Danube River, with the help of karstification, carved the gorge into these rocks.

CRETE'S ABUNDANT GORGES

The island of Crete has had an interesting geological history, which has led to the formation of over 100 limestone gorges. About 10 million years ago, the Mediterranean Sea rose over the ancient landmass of Aegeis and formed the island. Upheavals caused by the African tectonic plate colliding with the European Plate raised the island further. Continuous uplift cracked the earth, creating volcanic eruptions. Ensuing climate changes brought either glaciers or river torrents, which eroded the limestone and marble, creating the gorges. Many are in the Sfakia area, over a dozen of which run north–south into the Libyan Sea. These include the 11-mile (18 km) long Samaria Gorge, Crete's longest and most visited canyon. Composed mainly of dolomite and covered in cypress trees, the gorge was created during the Quaternary Period through glacial action and karstification. Samaria Gorge is 1,640 feet (500 m) deep and, at some points, only 7 feet (2 m) wide.

Other Cretan gorges are the 7-mile (11 km) long Imbros Gorge, the 2,300-foot (700 m) deep

ABOVE View over the Samaria Gorge and the surrounding mountains in southwestern Crete, Greece. At around 11 miles (18 km) in length, it is the second longest gorge in Europe. Its towering 1,600 feet (490 m) sides narrow to a gap less than 13 feet (4 m) apart at the Iron Gates.

Age that occurred at that time, water in the limestone froze, then thawed. Torrents of melted ice eroded the rock and it became a gorge.

Home to many rare plants and animals, the gorge shelters lime trees, hazel shrubs, and bluebells; endangered greater horseshoe bats; and ravens, tawny owls, and adders.

Another limestone wonder is the craggy Danube Gorge in the Swabian Alb—a low mountain range

Ha Gorge, and the Kourtaliotiko Gorge—the cliffs of which are roosting sites for the rare griffon vulture (*Gyps fulvus*).

The gorges of Crete contain 130 endemic plants including dictamo (*Origanum dictamnus*), which is a herb with medicinal properties used by the Ancient Greek philosopher Aristotle, and the evergreen Cretan plane tree. Endemic animals are equally diverse—there are Cretan subspecies of wild goat, marten, wildcat, badger, and golden eagle, among others.

METAMORPHIC TRANSFORMATIONS

Not all of Europe's gorges have walls of lime-stone—some are granite. The gorges of Dartmoor, in England, are composed of Dartmoor granite, a coarse igneous rock made of quartz, feldspar, bio-tite, and tourmaline.

During the late Carboniferous to early Permian periods, disturbances in the crust made Dartmoor's shale, sandstone, and limestone fold and fault. At the same time, heat generated by the disturbances created molten rock, which intruded into the other rocks and solidified, forming the Dartmoor granite. Starting about 280 million years ago, the rocks were metamorphosed by heat and pressure, causing concentration of metals such as tin and copper. Some granite also altered, its feldspar turning into china clay during these lengthy geological processes.

Continuing uplift of the land, together with water erosion, formed many gorges. Lydford Gorge is the most spectacular—a deep narrow canyon containing the 90-foot (27 m) high Whitelady Waterfall.

The island of Corsica in France is also composed of granite. The island's canyons are deep and rugged, with emerald green pools sparkling in the polished white or coral-colored rock. Two of Corsica's most spectacular canyons are the Gorges du Tavignano and the Gorges de la Restonica.

LEARNING ABOUT GORGE FORMATION—BLETTERBACH GORGE

The Bletterbach Gorge in Italy's South Tyrol is 5 miles (8 km) long and up to 1,300 feet (400 m) deep. Its strata can be clearly seen, so the gorge is now the central feature of a geological park that contains signs explaining its composition to visitors.

Hot ash and lava from volcanic explosions 260 to 280 million years ago hardened into the reddish-gray Bozen ignimbrite, which formed the base of the gorge. Toward the end of the Paleozoic Era, this rock was eroded to form the Gröden sandstone, 525 feet (160 m) thick, on top of the ignimbrite.

A shallow sea covered the area, retreating and returning. Gypsum built up in the mud, forming dark gray strata 130 feet (40 m) thick. Beginning about 248 million years ago, the advancing and retreating seas caused deposits of oolite, chalk, marl, and claystone to form, which have created strata about 1,300 feet (400 m) thick. This strata contains fossils of mussels, snails, and algae.

The summit of Blettebach Gorge is made of pale porous dolomite formed in an ancient tropical sea by algae.

ABOVE A griffon vulture takes flight in its search for food, scavenging for the remains of dead animals. An endangered species, it lives among the rocky crags and outcrops of Kourtaliotiko Gorge in Crete.

BELOW The rugged orange granite outcrops of Calanques de Piana in Corsica glow a vibrant red in the sunset. This steep-sided valley plunges 980 feet (300 m) to the sea, forming a gateway to the narrow Gorges de Spelunca, which is over 3,000 feet (900 m) deep in places.

Canyons and Gorges of Africa and the Middle East

Africa's canyons are carved mainly into dolomite. This pink or gray sedimentary rock is an enigma, as it changes once it has been deposited from a calcite- or aragonite-rich limestone to dolomite. African canyons contain a variety of exotic wildlife and fossils of dinosaurs and ancient humans. In the Middle East, the Jordan Valley–Dead Sea Rift Valley is dissected by canyons featuring unique flora.

ABOVE The quiver tree or koker-boom (*Aloe dichotoma*) has very rough bark that is used by native people to make quivers for their arrows. Not actually a tree, this plant grows in western southern Africa and many are found in Namibia's Fish River Canyon.

RIGHT The Blyde River carves its way through the Blyde River Canyon in Mpumalanga Province, South Africa. It is the third-largest canyon in the world and one of its main features is a number of massive potholes up to 20 feet (6 m) deep.

OPPOSITE, TOP RIGHT The Blue Nile flows swiftly through the Blue Nile Gorge in Ethiopia. Local stories tell of evil spirits that will reach up from the river and pull the unwary down into its treacherous rapids.

South Africa's Blyde River Canyon was formed where Madagascar and Antarctica tore away from Africa nearly 200 million years ago.

The land's torn edge was gradually tilted up-ward around a vast shallow sea that left behind thick layers of dolomite and red sandstone. Today these strata form the base of the canyon that was carved by the Blyde River as the land was uplifted. The canyon is now 16 miles (26 km) long and up to 5,250 feet (1,600 m) deep in places.

The canyon's wonders include the Three Rondavels, huge domes of dolomite decorated with bright orange lichen and topped with vegetation, and the Bourke's Luck Potholes, yellow and red, mainly sandstone cylinders that contain deep dark pools. The canyon's waterfalls, grasslands, and woodlands are home to somango monkeys, bushbabies, hippopotamuses, otters, and hundreds of birds.

A RAVINE CARVED IN DOLOMITE

Fish River Canyon in Namibia is one of southern Africa's most visited attractions. It is 100 miles (160 km) long, up to 17 miles (27 km) wide, and, in places, almost 1,800 feet (550 m) deep.

About 1.8 billion years ago, sandstone, shale, and lava layers became compressed and were heated to over 1,100°F (600°C) due to volcanic activity. Lava cut through these sediments about 900 million years ago, and soon after that erosion began to expose and level the rocks. These were covered by a shallow sea that left pebbles, black limestone, grit, shale, and sandstone.

Earth's crust fractured about 650 million years ago, depositing pink dolomitic sediments that can still be seen exposed in a north–south fault along which the canyon was formed. The canyon was widened by glacial action, wind, and erosion by the Fish River—a waterway that was once a mighty torrent but is now slow-flowing.

Namibia's national tree, the quiver tree—which can live for up to 300 years—grows in the canyon, as do succulents adapted to the semi-desert environment. Animals that can survive drought and hot weather live here, including leopards, klipspringers (rock-dwelling antelope), mountain zebras, and baboons. The canyon is also home to birds such as ostriches, fish eagles, and love birds, and reptiles like the black spitting cobra and the horned adder.

ETHIOPIA'S GRAND CANYON

The Blue Nile Gorge in Ethiopia is 250 miles (400 km) long, 15 to 20 miles (25 to 30 km) wide, and about 4,920 feet (1,500 m) deep—comparable to the depth of the Grand Canyon in the USA. It contains wild waterfalls and rapids, and exposes layers of rock laid down over hundreds of millions of years. The basement rock is 800-million-year-old granite, topped with 150-million-year-old sandstone and limestone that is more than a mile thick in places. Finishing off this rock sandwich is 20- to 30-million-year-old black basalt. This basalt makes the canyon seem dark and forbidding, especially where it is narrow and deep.

Crocodiles and fish-eating birds, such as kingfishers and eagles, live here as do vervet monkeys. Fossils have been found of *Allosaurus* and *Pliosaurus* dinosaurs, lungfish, and ancient turtles.

WILD CANYONS IN THE MIDDLE EAST

The Middle East's canyon gems include Gamla Gorge in Israel and Koprulu Canyon in Turkey, which has a wild river winding beneath its rocky crags. Oman's Wadi Shab flat desert landscape bursts into a sudden oasis of palms, waterfalls, and caves within Jebel Sham, the massive Canyon of Arabia.

Jordan's Dana Nature Reserve is criss-crossed with plunging canyons and contains peculiar white sandstone domes. Shag ar-Reesh or "the canyon of the feathers" is so named because of its feather-like rocks. Dana has abundant unique vegetation and the occasional wolf, golden jackal, striped hyena, red fox, or Nubian ibex can also be spotted there.

Also in Jordan, there are hot springs near a deep and narrow chasm in Wadi Mujib Reserve along the mountainous eastern side of the Dead Sea.

ABOVE Namibia's Fish River Canyon is the largest canyon in Africa and the second largest in the world. The flow of the river varies with the seasons. At the end of summer it floods, sending great torrents downstream; it later becomes reduced to a mere trickle then stops altogether, leaving behind still pools.

PETRA GORGE, JORDAN

There are over 800 caves and buildings in Petra, an ancient city carved into the rockface inside a sandstone gorge in Jordan. Wandering around these is like visiting a world where an artist has randomly thrown brightly colored paint. Splashes of contrasting colors—rose, orange, blue, ocher, and cream—abound.

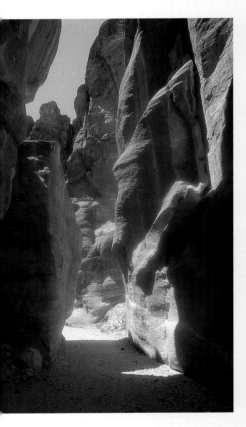

ABOVE The Siq is the path that leads between the high, narrow walls of reddish sandstone in Petra's gorge into the entrance to Petra, Jordan. In places it narrows to a just beyond a person's height with walls towering many hundreds of yards above.

OPPOSITE The facade of the Treasury in Petra, Jordan, one of many elaborate buildings carved into the gorge's sandstone walls by the Nabateans. These ancient people were skilled engineers who constructed ingenious hydraulics to create an oasis, and dams to divert flash floods.

The setting for the film *Indiana Jones and the Last Crusade*, Petra could have been created by a surrealist painter such as Salvador Dali. The multi-colored walls of Petra's sandstone gorges were carved into colossal and beautiful buildings more than 2,000 years ago by the Nabateans—a powerful people who controlled the trade routes through Arabia.

Petra (Arabic for "stone") was unknown to the outside world for centuries—although local Bedouin people lived and worked there. It was found in 1812 by a Swiss explorer, John Louis Burckhardt, who convinced the local people to lead him into Petra via the slot canyon, which is 1 mile (1.6 km) long, up to 650 feet (200 m) deep, and at times only 7 feet (2 m) wide.

Today, visitors can also enter Petra through the Siq, as locals call the canyon. Winding between its sheer walls, they suddenly glimpse—through a narrow space in the rock—the glowing pink facade of El Khazneh, or the Treasury. Visitors can also walk through other gorges, such as a ravine where goats climb and bushes cling to crevices in rock walls, to the massive, sand-colored, 150-feet (50 m) high Ad-Deir, or Monastery. They can also enter Petra through other narrow canyons cut through the sandstone by nature.

GEOLOGY OF A RAINBOW

Petra is situated near the Dead Sea Rift, a fault that formed 30 million years ago when the Arabian and African plates began to separate. Both plates are moving northward but at different speeds, causing stresses in the earth that create many earthquakes. Petra's gorges were formed through a combination of these disturbances in the earth and water erosion widening and deepening cracks in the rock. Other gorges in Jordan created in this way include the Mujib Canyon near the Dead Sea, into which a waterfall tumbles.

Petra's gorges are dry except during flash flooding and are mainly composed of sandstone. Much of this dates from the Cambrian to Silurian periods and is 408 million to 590 million years old. Some younger—early Cretaceous Period—sandstones are 100 to 140 million years old.

Petra's sandstone contains quartz that is colored by iron, aluminum, manganese, sodium, and lithium. Various combinations of these elements create the spectacular colors found in the sandstone. Oranges occur where there is a predominance of iron, yellows are produced by sodium and iron, pinks from the combination of sodium and lithium, and mauves are produced by manganese. Geologists call these wavy concentric patterns in the sandstone Liesegang banding. This banding was formed as deposits from mineral-rich water once flowed through the rock like ink through blotting paper.

Rock varnish can also be seen at Petra. This feature is a coating of manganese and iron hydroxides on the surface of the rock that makes it appear to glow. Petra's famous El Khazneh is coated with a manganese-rich varnish—which explains why its carved walls stand out so much.

PETRA'S WILDLIFE AND PEOPLE

Petra's harsh arid conditions favor birds that are desert specialists. The scarcity of food and unpredictability of rainfall mean they live at low densities and make full use of rain when it arrives. Some, like the 13 species of lark found at Petra, are sandy-colored to merge with the desert background. Some, like the bright pink Sinai rosefinch, are camouflaged against Petra's vibrant hues. Other birds found in the area include the Barbary falcon, desert lark, scrub warbler, Hume's tawny owl, and orange-tufted sunbird.

Reptiles found in and around Petra include rare Sinai agama lizards, the males of which have bright blue heads and necks.

The Bedouin are nomadic people who have lived at Petra for many years. While some have modern homes, others live in tents and caves—as they have for centuries—and preserve their traditional customs.

Canyons and Gorges of Asia

When it comes to canyons and gorges, Asia is a continent of superlatives. It has the deepest canyons in the world, some of which snake between the world's tallest and youngest mountains, the Himalayas. With walls composed of marble, limestone, granite, or gneiss, the gorges contain a kaleidoscope of diverse landscapes, rare plants, and endemic animals.

ABOVE Being sensitive to heat, the diminutive red panda (*Ailurus fulgens*) is found in the gorges of the Himalayas at altitudes where the temperature rarely rises above 77°F (25°C). The size of a domestic cat, it feeds mainly on bamboo.

Asia's canyons are a product of the continent's turbulent geological history. Fifty million years ago, the Indian tectonic plate, moving northward, collided with the Eurasian Plate. It pushed more than 1,200 miles (2,000 km) into Asia at a rate of around 2 inches (5 cm) per year, and is still moving northward today at a rate of approximately $\frac{4}{5}$ inch (2 cm) per year.

The collision between the plates and subsequent pressure led to buckling and folding of the crust, followed by upheavals. Large sections of Asia's crust soared skyward about 5 million years ago, creating mountain ranges. Rivers took advantage of the fault lines and cut into the bedrock. Subsequent glacial action and flooding, combined with constant earthquakes and overheating of the crust, formed deep narrow gorges.

WINDING GORGES OF THE HIMALAYAS

In the rain shadow of the Greater Himalayas in Nepal, the Kali Gandaki Gorge has been carved by a dark river. The dark coloring is acquired from the black stones of the river's bed and the sediments it contains. The gorge is dramatic, deep, and shady, running at one point between the Dhaulagiri and Annapurna ranges, which are both more than 26,250 feet (8,000 m) in height. The gorge itself is up to 19,000 feet (5,800 m) deep.

Kali Gandaki Gorge has walls of schist and gneiss that formed during compression of the tectonic plates. Rock slides have contributed to its geology—for example, during the late Pleistocene to the Holocene Epoch, between 5 to 10 feet (1.5 to 3 m) of broken rock, river sand, and lake sediments accumulated in the gorge.

Sheltering temperate broadleaf forests on its higher elevations, the gorge is home to the rare Namdapha flying squirrel, red panda, clouded leopard, and tiger. By restricting their territory size and reproductive rate, tigers have adapted to survival in these forests where prey is less available than in other areas.

In Pakistan, the 2,000-mile (3,200 km) long, pale blue Indus River separates the Himalayan and Karakoram ranges and has carved deep narrow canyons, including the Indus Gorge. This gorge, at 3,300 feet (1,000 m) above sea level, cuts through the western Himalayas' Nanga Parbat massif, which at 26,657 feet (8,125 m) high is the third highest mountain in Asia. The gorge's folded layers of dark gray and pale gray igneous and metamorphic rocks testify to its geological history. Gneiss and basalt from Asia's ancient continental crust combine with granites only 2 million years old from the young Himalayan mountain belt.

Granite is also exposed in the walls of Bhagirathi Gorge in India. Its river is one of the headwaters of the Ganges River and emerges from beneath a glacier that lies 13,000 feet (4,000 m) above sea level from where it falls into the gorge. The narrow gorge—between 100 and 130 feet (30 and 40 m) wide—runs between the Himalayas and is up to 13,000 feet (4,000 m) deep. The granite contains gouges made over time by catfish while feeding on mineral-rich algae. Sandy coves are home to basking shags and cormorants, flitting redstarts, and barking deer. Above them soars the lammergeier, or bearded vulture, with a wingspan of 9 feet (2.7 m) from tip to tip.

TORRENTIAL RIVERS, TIGERS, AND ROCKS FROM THE TETHYS SEA

The Three Parallel Rivers Protected Area in southwestern China covers more than 13,000 square miles (roughly 34,000 km^2) of snowy mountains, glaciers, karst formations, forests, grasslands, and spectacular gorges.

The three rivers—the Jinsha (a tributary of the Yangtze), Lancang, and Nu—flow through steep parallel gorges for almost 200 miles (300 km) between the Hengduan Mountains. In places, the gorges are less than 12 miles (20 km) apart but never meet. There are complex patterns of folded rock in their walls. These rocks were originally igneous rocks, part of the ancient ocean floor and the sediments that were deposited on it. These were crushed and uplifted during the closure of the Tethys Sea by the approaching Indian Plate.

These gorges include the 196-mile (315-km) long and up to 12,500-feet (3,800-m) deep Nu River Gorge, which is flanked by two 13,000-feet (4,000-m) high mountains, Gaoli and Biluo, and the Hutiao or Tiger Leap Gorge that was carved by the emerald green Jinsha River. This gorge is 10 miles (16 km) long, up to 12,800 feet (3,900 m) deep, and only reaches between 100 to 200 feet (30 to 60 m) wide—its name stemming from a legend that once a giant tiger jumped over it as it was so narrow.

All the gorges shelter semi-arid shrublands on their lower levels. Between 8,000 and 10,000 feet (2,400 and 3,000 m) there are coniferous and

ABOVE The Jinsha River in China flows through the Jinsha River Gorge, part of which is the Tiger Leap Gorge. This gorge is so deep the climate varies distinctly from top to bottom.

BELOW Vast areas around Mt Yaoshan near Guilin bloom with azaleas. Much-loved by gardeners, these flowering shrubs are native to China, and more than 300 species have been found in Yunnan alone.

deciduous forests. On the mountain slopes are alpine heathlands, grasslands, and tundra. Flowering plants such as poppies, rhododendrons, gentians, and orchids color the area. Snow leopards live in the mountains and Bengal tigers in the forests, and the blood pheasant and black-necked crane are sovereigns of the sky. Black snub-nosed and Yunnan golden monkeys live in the trees.

China's other marvels include the beautiful gorges of Yangshuo and Guilin. These gorges have been carved by the Lijiang River through karst limestone and dolomite, which was deposited between the Middle Devonian to Lower Carboniferous periods. The gorges snake between row upon row of tall narrow limestone pinnacles.

There are also the spectacular Three Gorges—located in China's Hubei Province—a 75-mile (120-km) long network of canyons carved by the Yangtze River. These gorges are being flooded by an ambitious dam project—currently the largest hydroelectric river dam in the world.

Away from mainland China, the geological history of the province of Taiwan has formed the beautiful black-and-white marble-walled Taroko Gorge. Taiwan was underwater 100 million years ago, when a layer of marine shells and coral was deposited and turned into limestone. Between 80 and 90 million years ago, intense pressure and heat changed the limestone into marble. Eventually, the Philippine oceanic plate collided with the Eurasian continental plate between 2 and 4 million years ago, and subsequent compression, folding, and faulting forced what is now the island of Taiwan upward and pushed the marble to the surface.

Taroko Gorge was later formed by the action of the Liwu River and its tributaries. The Liwu River continues to gnaw away at the resistant marble, and at the gneiss and schist of which the gorge is also formed.

Taroko Gorge is about 12 miles (19 km) long, and its walls reach up to 5,465 feet (1,667 m) above the river. This spectacular gorge is flanked by tropical red maples and sugar palms at lower elevations, while forests of cypress, spruce, and walnut are found at higher elevations. This diverse environment is home to a variety of animal and bird species, including Formosan rock monkeys, flying squirrels, blue magpies, Mikado pheasants, and many spiderss.

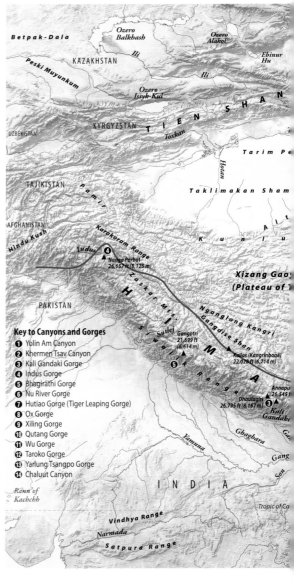

Key to Canyons and Gorges
1. Yolin Am Canyon
2. Khermen Tsav Canyon
3. Kali Gandaki Gorge
4. Indus Gorge
5. Bhagirathi Gorge
6. Nu River Gorge
7. Hutiao Gorge (Tiger Leaping Gorge)
8. Ox Gorge
9. Xiling Gorge
10. Qutang Gorge
11. Wu Gorge
12. Taroko Gorge
13. Yarlung Tsangpo Gorge
14. Chaluut Canyon

MARTIAN MONGOLIA

Scientists have found that the Mongolian Plateau (pictured) and the Tharsis Bulge on Mars have similar rift canyons that have been shaped mainly by water, although on Mars water has remained as ice except when disturbances in the crust have caused it to melt. Both areas contain evidence of glacial action and ice–magma interactions when volcanoes erupted under the ice. Both canyon systems can flood, due to melting ice in Mongolia and eruptions of groundwater on Mars.

Mongolia's spectacular gorges include Chulut Canyon, which has cut its way through basalts (dark igneous rocks, rich in iron and magnesium) originating from the late Tertiary to Quaternary periods. These basalts are 100 to 130 feet (30 to 40 m) thick and vesicular (porous, due to gas bubbles in the lava).

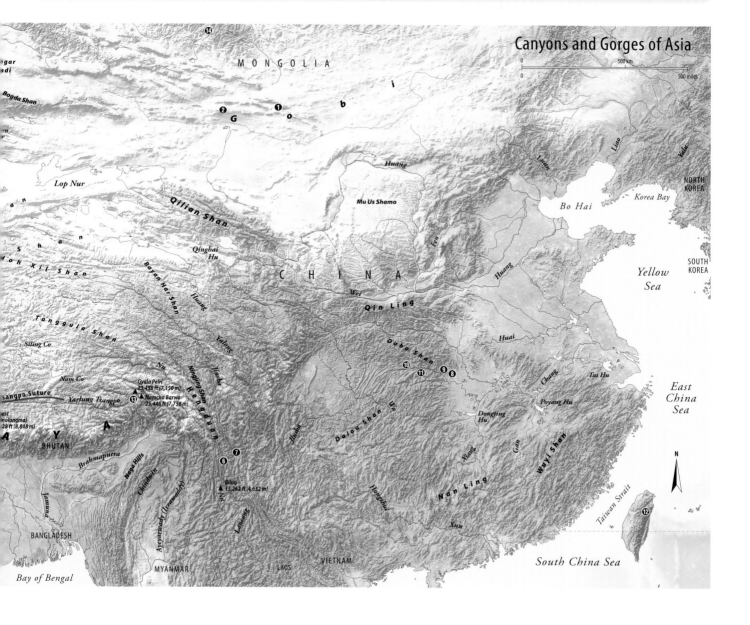

Canyons and Gorges of Asia

YARLUNG TSANGPO GORGE, CHINA

"A single drop of water or a blade of grass from this sacred place ... whoever tastes it will be freed from rebirth in the lower realms of existence," a seventeenth-century Tibetan Buddhist text says of the world's deepest canyon, the Yarlung Tsangpo Gorge.

ABOVE A clouded leopard (*Neofelis nebulosa*) hides behind a tree. These large cats inhabit deep tropical forests of parts of southeastern Asia. Their long tails and broad paws mean they are skilled climbers. They have such large canine teeth that they are sometimes nicknamed the modern saber-tooth.

RIGHT The Tsangpo River and Mt Gyala Pheri lie within the Yarlung Tsangpo Gorge. This is often abbreviated to Tsangpo, which roughly translates as "purifier." Its rugged and remote nature makes it somewhat inaccessible, the journey entailing a long trek.

Local people believe the gorge has hidden doorways to magical places and will be the last refuge of Buddhism when the world has ended. This remote beautiful place has been rarely explored and parts of it are still inaccessible.

Beginning at the village of Pe and stretching 150 miles (240 km) to the Indian border, the gorge is up to an astounding 19,700 feet (6,000 m) deep and 7,440 feet (2,270 m) deep on average. The world's highest river, the Yarlung, with an average elevation of 13,000 feet (4,000 m), flows through the gorge, dropping in height dramatically over the gorge's length, from around 10,000 feet (3,000 m) to under 1,000 feet (300 m).

HEATED PLATEAUS, PLUNGING TORRENTS

The gorge has been carved into the 16,000-foot (4,900 m) high Plateau of Tibet—the highest and largest plateau on Earth, created about 50 million years ago after the Indian and Eurasian plates collided. Buckling, folding, and upheaval lifted the plateau out of the Tethys Sea. The pressure of the continued northward motion of the Indian Plate created deep-seated thrust faults in the plateau, which is still subject to numerous earthquakes and landslides today. The gorge has been carved into a fault line called the Tsangpo Suture.

About 4 million years ago when ice covered the plateau, glaciers deepened the gorge, finally retreating only 10,000 years ago. Melting ice and rain became the Yarlung River. This powerful turquoise torrent has deeply eroded the gorge, carrying off vast quantities of rock over the last 3 million years.

The gorge contains Paleozoic Era limestones from when the area lay beneath the Tethys Sea, and Mesozoic and early Cenozoic era sedimentary rocks such as cherts. Eocene granite, gneisses from the Indian Plate, and metamorphic amphibolite, marble, and quartzite from the Eurasian Plate can all be seen crushed together in the walls of the gorge, as a result of the tectonic collision.

YETIS, TAKIN, AND TIBET'S LAST TIGERS

The gorge's climate ranges from Arctic to subtropical. It begins its journey through the Himalayas, bending around Mt Namcha Barwa, a 22,446-foot (7,756 m) high giant that is still rising at a rate of 1¼ inches (30 mm) per year.

This mountain has various ecosystems, from monsoon rainforest at its base to ice and snow on its peak, with broadleaf forest, pine forest, and alpine grasslands in between. Plants include endemic rhododendrons and orchids that can be found nowhere else, poisonous nettles, and tulips, blue pine and alpine oak trees. On the icy heights, snow leopards prey on sheep and the pika—a small furry animal that lives in alpine meadows and

stores food in its burrow to see it through harsh weather. Here too, local people believe the Himalayan yeti has its home. Below, clouded leopards prowl the forests, subtly camouflaged among the trees, while in the skies overhead the endemic Tibetan eared pheasant, black-necked crane, and Tibetan snowcock soar.

A few miles further on, the gorge passes by the Gyala Pelri peak, which is 23,458 feet (7,150 m) high. Next, steep spurs and drainages cut into the gorge from either side, glaciers abut it, and thunderous waterfalls crash into its depths. Two of the

largest of these waterfalls are Rainbow Falls and Hidden Falls, the latter being 100 feet (30 m) high.

Further along, there are lush hardwood forests where endangered deer species including the black muntjak, Tibetan red deer, and musk deer roam. Below, gorals—goat-like animals with ruddy coats—and takin—stout bovine animals—graze on the hillsides.

As the canyon descends, bamboo groves line the banks. Dense subtropical jungle that shelters Tibet's last tigers, red pandas, and golden monkeys marks the gorge's end.

ABOVE The Yarlung Tsangpo River is the world's highest major river, at an average elevation of about 13,000 feet (4,000 m), before it cuts through Yarlung Tsangpo Gorge and eventually empties into the Bay of Bengal. The green in the bottom right-hand corner reveals the region's rich vegetation, which includes over 400 species of orchids.

Canyons and Gorges of Australia

Australia's sandstone gorges are ancient, in some cases over 100 million years old. Some contain dried-up riverbeds peppered with red boulders and surprisingly deep waterholes, plants from the age of dinosaurs, and eroded scarred walls down which waterfalls tumble.

ABOVE Windjana is one of many gorges in the Kimberley, a remote region of Western Australia, that slowly carved into an ancient reef. This massive barrier reef formed the Napier Range during the Devonian Period.

As the world's oldest, driest, and flattest continent, Australia is geologically relatively stable, so most of its gorges have been formed over millions of years by wind and water. Spectacular gorges are peppered across the country—in Far North Queensland, in Australia's northeast, Carnarvon, Mossman, and Lawn Hill gorges can be found, and in the remote northwest are Geike, Windjana, and Mimbi gorges. In the vegetated wet tropics of the Top End are the gorges of Kakadu; a little further south, Katherine Gorge is hewn into the plateau. Some of the most fascinating gorges are found in two major mountain ranges, the MacDonnell Ranges in Central Australia and the Great Dividing Range running down the east coast.

FLAMING GORGES OF THE RED CENTER

The MacDonnell Ranges are 100 miles (160 km) long and formed 350 million years ago when volcanic activity created a range of scarlet iron oxide-saturated quartzite. The range's sheer walls are scarred by the actions of inland seas, wind, and intense heat. Along the arid valley floor, the 100-million-year-old Finke River—one of the world's oldest rivers—has carved out numerous gorges. Here, moister sheltered environments protect species such as the MacDonnell Ranges cycad (*Macrozamia macdonnellii*), a relic from the wetter days of the late Cretaceous Period, some 65 million years ago.

All of MacDonnell's gorges shelter species that have adapted to the climatic variations—which can range from 122°F (50°C) in summer to 18°F (−8°C) in winter—and to its infertile soil. Many are highly specialized and endemic to the area, such as the red cabbage palm that is only found in Finke Gorge, and the critically endangered central rock rat, a furry cream-colored marsupial, 5 inches (12 cm) long, with a tail the same length.

Ormiston Gorge lies along a tributary of the Finke River, and is 1,000 feet (300 m) deep in some places. The scarlet cliffs on either side of it are scarred with grooves from erosion. A permanent 46-feet (14 m) deep waterhole houses up to nine species of fish that are stranded until the river floods maybe once a year. Huge boulders striped with purple manganese litter the mainly dry, sometimes puddled, riverbed. River red eucalypts that tap into groundwater shelter colorful ringneck parrots. Black-footed rock wallabies warm themselves on the rocks during the early morning then shelter beneath them when the sun is ferocious.

Canyons and Gorges of Australia

0 — 500 km
0 — 500 miles

Key to Canyons and Gorges

1. Murray Canyons
2. Kings Canyon
3. Grand Canyon
4. Claustral Canyon
5. Thunder Canyon
6. Carnarvon Gorge
7. Mossman Gorge
8. Lawn Hill Gorge
9. Geike Gorge
10. Windjana Gorge
11. Bell Gorge
12. Twin Gorge
13. Koolpin Gorge
14. Barramundi Gorge
15. Katherine Gorge
16. Finke Gorge
17. Ormiston Gorge
18. Dales Gorge
19. Knox Gorge
20. Brachina Gorge
21. Chambers Gorge
22. Parachilna Gorge
23. Cataract Gorge
24. Loch Ard Gorge
25. Glenbrook Gorge
26. Cathedral Gorge
27. Piccaninny Gorge

UNDERWATER MAJESTY

Australia's deepest canyons lie underwater. Stretching for more than 93 miles (150 km) off the South Australian coast, the Murray Canyons descend 15,000 feet (4,575 m) below sea level. Core samples taken here are reminiscent of the geological structure of Antarctica—a reminder of when Australia was joined to that landmass over 50 million years ago. These canyons shelter animals such as blue whales, great white sharks, and Risso's dolphins.

BLUE MOUNTAINS GORGES AND WATERFALLS

The Great Dividing Range is a relative youngster, thrown up during dramatic volcanic activity around 170 million years ago, and made to rise further when the landmass comprising New Zealand, New Caledonia, and offshore islands separated from the mainland 80 million years ago.

The Blue Mountains are part of the Great Dividing Range, following the orientation of a 60-mile (100 km) long fault line. The mountains were formed when the Hawkesbury sandstone of the Sydney Basin was thrust up and lifted a plateau rising to 3,600 feet (1,100 m) above sea level. Deep valleys and gorges have been cut into this plateau. Running water was largely responsible for this dissection. Weathering also gradually weakened the shales and coal beneath the sandstone, causing angular sections of the cliffs to break away.

The gorges of the Blue Mountains vary in depth, the deepest being 2,500 feet (760 m) deep. The cool moist environment shelters tree ferns and provides a home for birds such as satin bowerbirds, eastern whipbirds, and lyrebirds, the males having a decorative harp-shaped fan of tail feathers.

ABOVE LEFT The Katherine River cuts through a sandstone plateau in Nitmiluk National Park, Northern Territory, along Katherine Gorge. Freshwater crocodiles inhabit this gorge.

BELOW The Three Sisters in Katoomba overlook the spectacular Blue Mountains, which are sculpted by numerous gorges and slot canyons.

KINGS CANYON

The sandstone forming Kings Canyon was laid around 440 million years ago during the Silurian Period, when the interior of Australia was covered by a vast inland sea. Since then, the action of the elements has formed a paradise for 600 highly specialized species of plants and animals, and formed a canyon that rises 985 feet (300 m) above Kings Creek.

ABOVE Australia's largest lizard, the perentie (*Varanus giganteus*) mainly inhabits the rocky terrain of Australia's arid interior but is also found in sandy deserts.

BELOW One of the permanent pools is in the Garden of Eden, about halfway along the walkway around Kings Canyon on its eastern side. Lush riverside plants such as palms and ferns grow nearby.

Four hundred million years ago, during the Devonian Period, the inland sea levels dropped. Two layers of rock formed, made up of the red Carmichael sandstone and the white Mereenie sandstone. About 300 million years ago, volcanic activity and earthquakes forced both sandstone layers to buckle and be pushed up, forming the George Gill Range, which rises over 330 feet (100 m) from the surrounding desert bush. Fractures occurred in the sandstone, and eventually a ravine was formed, now called the Garden of Eden, which shelters Kings Creek. Here, deep waterholes and temperamental waterfalls plunge into rock pools or dry up completely during the summer heat.

Above the creek, huge boulders fell from the range, weathering to beehive-shaped domes, now called the Lost City. These domes of burnished red stone rest on the ancient seabed, resplendent with marine fossils.

From the Lost City, visitors can peer down into the ravine that forms part of the Garden of Eden. The walls of the ravine are a rich red color that glows at sunrise and sunset. The vibrant red coloring is due to iron oxide staining the sandstone, aided by fungi that thrive on the iron, silica, and rainwater.

UNUSUAL PLANTS AND ANIMALS

The canyon and its surrounds shelter endemic plants and animals that have adapted to the extreme summer sun, freezing winters, and lack of ready water. The marsupial mole, burrowing frog, and mulgara (a small furry marsupial with a pointed nose and long tail) live in burrows underground. Red kangaroos, dingoes, and an assortment of reptiles can obtain most of the moisture they need from their food. Desert oaks, river red gums, spear grasses, and the witchetty bush—source of the Aboriginal bush food, the witchetty grub—are some of the region's unique plants.

The banks around the Lost City are dotted with spinifex—a tall spiky grass that can survive in infertile conditions. Here, three animal species thrive that are specially adapted to the environment and two of them take their name from the grass—the spinifex hopping mouse, which shelters in the vegetation, and the spinifex pigeon, which feeds on the seeds. The rufous-crowned emu-wren is also found here.

Around the Garden of Eden, pelicans and black swans swim in the waterholes where yabbies—large shrimp—lurk. Tree frogs live in the nardoo ferns, cycad palms, and eelweed grass. The perentie—a large lizard up to 10 feet (3 m) long—may be found sheltering under rocks.

PEOPLE OF THE CANYON

Kings Canyon is the home of the Luritja people, who have lived here for over 20,000 years and named the surrounding national park Watarrka, for the umbrella bush that thrives here.

The explorer Ernest Giles stumbled on the canyon in 1872 and named it for his sponsor Fieldon King when he was fortunate enough to find water there that saved his life. Less fortunate was a 47-year-old English woman who suffered a heart attack in the late 1990s while attempting the four-hour circuit walk around the canyon in the broiling 104°F (40°C) heat of a Central Australian summer's day.

The walking trail takes in all the canyon's major features. When the sun is high, helicopters thrum overhead, the ravine echoes with the laughter of busloads of visitors, and this geological marvel loses its luster. The best time to visit is at sunrise, while kangaroos are feeding, the canyon walls glow a deep crimson, and everyone else is tucked up in bed. Then, the magic of this special place can be sampled to the full.

RIGHT The cooler hours and gentle light of dawn is the best time to view the beehive-shaped eroded sandstone domes of the Lost City in Kings Canyon, Watarrka National Park.

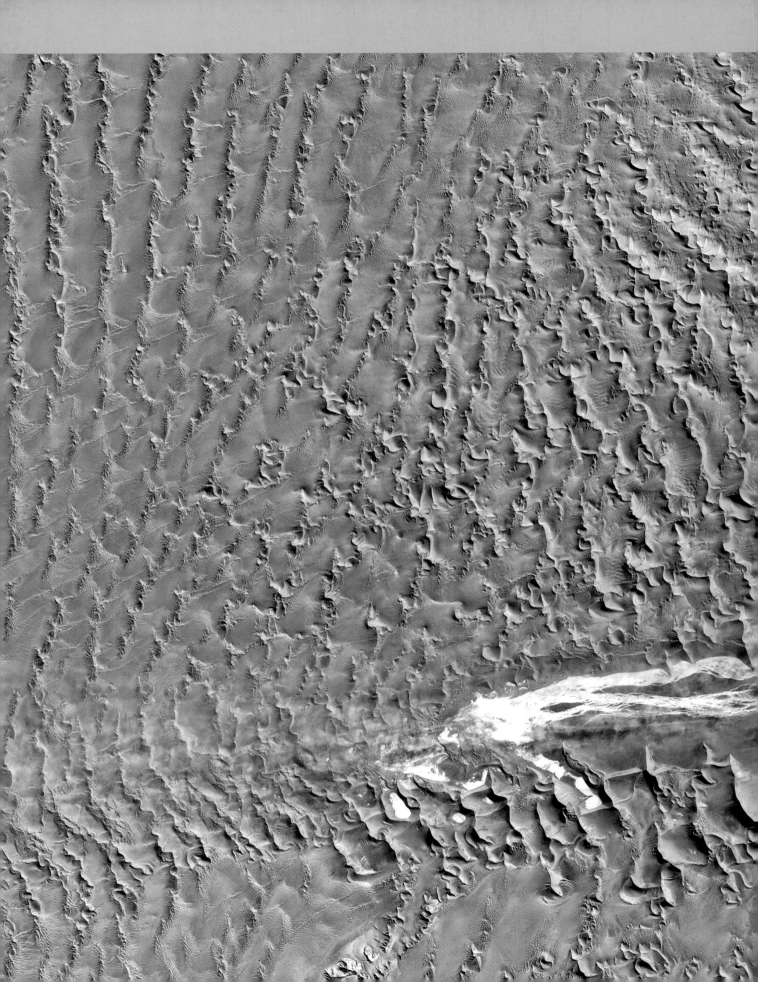

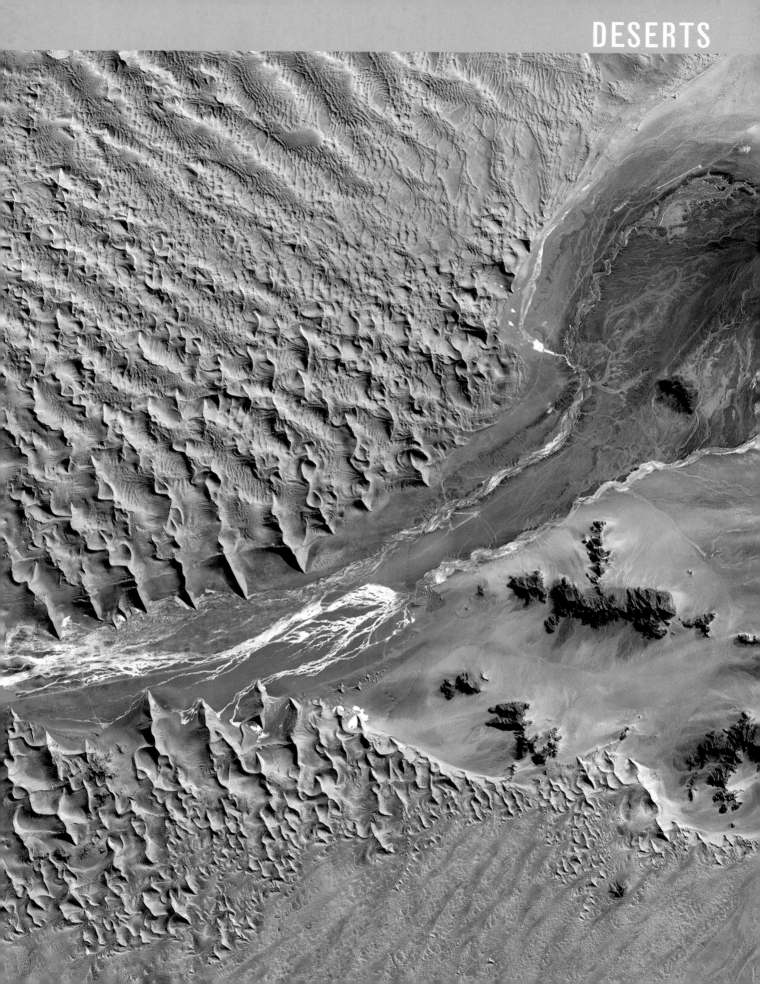

D eserts are among our most intriguing but least visited landscapes. Most people have never experienced the contrast of the desert's soaring daytime temperatures accompanied by freezing temperatures at night. Evaporation rates are so high that perspiration in humans is minimal, yet even a short time without water is enough to create the light-headed euphoria of dehydration. Remote and hostile to human life, deserts remain one of Earth's last unpopulated frontiers.

ABOVE Despite being one of the driest areas on Earth, the Namib Desert in Namibia supports a rich variety of plant and animal life, including gemsbok.

RIGHT The Libyan Desert represents the northeastern section of the Sahara Desert, and crosses the countries of Egypt, Libya, and Sudan. Few people venture into this harsh and arid place.

Deserts are defined as regions of Earth that receive extremely little precipitation—arbitrarily this boundary is set at below 10 inches (250 mm) of rainfall per year. By this definition, the cold dry regions of the world are also defined as deserts and sometimes called polar deserts. With an area of 5,400,000 square miles (14,000,000 km²), the Antarctic Desert, an ice desert, is by far the world's largest. The Sahara Desert is most certainly the world's largest hot desert, with an area of 3,320,000 square miles (8,600,000 km²) covering Egypt, Libya, Chad, Mauritania, Morocco, and Algeria. At no more than 150,000 square miles (400,000 km²), Australia's Great Sandy Desert and the Karakum Desert in Turkmenistan are the two smallest of the world's ten biggest deserts.

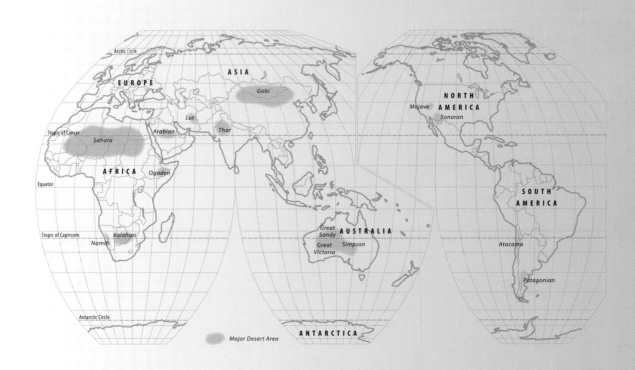

LATITUDINAL BELTS OF DESERT

About one-fifth of Earth's land surface is desert. Several factors influence desert locations. The major deserts, such as the Sahara and the Great Australian Desert, occur in bands around the globe outside of the tropics and between latitudes of about 20 to 35 degrees north and south of the equator. This is a zone of low winds and high pressure, also known by becalmed sailors as the Horse Latitudes. Here, airmasses from the equatorial and temperate convection cells descend and become dry and warm, so there is very little chance of rain.

Deserts tend to be located toward the center and western side of large continental landmasses far from the reach of moisture-laden easterly trade winds (strong winds driven by Earth's easterly rotation). Furthermore, deserts will also form on the downwind or rainshadow side of mountain ranges because all moisture is extracted from the prevailing winds by uplift and cooling on the mountain's windward side. An unfortunate mix of these factors can produce some of the driest deserts on Earth, such as the Atacama Desert of Peru and Chile.

TYPES OF DESERT LANDSCAPES

Not all desert landscapes are vast oceans of wind-blown sand or sand deserts, such as the Empty

Quarter of Saudi Arabia. There are also stony deserts, such as the Tirari-Sturt Stony Desert, a vast plain of gravel and wind-polished pebbles or gibbers, named after explorer Charles Sturt who crossed it trying to reach the center of Australia in 1844. The gibbers are so close together that they form a hard flat pavement and protect the underlying sand from erosion. Rocky deserts have shallow bedrock not covered deeply by talus or soil, or large pavements of bare rock swept clear of sand or gravel by wind. Extensive areas of this type of desert are found in the Libyan Sahara and are

ABOVE The Sahara Desert is an enormous region of elevated plateaus of rock and stone called *hammada*. The plateaus can be as much as 11,000 feet (3,400 m) high in some places.

LEFT Saguaro and cholla cacti thrive in the Sonoran Desert in Arizona, USA. The Sonoran is famous for its saguaro cacti (*Carnegia gigantea*), which can live for up to 200 years.

RIGHT Salar de Uyuni in Bolivia is one of the world's famous salt deserts. It formed around 40,000 years ago when prehistoric Lake Minchin began to evaporate.

OPPOSITE These sand dunes in the Namib–Naukluft National Park area of the Namib Desert, are transverse or crescentic dunes. Their burnt orange hue signifies their age; the brighter the color, the older the dune.

ABOVE Much of the Atacama Desert in Chile is devoid of vegetation. However, in the section of desert south of the Tropic of Capricorn, coastal mists known as *camanchacas* bring the humidity needed for wildflowers to grow when rain does fall.

known as *hammada*. Plateau deserts occur on flat, bare tablelands, such as Golan Heights on the border of Israel, Lebanon, Jordan, and Syria. These rugged plateau lands are often deeply dissected by steep-sided narrow ravines or wadis. Mountain deserts feature bare and dry rock peaks such as seen in the Tibesti and Ahaggar ranges of the Sahara, or the MacDonnell Range in central Australia.

DESERT SAND DUNES

The lack of soil-stabilizing vegetation in deserts means that loose sand is common, and this can be blown into a variety of differently shaped sand dunes depending on the prevalent wind direction and velocity. Large fields of sand dunes are known as ergs in the Sahara.

The most common sand dune is the transverse dune. These are also known as a barchan or crescentic dunes because of their elegant crescent moon shape with their two arms or horns pointing downwind. Barchan dunes form in unidirectional winds of moderate strength. Sand grains are blown up the windward slope of the dune and then roll down the downwind side. In this way the dune slowly creeps forward, grain by grain. Barchan dunes have been measured in China as traveling at about 300 feet (90 m) per year, and they can bury trees, houses, or anything else in their path. In regions where the wind is constantly changing, the sand will not migrate in one direction and star dunes form. The star shape is actually formed by a series of barchan dune horns at different orientations. These dunes tend to pile upward, and in the Badain Jaran Desert of China, star dunes have been recorded reaching heights of around 1,600 feet (485 m).

Longitudinal or seif dunes form where winds are strong and unidirectional. These shoestrings of

sand lie parallel to the wind direction, and present an amazing image. In the Sahara Desert they may achieve lengths of 200 miles (320 km) and heights of 900 feet (275 m). Rocky or clay windswept corridors separate the parallel seif dunes.

Parabolic dunes occur in destabilized areas of vegetated dune systems, hence the name "blowout dune" is also commonly used. Once the sand is free, it is pushed downwind into a parabolic arc in front of a wind-scoured corridor. The parabola's arms point upwind, back to their point of attachment with the vegetated part of the dune. Parabolic dunes are common in coastal areas and can be caused by any natural or human disturbance. For example, the activity of off-road vehicles cutting a track to the beach across a vegetated coastal foredune can be enough to initiate a blowout dune.

ANCIENT SAND DUNES

Sand dunes are preserved in the geologic record when they become lithified or turned to rock. This process takes place when the dunes are buried and mineral-saturated groundwater deposits natural cement between the sand grains to form hard sandstone. Evidence of the former dunes remains in a series of inclined beds or cross-beds that mark the slope of the downwind dune face. The cross beds are usually inclined at the maximum angle at which free, unconsolidated, dry sand can lie before collapsing. Magnificent dune cross beds can be seen in the Navajo Formation sandstones in Zion National Park, Utah.

SALT FLATS AND PLAYA DESERTS

Salt deserts occur in enclosed dry basins where drainage does not flow toward the coast but inland toward the center of the basin. Any rainfall that does occur in these deserts carries soluble salts

inward to temporary lakes or playa lakes that form at the basin's lowest points. These lakes may only have water in them once every few years or even decades. High evaporation rates mean that they quickly dry out, leaving vast shimmering white flats of salt and other minerals such as gypsum and calcite. Some famous salt deserts include Salar de Uyuni in Bolivia, Dasht-e Kavir in central Iran, Australia's Lake Eyre Basin, and the Great Salt Lake of Utah, USA. The extremely flat surface of the Great Salt Lake of Utah is perhaps best known as the site of many world land-speed record trials.

PLANTS OF THE DESERT

The desert is an extremely harsh environment so the plants and animals that survive there have evolved a number of water conserving strategies. Xerophytes are plants that efficiently collect and store water and have features that reduce water loss. Water loss limiting strategies include small waxy leaves, less surface pores (stomata), or no leaves at all. Long hairs on the leaf, known as trichomes, assist by providing shade, breaking up wind flow over the leaf surface and hence moisture loss, and by acting as water collectors during morning cloud mists. The cactus can close its stomata during the day to reduce transpiration. Other xerophytes minimize water loss by shedding their leaves during the dry season. Many xerophytes have either succulent fleshy stems such as cacti, or a large fleshy underground tuber, enabling them to store large amounts of water. Spines or prickly hairs discourage animals from dining on the plant's succulent flesh in order to quench their thirst.

The root system of xerophytes can be extensive, shallow, and designed to maximize water uptake during brief sporadic rains, as in the case of the cactus. Or, the root systems can be deeply penetrating in order to reach a permanent subsurface water supply, as in the case of the acacia and the oleander. Desert palms have also developed long taproots that enable them to survive during the driest of surface conditions.

Short-lived ephemerals avoid drought by rapidly growing, flowering, and seeding only during brief rainy periods. Their seeds lay dormant for years until the opportunity to sprout in the next rains. Other plants, known as drought evaders, shoot from underground bulbs whenever it rains.

With a pair of human hands and a bit of skill, it is possible to get past the fine, hair-like spines on the plump purple fruit of the prickly pear (genus *Opuntia*) and enjoy its sweet crimson flesh. Local indigenous people have long eaten the fruit (which they call tuna) from this native of the tropical Americas; and they know how to get rid of the needles by briefly roasting the fruit in the flames of a fire or shaking them in a sack of hot coals. This fruit is also made into juice, jellies, and jam. Considered a pest in many countries outside the Americas, this cactus, which can grow into a sizable tree, can be seen planted in rows and lovingly tended to in the Mexican countryside. Mexicans also enjoy eating the flat leaf-shaped paddles, which are sliced and fried with eggs or used in salads, and are known as *nopales*.

DESERT ANIMALS CONTEND WITH THE HEAT

A look into the desert from a car window will not reveal anything but a lifeless landscape. The hostile environment of intense daytime heat and searing sun mean that most animals here are nocturnal. Sleeping during the day, they come out at night to hunt and eat. Some, like the rodents, even block the entrance of their burrows to keep out the hot dehydrating air. Some animals never drink, but get their water from plants and seeds.

Animals, such as the desert toad, spend most of the year deep underground and remain dormant until seasonal rains create temporary ponds. During this brief period they come out to eat, drink, breed, and lay eggs before retreating underground for another long period. Arthropods, such as brine shrimps, survive as eggs, and will only complete their life cycles when sporadic rains fill saline ponds and temporary lakes.

Most animals have evolved a means to avoid overheating and water loss, and to dissipate heat. Some desert mammals, such as the kangaroo rat, have extra-efficient kidneys that can extract more

water from their urine, which is then recycled into their blood. Specialized organs in their nasal cavities also recapture much of the moisture that would normally be exhaled during breathing. Large ears loaded with tiny blood capillaries, such as those of the jackrabbit, will dissipate heat effectively. Pale colored fur or skin also minimizes heat uptake.

PEOPLE OF THE DESERT

The people of the desert have traditionally been nomadic. They move around, either daily, monthly, or semi-annually. Hunter-gatherers move from waterhole to waterhole in search of food animals and plants. Pastoralists also move their herds of domesticated animals—such as camels, goats, or sheep—from place to place depending on their needs. They mostly follow predictable seasonal pasturelands and a reliable water supply. Some groups also cultivate crops and may trade in order to survive. Typical clothes of desert people include long, flowing robes and headgear that protects the skin from the sun. Veils also act as masks to block out sand and dust. These garments may seem hot, but they are loose fitting and allow air to cool the body while slowing down sweat evaporation and dehydration. At night, when it is cooler, this style of clothing is very warm, acting like a blanket.

While some desert dwellers live in permanent homes, the nomads have developed a variety of rather ingenious, comfortable, and lightweight portable homes. The tent had its origins with the nomad. In the desert, tents must be well insulated from extremes of heat and cold, and they must be waterproof. Traditionally the woven hair and skins of the herd animals, waterproofed with fat, were used for this purpose.

ABOVE Desert palms are well adapted to their environment. They send down a long taproot to give them access to water below the dry desert surface.

THE WORLD'S TEN BIGGEST DESERTS

	NAME	AREA (SQ. MI.)	AREA (KM²)	COUNTRIES
1	Antarctic Desert	5,400,000	14,000,000	Antarctica
2	Sahara Desert	3,320,000	8,600,000	Egypt, Libya, Chad, Mauritania, Morocco, and Algeria
3	Arabian Desert	900,000	2,331,000	Saudi Arabia, Jordan, Iraq, Iran, Kuwait, Qatar, United Arab Emirates, Oman, and Yemen
4	Gobi Desert	500,000	1,300,000	Mongolia and China
5	Patagonian Desert	260,000	673,000	Argentina and Chile
6	Great Victoria Desert	250,000	647,000	Australia
7	Great Basin Desert	190,000	492,000	United States
8	Chihuahuan Desert	175,000	450,000	Mexico and United States
9	Great Sandy Desert	150,000	400,000	Australia
10	Karakum Desert	135,000	350,000	Turkmenistan

Deserts of North America

The four predominant deserts of North America are the Mojave, Chihuahuan, Sonoran, and Great Basin. All are small and geologically young when compared to other deserts in the world, and are located in a broad band running down the western United States between the Sierra Nevada and Rocky Mountains.

The aridity and patterns of precipitation in each of these deserts have contributed to the creation of four very different biotic environments. All are classified as hot deserts, with the exception of the Great Basin on the high Colorado Plateau, which experiences long cold winters with much of its precipitation falling as snow.

THE MOJAVE DESERT

America's smallest desert is located in southern California and extends into Arizona, Utah, and Nevada. The Mojave's basin and range topography of steep-sided mountain ranges alternating with flat sediment-laden valleys is typical of the American southwest and the product of 6 million years of crustal stretching.

Though difficult to define in biological terms due to its variety in elevation, vegetation, and climate, the Mojave is generally considered a shrub-dominated desert with a quarter of its 2,000 species of plants endemic to the region. It receives less than 6 inches (150 mm) of rainfall a year, with dry lakes a common feature of the region.

From Telescope Peak's lofty 11,056-foot (3,372 m) summit to −282 feet (−86 m) at Badwater, the lowest point in the western hemisphere, the Mojave's arid Death Valley represents

ABOVE Ribbons of color streak the badlands of the Painted Desert in the Great Basin Desert. These are caused by different mixes of minerals and decayed vegetation seeping into the strata and tinting the many distinctive sedimentary layers.

the greatest topographic relief in the conterminous United States. Millions of years of earthquakes and fracturing have built its mountains and lowered its basins, making it one of the hottest and driest regions in North America.

THE CHIHUAHUAN DESERT

The Chihuahuan Desert covers a large portion of northern Mexico and extends into the southern United States. The southernmost of the North American deserts, its high elevation with extensive areas above 4,000 feet (1,200 m) exposes it to Arctic winds resulting in thin gravelly soils and a prevalence of perennial grasslands dominated by agaves, leaf succulents, and small cacti. Large trees and columnar cacti are missing in a desert where 80 percent of the soil is calcium-rich and born from limestone beds, evidence that the area was once submerged.

The Chihuahuan Desert is rich in bat fauna with over 18 species cataloged, including the Brazilian free-tailed bat, the pallid bat, and the western pipistrelle.

THE GREAT BASIN DESERT

One of the last great open spaces still found in North America, the Great Basin is an arid high altitude desert covering 200,000 square miles (520,000 km^2) that stretch across much of Nevada into Utah and northern Arizona. Its somewhat isolated mountain ranges and broad plains are still in the process of stretching and cracking.

PETRIFIED FOREST NATIONAL PARK

The 218,500 acres (88,500 ha) of Petrified Forest National Park are among the largest areas of intact grassland remaining in the American southwest. Located in southern and southwestern Arizona, alongside the Painted Desert, the park is home to one of the world's largest concentrations of petrified wood. These fossilized remains are the remnants of three now-extinct trees—*Woodworthia arizonica*, *Araucarioxylon arizonicum*, and *Schilderia adamanica*—common during the Triassic Period (250 to 200 million years ago) when the Colorado Plateau was located near the equator and covered in a 200-foot (60-m) canopy of conifers and tropical flora.

These ancient trees were deposited by floodwaters within the sediments of rivers and buried in stream channels. They slowly became altered as volcanic ash infiltrated the sedimentary layers. The groundwater that dissolved the silica in the ash carried it through the logs, replacing the cell walls and ultimately crystallizing them.

The cell walls of these ancient trees crystallized into mineral quartz, resulting in the exquisite coloring we see today.

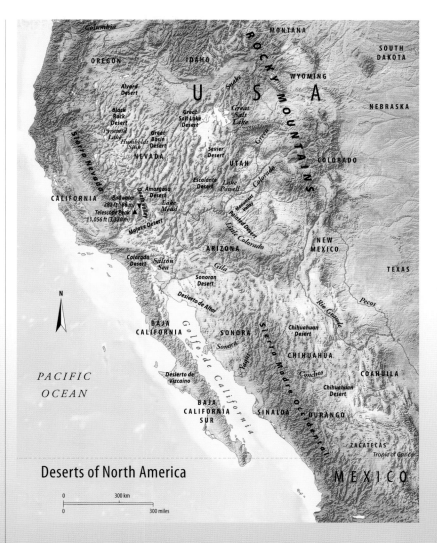

Deserts of North America

```
0          300 km
0                    300 miles
```

Great Basin is characterized by a process called internal drainage, where its meager precipitation, lacking any outlet to the sea, collects in low-lying areas such as Pyramid Lake or Humboldt Sink.

Its particularly thin crust and low temperatures result in a limited variety of plant and animal life. The Great Basin bristlecone pine, the world's oldest living tree, can be found within its higher elevations and is easily distinguished by its gnarled twisted trunks and vibrant orange-yellow bark.

The Painted Desert is located along a 160-mile (250 km) crescent-shaped arc in the Great Basin's southernmost boundary. A region of flat-topped mesas and buttes, its name refers to the rainbow-like hues found within the sandstone and mudstone layers of the Chinle Formation.

BELOW Looming thunderclouds predict rain in Antelope Valley, a high desert region in the Mojave Desert. Wildflowers such as California poppies bloom here during the spring season, mid-March to mid-May, along with a variety of desert grasses.

SONORAN DESERT

Exploring the Sonoran Desert in 1699, Captain Juan Mateo Manje heard a terrifying tale from the Pima Native Americans. They told of a giant man-eating bird that roamed the skies, catching as many Pimas as it could eat. One day a group of warriors followed it back to its cave and, while it was sleeping, closed off the cave's entrance with wood and set it on fire. The bird, growling fiercely, eventually died, asphyxiated by the flames and smoke.

OPPOSITE The typical landscape of the Sonoran Desert in New Mexico features vast forests of saguaro cacti and other desert plants. To the north, the desert features riparian forests, while to the south it gradually merges into thornscrub in the Mexican states of Sonora and southern Baja California.

Despite the demise of the Pima's avian man-eater, the Sonoran Desert is still home to a vast array of birdlife that depends upon its varied ecosystem for survival. "Sky islands" of coniferous forests bring birds normally associated with cooler northern environments into proximity with desert and tropical species, creating an abundance of birdlife rarely seen in a desert environment.

A UNIQUE DESERT LANDSCAPE AND CLIMATE

The Sonoran is a desert of high mountains and deep broad valleys underlain by thick deposits of sand, silt, and gravel. Its characteristic basin and range topography was born from widespread volcanic activity occurring 20 to 40 million years ago, which formed massive calderas and produced intense heat. Subsequent geological activity resulted in the distinctive parallel mountain ranges, sandy beaches, broad plains, lava sheets, and playas. The sedimentary, igneous, and metamorphic rocks in its strata vary in age from Precambrian outcrops born when the continental crust below Arizona was forming 2 billion years ago, to recent pockets of volcanism dating to as recently as 700 CE.

The Sonoran's low latitudes and relatively low elevation combine to make it not only North America's warmest desert but also its wettest, with a biseasonal rainfall regime providing a lushness and diversity in both habitats and species unmatched by other deserts of the region. Unlike other deserts, it has mild winters and rarely experiences frost. The two rainy seasons come in the form of heavy summer thunderstorms from July to September, with frontal storm remnants from the Pacific Ocean bringing light widespread winter rains from December to March.

ABUNDANT DESERT LIFE

The Sonoran's diversity rivals that of any other terrestrial ecosystem. Its 100,000 square miles (260,000 km²) are home to more than 60 species of mammals, 35 species of native fish, and over a hundred reptiles. Its 350 species of birds do not, thankfully, include any mythological man-eaters.

Half of its biota is tropical in origin with tree-sized columnar cacti growing densely enough in places to be called forests. One cactus exclusive to the Sonoran Desert is the organ pipe cactus (*Stenocereus thurberi*). Rather than branches growing out from a central trunk, its many branches rise from a singular base at ground level. It has a juicy edible fruit called pithaya that matures each July. Birds puncture the pithaya's skin, exposing a bright red pulp and tiny black seeds that are, in turn, eaten by other birds and animals.

Coniferous forests can be found here in the high mountain ranges where precipitous elevations provide for radically differing climatic conditions. Cool moist summits—some as high as 8,800 feet (2,700 m)—provide life zones more akin to Canada than the southwestern United States. Temperate deciduous forests are seen in scattered groves of aspens and ribbons of riparian trees, with forests of oak and pine from the Sierra Madre Occidental extending as far north as central Arizona.

THE SAGUARO CACTUS

The Sonoran's signature cactus, the saguaro (*Carnegiea gigantea*), can grow as high as 50 feet (15 m), weigh several tons, and live for up to 200 years. A horizontal root structure easily absorbs whatever moisture is available, which is stored in its stems and branches. An accordian-like external pleating expands to collect precipitation and closes during periods of drought. It blooms from May to June with creamy white blossoms opening in the evening and attracting nocturnal pollinators such as bats. Gila woodpeckers make holes in saguaros for nests and, when abandoned, they become homes for small birds such as the tiny elf owl. The local Tohono O'odham Native Americans harvest the fruit, grinding its seeds into a rich butter-like substance, which is considered a delicacy. Some of its juices are set aside to make a sweet syrup, or fermented as wine for use in traditional ceremonies that beckon summer monsoons.

The flowers of the saguaro cactus provide an excellent food source for gila woodpeckers (*Melanerpes uropygialis*).

Deserts of South America

Chile's Atacama Desert and the Patagonian Desert in Argentina are the two principal deserts of South America, yet it would be difficult to find two more diverse desert landscapes.

The Atacama Desert has a history of hyper-aridity stretching back 20 million years and is 50 times as arid as California's Death Valley. Devoid of vegetation, its riverbeds have not seen rain for 120,000 years and there are areas where not a single drop of precipitation has ever been recorded.

The cold desert scrub of the Patagonian Desert, by contrast, receives between 20 and 24 inches (510 and 610 mm) of rain annually and is home to a mix of grasses, low shrubs, and deciduous thickets typical of a cold desert–scrub biome.

WHY THE ATACAMA IS SO DRY

Studies of the Atacama trace its arid climate back to the Miocene Epoch, 23 to 5 million years ago, though sediments from the late Triassic Period 200 million years ago also show traces of an abnormally high evaporitic environment.

Beginning at the Peruvian border, the Atacama stretches south along a 600-mile (970 km) corridor between the Pacific Coast Range and the Andes Mountains, forever trapped in a double rain shadow that limits rainfall to infinitesimal levels. Organic materials in its soil are so low that microorganisms are unable to grow in it, even under laboratory conditions. In the 1970s, the Atacama Desert was chosen by NASA to conduct microbe

BELOW Despite parts of the Atacama Desert not receiving rain for thousands of years, there are a dozen lakes in this extremely dry desert. The high evaporation means all have a high salt concentration.

ANCIENT ART OF THE ATACAMA

A geoglyph is a prehistoric type of art that uses the landscape as its canvas. Depictions of geometric shapes as well as humans, animals, and birds are created either by scraping away the desert varnishes to expose the lighter subsoil beneath, or by placing rocks and other objects onto the landscape.

Over 5,000 individual glyphs have been cataloged in the Atacama across an area of 58,000 square miles (150,000 km²). They were built from 600 to 1500 CE possibly as a form of cultic worship to Andean deities, or as beacons for safe passage for llama caravans, providing information on salt flats or water sources.

hunting experiments for its new Mars *Viking* rover due to this desert's Martian conditions. No evidence of life was found.

Because it has an average elevation of 7,900 feet (2,400 m), the Atacama Desert is not a hot desert. Its daytime temperatures range from 32°F up to just 77°F (0° to 25°C) in a mountainous and volcanic landscape with plateaus interspersed by great saline lagoons and dry salt beds. As it nears the Pacific Ocean, the land descends into large basins that open out to gently sloping, sandy,

nitrate-filled plains, ending in high bluffs ranging in height from 800 to 1,500 feet (244 to 458 m).

In the absence of moisture, nothing decays. Anything left behind in the Atacama will eventually become an artifact. Even people. In 1983, the oldest mummies in the world were discovered by workmen in the northern Atacama city of Arica. Ninety-six bodies were unearthed, belonging to the Chinchorro culture, which lasted from 7,800 to 3,800 years ago. The oldest was dated to over 7,000 years, two millennia older than the oldest Egyptian mummies, preserved by a combination of sophisticated mummification techniques and the aridity of the driest desert on Earth.

THE PATAGONIAN DESERT

One of the world's major deserts, the Patagonian descends from 6,500 feet (2,000 m) high in the foothills of the southern Andes down to the Atlantic Ocean, and from the Straits of Magellan in the south northward to the Valdez Peninsula.

Lying within 200 miles (300 km) of both the Atlantic and Pacific Oceans would normally guarantee a region a cool moist climate, but the Patagonian Desert is, in fact, the world's outstanding example of a rainshadow desert, with the Andes mountains shutting out rain-bearing westerly winds from the South Pacific.

Nevertheless, this diverse landscape with its year-round frosts, sparse precipitation, and unrelenting winds is home to a wide variety of flora and fauna. Semi-forested Andean slopes give way to deciduous thickets and tuft grass tundra that are home to endemic species such as bunchgrasses and the Patagonian opossum.

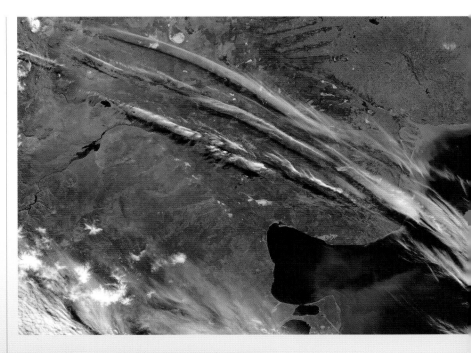

ABOVE A satellite image shows the northern border of the Patagonian Desert. The Valdez Peninsula, in the bottom right, is an area of sub-desert scrub.

LEFT In the high desert region of southern Peru near Arequipa lies a sea of perfect barchan dunes. These crescent-shaped transverse dunes are created by the prevailing winds blowing consistently in one direction.

Deserts of Africa and the Middle East

From the Kalahari with its monsoonal rains to the arid Sahara and the ancient strata of the Namib, the world's oldest desert, Africa's deserts cover over a quarter of the continent. The deserts of the Middle East are largely vast expanses of arid shrublands, and they extend from Saudi Arabia, Yemen, and the United Arab Emirates on the Arabian Peninsula westward through Iran, Iraq, Jordan, Syria, Israel, and into Egypt.

ABOVE Meerkats (*Suricata suricatta*) live in southern Africa's Kalahari Desert. These pack animals take turns standing on guard, looking out for the group and will bark to alert of danger.

BELOW The towering dunes of the Namib Desert have been sculpted by winds. The rich orange color is caused by the iron in the sand being oxidized, in the same way that metal gets rusted. Thus, the mostly brightly colored dunes, being the most oxidized, are the oldest.

Africa's principal deserts are the Kalahari and Namib deserts in Namibia, Botswana, and South Africa, and, of course, the best-known desert—the vast sprawling Sahara Desert in the north, the world's largest desert at 3.5 million square miles (9,000,000 km²), its sand seas and gravel plains enveloping all or part of 11 African nations.

The Middle East covers over 2.8 million square miles (7,300,000 km²) of which nearly two-thirds is classified as arid to semi-arid regions. Its topography varies from the Taurus and Elburz mountains of Turkey and Iran in the north to the flat plains and plateaus of the south, stretching from the Libyan Desert in North Africa to the Ar Rub'al-Khali Desert on the Arabian Peninsula.

THE NAMIB DESERT, NAMIBIA

The Namib is the world's oldest desert and is also one of the driest regions on the planet, with a history of aridity going back 50 million years. Despite this desert receiving less precipitation than the Sahara, it nonetheless supports a rich diversity of life, sustained by mild coastal temperatures and moisture-laden Atlantic fogs.

With an average width of just 60 miles (100 km), the Namib parallels Namibia's Atlantic coastline for 1,200 miles (2,000 km), as it stretches ribbon-like from the mouth of the Orange River in South Africa's Northern Cape Province north

THE NAMIB'S *STENOCARA* BEETLE

In what is very likely the hottest terrestrial environment on Earth, the Namib's tiny endemic *Stenocara* beetle possesses a uniquely adapted exoskeleton along its back and wings, which it uses to harvest moisture from the desert's occasional early morning fogs. Tilting its body into the wind, water droplets from fogs normally so light that condensation cannot occur soon begin to coalesce around a series of rounded hydrophilic nodules on its wings. These are arranged in hexagonal concentrations and embedded in a bumpy matrix. Upon reaching ⅕ inch (4 mm) in diameter they roll down the *Stenocara*'s back and into its mouth in a self-replenishing stream.

through Namibia to Mossamedes in Angola. Located along a broad plain eroded into monotonously flat bedrock, it rises gradually on a gradient reaching inland from the coast to a height of 3,000 feet (1,000 m) at the Great Western Escarpment, which marks its eastern boundary.

The escarpment's increased humidity and annual rainfall of 8 inches (200 mm) allows for thin to moderate coverings of annual grasses in an otherwise barren expanse of gravel flats and sand dunes. Dry riverbeds etch their way across the Namib as linear oases, providing conduits for surface water, groundwater, and flood sediments.

In the central Namib, prominent steep-sided hills of granite called inselbergs rise from the surrounding low-relief plains. Spitzkoppe, a granite massif between Usakos and Swakopmund, is the country's most recognizable mountain, often called the Matterhorn of Namibia. It rises roughly 2,300 feet (700 m) above the surrounding plains despite its summit being 5,835 feet (1,784 m) above sea level. Part of the Erongo Mountains, it is one of many igneous complexes that formed in Namibia's Damaraland during magmatism resulting from the opening of the South Atlantic.

THE KALAHARI DESERT, SOUTHERN AFRICA

Derived from the Tswana word *keir* meaning "great thirst," the Kalahari semi-desert has one of the largest sand mantles in the world, covering 100,000 square miles (260,000 km²) of the southern African nations of Namibia, South Africa, and Botswana.

Its expanses of red-brown sands possess little surface water, with drainage provided by seasonally inundated pans, dry valleys, and saltpans. A vast plateau ranging in height from 1,600 to 5,000 feet (500 to 1,500 m) above sea level, its abundant vegetation includes tall savanna woodlands of sycamore, fig, ebony, and baobab in the north sustained by a summer wet season, through to scrub areas and grasslands in its arid southern boundaries.

Rainfall varies from 5 inches (130 mm) in the southwest to 20 inches (500 mm) in the north, contributing to the Kalahari's reputation as one of Africa's last true wildlife refuges. It is home to meerkats, lions, brown hyenas, and antelope, while hundreds of sparrow-like weaver birds construct huge communal nests up to 7 feet (2 m) in diameter in its camel thorn and acacia trees.

The Kalahari is also the ancestral home of the San (Bushmen), remnants of Africa's oldest cultural group and thought to be the descendants

of southern Africa's original inhabitants. Their society, rather than structured and hierarchical is instead built on a family culture where decisions are made collectively through open discussion. Evidence suggests they have been a hunter-gatherer society for over 20,000 years. Predominantly vegetarian, 80 percent of their diet consists of plant food such as berries, roots, melons, and nuts.

THE SINAI DESERT, EGYPT

Egypt's Sinai Desert lies at the crossroads of Asia and Africa, a 24,000-square-mile (70,000 km²) triangular landmass of jagged mountains scoured by wadis. Here lie ancient Precambrian Era rocks and gravel and limestone plains, surrounded by the saline coastal mudflats of the Gulf of Suez to the west and Israel's Negev Desert and Gulf of Aqaba to the east.

The Sinai's geology and topography allow it to easily be divided into a northern and southern region. In the north, elevations are far lower than those in the central and southern peninsula, with expanses of flat-lying Paleozoic sediments, wadis, fossil beaches, and extensive areas of mobile sands.

TOP An oasis in the Sinai Desert indicates fresh water is nearby. Oases have for centuries served as stopovers for Bedouin herding livestock or trading goods, transporting them across the desert in camel caravans.

ABOVE Antelopes such as the gemsbok (*Oryx gazella*) roam the dunes of the Kalahari and Namib deserts. Highly adapted to deserts, they can survive long periods without water.

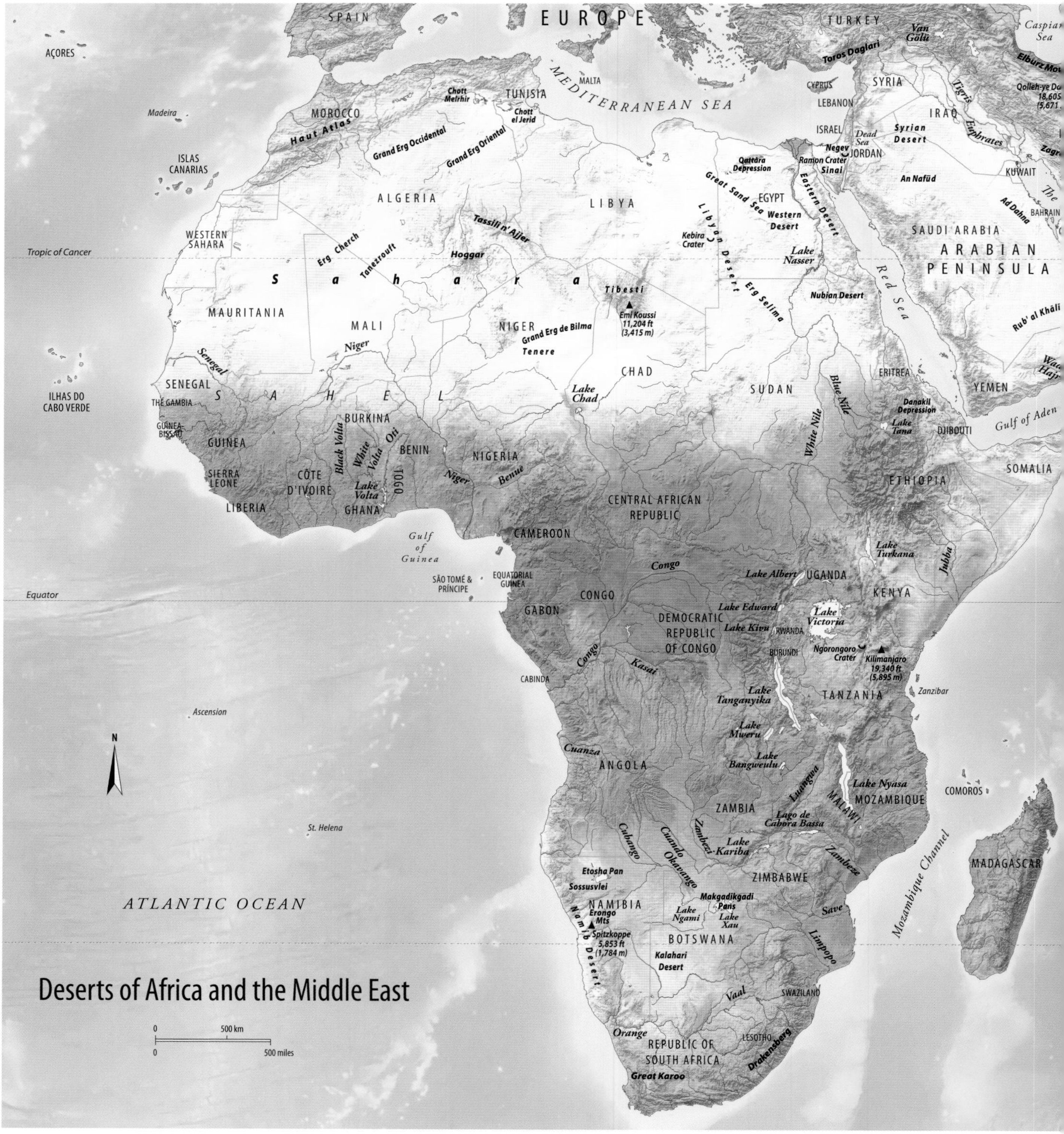

Deserts of Africa and the Middle East

0   500 km

0   500 miles

Post-folding erosion has led to the formation of extensive limestone cuestas—ridges with gentle slopes on one side and cliffs of the other.

The central Sinai is largely comprised of the El-Tih Plateau, a region of limestone formed during the Tertiary Period. In the southern Sinai,

mountains composed of acid plutonic and volcanic rocks slope steeply to the gulfs of Aqaba and Suez producing steep ephemeral wadis flowing to the sea. Here, nine peaks rise over 6,500 feet (1,983 m) in height, including Mt St Catherine, Egypt's highest peak at 8,639 feet (2,639 m).

The southern Sinai is rich in plant diversity. Over 700 vascular plant species have been catalogued including 28 endemic species mostly found in the gorges of the St Catherine Mountains.

THE NEGEV DESERT, ISRAEL

The diminutive 5,000 square miles (13,000 km²) of Israel's Negev Desert comprises several climatic zones. The northern Negev has a Mediterranean climate with fertile soils and annual rainfall exceeding 12 inches (300 mm), while its western region is characterized by light, sandy soils and dunes that reach heights of 100 feet (30 m). The central Negev's high plateaus rise to 1,800 feet (550 m) above sea level and receive only 4 inches (100 mm) of rainfall a year. To the east, the Negev Plateau abruptly plummets into the arid Arava Valley, which stretches from the port city of Eilat in the Negev's extreme south, northward in a 100-mile (160-km) depression along the Jordanian border to the Dead Sea.

In the Negev's highlands can be found leopards, gazelles, wolves, red foxes, and hyenas, though the region's inferior salty soils provide little in the way of floral diversity.

The Ramon Crater is the world's largest karst erosion cirque and one of the Negev's primary geological features. Twenty-four miles (40 km) long, 6 miles (10 km) wide, and up to 1,500 feet (500 m) deep, ancient rivers carved and shaped this crater over millions of years. Water eroded the crater floor, exposing increasingly ancient rock strata that are up to 200 millions years old.

THE SYRIAN DESERT, SYRIA

The Syrian Desert is a combination of steppe and true desert covering nearly 60 percent of the country and extending into neighboring Jordan

and Iraq. Comprised primarily of both rock and gravel this geographic northward extension of the vast Arabian Desert plateau is a region of undulating plains characterized by low and erratic rainfall, with elevations ranging from 900 to 1,600 feet (300 to 500 m).

Overexploitation and habitat destruction have combined with huge increases in domestic livestock over the past 50 years to wreak havoc on the region's endemic biodiversity. Over a dozen native animals have become extinct across the Syrian steppe, including the Syrian wild ass and the Syrian ostrich.

Its sparse vegetation includes *Artemisia* shrub-steppe and scattered *Pistucia* trees. Bird species and densities are relatively low, although the region's semi-natural and cultivated steppes provide important winter breeding grounds for various species of wetland birds including the endangered *Tetrax tetrax*, a large bird in the bustard family.

LEFT Camels are the main form of transport in Middle Eastern deserts. For the nomadic desert inhabitants, camels not only provide transport but are also a source of meat, creamy milk, wool, leather, and even offer shade from the intense heat. They are able to cover terrain impenetrable to vehicles.

BELOW This bizarre sandstone rock formation in Timna Park, in the southern part of Israel's Negev Desert, is known as "the mushroom". Set in a desert wadi, surrounded by cliffs, Timna Park features a variety of landscapes. It may have the first-known copper mine; some believe it dates back to King Solomon's time.

SAHARA DESERT, NORTH AFRICA

The Sahara Desert in North Africa is the largest hot desert on Earth, a vast region of sand seas, elevated plateaus, gravel plains, and depressions covering over 33 million square miles (9 million km²). It stretches across the continent from Morocco's Atlantic coastline to Egypt's Red Sea.

BELOW The Sahara is known for its seemingly infinite stretches of sand. However, it is really a diverse landscape of varying features, including volcanic remnants. The dark rocks in Chad's Tibesti Plateau (seen below left) are the cones and calderas of ancient volcanoes. Dark-colored lava beds to the north of Tibesti, in Libya, are also visible.

Gravel plains and sand dunes make up a massive 90 percent of the Sahara, with three-quarters of its inhabitants living in scattered oases that total a mere 800 square miles (2,000 km²). The highest shaded temperature in the world, 136°F (58°C), was recorded here. There was a time though, when the Saharan climate was very different.

IN THE ANCIENT PAST, A PLACE OF DEEP LAKES

During the Halocene Epoch, starting 11,000 years ago, the Sahara was home to enormous freshwater lakes up to 1,500 feet (500 m) deep, such as the paleolake Megachad, larger in area than today's Caspian Sea. Nine thousand years ago, the climate of North Africa began to change. Summer temperatures increased and precipitation decreased, devastating the socio-economic systems of the civilizations that had dwelt along these ancient shorelines for almost 500,000 years.

Changes in Earth's orbit and the tilt of its axis abruptly altered the Sahara. In only 300 years it evolved from a region of annual grasses and low shrubs, as found in fossilized pollen, into the desert we know today.

The Sahara is an extremely stable environment with a widely varying topography characterized by high levels of aridity and desiccated dust-laden winds called sirocos that often reach speeds of 60 miles per hour (100 km/h), capable of eroding rocks and reducing visibility to zero.

Petroglyphs of river animals such as crocodiles have been found in the Tassili n'Ajjer in southeast Algeria. The Messak Plateau petroglyphs of the central Sahara depict horses, camels, and two-wheeled carts dating to 1500 BCE alongside crude portrayals of ostrich and giraffe circa 3000 BCE, a pictorial testimony of a more temperate past.

REGS, ERGS, HAMADAS, AND CHOTTS

Endless expanses of sand mixed with black, red, or white gravel known as regs account for 70 percent of the Sahara's surface area. These remains of ancient seas and rivers can stretch for thousands of miles, such as the Tanezrouft Reg to the west of the Hoggar (Ahaggar) Mountains that covers 200,000 square miles (500,000 km²).

Ergs or sand dunes cover one-fifth of the Sahara, and can stretch over hundreds of miles and reach heights up to 1,000 feet (300 m). The Selima Erg in Sudan covers 3,000 square miles (7,800 km²) and the Grand Erg Occidental and Oriental envelop most of Algeria. The Sahara's sand sheets and dunes come in many forms—transverse, blowout, tied, and barchan, as well as the rare pyramidal dunes.

Elevated plateaus of rock and stone called hamadas reach over 10,000 feet (3,000 m) in the central Tibesti Mountains of northern Chad and

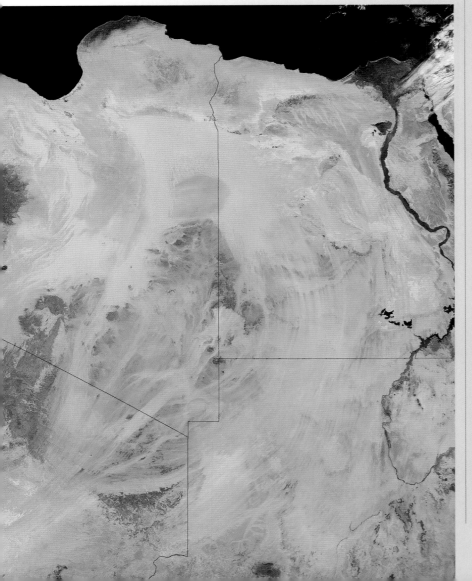

southern Libya, a region of dormant volcanoes across a broad uplifted area that contains the Sahara's highest peak, Emi Koussi, at 11,204 feet (3,415 m).

Chotts are inland saltwater wetlands with no connection to water sources. They occur throughout the Sahara in depressions where water collects from underground sources and seasonal rainfall. The three major Saharan wetlands are the Chott el-Jerid in Tunisia, the Chott Melrhir in Algeria, and the 7,000-square-mile (18,000 km^2) Qattara Depression in northwestern Egypt.

ONE DESERT, THREE LANDSCAPES

The Sahara can be divided into three distinct geographic regions. The Western Sahara is an expanse of rocky plains and sand deserts largely devoid of rainfall and surface water, though underground rivers originating in Morocco's Atlas Mountains carry groundwater to oases, producing fertile soil and good crops. The Central Plateau runs 1,000 miles (1,600 km) on a northwest–southeast axis, varying in height from 2,000 to 2,500 feet (600 to 800 m). Despite limited rainfall, several central Saharan peaks occasionally receive brief snowfalls. To the east, the Libyan Desert is considered the Sahara's most arid region, with little or no moisture and few oases. Its sandy wastes, boulder plains, and dunes up to 400 feet (120 m) in height are spread across a region of basement rocks and horizontally bedded sediments.

The Sahara possesses little perennial vegetation with large expanses of regs and ergs devoid of any life for years at a time. Its fauna, however, is richer than generally perceived, with over 70 species of mammals, 100 species of reptiles, and 90 species of birds cataloged.

ABOVE A nomad walks along the sand dunes of the Sahara Desert. Most of the people who live in this desert are nomadic including the Berbers, the original dwellers, who now comprise a large proportion of the populations of Morocco and Algeria.

LEFT Most of the Sahara's oases occur in places below sea level, where groundwater seeps up from underground basins or natural springs. The Qattara Depression In Egypt is the lowest point, at 425 feet (130 m) below sea level. Fruit such as dates are grown in these oases.

ABOVE A deadly fat-tailed scorpion (*Androctonus australis*) is nearly as toxic as a cobra. Here in the Touggourt, in the Sahara's Great Sea of Sand, in Algeria, daytime temperatures can reach 172°F (78°C). This animal survives by burrowing during the heat of day and being active at night.

SALT DESERTS

Salt deserts, also referred to as salt flats or playas, consist largely of alkaline salts and fine-grained sediments. Often located in regions with little or no rainfall, high levels of evaporation produce excesses of dissolved salts, leading to the formation of vast salt crusts.

The world's largest salt desert is the Salar de Uyuni. It encompasses over 4,000 square miles (10,400 km²) of Bolivia's southwest Potosi region and is estimated to contain a massive 10 billion tons (11 billion tonnes) of salt. Other significant salt deserts are the Dasht-e Kavir in Central Iran and the Great Salt Lake in Utah, USA.

THE DASHT-E KAVIR

The Dasht-e Kavir is a 10,000-square-mile (26,000 km²) basin in the middle of the Central Iranian Plateau, stretching from the Dasht-e Lut "Empty Desert" in the southeast, to the Alborz Mountains in the country's northwest.

BELOW Like swirls of paint on an enormous canvas, shallow lakes, mudflats, and salt marshes share the sinuous valleys of Iran's largely uninhabited Dasht-e Kavir, or Great Salt Desert. This satellite image was made using enhanced thematic mapping.

A rainless arid expanse with scattered wadis, lakes, and marshes, it is named after its constituent kavirs or salt marshes, such as the Darya-ye Namak, a salt lake with plates arranged in mosaic-like patterns, and the Rig-e Jenn, a forbidding expanse of dunes believed by local caravan traders to be inhabited by evil spirits.

Many kavirs possess properties akin to quicksand, making the Dasht-e Kavir a treacherous and still largely unexplored region with habitation restricted to its surrounding hills and mountains.

Summer temperatures can often reach as high as 140°F (60°C), causing extreme levels of evaporation and the formation of salt crusts over its mud grounds and marshes. Vegetation is limited to plants that can survive its saline soils, such as mugwort, a herbaceous perennial plant with reddish purple stems and tiny yellow petals that thrives in nitrogenous soils.

THE SALAR DE UYUNI

High in the Bolivian Altiplano near the crest of the Andes lies one of the world's largest salt deserts, the Salar de Uyuni. Formed after the evaporation of prehistoric Lake Minchin began some 40,000 years ago, the salar covers over 3,000 square miles (7,800 km²) of a 12,000-foot (3,700-m) high plateau. Surrounded by mountains in an internally drained basin, its climatic history of precipitation and aridity lies trapped within its thick Quaternary sediments, providing a valuable paleoclimatic history of the region.

Periodic flooding and evaporation have produced widespread calcareous lake sediments and salt deposits over 300 feet (100 m) deep, in addition to large reserves of gypsum and halite or

Minerals such as halite and gypsum form beds of sedimentary rocks when seawater trapped in isolated lakes or shallow seas evaporates. Thick layers build over time as a lake is repeatedly replenished with seawater then dried out to a salt pan. Evaporites are mined for halite and potassium minerals, such as sylvite and carnallite, in ancient seas such as the Dead Sea and Aral Sea. Evaporite beds buried beneath later sandstone and shale layers may rise upward into huge salt domes due to their lower density.

common table salt, as well as brines of lithium and boron. Coral is also present, trapped in limestone layers beneath the desert's paleo-lakeshore terraces.

A dormant volcano, Mt Tunapa, sits in the salar's north, its summit supporting a small glacier with precipitation from its windward slopes supporting scattered communities along its base.

THE GREAT SALT LAKE DESERT

The Great Salt Lake Desert stretches west from the shores of Utah's Great Salt Lake to the Nevada border. Its 4,000 square miles (10,400 km^2) are blindingly white due to large concentrations of salt in its soil left over from evaporite deposits from the extinct Lake Bonneville, an ancient

lake that covered around 20,000 square miles (520,000 km^2) of western Utah between around 15,000 and about 30,000 years ago, before erosion caused it to drain north into the Snake River.

A remnant of Lake Bonneville, the Great Salt Lake itself is a terminal lake with no surface outlet, receiving water from four surrounding rivers with a combined drainage basin of over 21,000 square miles (54,400 km^2). Water loss is due to evaporation. Average depth is just 30 feet (10 m). There are over 11 recognized islands found throughout its more than 1½ million square miles (4,000 km^2) including Antelope Island, inhabited since pioneering times and possessing the state's oldest Anglo-built structure still set on its original foundations.

TOP Salt beds in the Great Salt Lake Desert form an expanse of salt-encrusted land. Part of the lake is used as a racing track to break speed records.

ABOVE LEFT Like sentinels guarding a lonely desert outpost, giant cactus dominate the shores of Salar de Uyuni's salt lake. Hardy and salt-tolerant, they can reach 20 feet (6 m) tall, and are known to live for over 1,000 years. The rock salt in the lake is cut into blocks and used for building.

Deserts of Central Asia

Central Asia is the region between the eastern shore of the Caspian Sea east to the Altai Mountains in Mongolia. Its primary natural features include the Pamir and Tien Shan Mountains, the Caspian and Aral Seas, the Amu-Daria River and its twin deserts, the Kyzyl Kum ("Red Sands") Desert in north-central Uzbekistan, and the Kara Kum ("Black Sands") Desert that occupies over 80 percent of Turkmenistan.

Broader definitions of the region that are based on climate encompass far larger borders which include Mongolia's Gobi Desert and the Taklimakan and Ordos deserts of western China.

TOPOGRAPHY OF KARA KUM AND KYZYL KUM

Typical of the mosaic of clay, stone, salt, and sandy deserts common to Central Asia, the Kara Kum (Garagumy) is a large denuded desert plain with abundant salt flats, soft sediment bluffs, and karst topography. It is separated from the neighboring Kyzyl Kum Desert to the east and its eolian sand fields by the 1,500-mile-long (2,400 km) Amu Darya River, the longest river in Central Asia.

The Kyzyl Kum and Kara Kum deserts possess the sort of common geological traits one would expect from having evolved in an arid environment. The soils of both deserts are characterized

BELOW The Gobi Desert in Mongolia looks quite unlike an arid region. Much of its landscape is rocky and vast areas of the Gobi include steppe and semi-arid areas. Gorges in the Gobi have yielded some important Cretaceous dinosaur fossils, including eggs and trace fossils.

THE MONGOLIAN DEATH WORM

In his 1926 book *On The Trail of Ancient Man*, American paleontologist Roy Andrews wrote an intriguing account of the elusive Mongolian Death Worm. Apparently able to spray acid and kill from a distance using an electrical current, Mongolian tribesmen know it as the *allghoi khorkhoi*, or "intestine worm," a sausage-like creature as thick as a man's arm covered in dark blotches with spikes at both ends. No discernible facial features makes it impossible to distinguish its head from its tail. Resembling a cow's intestine, its venom is believed to corrode metal. Most of the year it hibernates beneath the dunes of southern Mongolia, only surfacing after heavy rains when the soil is moist. It's impossible to find a nomadic herder who doesn't believe it exists, and many claim to have seen it, including a former Mongolian Premier.

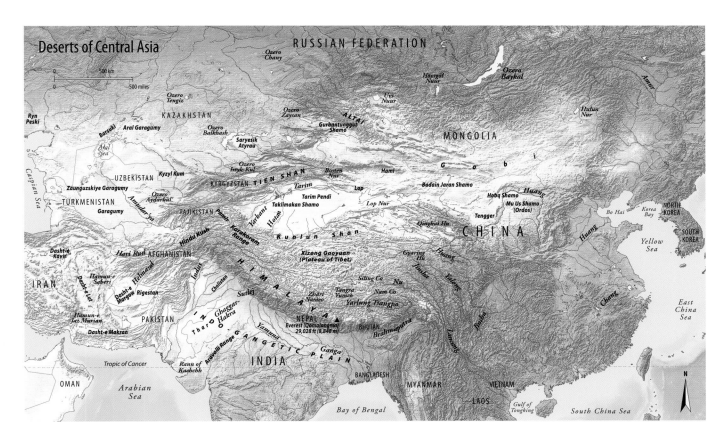

Deserts of Central Asia

by low amounts of organic matter along
with high alkalinity and salinity levels. Annual
rainfall can occasionally rise as high as 5 inches
(130 mm), making the Kara Kum and Kyzyl Kum
deserts the richest desert complex in Eurasia. It is
home to desert mammals such as the long-eared
hedgehog, the rare honey badger, and sand lynx,
along with a variety of nocturnal burrowing
rodents such as gerbils and jerboas.

VAST PLATEAUS OF THE GOBI DESERT

One of the great deserts of the world, the Gobi
Desert stretches over 1,000 miles (1,600 km)
from the Da Hinggan Mountains in Mongolia
to the Tien Shan Mountains in northern China
in a vast arc across Central Asia.

Although the Chinese call the Gobi *sha-mo*
(sand desert), only around 3 percent of the desert
is sandy, with the remainder covered by a thin
layer of gravel over a base of metamorphic rock.
Its plateaus consist of bare rocks with occasional
patches of shifting sand, the region remaining true
to the Mongolian definition of *gobi*, a place where
habitation is scarce and where gravel and rocks
cover the landscape.

Rainfall varies from 3 inches (80 mm) in the
west to 8 inches (200 mm) in the east. Winter
temperatures can plummet to −40°F (−40°C) in
winter and climb to 113°F (45°C) in summer.

WESTERN CHINA'S WINDSWEPT ORDOS DESERT

The Ordos Desert lies along a high, arid, windswept plateau on the southern boundary of Inner
Mongolia. Almost encircled by the great northern
bend of the Yellow River to the north, west, and
east, the Great Wall of China separates it from fertile loess lands to the south. The Ordos Plateau
(Mu Us Shamo) is characterized by bitterly cold
winters, hot summers, and chronic droughts. Its
abundance of salt lakes have intermittent streams.
An annual precipitation rate of less than 10 inches
(250 mm) combines with alkaline soils to produce
a landscape consisting largely of montane grass and
shrubland in a vast wilderness area.

ABOVE The World Heritage-listed
Mogao Caves lie along an oasis
on the ancient trading route of
the Silk Road near Dunhuang
in China. On the edge of the
Dunhuang Desert, carved into
the barren ground is a labyrinth
of ancient grottoes featuring an
extensive array of early Buddhist
art and shrines. Hundreds of
thousands of square feet of the
walls are covered in murals.
Many ancient manuscripts have
also been recovered from the site.

TAKLIMAKAN DESERT, CHINA

At times known as the "Desert of Death" and the "Place of No Return," the forbidding Taklimakan Desert in the northwest Xinjiang Uygur Autonomous Region of western China is one of the most barren and inhospitable deserts on Earth.

Along the Taklimakan's north-western edge on the Silk Road (below) lies a string of oases fed by glacial waters. In the hills lining the Tarim Basin, the intense aridity of the desert is less severe but can dry the earth (above).

Estimated to be some 4.5 million years old, despite the absence of any universally accepted theory as to its origins, an ongoing debate continues as to whether the Taklimakan is a rainshadow desert or simply a land-locked region too far removed from any meaningful sources of precipitation.

One of the world's great deserts, it covers over 100,000 square miles (275,000 km²) of the Tarim Basin (Taklimakan Shamo), the largest basin in the world. Two arms of the legendary Silk Road traverse its northern and southern boundaries. The glacier-fed streams of the Pamir Mountains to the west, the Tien Shan Mountains to the north, and the Kunlun Shan Mountains to the south emerge from alluvial fans but soon disappear into an endless expanse of loose sands and fine-grained alluvial clays. The only rivers to cross it are the

Tarim and the Ho-t'ien, though their flow is anything but constant and only during rare periods of flooding do they carry water along their entire course.

THE TAKLIMAKAN'S SAND DUNES

The Taklimakan's crescent-shaped sand dunes often reach heights of over 300 feet (100 m) and are shaped across elliptical, cauldron-like basins that generally follow a northeast–southwest line, and whose crest-to-crest widths can measure over 2 miles (3.2 km)—the largest on Earth.

In the west, between the towns of Kashgar and Yarkand, strong winds aid these dunes in traveling 13 feet (4 m) annually, and the sands along its 1,100-mile (1,800-km) southern edge insidiously swallow up cultivated land at the rate of 30 feet

(10 m) a year. Several theories on the source of the Taklimakan's vast deposits of sand range from them being the remnants of the alluvial clays of the desert itself through to being the remains of a once-great Central Asian Mediterranean–type sea.

A FORBIDDING CLIMATE

This is an arid environment that receives less than 1½ inches (40 mm) of rain a year. The winters are cold with temperatures averaging –4°F (–20°C) while summers are relatively mild and rarely exceed 102°F (39°C). In spring, high winds can result in choking impenetrable dust storms capable of smothering local fauna and which can cause the temperature to plummet to below zero.

This is a sparsely settled region where plant and animal life is scarce outside cultivated areas. Tamarisk trees and steppe shrubs such as kamish compete for moisture with hares, rats, foxes, and the occasional wild camel. Poplar trees in varying stages of decay testify to the area's dwindling reserves of water. The Chinese government is in the midst of a 10-year plan to encircle the Taklimakan Desert's 1,900-mile (3,000 km) circumference with 1.56 million acres (630,000 ha) of trees and grasses designed to alleviate damage from sandstorms and slow desertification.

THE TAKLIMAKAN MUMMIES

Since the late 1970s, hundreds of Caucasian mummies with distinctly European features have been unearthed in western China in the Tarim Basin. Commonly referred to as the Taklimakan mummies, they became one of the great mysteries of the twentieth century. Many of them are on display in Xinjiang's museums.

Preserved for thousands of years by the region's exceptionally arid alkaline soils, these non-Mongoloid people had high-bridged noses and fair hair. They apparently lived unchallenged in the Tarim Basin for over a thousand years, long before the first recorded appearance of East Asian peoples.

One of the oldest finds is a 3,800-year-old 6-foot (1.8-m) Caucasian female nicknamed Beauty of Loulan. She has a facial bone structure of unmistakable Indo-European descent, and was buried wearing ankle-high moccasins, a knee-length overwrap, and a skirt with the fur turned inward for warmth. Another famous mummy is Cherchen Man, who had red hair, and a blond baby.

These desiccated corpses have given historians a new perspective on Bronze Age people and given credence to ancient Chinese texts dating to 500 CE describing legendary figures "of great height" with deep-set blue-green eyes and blond hair.

BELOW A vast alluvial fan spreads out over the desolate landscape near the southern border of the Taklimakan Desert between the Kunlun and Altun mountain ranges. The left side is the active part of the fan, which shows up as a vivid blue where water is flowing out of the mountains at the bottom right corner. It fans out into scores of intricate braided channels that disappear into the desert. The silvery braidings at lower right are dry channels.

THAR DESERT, INDIA AND PAKISTAN

The Thar Desert in western India (also known as the Great Indian Desert), covers an area of 90,000 square miles (235,000 km²), crossing the border into the eastern Pakistani provinces of Sind and Punjab. Here it is known as the Cholistan Desert, whose barren landscapes, once watered by the-dried up Ghaggar River are now home to over 400 archeological sites dating to the Indus civilization.

BELOW A cameleer leads his camel through the rippling sand dunes of the Thar Desert. Camel is the best means to traverse the shifting dunes of a desert that has no oases. Travelers must rely on hand-dug wells that can vary in depth and quality and the water needs boiling as it is frequently infested with parasites.

An expanse of thornscrub encircles the Thar Desert, which, together with the desert itself, has a combined area of over 280,000 square miles (725,000 km²). Of this, only 10 percent consists of shifting sand dunes, with the remainder a mix of salt pans, fixed dunes, and gravelly plains.

It is a desert of myth. In the allegorical Indian Sanskrit epic *Ramayana*, Lord Rama, in pursuit of his wife Devi Sita who was abducted by the demon–king Ravana, prepared to use a fire–weapon to dry up the ocean separating him from Ravana's kingdom. All the creatures of the ocean pleaded with Lord Rama to spare their lives, and so he instead aimed the weapon at a distant inland sea in western India, and thus the arid and desolate Thar Desert was born.

Throughout the Thar's 92,000 square miles (240,000 km²) not a single oasis can be found, and artesian wells are rare due to the high percentage of soluble salts turning well water toxic. Since the

1600s, local communities have built saucer-shaped catchments into the land, packing them with pond silt or gravel to collect the less than 10 inches (250 mm) of rainwater that falls annually. The fire-weapon of Lord Rama apparently achieved its aim.

THE THAR'S INCREASING ARIDITY

Drainage systems in the Thar Desert are poorly organized, small in size, and easily lost to the surrounding sand dunes and aeolian plains. Neo-tectonics have played a sizeable role in modifying old drainage courses, forcing tributaries to alter directions that have long since resulted in drying up rivers such as the Ghaggar and Yamuna.

On average, about four out of every 10 years are drought years here, with human habitation adding to the vagaries of nature; in combination these give the Thar the largest human and live-stock population among all the world's deserts, along with the resultant high pressure on grazing lands that further increase the region's aridity.

FURTHER DESERTIFICATION

One third of India's arable land is susceptible to deser-tification, with the Thar now beginning to threaten its most significant geological barrier, the Aravalli Range. These mountains to the south-east have long prevented the encroachment of the Thar toward the east and also the Ganges Valley, but they too are fast losing their vege-tation to the relentless processes of aridity.

The region's landscape is generally characterized by sta-bilized sand plains with longi-tudinal ridges in the south and transverse ridges to the north that are mostly calcareous and rich in minerals such as quartz. Soil quality improves outward to the east, consisting largely of deposits from eroded rock blown in from the coast and from the Indus Valley.

The Thar's shifting sand dunes come in three types: longitudinal parabolic (parallel to the prevailing winds); trans-verse (aligned across the wind direction); and barchan dunes, arc-shaped sand ridges with

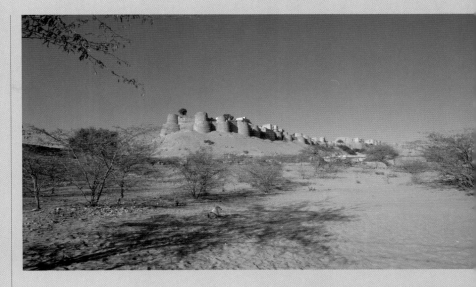

concave sides facing the wind. Temperatures vary from below freezing in winter to as high as 125°F (52°C) in summer and rainfall is precarious and erratic, ranging from 5 inches (130 mm) in the west to 15 inches (380 mm) in the east.

DIVERSE PLANT LIFE, DECREASING WILDLIFE

The combination of hillocks and sandy and gravel plains have, despite a general lack of precipitation, resulted in a great diversity of vegetation. Over 700 species of plants grow in the Thar, of which 107 are grasses that are nutritious and rich in minerals and trace elements. Shrubs here are deep-rooted, drought-tolerant, and able to gain biomass rapidly when rain is plentiful.

Wildlife that is fast disappearing elsewhere in India and Pakistan can be found here in abun-dance. Wild ass, blackbuck antelope, and the great Indian bustard have evolved excellent survival strategies, such as being smaller in size than their namesakes elsewhere and their predominantly noc-turnal habit allows them to avoid the heat. They also may be protected from the elements by the people within various local communities, such as the Bishnois—a religious sect in western Rajasthan whose devotees worship nature.

ABOVE The desert villages of Rhajasthan, India, are largely home to agriculturalists. Many of their dwellings are built from mud or dung, hay, and sand. Their thick walls keep the buildings cooler than brick.

TOP The fortified medieval city of Jaisalmer lies in the north-western part of the Thar Desert, Rajasthan. Also known as the Golden City, it contains many finely sculpted temples.

LEFT Sparring male blackbucks (*Antilope cervicapra*) use their horns in an attempt to assert dominance, defend territory and gain access to females by indirectly controlling the females' access to resources.

Deserts of Australia

Over 70 percent of Australia is classified as semi-arid to arid, with the world's driest continent after Antarctica being home to the largest expanse of desert in the southern hemisphere. These desert regions are found mainly throughout the western plateau and interior lowlands.

ABOVE The Simpson Desert is Australia's best-known desert owing to its extraordinarily long, brilliant red sand dunes.

BELOW Tall pillars rise up from the desert floor in the Pinnacles Desert, Western Australia. This is a small desert, covering just 1,000 acres (400 ha).

They vary from traditional sand deserts, such as the Simpson Desert in South Australia with its signature hundred-mile-long (160-km) dunes, to stony deserts like the Tirari Desert and Sturt's Stony Desert in south-central Australia, known for their red soils and "gibber plains"—vast deposits of rock and pebble-littered silica.

THE SIMPSON DESERT

The Simpson Desert is the world's largest and youngest parallel sand dune desert, covering an area of 68,000 square miles (175,000 km²), straddling the corner of the Northern Territory, Queensland and South Australia. It is bordered by Lake Eyre to the south, the Finke and Todd rivers to the west, and the Diamantina, Warburton, and Georgina rivers to the east.

Its sand dunes are stationary, fixed in place by vegetation. Reaching heights of 120 feet (40 m) and separated by as much as 450 feet (140 m), they run parallel on a northwest–southeast axis with inter-dune corridors populated by a mix of acacias, eucalypts, and occasional large stands of bloodwood trees and ghost gums.

Rainfall averages 8 inches (200 mm) annually and temperatures can reach 122°F (50°C) in summer and can fall to below 32°F (0°C) in winter.

Some of Australia's rarest marsupials can be found here including the fat-tailed mouse. Vegetation that has managed to adapt to this harsh environment includes sandhill canegrass, which helps stabilize loose sand along dune crests while providing a home for the rare Eyrean grasswrom. Lobed spinifex, a perennial grass well-adapted to arid environments, its leaves curled to reduce surface area with wax-like cuticles to limit water loss, provides lizards and small mammals with shelter from airborne predators.

THE PINNACLES DESERT

Rising from the yellow sands of Western Australia's Nambung National Park are thousands of pillars of Tamala limestone. These are known to geologists as the by-product of shelly sand dunes cemented by eons of percolating rainwater, and collectively called The Pinnacles. These shell sands had their genesis 15,000 years ago during the last Ice Age, evolving into layers of calcareous deposits that formed over the limestone, which eventually leached away, leaving the columns we see today. The remnants of three ancient systems of sand

dunes that long ago paralleled the West Australian coastline, they vary in size from around 3 feet to over 10 feet (1 to 3 m) in height.

THE TANAMI DESERT

The 71,000 square miles (184,000 km^2) of the Tanami Desert in the Northern Territory is one of the most isolated and arid areas on Earth, not fully explored until well into the twentieth century. It has an arid subtropical climate and is characterized by gently undulating sand plains, broad drainage depressions, and depositional plains that support trees and steppe shrubs, open hummock grasslands, and low woodlands of mulga and eucalypts.

Its widely spaced dunes are parallel on a largely featureless plain. Ancient streambeds traverse the region over a layer of Cambrian Period sediments and scattered outcrops of Proterozoic Era granite.

The Tanami's isolation has spared it the ravages born from feral animals such as foxes and rabbits, giving it possibly the most diverse assemblage of native desert fauna in Australia, including the nocturnal mole, the bilby, and the great desert skink.

AUSTRALIA'S STONY DESERTS

Rock samples collected from Australia's stony deserts indicate they were formed as early as 4 million years ago. They are characterized by stones buried under deep loam soils with exposed coverings of silcrete, a resistant rock formed by groundwater silicification of sedimentary layers. Called gibbers, these ancient landforms cover 10 percent of the vast Australian continent.

Sturt's Stony Desert was named by explorer Charles Sturt in 1844 during his unsuccessful attempt to locate an inland sea. It is an endless landscape of stony plains and gibber, with very few linear dunes, dotted with saltbush scrub that thrives in what is a very harsh, dry, and salty environment.

THE GREAT SANDY DESERT

The Great Sandy Desert is Australia's second largest desert. A wilderness of red sand dunes, salt marshes, and sand hills, it covers 110,000 square miles (285,000 km^2) of northern Western Australia and the Northern Territory, from Eighty Mile Beach on the Indian Ocean through the Pilbara and Kimberley ranges to the Gibson Desert in the south.

A sparsely populated region with no significant human habitation, its remarkably parallel linear sand dunes are covered primarily in spinifex and desert oak in a landscape of tree steppe, bloodwoods, acacias, and grevilleas amid emergent sandstone hills and sandy plateaus dotted with salt marshes and dry lake beds.

Rainfall is patchy. Its northerly region along the West Australian coastline and throughout the Kimberleys benefits from monsoonal rains and tropical cyclones with rates of precipitation of up to 12 inches (300 mm) annually. Even its more arid southern regions receive 10 inches (250 mm) a year—high by desert standards though massive evaporation rates result in little real benefit.

Summer daytime temperatures can vary from 98°F to 100°F (37°C to 38°C) in the north near Halls Creek in the Kimberleys to 100°F to 108°F (38°C to 42°C) further south. Winters are short and warm with only the occasional frosts along its southern boundary with the Gibson Desert.

ABOVE Rock samples collected from Australia's stony deserts indicate they were formed as early as 4 million years ago. Sturt's Stony Desert (above), like others, is characterized by stones buried under deep loam soils with exposed coverings of silcrete, a resistant rock formed by groundwater silicification of sedimentary layers. Called gibbers, these ancient landforms cover 10 percent of Australia.

Ice Deserts of Antarctica

Deserts are generally defined as regions that receive less than 10 inches (250 mm) of precipitation a year. By this definition, Antarctica, the fifth largest of the world's seven continents, sits very comfortably alongside its warmer sandier cousins. Such deserts are sometimes called ice deserts.

ABOVE Katabatic winds are characteristic of Antarctica. These are dry winds that spill down into the valleys then head toward the coast. They quickly evaporate any ice in Antarctica's dry valleys almost as fast as it forms. This wind can sculpt the ice into unusual shapes before it is completely eroded away.

The interior of the continent receives less than 2 inches (60 mm) of rainfall a year, less than the Sahara. With virtually no evaporation due to its cold climate, rainfall and snow have built up over millennia into the massive ice sheets we see today. Antarctica was not always a land of ice sheets and permafrost. Three hundred miles (500 km) from the South Pole, sandstone beds lined with coal deposits have been found, laid down in marshy conditions under a cool moist climate.

ANTARCTICA'S DRY VALLEYS

A series of dry valleys throughout the continent receive virtually no rainfall. These ice-free valleys are generally 3 to 6 miles (5 to 10 km) in width between ridge crests and up to 30 miles (50 km) long, with permafrost patterns common to perennially frozen ground seen in their loose sand and gravel surfaces. Principal ice-free valleys include the Victoria, Taylor, Wright, and Balham valleys. The Taylor and Wright valleys have glaciers at their heads that connect to the ice of the polar plateau. Vertical terminus walls, little or no rock debris, and a general absence of crevasses are characteristics of dry valley glaciers.

Lakes found within these dry valleys possess saline levels three times as high as seawater, and temperature inversions from solar heating of water through the ice have resulted in deep water temperatures as high as 77°F (25°C).

LIFE ABOVE AND BELOW THE ICE

Antarctica's massive ice sheets form an effective barrier to two widely varying bioregions. Below the ice, life is rich and complex, while above the ice, it is restricted to some 350 species of lichens, mosses, and algae. Algae and lichens have been found growing inside sandstone protected from

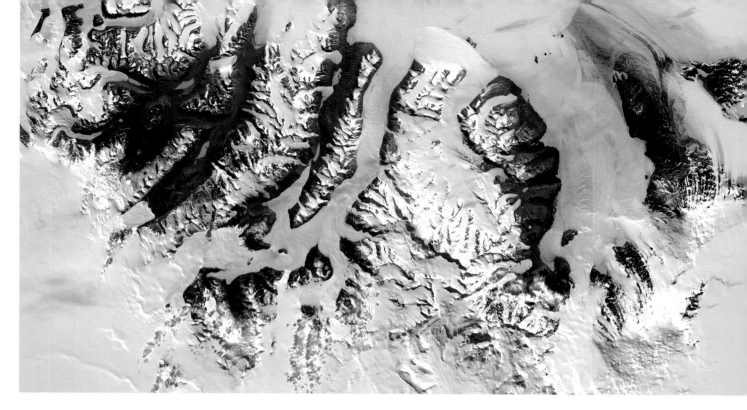

the harsh climate beyond, while dry valley soils contain various bacteria, fungi, and algae. There are no trees or shrubs, and the continent supports no permanent land-based vertebrate animals though emperor penguins do breed on the surrounding ice in winter. Various invertebrates, no more than a tiny fraction of an inch in length, such as mites, ticks, and nematode worms live here. Their bodies contain a type of anti-freeze, surviving Antarctica's long winters by lying insulated and dormant within ice and rock crevasses, becoming active once temperatures rise in summer.

Below the ice, life is abundant. Four times more productive than the other oceans of the world, Antarctica's waters possess an immense food chain that begins with microscopic algae and zooplankton, such as krill through to squid, octopus, seals, fish, penguins, and whales.

If Antarctica's ice sheets were to melt, all the oceans would rise over 200 feet (60 m). In some places the ice is 3 miles (5 km) deep, giving Antarctica the highest average elevation of Earth's continents. Ice covers 5.3 million square miles (14,000,000 km²) or 97 percent of its surface area and contains over 90 percent of all the world's ice. Mean winter temperatures average −76°F (−60°C). The lowest ever recorded was −129.3°F (−89.6°C) at the Russian Vostok station.

ABOVE As one of the few areas of Antarctica not covered by thousands of feet of ice, the McMurdo dry valleys stand out starkly against the glaciers in this satellite image. For a few weeks each summer, temperatures are warm enough to melt glacial ice, creating streams that feed freshwater lakes that lie at the bottom of the valleys.

EMPEROR PENGUINS

The emperor penguin (*Aptenodytes forsteri*) is the world's largest penguin, standing nearly 4 feet (1.2 m) tall and weighing as much as 90 pounds (41 kg). It is estimated some 200,000 breeding pairs live in temperatures as low as −76°F (−60°C) on the fast ice surrounding the Antarctic continent.

Their tiny wings allow them to dive to depths of up to 900 feet (275 m) in search of food, and they have the ability to swim at speeds in excess of 9 miles (15 km) an hour, enabling them to evade their primary predator, the leopard seal. A combination of body oils and a thick layer of downy feathers traps body heat and keeps water out.

Females lay a single large egg that is immediately rolled onto the male's feet and cared for within the folds of its skin. Groups of males then huddle together incubating the eggs for over 2 months, in the process losing half their body weight, while the females return to the sea to feed. The chicks become fully grown in 6 months, returning to the ocean with their parents at the start of the Antarctic summer.

OPPOSITE, FAR LEFT Looking directly along the dry valleys of Antarctica, the extent of exposed rock is revealed. These vast expanses of ice-free rock are sometimes called Antarctic oases. The exposed rock strata make fossil collection possible. Fossil leaves and wood found in Antarctica indicate it was once warmer and forested.

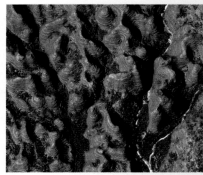

ABOVE The beehive-shaped Bungle Bungle mountains rise up to 1,000 feet (300 m) above a plain of forest and grass in Western Australia's Kimberley region. Ancient rivers shaped these mostly sandstone mounds.

RIGHT People have made homes in these soft volcanic tuff cones near Goreme, Cappadocia, on Turkey's Central Anatolian Plateau. Apart from this small area of fertile soil, Cappadocia features red sandstone and salt deposits.

RIGHT The Kata Tjuta (Olgas) rock formation is the remains of an alluvial fan that was created by sediment washed away from a mountain hundreds of millions of years ago. Kata Tjuta consists of conglomerate rock: Gravel bonded by sand and mud.

A broad range of landforms can be found within large continents. Inland landforms are often very distinctive topographic features, having been created by a unique combination of tectonic activity, erosion, and climate. Uplift of large areas of continental crust is occurring almost continually, as these thick landmasses act like huge lids on top of the heat escaping from Earth's interior.

Rising heat causes large areas of continental surface to dome upward, like boiling soup lifting the lid of a saucepan. It may even lead to the eventual tearing and rifting of the crust. Once raised high above sea level, landforms develop that are typical of that region. In many cases, these features are so spectacular and interesting that they are protected as national parks, or achieve World Heritage status.

Inland topographical features include: plains, plateaus, mesas, buttes, pinnacles, arches, folded and buckled rocks, domes, basins, different types of faults, cuestas, and razorback or hogback ridges. Many of these awe-inspiring natural phenomena have been known to, and are often significant to, indigenous populations; over the past several hundred years, colonizers have been amazed to "discover" these ancient wonders.

SEDIMENTARY STRUCTURES:
HARD AND SOFT SEDIMENTARY BEDS

Some of the most unusual landforms occur in sedimentary rocks and are caused by the fact that sedimentary rocks are made up of planar units, or strata, of differing resistance to weathering and the forces of erosion. Conglomerates and sandstones can be quite strong and difficult to erode provided that there is strong cement, such as silica, holding the pebbles, cobbles, and grains of sand together. Silica cement is reasonably common. Often it is derived from the quartz sand grains themselves, which dissolve very slightly at their points of contact with surrounding grains during the process of burial and compaction.

The term lithification is used to describe the process of sediments turning into rock. Limestone strata are also generally very strong physically; however, these behave quite differently in different climates, depending on the temperature and rainfall. Limestones will form resistant cliffs that seem to last forever in arid climates, yet under wet tropical conditions the same rock will rapidly dissolve and become riddled with caves. Shales and siltstones, because of their higher clay content, are generally much softer and weaker physically, and prone to break down very quickly when exposed

to the weather. When wet, clay minerals absorb water into their molecular structure, they expand very slightly then shrink again when they dry out. This constant process of swelling and shrinking rapidly weakens the integrity of the rock by creating micro-fractures between the grains, which in turn allows more water to enter. The process continues quickly, with exposed rock surfaces crumbling away. Anyone with a house built on clay or shale will testify to the need to repair cracks that have been created by slight movements of the ground beneath the foundations.

ABOVE The Three Sisters in the Blue Mountains, New South Wales, Australia, are composed of strong sandstone that was able to resist the forces of wind, rain, heating and cooling, and, most significantly, gravity.

LEFT The Autana Tepui is one of more than a hundred sandstone–quartz mesas in La Gran Sabana, southeastern Venezuela. This tabletop mountain rises abruptly 4,600 feet (1,400 m) above the Amazon rainforest.

PLAINS, PLATEAUS, MESAS, BUTTES, AND PINNACLES

A large, flat, unbroken expanse of land is called
a plain. At sea level, a sluggishly meandering river
has no energy to cut down into its floodplain,
but the river can become rejuvenated if the land
is uplifted. An uplifted plain is called a plateau.
Resistant geological strata often form a protective
cap over weaker rocks beneath, creating an array
of spectacular landforms. Once streams and rivers
cut into the resistant top of a raised plateau, it
quickly becomes dissected into a series of flat tabletop mountains with steeply cliffed edges. These are
known as mesas. The tepuis of Venezuela, capped
by an ancient layer of resistant sandstone, are a
spectacular example of this type of landform.

The weaker layers in a horizontal stack of sedimentary rocks are rather like its "Achilles heel,"
because they allow weathering and erosion to penetrate and undermine the stronger layers. Without
support, the resistant layers eventually break away
and collapse, forming a sloping apron of debris
around the base of the mesa called a talus apron.
Erosion thus takes place around the weakened
sides of the landform in the form of parallel retreat
of its cliff faces, rather than downward through the

protective cap rock. The result is that the mesas
do not shrink in height but rather become aerially
smaller and smaller. When the diameter of the flat
tabletop eventually becomes less than the height
of the mesa above the plains level then a different
term is used—they are called buttes. Even more
dramatic is when the cap rock shrinks to almost
nothing more than a large boulder protecting a
thin column, or pinnacle, of weaker rock. Some
of these look so precarious, as if they could be
toppled with just one push, and they are obviously
the final stage before the complete disappearance
of the landform. The colorful and bizarrely shaped
pillars of Bryce Canyon, Utah, display the different
stages of this process beautifully.

NATURAL ARCHES

Arches are very temporary geological structures.
They represent an unusual situation where closely
spaced jointing has allowed erosion to carve thin
fins or vertical slices of rock, which are then eroded
through from one side to the other to form an
arch beneath a more resistant cap rock layer. Water
and wind are the main agents of erosion in inland
settings. Arches National Park, Utah, is a unique
place with some 2,000 documented natural arches

ranging from tiny holes to the broad paper-thin Landscape Arch. This arch is one of the main draw-cards to the park. It probably took some 100,000 years to carve but is now very close to the point of collapse. Sea arches, found around the coastlines, are similar in structure but have been created by wave action eroding the base of cliffs.

FOLDED AND BUCKLED ROCKS

Sedimentary rocks can be slowly rumpled into a series of folds, much like when a floor rug has been pushed up against a wall. The upper parts of the folds, or hills, are called anticlines while the bottoms or troughs are termed synclines. Because the process takes place very slowly, most of the sedimentary strata accommodate the bending. However, the very resistant layers will generally develop a series of parallel tension cracks along the tops of the folds (at the point of maximum bending)—much like those you can see on a small scale if you bend a fruit candy bar. These cracks weaken the layers and, as a result, the tops of anticlinal hills will quickly erode into a series of valleys that may eventually become deeper than the adjacent synclines. This is referred to as topographic inversion. Many fine examples of synclinal folds can be seen throughout the Rocky Mountains, perched high on the sides of mountain faces.

BASIN AND RANGE COUNTRY

Interesting landforms arise when sedimentary strata are tilted at various angles by faulting and are then exposed by erosion. On a large scale, in areas of extensional stress, blocks of Earth's crust can be rotated along a series of parallel normal fault planes. The edges of the up-side block form a series of parallel ridges adjacent to a series of long straight valleys. The Basin and Range Province, which covers much of the southwestern United States and northwestern Mexico is a typical example. Here,

parallel, hot, dry valleys separate a series of elongated north–south trending mountain ranges. Death Valley, California—one of the hottest, driest, and lowest places in the USA—is a classic example of a basin and range valley. It was named Death Valley in 1849 by the California Gold Rush immigrants who had to cross its hostile flats on their way to the goldfields. Death Valley was not always as dry as it is now. During the late Pleistocene Epoch (around 1.8 million to 10,000 years ago), Lake Manly, a prehistoric lake, filled the valley.

DOMES AND BASINS

If folding occurs in more than one direction, it will create interesting closed dome and basin structures in the sedimentary rocks. Oil exploration companies eagerly explore for anticlinal domes below the land surface and sea floor, because these structures can trap and hold rising oil and gas. When these geologic structures erode, the relative positions of

ABOVE Death Valley in California, USA, is a model basin and range valley. This view of Zabriskie Point (with the Panamint Range in the background) illustrates what is known as badlands topography—arid terrain with soft clay soil that has been extensively eroded.

BELOW Water and wind have formed at least 2,000 sandstone arches in Arches National Park, Utah, USA. Here we see Turret Arch through North Window. The natural process is ongoing; old arches are collapsing and new arches are being created.

STONE TOOLS, OCHER, AND PIGMENTS

The art and traditions of native peoples are strongly influenced by the geology around them—namely the availability of certain rocks, ocher, and pigment. Compact hard stones, which could be chipped and flaked, were highly sought after for tools and weapons. These included grinding stones, knives, axes, arrowheads, and spear points. The glassy igneous rock, obsidian, was prized in the Americas. The Aztecs particularly favored obsidian for carving, and even polished it into perfectly smooth disks to use as mirrors.

The most common color pigments used by native peoples were white (from clay); reds, browns, and yellows (from oxides of iron); and dark brown and black (manganese oxides). Blues and greens were rare but could be extracted from the minerals found around copper ore bodies. Colors available in one area were often a valuable commodity to be traded with groups from other localities.

Clearwell Caves in Gloucestershire, Britain, have been hand mined for their natural earth pigment for more than 7,000 years. The mines were a source for most colors, including the rare purple. Later, many cultures learned that they could apply certain pigments to the surface of pottery. When the item was fired in a kiln, these pigments would melt to form a waterproof glaze, allowing the pottery to be used to store liquids. The glaze was often surprisingly colorful and was also applied for purely decorative purposes. Porcelain ware still plays an important part in our daily lives.

the harder and softer layers determine the appearance of the final landform. If the core of the dome is more resistant to erosion then an upland area will result with radial drainage flowing outward; whereas, if the central core is a weaker rock type then a central depression will form with drainage flowing inward. Resistant rock units can stand out as striking circular or elliptical natural walls that create completely enclosed valleys. Wilpena Pound in South Australia is a spectacular example of such a structure. On a smaller scale, eroded elliptical synclinal basins can look very much like the jagged stone hulls of ancient boats.

NORMAL, REVERSE, AND THRUST FAULTS

Compressional or tensional forces in the crust can often cause the rocks to crack, break apart, and eventually slide along those broken surfaces. This is called faulting. Under tensional stress, blocks of crust will drop down in a series of steps along parallel fault planes, dipping steeply toward the

LEFT Following millions of years of erosion, the resistant Precambrian sandstone at Wilpena Pound in the Flinders Ranges National Park, South Australia, has formed an enormous protective wall around this natural basin.

down-dropped block. These are known as normal faults and they are common in the initial stages of formation of a rift valley. Under compression, the movement along the fault is in the opposite direction and these faults are known as reverse faults. In cases of extreme horizontal compression, huge segments of crust may push over the top of adjacent crust along shallowly dipping plains called thrust faults. In smaller faults exposed in cliffs or roadside cuttings, it is often possible to determine the direction of movement by matching strata on one side of the fault to those on the other.

CUESTAS, RAZORBACKS, AND HOGBACKS

In a folded or faulted sedimentary rock sequence, a series of extremely interesting landform shapes can develop, depending on the angle at which the bedding planes have been tilted. At shallow angles, features that look very similar to mesas can develop. They differ in that their tabletops have a distinct inclination, giving the feature a rather strange lop-sided appearance. These are known as cuestas, and indigenous American tribes sometimes used their cliffed edges as "buffalo jumps" to drive herds of bison to their deaths.

At higher angles, the resistant strata form much steeper ridges called razorback or hogback ridges. Each side of the ridge has rather a distinctive appearance. The side where the planar beds of resistant strata dip steeply out of the ground is called the dip slope, while the other side, in which all of the sedimentary beds are exposed in cross-section, is called the scarp slope. When the bedding is inclined to near vertical, the profile of the ridges is quite symmetrical and the resistant strata can look very much like vertical dikes protruding from the ground. Uluru (Ayers Rock) in central Australia is one good example of near-vertically dipping sedimentary rocks protruding from the floor of the desert. The resistant sandstone beds have been rounded by erosional forces, giving the monolith its characteristic shape.

ABOVE The Australian monolith Uluru is made up of sedimentary rock that dips almost vertically. Shown here is the top of the huge sandstone slab—the rest of it extends at least 3 miles (5 km) below the ground.

Head-Smashed-In Buffalo Jump, Alberta, Canada

Near the Porcupine Hills in southwestern Alberta, Canada, lies one of the world's most significant prehistoric hunting sites—Head-Smashed-In Buffalo Jump. Its significance in buffalo hunting for almost 6,000 years led to this escarpment being made a World Heritage Site by UNESCO in 1981.

ABOVE A hoary marmot (*Marmota caligata*) rests on a rock at Head-Smashed-In Buffalo Jump. It is the largest North American ground squirrel.

BELOW A herd of buffalo graze on grasslands in southwestern Alberta. Numbers were estimated to be around 30 million wild animals prior to the 1800s.

Located on the edge of a landscape of highlands and hills cut by a series of natural passes, this jump site is a complex series of interconnected geological features with a kill site far larger than comparative jump sites found in Solutre in France, where wild horses were led to their fate, and Vistonice in the Czech Republic, where young mammoths plunged to their death.

Today, the members of the Blackfoot Indian Nation are the custodians of the site, which their ancestors used as a kill site for thousands of years. The Blackfoot Indian people provide interpretive tours of the area and consider it to be of great cultural and spiritual significance.

GEOLOGY OF THE JUMP

The jump itself is composed of Paskapoo sandstone, which forms an escarpment almost $1\frac{1}{4}$ miles (2 km) in length running along a northeast–southwest axis. The kill site below is situated on a slumped rock terrace at the end of a 22-square mile (57-km²) Gathering Basin, the banks of which are cut by six valleys where the buffalo were herded prior to beginning their stampede.

Surrounded by drive lines—called "cairns"—within the Gathering Basin, the buffalo were herded toward the jump, a 1,000-foot (300-m) long section of sandstone that broke away from the cliff 6,000 years ago to form a relatively flat terrace. Two thousand years ago a second section collapsed, raising the level of the kill site and reducing the depth of the jump to the 40 feet (12 m) we see today.

Some sharp dips in the landscape approaching the jump, which are known as "coulees," in conjunction with the placement of trees and logs covered in buffalo hide, helped to funnel the buffalo over the precipice. Those that did not perish on impact were slaughtered by the use of clubs, knives, and spears.

DISSECTING THE CARCASS

Women and children then began the process of skinning, dressing, and separating the meat. Flesh and bone marrow were eaten or made into a food concentrate called pemmican for winter. Hides were tanned and used for clothing and shelter. Tails were used as fly swatters, while the buffalo's coarse hair was used for decorative purposes and to stuff pillows and blankets.

HISTORIC KILLING SITE

The oldest evidence of human habitation here can be seen in the discovery of two Scottsbluff spear points that date back to almost 9,000 years ago.

Archeological excavations have since established the presence of successive cultures that have used the site for the slaughter of buffalo, and they date back as far as 5,600 years ago. Over 1½ million cubic feet (400,000 m³) of old kill deposits have been discovered here in almost pristine condition. The site is spread through many undisturbed stratified layers, where bones and cultural artifacts, such as tools and implements, point to extended periods of use and complex social networks, interrupted only by a single period of non-use.

The earliest group to use the site were the Mummy Cave people (3600 to 2100 BCE), whose stone tools recovered from the site's deepest layers represent a limited range of style and sophistication.

The jump remained idle for a period of over a thousand years. The local Pelican Lake people (900 BCE to 300 CE) used a variety of stone tools that originated from the Kootenay Lakes and the Rocky Mountains of British Columbia to as far south as present-day Montana in the United States. Larger kills are attributed to the Avonlea Phase (300 to 850 CE), culminating in the

final period, which is known as the Old Women's Phase (850 to 1850 CE). Frequent and massive kills are characteristic of this period and include tools made from petrified wood, as well as tools typical of the plains Indians and the local Piegan people.

The communal activity of buffalo-jumping became unnecessary with the arrival of European weaponry. Combined with the introduction of disease and the mass hunting of buffalo for sport, the impact decimated the vast herds, depleting Head-Smashed-In Buffalo Jump not only of buffalo but also of the people who called it home.

ROCK FEATURE: SANDSTONE

Sandstone is a sedimentary rock made up of sand-sized grains cemented together by calcite, quartz, or iron oxide. Grains are usually quartz, although sandstones can also be made up of feldspar and other minerals. The sand would have been originally deposited in a moderately high-energy environment such as a river delta, beach, or desert. The grains may be angular or rounded depending on the amount of transport that they have undergone.

John Day Fossil Beds, Oregon, USA

The John Day Fossil Beds National Monument was established in 1975. It is world-renowned for possessing an unbroken sedimentary accumulation of ancient soils, rivers, floodplains, and forests, as well as fossilized remains of plants and animals that span 40 million of the 65 million years of the Cenozoic Era.

The monument is spread across 14,000 acres (5,700 ha) along the John Day River in north-central Oregon and is split into three distinct and separate units—the Painted Hills Unit, located 9 miles (15 km) northwest of Mitchell; the Clarno Unit, 20 miles (34 km) west of Fossil, and Sheep Rock Unit, 6 miles (10 km) west of Dayville.

BELOW An aerial view of the John Day Fossil Beds in Oregon, USA, reveals bands of color in the claystone, which is made up of various different minerals tinting each sedimentary layer of rock.

WHO WAS JOHN DAY?

John Day was a backwoodsman from Virginia who was engaged as a trapper and hunter by the Pacific Fur Company in Astoria, Oregon, in the fall of 1810. Crossing the Blue Mountains near the Columbia River in early 1812 he was attacked by a band of Native Americans and left to die near the mouth of the Mau Mau River, 30 miles (48 km) east of Dalles City, Oregon.

In May of 1812, he miraculously appeared in the coastal town of Astoria, and people soon began referring to the Mau Mau River as the John Day River. Eventually, a valley, two cities, a dam, and the John Day Fossil Beds all acquired his name. Four differing accounts of his death persist to this day. There is no record of his ever passing

through the area now known as the John Day Fossil Beds, and his precise association with one of America's most unique National Monuments remains a mystery.

EARLY DOCUMENTATION OF THE FOSSIL RECORD
The first mention of fossiliferous deposits in the John Day Basin appeared in a paper presented to the Philadelphia Academy of Sciences by Professor Joseph Leidy in October 1870, in which he cataloged remains of mammals presented to the Smithsonian Institute by a Reverend Thomas Condon of Dalles City.

In 1873, an exploratory party led by Professor Marsh described several fossil mammals dating to the Miocene and Pliocene epochs, thus making the first statement as to the age of the beds.

THE PAINTED HILLS UNIT
The Painted Hills Unit is an area of scenic beauty unique in the Pacific Northwest, made up of eroded remnants of the lower John Day Formation.

Composed predominantly of claystone, its colorful layers are an unusual collection of naturally occurring oxides. Lignite produces its black soils; its bands of red are laterite soils formed by floodplain deposits, and its muted grays are a collection of siltstone, mudstone, and shale.

Climbing on the hills is discouraged, to prevent erosion and to protect the fragile bee flower that grows in the dry ravines of the region.

THE CLARNO UNIT TREASURE TROVE
The 2,000 acres (800 ha) of the Clarno Unit are characteristic of central Oregon's semi-desert biome. Its primary landform is the cliffs of the Palisades, which include numerous clast-rich, channelized debris flows and hyperconcentrated flood flow deposits. Its rocks, born of volcanic lava flows, ash, and mudflows, have preserved an extraordinary record of the plants and animals that lived here 50 million years ago.

A selection of Clarno's treasures includes exquisite fossil leaves, an early cat-like predator known as *Patreofelis*, miniature four-toed horses and large rhino-like animals called brontotheres.

SHEEP ROCK UNIT: GORGES AND SUMMITS
Named after the 1,100-foot (335-m) high Sheep Rock Peak that rises over the John Day River valley, the eroded lava beds and exposed claystone layers of Sheep Rock Unit date back 28 million years, and continue to expose new fossils through the processes of natural erosion.

Sheep Rock Peak is the unit's most prominent landform and represents one of the John Day formation's most complete stratigraphic sections. From the Picture Gorge basalt and Late Tertiary Rattlesnake formation of its summit to the volcanic tuffs of ignimbrite stretching diagonally across the mountain, Sheep Rock Peak is one of the most complete and accessible examples of Tertiary Period deposits to be found anywhere in the western United States.

ABOVE A closer look at the bands of color that line the folds of the Painted Hills Unit in John Day Fossil beds reveals layers of vivid reddish orange. This is caused by staining with iron-rich minerals that "rust" when exposed to air.

ABOVE Wildflowers line the semi-arid hills of the fossil beds. This region was a completely different landscape 50 million years ago, when the same area was a lush tropical forest inhabited by palm trees and crocodiles. At this time, the first horses evolved in North America—14 horse genera have been found at the site.

Monument Valley, Arizona and Utah, USA

Monument Valley is not, in fact, a valley but rather a wide flat landscape of buttes and pinnacles ranging in height from 400 feet to 1,200 feet (120 m to 360 m). Composed of 270-million-year-old Permian Cedar Mesa sandstone, this ancient lowland basin represents one of the most enduring and definitive images of the once wide untamed American frontier.

ABOVE A coyote (*Canis latrans*) running across sandy terrain in Monument Valley. They favor arid grasslands and high mesas.

BELOW The Left and Right Mittens (left and center) and Merrick butte (right) as seen from the visitor center may be the world's best-known buttes.

The 30,000 acres (12,000 ha) of Monument Valley Navajo Tribal Park was established in 1958 and sits at an elevation of 5,600 feet (1,700 m) on the Utah–Arizona border. A loop road passes some of Monument Valley's more recognizable landforms such as Left (East) Mitten, Right (West) Mitten, and Merrick Butte, though the majority of the park can only be seen with official Navajo guides.

MONUMENT VALLEY'S DISTINCTIVE LAYERING

Monument Valley extends along the crest of a wide anticline. During the Eocene Epoch, vast layers of rocky mountain sediments, siltstone, and shale were deposited here, while continuing uplifting and folding exposed the lower levels of shale to erosion, so that the overlying sandstone was undermined. As huge blocks of rock fell away from the cliffs, the plateau began to erode into isolated plateaus, buttes, and pinnacles.

These buttes and pinnacles are composed of three easily recognizable and distinct layers. The butte's bottom layers are composed of Organ Rock shale, with the middle layers made up of the harder De Chelly sandstone.

The uppermost parts of the buttes and pinnacles consist of Triassic Period shale known as Moenkopi and capped with Shinarump siltstone. This layer creates a process known as "reverse armoring" with the denser shale protecting the De Chelly sandstone beneath from rain and its associated effects of weathering. Once the Moenkopi layer is eroded, weathering commences through pre-existing joints in the De Chelly sandstone, eventually producing the myriad butte formations seen today.

PLATEAUS, MESAS, BUTTES, AND PINNACLES

The bases of many buttes are surrounded by broad gentle slopes of sediment that will in time consolidate and provide new surfaces for erosion.

Many majestic mesas, buttes, and pinnacles can be found throughout the park, and although they may seem fixed in time and space, in geological terms, no landscape is ever "finished." Monument Valley has been in the process of eroding for the past 50 million years.

When plateaus—vast expanses of land that are bounded on at least one side by steep-sided cliffs—erode away they eventually become mesas.

MONUMENT VALLEY AND THE NAVAJO NATION

Monument Valley is home to some 300 Navajo people. They live in the traditional "blessing way" in six-sided mud homes called hogans, their doors facing east to greet the morning sun. They cook over wood fires and live without the comfort of electricity or running water.

Within the boundaries of the park they have managed to maintain their cultural and artistic traditions, their connection with the land, and their pastoral way of life.

Today the Navajo Reservation is spread over 27,000 square miles (70,000 km²), spilling over into Arizona, Utah, and New Mexico.

A hogan, a traditional Navajo dwelling, in Monument Valley Navajo Tribal Park.

As the cap rock of a mesa, more resistant to erosion than the rock beneath, begins to erode, rocks fracture and fall away, with the mesa transforming into a butte when its width becomes less than its height.

Buttes eventually erode still further, leaving just a slender stone monument, or pinnacle that itself will one day collapse, becoming part of the sediments of the valley floor.

The floor of Monument Valley is composed largely of Cutler Red siltstone and sands left behind by the numerous rivers that cut the valley. Reddish hues found throughout the sands and rocks of the valley are the result of deposits of iron oxide, while black vertical streaks called "desert varnish" are the product of manganese oxide.

SURVIVORS IN A HARSH LANDSCAPE

Due to the valley's extreme dryness and lack of moisture, few trees grow here. Junipers can be glimpsed along the park's boundaries, and when precipitation occurs, rabbitbrush, snakewood, and cliffrose will appear. Purple sage, a silvery-leafed Californian herb with purple flowers, grows abundantly throughout the park.

ABOVE LEFT Monument Valley has many natural arches, and while this feature is often called Teardrop Arch, it is actually not an arch, but a natural crack in the rock among the cliffs in southeastern Utah near Gouldings.

The Tepuis, Venezuela

Sir Arthur Conan Doyle's 1912 fantasy novel *Lost World* told of a land of dinosaurs and various
other evolutionary leftovers that inhabited the tops of sheer-walled, heavily forested plateaus
deep in the South American jungle. Venezuela's tepuis was the lush setting that inspired him.

ABOVE Poison arrow or dart
frogs (*Dendrobates* species) live
in the rain forests of Venezuela
and other surrounding countries.
They are toxic to predators.

Cut off from the world and in enforced compli-
ance to the Darwinian principles of natural selection, the creatures in Doyle's novel continued to
evolve in isolation and on their own terms in a
primeval environment of exotic tropical foliage,
mists, and waterfalls. It's easy to see why the tepuis
of southern Venezuela provided Doyle with the
ideal setting for his story. He called them "islands
in the sky"—a land-locked Galapagos.

The sandstone table mountains of the
Venezuelan tepuis sit astride the Precambrian
Guayana Shield, an almost 1 million square mile
(2.5 million km²) expanse of igneous and metamorphic rocks nearly 2 billion years old. The
region has more than 25 percent of Earth's humid
tropical forests and 20 percent of its fresh water.

The Guayana Shield's granitic base, isolated by
denudation due to the Gondwana breakup and
subsequent formation of the Orinoco and Amazon
river basins, became overlain with layers of sand
that over time compressed into quartzitic and
sandstone rocks. Uplifting and erosion led to
the separation and isolation of the tepuis plateaus
and the formation of the Guayana Highlands.
Ranging in height from 3,000 to 10,000 feet (1,000
to 3,000 m), they are today spread primarily across

a 500-mile (800-m) expanse of rainforest and tropical savanna in southern Venezuela and extend into neighboring Brazil, Guyana, and Suriname.

A DIVERSE ARCHIPELAGO

Despite having many geographical and floristic features in common, there is nevertheless a surprising diversity among the tepuis plateaus.

Some possess more irregular and undulating summit contours than others, and each plateau has its own characteristic shape, floristic complexity, and elevation. Depending on a tepuis' elevation, temperatures may be as low as 32°F (0°C), and rainfall can vary from 100 to 160 inches (2,600 mm to 4,000 mm) annually, depending on the local orographic features.

Cerro Ichun on the Brazil–Venezuelan border has by far the largest surface area at 1,250 square miles (3,200 km²). The highest peak is Pico da Neblina in Brazil, 9,888 feet (3,014 m), followed by Pico Phelps in Venezuela at 9,816 feet (2,992 m).

The tepuis ecoregion is split into four distinct floristic and geographic districts. The Southern District along the Brazil–Venezuela border possesses the highest number of endemic flowering plants in a landscape of meadows and low forest. The Western and Jaua-Duida Districts are dominated by granite and sandstone mountains, while the Eastern District is characterized by the endemic treelet *Bonnetia roraimae*.

FLORA AND FAUNA

Vegetation in the tepuis shows an extremely high rate of endemism. Over 900 species of plants have been cataloged on the summit of Auyan-tepui alone, of which 10 percent are endemic to the massif. Fauna is diverse, but not particularly abundant; however, perhaps in a nod to Conan Doyle, the summits are rich in reptiles and amphibians, including boa constrictors, coral snakes, iguanas, and tegus lizards.

LEFT The sheer sandstone walls and corrugated tops of Auyan-tepui rise above the cloud-studded valley. One of the largest of Venezuela's tepuis, its name means "devil's mountain." It is also called "roaring canyon" by the local Pemon Indians.

ANGEL FALLS: THE WORLD'S TALLEST WATERFALL

In the 1920s a Missouri-born bush pilot named Jimmy Angel claimed to have plucked around 75 pounds (34 kg) of gold nuggets from a tepui-top stream in just a few days. He spent years trying, in vain, to relocate the stream.

In 1935, his radial-engine monoplane became mired in mud on the tundra-like surface of the tepui mountain of Auyan-tepui. It remained there for 33 years. On his long walk back to civilization, Jimmy noticed a waterfall. It emerged from a slot canyon and cascaded down steep talus through some heavily fractured and jointed sandstone before beginning its final dramatic descent 2,647 feet (807 m) to the valley below. It was 20 times as high as Niagara Falls, and the highest waterfall in the world.

Known to the local Pemon Indians as Kerepakupai-meru—"fall from the deepest place"—it soon became known to the rest of the world as Angel Falls.

Roussillon Ocher, France

The natural ochers of Vaucluse were formed long ago at the end of the Cretaceous Period. As the climate became tropical the emergent thick sands began to solidify, their mineral compositions altering into iron hydroxide and clays that have since resulted in the 17 to 25 different tints seen throughout the region today.

OPPOSITE The vibrant hues of the ocher cliffs around the old Roussillon quarry during sunset. Before paints were manufactured artificially, only natural pigments were available to artists and these ocher cliffs were at times mined for as many as 25 variations in tint.

ABOVE The quaint village of Roussillon sits high atop the ocher cliff tops of the Plateau de Vaucluse in Provence. It is often called the "red village" because so many of the local buildings are built from the red rock that was mined at the local quarry.

According to legend, Sermonde, wife of the Lord of Roussillon, once fell in love with a traveling troubadour. Upon hearing of his wife's dalliance, Sermonde's husband, filled with jealousy and rage, murdered her lover with a dagger.

After arranging for her lover's heart to be removed and cooked, it was served to the unsuspecting Sermonde for dinner. When told the truth after her meal had digested, Sermonde threw herself to her death from an upper window, her blood forever turning the soils of Roussillon a deep and enduring red.

THE OCHER LANDSCAPE
In the region between St Pantaleon and Gignac, there is a 15-mile (24 km) wide band north of Apt where the ocher lies in strata up to 50 feet (15 m) wide. The Ocher Trail outside Roussillon begins in the east near the town of Rustrel, the "Provencal Colorado," and the most spectacular of France's ocher sites. A stream from Gignac to Apt eroded Roussillon's ocher valleys, exposing the white limestone, green clays, and banks of ocher in tortured tormented shapes rarely seen anywhere else.

OCHER'S HUMAN HISTORY
Once believed to be the "red earth" used by God to create Adam in the Book of Genesis, ocher has been used by humans since prehistoric times for coloring earthenware, for household utensils, and for decorative purposes. The oldest evidence of ocher mining stretches back over 40,000 years with the discovery of Swaziland's Lion Cave, and ocher has been found in various Neanderthal burial sites.

Ocher was used by the Romans and rediscovered in the Vaucluse in 1780, when it was used in the flourishing textile and soap-making industries.

ROUSSILLON'S FLORA AND FAUNA
The soils of Roussillon's Ocher Trail are home to parasol pines and chestnut trees that grow in the coolness of the valley floors. In fall, shrubs such as broom heather produce masses of bright pink flowers shaped like tiny bells.

Twenty-six species of wild orchids thrive in the chalky red soil, while white oaks, holm oaks, Scotch pines, and boxwood—species that are normally common to the region—grow poorly in Roussillon's siliceous soils.

OCHER AND PIGMENT CONSERVATORY
In the midst of a pine forest on the outskirts of Roussillon, the Conservatory of Applied Ocher and Pigments is dedicated to perpetuating the traditions involved in ocher production. Excess sands are filtered through stone pipes and the purified ocher is dried in settling basins. It is then cut into briquettes, fired, crushed, and filtered again.

The only corporation in France today dedicated to the continued marketing of ocher is the Société des Ocres de France in the Provencal town of Apt. Founded in 1901, it specializes today in limewater color-washes imbued with ocher.

WHAT IS OCHER?
Ocher is essentially a mix of silica and clay tinted with naturally occurring pigments of hydrous forms of iron oxide such as yellow-brown limonite, goethite, gypsum, and manganese carbonate. To be considered a true ocher, the iron oxide content must be a minimum of 12 percent, though it can be as high as 70 percent. Colors can range from yellow through to crimson, red, and brown. The preparation of red oxide pigments involves extensive milling and separation of its coarse particles by elutrition before being either dried or burned to improve its iron content and color.

Ocher is used in color washes, and to make colored paper; it is among the most permanent colors in the artist's palette. Red oxide pigments are widely used in the painting of automobile bodies, structural steel, and ships' hulls. The quality is judged according to the pigment's brilliance, fineness of texture, and staining power.

Sachsische Rock Formations, Europe

Saxon Switzerland National Park is a geological Noah's Ark in the midst of central Europe, a largely unpopulated expanse of chalky sandstone cliffs, deeply incised valleys, table mountains, and gorges 12 miles (20 km) southeast of Dresden in eastern Germany. The Sachsische rock formations lie within this park.

More a deeply cut plateau than a mountain range, the park protects over 89,000 acres (36,000 ha) of Elbe sandstone massif, a typically eroded sandstone landscape carved out by the Elbe River and its tributaries over 100 million years ago.

TWO PARKS WITH A COMMON PAST

One hundred million years ago there was little here save for a lake with a chalk bed over a sandstone slab some 2,000 feet (600 m) in depth. Erosion gradually carved out crevasses and notches into which the Elbe and other smaller streams flowed, creating the bizarre formations of friable sandstone known as Elbe sandstone.

The natural landscape of Saxon Switzerland National Park flows into the Czech Republic where it becomes the Bohemian Switzerland National Park. Together these two parks constitute a trans-boundary national park and protected landscape area, which encompasses over 270 square miles (700 km^2) along the border between Germany and the Czech Republic.

A FORTRESS ON THE PLATEAU

One of the Elbe sandstone mountain's most prominent features is historic Konigstein Castle, a mountaintop fortress built along a sheer 750 feet (230 m) sandstone formation overlooking the Elbe River and the medieval town of Konigstein.

Constructed by the kings of Bohemia in the mid-thirteenth century, in 1408 Konigstein passed into the hands of the Saxons, who often took refuge in it during times of crisis. Always considered to be invincible, the castle has never been taken in battle.

Konigstein is an architectural ensemble, with a 750-year history reflecting styles from late Gothic through to Renaissance, baroque, and nineteenth-century influences.

Covering an area of 58 acres (23.5 ha), Konigstein's principal points of interest are its main entrance dating to 1590, the Old Arsenal (1594) with its groined vaulting and massive columns, and the well house (1735) with its 500-foot (150-m) deep well.

The Elbe Sandstone Mountains cover 36 square miles (93 km²), stretching along the right bank of the Elbe River. The region is at 2,400 feet (730 m) above sea level and consists of three strata of sandstone overlaid with Ice Age sediments. These have been eroded into more than 1,100 random, twisted pinnacles with heights of up to 1,500 feet (450 m), set among a landscape of conical basalt hills with granite cliff faces along its boundaries.

A HISTORY OF ENVIRONMENTAL PROTECTION

From 1873 through to 1930 a series of successful appeals were initiated against the construction of mountain railways to the Bastei rocks and Lilienstein. In 1911, an area near Hohnstein was set aside for preservation and in 1938–40 the Bastei and Polenztal were declared protected. In 1956, Saxon Switzerland was officially declared an area of outstanding natural beauty, and in 1990, it finally achieved the status of national park. Over 4,000 observation areas are scattered throughout the park, monitoring everything from the lifestyle of the bark beetle in a coniferous forest environment to the influence of deer on plant undergrowth.

THE ELBE: AN "ARK" OF BIODIVERSITY

The park's moist gorges and valleys provide an ideal environment for a wide variety of ferns and mosses, and a bright yellow lichen and the claspleaf twisted stalk traditionally found in far higher elevations. Its streams are home to brown trout and fire salamanders, while dry ledge pinewoods and their surrounding heathlands provide a habitat for owls and cave-dwelling woodpeckers.

While peregrine falcons and black storks may be somewhat of a rarity, vertebrates such as deer, lynx, and otters can often be seen at dusk.

Beeches are found in profusion amid the park's basaltic outcrops and display a profusion of nine-leaf coral root and wood anemone blossoms in the spring. The beeches also provide shelter for black woodpeckers and stock doves.

ABOVE The vibrant sulfur-yellow tones of lichen stand out starkly against the dark hues of the surrounding rock. Lichens are not actually plants but a combination of fungi and plant cells that work together to create a single organism. It is the fungus cells that absorb the minerals from the rocks the lichen lives on.

LEFT A light dusting of snow gently coats the peaks of the rock formations in Saxon Switzerland National Park.

Ethiopian Highlands

Among the most densely populated agricultural regions in Africa, the Ethiopian Highlands cover an area of over 200,000 square miles (520,000 km²) and include over 80 percent of Africa's tallest mountains. Known as the "Roof of Africa," many of the area's summits reach heights of between 15,000 and 16,000 feet (4,600 and 4,900 m), with little of its surface area falling below an elevation of 5,000 feet (1,500 m).

ABOVE Geladas sit on top of a cliff in the highlands. Often called gelada baboons, they are not true baboons, though they are related to them.

BELOW These highlands in Shoa Province are in the northern part of the country. The devastating droughts during the last few decades along with landuse activities have severely degraded the lowlands, and now over 90 percent of the population lives in the highland regions.

This region of rugged craggy peaks is divided along a northwest–southeast axis by Africa's Great Rift Valley. Within the northwest portion is Ras Dashan, at 15,158 feet (4,620 m) Ethiopia's tallest peak, and Lake Tana, the legendary source of the Blue Nile. The southeast is home to the Bale Mountains, habitat for many of Ethiopia's endemic species, and the largely unexplored Harenna Forest.

FROM CONTINENTAL PLATE TO ROOF OF AFRICA

Seventy-five million years ago magma began to uplift a region of the African craton, a large portion of Africa's continental plate that had been relatively dormant since Precambrian times.

Eruptions of lava flowed over underlying Cretaceous Period rock, which led to a build-up of massive basaltic layers up to 10,000 feet (3,000 m) deep. Folding of these deposits resulted in the formation of the Ethiopian dome, which was in turn split into two great massifs during the Miocene and Pliocene epochs.

The Ethiopian Highlands are young in geologic terms, with volcanic activity as recently as 5 million years ago continuing to shape the Ethiopian dome. Fourteen thousand years ago, the region's flora of dry montane forest, open grasslands, and heath were still affected by periods of glaciation.

HIGHLAND HABITATS

Montane grasslands and shrublands dominate the eastern and western Ethiopian Highlands and are home to a vast array of rare birds and mammals including 250 of the endangered Walia ibex, a species of wild goat with enormous curved horns.

Geladas (*Theropithecus gelada*), which were quite common throughout Africa in the Pleistocene Epoch, are now restricted to the highlands. They live a sophisticated communal lifestyle, often numbering in the hundreds, and they huddle together at night for warmth. They feed on native grasses while seated, and are often observed shuffling from one meal to the next without getting up. The cleared patches of grasslands left in their wake are known as "gelada fields."

Birds endemic to the region include blue-winged geese (*Ankober serins*) and the rather rare Ruppell's chat (*Myrmicocichla semirufa*). A black bird

THE ETHIOPIAN WOLF

The Ethiopian wolf (*Canis simensis*) is the rarest canid on Earth and one of the world's rarest mammals. Possessing a distinctive chestnut-colored coat, elongated legs, and slender snouts, these wolves inhabit scattered pockets of grassland above 10,000 feet (30,000 m) in dens consisting of a complex system of burrows usually located beneath rock overhangs and cliffs. Despite being observed roaming in packs of up to 13 adults, they are primarily solitary animals that forage alone on giant mole rats and various small rodents, which together account for 90 percent of its prey. Endemic to Ethiopia, the Ethiopian wolf is presently in decline due to habitat loss and overgrazing.

The highlands have seven isolated populations of the Ethiopian wolf, with the biggest concentration in the Bale Mountains.

BELOW Craggy peaks rise from the valleys beneath the highlands in Simien Mountains National Park. A rich diversity of flora occurs over several climatic zones, changing with rising altitude from mountain woodlands (now felled) to alpine moorland.

with a white patch on the inner surface of its wings, Ruppell's chats inhabit the edges and sides of the highland's extensive precipitous gorges amid associated bare rocks and waterfalls at elevations above 6,000 feet (900 m).

SIMIEN MOUNTAIN NATIONAL PARK

The 52 square miles (136 km²) of Simien Mountain National Park lies along a broad undulating plateau situated on the northern edge of the Ethiopian Amhara Plateau and bounded to the south and northeast by the deep valleys of the Tacazze River.

Part of the 25-million-year-old Simien Massif, the park's igneous basalts have long since been eroded to form impressive gorges reaching up to 5,000 feet (1,500 m) in height and escarpments such as the north scarp that extends a distance of 21 miles (35 km). Its famous pinnacles are volcanic necks of solidified lava, remnants of volcanic activity stretching back 40 million years.

Inland South Africa

The South African interior is dominated by a vast region of plateaus with unwarped rims. This plateau area is enclosed by the coast-hugging Great Escarpment, an elongated erosional feature comprising southeastern and western coastal belts, the Cape Ranges to the south and southwest, and the Guateng Lowveld to the northeast.

OPPOSITE The sun gives a golden glow to the sandstone cliffs in Golden Gate Highlands National Park, hence the park's name. The park is home to many mammal species, including the springbok (*Antidorcas marsupialis*).

BELOW A river meanders across the floodplain in KwaZulu-Natal Province. This province is located in the higher altitudes of the Great Escarpment, which separates South Africa's coast from its interior region.

The plateau area in the interior of South Africa descends from a height of 8,000 feet (2,500 m) in the basaltic Lesotho region in the east to 6,500 feet (2,000 m) in the lowlands of the Kalahari Basin. Its central region, the Highveld, is a flat expanse that ranges in height from 4,800 to 5,000 feet (1,500 to 1,600 m), which features species-rich grasslands traversed by meandering rivers.

THE GREAT ESCARPMENT

The uplifting of the Great Escarpment began in the Cretaceous Period during the breakup of Gondwana. The escarpment is South Africa's longest and most spectacular topographic feature. It runs almost the entire length of the country's 1,800-mile (3,000-km) coastline, dividing South Africa's narrow coastal belt from its vast interior.

The escarpment is at its most spectacular in the northeast where basaltic rock provides sheer vertical elevations of up to 6,000 feet (1,800 m). Some of the country's highest summits, such as Mont-aux-Sources 10,823 feet (3,300 m), are located here.

Traveling southwest and paralleling the coast, the escarpment provides the boundaries between KwaZulu-Natal and Free State provinces as well as between KwaZulu-Natal and Lesotho before turning westward and descending to lesser altitudes of between 5,000 and 8,000 feet (1,500 and 2,400 m) through the Eastern Cape Province.

THE CENTRAL AND SOUTHERN KAROO BASINS

The ancient sedimentary rocks of the flat and arid Central and Southern Karoo basins are found within the Highveld interior plateau region, which encompasses about two-thirds of the total land area of South Africa. The rock sequences of the Karoo Supergroup date from the late Carboniferous Period (320 million years ago) through to the late Triassic Period (208 million years ago).

The otherwise monotonous expanse of the Karoo Basin, meaning "great thirst," is occasionally broken by inselbergs and kopjes. Prominent conical, fynbos-clad, erosion-resistant mountains rise above their surrounding plains, and range in elevation from Warmwaterberg at 4,400 feet (1,300 m) to Kamanassieberg at 6,400 feet (2,000 m).

The undulating sandy plains of the Karoo are also home to one of the world's finest assemblages of fossils, which catalog a remarkable sequence of terrestrial tetrapod development and the world's most extant example of the development of early mammalian life. Fossil remains of Africa's oldest known tortoise can be found in the rocks of the Karoo's Elliot Formation, while the rocks of the Molteno Formation provide unique glimpses into the biodiversity of the late Triassic Period.

The rocks of the Karoo's Beaufort Group also provide our most complete window into the great end-Permian extinction of 251 million years ago. This resulted in a mass extinction of life forms and loss of up to 95 percent of species on Earth.

For semi-arid vegetation, the Karoo Basin has the highest floristic diversity in the world, with more than 5,000 plant species so far cataloged. More than 50 percent of these species are endemic to the region and receive the bulk of their precipitation in the form of coastal fog.

The climate and vegetation of the region are largely determined by the Great Escarpment and Drakensberg mountains, whose extreme elevations prevent any significant precipitation from reaching the country's interior.

THE KALAHARI CRATON

The Kalahari craton is a stable, undisturbed segment of Earth's continental crust that has experienced little internal deformation and is comprised of low-density, felsic rock (igneous rock that has a very high silica content). It stretches from the Zaire–Zambezi watershed through to South Africa's Orange River. In Limpopo Province, the craton possesses a Precambrian granitic basement that contains some of the world's oldest known rocks: migmatites, granites, and gneisses as much as 3.4 billion years old.

The region has been stable for approximately 3 billion years, and has since been overlaid by lavas, shallow-water sediments, and igneous intrusions. It has a series of well-preserved rocks

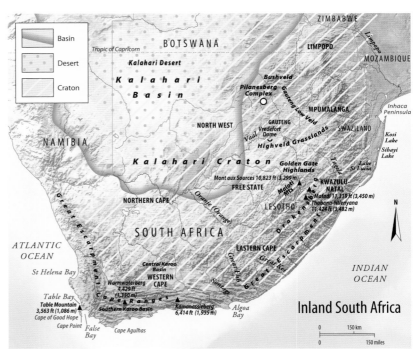

Inland South Africa

that contain deposits of gold, platinum, titanium-iron, and chromium, which provide the basis of South Africa's mineral wealth.

THE VAST KALAHARI BASIN

The Kalahari Basin or depression is a lowland region covering an area of two and a half million square miles (6.5 million km^2), which encompasses the Kalahari Desert and almost all of Botswana and neighboring Zambia, Angola, Namibia, and Zimbabwe, and extends into South Africa's Northern Cape Province.

Overlaying a very thick geological base of Precambrian rocks, the massive sediments of the Kalahari Basin, also known as the Mega Kalahari, began forming 120 million years ago after the breakup of Gondwana. When the Great Escarpment began to develop, the interior of southern Africa began a period of subsidence, leading to the creation of a basin that was highly prone to sedimentation, which later became the Kalahari–Cubango–Congo Basin.

The southern Kalahari region is fondly known to some South Africans as "Agterland," a word with similar connotations to the Australian term "Outback." It possesses tan-colored, calcareous rocks that are best exposed along the area's ancient riverbeds. It also harbors diamonds, copper, and coal below its nutrient-poor sands.

The formation of its characteristic longitudinal 100-feet (30-m) high sand dunes can be traced back to between 16,000 and 20,000 years ago.

THE PILANESBERG COMPLEX

Just 75 miles (120 km) northwest of Johannesburg in North West Province, the Pilanesberg Complex is the third-largest complex of alkaline rings in the world. The complex is set within a series of concentric hills covering 300 miles (500 km) and located within a long-since eroded volcanic caldera, the remnant of an ancient eruption that occurred around 1.3 billion years ago.

Its most prominent features are a series of ring dikes and cone sheets formed by resurgent magma that intruded along fractures during the caldera's collapse, with overlying layers of alkaline pyroclastic rocks sitting atop an intrusive base of syenite and foyaite (unusual rock types that are high in potassium and sodium but low in silica). The complex has been heavily distributed by faulting. Pilanesberg National Park sits within the complex.

THE HIGHVELD GRASSLANDS

The once-expansive grasslands of the Highveld, whose peat purifies the waters of the plateau's westward flowing rivers, have suffered severe fragmentation from agriculture and water projects that were constructed to ensure a continuity of supply for the population of greater Johannesburg.

The Highveld Grasslands are the largest remaining expanse of grassland in southern Africa, and have been present since the early Holocene Epoch approximately 11,000 years ago. The area consists of two types of grassland: the sweet grasslands where the grasses retain moisture in their leaves during winter and possess low fiber content; and the shorter, denser sour grasslands where the grasses have a high fiber content.

The region supports several species of large mammals rarely seen in southern Africa, including the brown hyena and the honey badger. A rich avian diversity includes the whitewinged flufftail and the endemic Botha's lark.

Remains of pristine grassland are now found predominantly within nature reserves, such as the Koppies Dam Nature Reserve and the Ermelo Gamepark, although the total percentage of grassland under protection in

GOLDEN GATE HIGHLANDS NATIONAL PARK

Proclaimed a national park in 1963, the 28,660 acres (11,600 ha) of Free State Province's Golden Gate Highlands National Park is home to some of South Africa's most dramatic examples of exposed sandstone cliffs. Nestled in the foothills of the Maluti Mountains, Golden Gate's distinctive sandstone outcrops rest on a base of 200-million-year-old Elliot mudstone, overlain with layers of clarens and calcified sandstone and capped with thin layers of quartzite and Drakensberg basalt, the product of molten rock forced upward when the area was volcanically active around 183 million years ago.

Examples of geological formations within Golden Gate Highlands National Park include the Molteno Formation, the Drakensberg Formation, and the fine-grained sedimentary rocks of the mudrock-rich Elliot Formation.

Golden Gate is a classic highland environment and home to mammals such as wildebeest, springbok, and zebra, and birds such as the bearded vulture and the bald ibis. Its montane grassland flora includes many species of rare and indigenous flowers.

these and other parks amounts to less than one percent of the grasses of the whole region.

VREDEFORT DOME

Vredefort Dome, 75 miles (120 km) southwest of Johannesburg, is the world's oldest, deepest, and largest astrobleme (meteorite impact crater) with a radius of 120 miles (190 km). The astrobleme dates back over 2 billion years, and represents the largest single release of energy in recorded history.

Vredefort Dome's geological sequence begins at its granitic core with its coarse-grained quartz-feldspar and gneiss rock. Granophyre dikes are also visible and are representative of the original impact melt. Shatter cones, which are characteristic features of meteoric impact formed during episodes of enormous compression, are also evident.

It is estimated that the meteorite was between 6 and 9 miles (10 and 15 km) across and impacted at a speed of about 9 miles per second (15 km per second). The crater would then have been almost 180 miles (290 km) wide and 3 miles (5 km) deep.

ABOVE LEFT Red sand dunes are a major attraction for visitors to Kgalagadi Transfrontier Park. In the late 1990s, South Africa's Kalahari Gemsbok National Park joined with Botswana's Gemsbok National Park to create this enormous 8.9 million acre (3.6 million ha) wildlife sanctuary.

LEFT Golden Gate Highlands National Park lies in the foothills of the Maluti Mountains. As the sun goes down, a hazy purple, red, and gold hue blankets the mountains, particularly the highest peak called Generaalskop.

Cappadocia, Turkey

Known throughout Turkey as "a picture etched upon the earth by nature," the region of Cappadocia (Kapadokya in Turkish) is located in the Anatolian highlands of central Turkey. It lies between the towns of Nevsehir, Urgup, and Avanos.

ABOVE A close-up of the unique rock formations known as "fairy chimneys" in Cappadocia, which are formed from weather-worn tufa, a type of volcanic rock. These unique formations have historically been used for housing in this valley.

Cappadocia is an open-air museum, a geologic wonderland of twisted, Gaudi-like, conical and pyramidal rock formations with a morphology that is the result of thousands of years of continual erosion. After nature finished with Cappadocia, Anatolian man took over, using simple tools to construct more than 120 underground communities within the porous tufa—including the city of Ozkonak that has a population of over 60,000.

MT ERCIYES—THE "FATHER OF CAPPADOCIA"

In the early Miocene Epoch, between 10 and 7 million years ago, Cappadocia was in the midst of a period of intense volcanic activity. Volcanic ash (soft tufa) descended onto the steppes of Cappadocia in depths up to 500 feet (150 m) from the 12,847 foot-tall (3,916 m) Mt Erciyes and the surrounding volcanoes, Hasandagi and Golludag.

During the mid-Pliocene Epoch, 2.9 million years ago, a large caldera formed in eastern Cappadocia and agglomerates such as basaltic lava began to accumulate alongside the soft tufa. These soft tufa deposits have since been eroded and sculpted by steam, wind, and rain into the conical "fairy chimneys" we see today.

CAPPADOCIA'S FAIRY CHIMNEYS

Cappadocia's climate can be extreme, with hot dry summers and cold damp winters. Melting snow in the spring, heavy rains, and sudden changes in temperature combine to fragment and erode its rocks. Fairy chimneys begin to form when lava covering the tuff gives way along pre-existing cracks to produce isolated pinnacles.

These pinnacles can reach heights up to 130 feet (40 m), are generally conical in shape, and possess caps of harder rock such as lahar or

THE CHURCHES OF CAPPADOCIA

There are over 600 rock-cut churches throughout Cappadocia, constructed primarily from the seventh to the thirteenth centuries. They range from simple designs featuring rectangular vaulted naves, with apses covered by a projecting arch, through to cross-planned churches with triple apses and domes. Original frescoes, predominantly in red ocher, were painted directly onto the rock or over thin layers of plaster, and have been vividly preserved in the cave's filtered light. The majority of frescoes date from the eleventh and twelfth centuries.

Many of Cappadocia's underground churches possess Byzantine architectural features such as arches, columns, and capitals, all of which are there purely for aesthetic purposes, as the churches are hewn from tufa with no supportive elements required. The best-preserved frescoes are in the Goreme Valley's "Dark Church," while the yellow ocher of the region dominates the vaulted cross-shaped interior of the eleventh-century Elmali Church. At its height, Goreme was reputed to have had a different church for every day of the year.

Some of the churches, dwellings, and fortresses carved into the rock formations of the Goreme Valley appear to be inaccessible.

ignimbrite that protect the softer tuff below from erosion. Cappadocia's fairy chimneys are most commonly found near the town of Cat in Nevsehir, in the Soganli Valley, and within the valleys of the Uchisar-Urgup-Avanos triangle.

GOREME NATIONAL PARK

Goreme National Park was inscribed as a World Heritage Site in 1985 and has been a center of continuous human habitation for over 16 centuries. The Goreme Valley, with a circumference of 25 miles (40 km), is primarily an agricultural region characterized by orchards, vineyards, and farmland, an area where man-made works blend seamlessly into the landscape. Rock dwellings here have been dated to 4000 BCE and the town of Goreme itself is set amid innumerable cones and fairy chimneys. A monumental twin-pillared Roman tomb hollowed into a fairy chimney in the center of Goreme is proof the town was once used as a necropolis by the people of nearby Avanos during Roman occupation. The Goreme Valley emerged as a significant monastic center after the Iconoclast Period of the eighth and ninth centuries, when iconic and various other figurative representations were prohibited.

ABOVE Looking down along the floor of the Baglidere Valley in Cappadocia, the vast area the strange rock formations cover is most apparent. They are essentially the erosional remnants of ancient volcanic ash.

Masada, Israel

Masada has played a significant role in the history of Israel and is today regarded as a national shrine. Its geographic isolation and steep vertical cliffs saw it used as both the last stronghold of the zealots who defied the Roman legions of Flavius Silva in 70 CE and the site of an opulent palace complex constructed by Herod the Great in classic Roman style.

BELOW An aerial view of the ancient ruins of Masada reveal its massif lies on top of an enormous mesa. The eroded rocks were originally sediment deposited in an ancient lake bed.

Masada is a partially isolated massif that is located near the Dead Sea in southern Israel, 11 miles (18 km) south of En Gedi on the eastern fringes of the Judean Desert. This massif is the primary feature of the 682-acre (276-ha) Masada National Park, towering above a geographic transition zone that includes steppe, intermixed desert, and Mediterranean elements. The site was inscribed on the World Heritage List in 2001.

MASADA'S MESA AND HANGING WALL
Masada's rhomboid mesa top covers 20 acres (8 ha) and rises as much as 1,300 feet (400 m) above the surrounding terrain. It sits on a fault-bounded up-lifted portion of the Earth's crust in a rift valley that is an extension of the Syrian–African Rift Valley system. This system stretches from northern Syria south through Lebanon's Bekaa Valley to the Gulf of Aqaba and southeast to the Indian Ocean.

Massive dolomites and limestone of marine origin overlying less-resistant limestone form the bulk of Masada. It is bounded to the east by lacustrine silts, gravels, conglomerates, and sandstones up to 80,000 years old—once part of a huge lake, the forerunner to the Dead Sea—and to the west by the terraces, hills, and wadis of the Judean Plateau.

On Masada's western slopes, a naturally occurring spur offers the most convenient path to the mesa top. Part of the downfaulted hanging wall of the West Masada fault, it is over 700 feet (215 m) long at its base and up to 650 feet (200 m) wide at its widest point. Today it ascends to just 40 feet (12 m) below the summit and is composed of almost 9 million cubic feet (255,000 m³) of rock and earth in neatly layered strata of local bedrock, topped with talus (rock debris) and the remains of Roman earthworks. These earthworks were used to build a ramp along the spur to pierce Masada's defenses and end the Jewish revolt in 73 CE.

A BEDROCK BASE

Masada lies along the northern edge of a horst formation, ranging between 1,600 feet and 3,000 feet (500 m and 900 m) in height, which extends on a north–south axis for 6 miles (10 km) to the south. It is isolated from the southern portion of the horst because of the extensive down-cutting by streams through the horst block Nahal.

Masada sits atop three distinct layers of bedrock. The youngest and uppermost layer is known as the Mishash Formation (dated at 83 to 71 million years ago) and possesses thin layers of dark gray flint, soft chalk, and marl. The white chalks of the older Menuha Formation (dated at 86 to 83 million years ago) are up to 150 feet (45 m) thick. Easily visible in the landscape, this layer is a particularly important stratigraphic marker that assists in unraveling the complex geologic history of Masada. The Bina Formation (dated at 93 to 89 million years ago) is the oldest and deepest of these layers, with a massive 500-foot (150-m) thick layer of dolomite and limestone.

ABOVE The crumbling ruins of one of Masada's forts have been painstakingly excavated by archeologists. The site was initially a palace, then a Roman garrison, and finally a refuge for the religiously persecuted.

A PLACE OF REFUGE FROM THE ROMANS

When the Romans attacked Jerusalem in 70 CE, 960 zealots retreated to Masada where, for three years, they held off the Roman Tenth Legion under the command of Flavius Silva.

At the time of the siege, Masada had a casement wall 4,600 feet (1,400 m) long and 13 feet (4 m) thick along the edge of the plateau, while large rock-hewn cisterns on the northwestern side provided ample supplies of fresh water. Buildings included bathhouses, storerooms, living quarters, and a synagogue. On its northern edge stood the elegant three-tiered palace of King Herod with its floors of cut marble, black-and-white mosaics, and fluted plaster columns. Rather than be taken alive and sold into slavery and prostitution, the zealots took their own lives as the Romans bludgeoned their way through the plateau's ramparts. The historian Flavius Josephus wrote that everything had been burned by the zealots except their stores of food, to let the Romans know it was not hunger that led them to suicide.

A bathtub in a ruined bathroom in one of the buildings at Masada.

Uluru and Kata Tjuta, Central Australia

Near the geographic center of Australia lie two of Australia's most prominent and iconic landforms—Uluru, formerly known as Ayers Rock, and Kata Tjuta, a local Anangu word for "many heads," and commonly known as The Olgas. They rise up almost alongside one another out of a vast, flat, arid plain of vivid red soil.

Together they form Uluru–Kata Tjuta National Park, an Aboriginal-owned World Heritage Site, inscribed in 1987 and jointly managed by the original owners and Parks Australia. Uluru and Kata Tjuta are set in a relatively flat, sand-plain environment dominated by spinifex, desert oaks, and large stands of mulga woodland and other low shrubs.

MT CONNOR: THE FORGOTTEN MESA

Southeast of Lake Amadeus and west of Uluru, in a direct line with Kata Tjuta, is the forgotten landform of Central Australia—the 700-million-year-old mesa, Mt Connor. Often mistaken for Uluru and considered by many locals to be larger than its famous cousin, Mt Connor is a horseshoe-shaped mountain on the fringes of the vast Curtin Springs cattle station. This conglomerate and quartzite formation is immersed in a talus (scree) slope that extends over its lower 500 feet (150 m). Above are sheer cliffs that reach a further 300 feet (90 m) to its summit. Mt Connor is 2 miles (3.2 km) long and ¾ of a mile (1.2 km) in width with sandstone and limestone ridges radiating up to 1½ miles (2.4 km) from its base. Local Aborigines call it "Artilla," believing it to be the abode of the "icemen" who created cold weather.

ULURU'S UNUSUAL COMPOSITION

Uluru rises 1,130 feet (345 m) above its surrounding plains, covers 1¼ square miles (3.3 km²) in area and possesses a circumference of just over 5½ miles (9 km) at its base. It is composed of sedimentary arkosic sandstone made of up 75 percent lithified sand and 25 percent feldspar, deposited in two massive alluvial fans of sand and rock—up to 1½ miles (2.4 km) thick—by erosion from the nearby Petermann Ranges 550 million years ago.

Immersed in the sediments of an ancient sea, the sands became sandstone and the rocks compacted into conglomerate. Further faulting and folding of the Alice Springs orogeny 400 to 300 million years ago resulted in the horizontal strata of arkose sandstone being rotated vertically 90 degrees from their original bedding plains.

The arkose sandstone of Uluru is a semi-metamorphosed coarse-grained sandstone. It is extremely resistant to erosion, permitting the rock to withstand millions of years of weathering. Previous deposits over this ancient sandstone have long been eroded, leaving exposed Uluru's flaky red surface born from the rusting of iron in the arkose.

The arkose's original rust-free grays can still be seen in Uluru's myriad caves, caused by uneven weathering and water erosion when the surrounding landscape was far higher than it is today.

LEFT As the sun rises and hits Uluru, the color slowly changes from a soft reddish brown to a deep red, then a vibrant glowing orange. Depending on the time of day, season, and weather conditions the hues can vary startlingly from purple to soft gray.

ABOVE Emus (*Dromaius novae-hollandiae*) are sometimes seen around Uluru. They inhabit a wide range of habitats, from some arid inland areas to tropical woodlands, and they occur in a broad range across Australia.

Surface features of Uluru include many examples of sheet erosion—deep parallel fissures as well as overhangs at the rock's base formed by sand blast erosion and chemical degradation.

THE "MANY HEADS" OF KATA TJUTA

The 8,650 acres (3,500 ha) of Kata Tjuta consist of 36 rock domes with near-vertical sides and hemispherical summits, the highest of which is Mt Olga at 1,640 feet (500 m). Kata Tjuta's Mt Currie conglomerate is a mix of cobbles and pebbles of granite and basic rocks, with isolated pockets of lithosols, calcareous red earth soils, and sands found on its surrounding alluvial fans and scree slopes. Enormous flocks of green and yellow budgerigars (*Melopsittacus undulatus*) often flit between the rock boulders.

Like Uluru, Kata Tjuta is also a product of folding and its subsequent erosion. It possesses a far coarser mix of conglomerate and sandstone than the finer arkose sandstone of Uluru. Both landforms represent only the exposed tips of massive rock slabs that descend many miles (several kilometers) below ground level.

Fractures caused by faulting have permitted water to penetrate and erode the rock, forming gorges and splitting them into blocks. Curved cracks, known as topographic joints, then began to develop. Combined with erosion, the topographic cracks have produced the distinctive shapes of Uluru and Kata Tjuta that we see today. Such variations in their shapes are prime examples of how the degree of folding can have profound effects on the shape a landform may take.

BELOW The massive boulders of Kata Tjuta can easily be seen from Uluru, with just a vast expanse of flat arid land in between. Between the domes is the Valley of the Winds, an undulating bushwalk.

Purnululu National Park, Western Australia

The 590,000 acres (240,000 ha) of Purnululu National Park lies 180 miles (300 km) south of Kununurra in Western Australia's Ord River region and consists of four primary landforms—the Ord River Valley; a series of limestone ranges to the park's north and west; wide sand plains; and the deeply dissected plateaus of the Bungle Bungle massif.

OPPOSITE From the air, every tiger-striped layer of sediment in the rock strata of the Bungle Bungle beehives is visible. The less-permeable orange bands are crusts tinted by manganese and iron oxide, while the slate gray and black bands are the more porous layers, which have a higher water content and are stained by the algae and lichen that grow on their wetter surfaces.

BELOW During the dry season, the creek bed running through the main ravine of Piccaninny Gorge appears bone dry; however local springs keep the gorge's waterholes filled year-round. During the wet season, it is flush with water and at this time the Bungle Bungle gorges can become inaccessible.

In the midst of a savanna biome and overlying a stable sedimentary landscape, the Bungle Bungle's sandstone karsts are the result of sand and other fine granular materials deposited into the Ord Basin, a process that began during a period of active faulting 360 million years ago. Today, these beehive-shaped cone karsts represent one of the most extensive examples of sandstone tower karst in the world, unrivalled in their volume, complexity, and grandeur.

THE BANDED DOMES OF THE BUNGLE BUNGLES

Set on a dramatic plateau of Devonian quartz sandstone, the steep-sided cone towers of the Bungle Bungles were shaped by complex processes of uplift, compaction, and sedimentation that began when the ancient landmasses of Gondwana and Laurasia collided some 350 million years ago.

Unique in their lithology and geomorphic evolution, the Bungle Bungle massif consists primarily of Glass Hill sandstone, a well-sorted, clean, porous, white quartz-rich sandstone. Its crossbedded and ripple-marked appearance are two characteristic features of a dry depositional environment.

The domes' distinct alternating bands of orange and dark gray cynobacterial crusts mixed with iron oxide help to stabilize their fragile outer surfaces. They provide an outstanding example of how dissolutional weathering can affect land formations through the erosional power of sand, wind, and rain—and its associated sheet wash—on slopes. The darker bands overlie more permeable rock layers. Water penetration through these layers occurs easily, promoting dark algal growth, with denser layers in between covered in a mix of manganese and iron creating the orange banding that protects the base of the towers from erosion. These towers occur in a random mix of clusters and ridges, as well as freestanding forms.

The cone towers of the Bungle Bungles join 15 other areas throughout the world recognized principally for their karst values. This area is one of three World Heritage sites where a karst landscape is present in non-carbonate rocks, the other sites being Wulingyuan in China and Canaima National Park in Venezuela.

LANDFORMS WITHIN THE BUNGLE BUNGLES

The Bungle Bungles provide a singularly rare insight into the various stages of Earth's geologic history, including the ongoing processes in the development of landforms and other significant geomorphic events.

Essentially a large plateau bounded by steep cliffs and incised with numerous joint-bounded valleys, it can be divided into the following separate landform units—incised valley forms; large incised valley forms; the plateau surface; the cliff face; and domes, towers, and ridges.

The incised valley forms are narrow gorges and sheer-sided canyons in the northeast and west of the massif that extend over half a mile (0.8 km) into the massif. These are the result of stream dissection. The large incised valley forms are flat-bottomed valleys several miles (kilometers) long and rising to 650 feet (200 m) in height on

RIGHT The Bungle Bungle massif supports a sparse tree steppe occasioned with grevilleas and acacia trees. Stores of permanent water are a rarity within its cliffs and chasms, so closed forests such as those found in the Osmond Creek Valley to the north cannot be supported.

ABOVE A group of rainbow bee-eaters (*Merops ornatus*) settle on a tree branch. During the winter months these highly social birds migrate north to the Kimberley and other warm to hot regions of Australia. They prey on flying insects, particularly bees.

either side. They are found mainly in the central south as well as to the southwest of the massif.

The plateau surface is best represented in the western and central regions of the massif. Lying some 650 feet (200 m) above the surrounding plains, this relatively flat expanse possesses distinct benches related to rock type and jointing. In particular, the Piccaninny structure contains deformed rocks and evidence of alteration suggesting the possible impact of a meteorite. The cliff face is a linear sheer-cliffed plateau edge that runs along the western boundary of the massif.

The distinctive domes, towers, and ridges were formed by the silicification of joints. The famous beehive-like sandstone karsts of the Bungle Bungles can rise up to 500 feet (150 m) in height and are mostly found in the southern and eastern regions of the massif. Their orange and gray-black banding is the product of silica, lichen, and bacteria skins that have developed over the sandstone.

A RECENTLY ESTABLISHED NATIONAL PARK

Prior to 1982, the Bungle Bungles were unknown to the Australian public and were familiar only to the traditional Aboriginals of the region and various pastoralists, stockmen, and locals. It only took until 1987 for the region to be given national park status, and in 2003 it joined the world's conservation elite when it was added to the World Heritage List by UNESCO for its representation of ongoing geological processes as well as its superlative natural beauty.

The Bungle Bungles lie wholly within Western Australia's Purnululu National Park. To the south and east, Purnululu's sparse low woodlands and mid-height grasslands are bounded by the Ord River Valley, while limestone ridges dominate to the west. The Osmond Range, a series of fault-bound ridges, lies to the north with its stromatolites and rocks of the Cambrian Period dating from 550 to 500 million years ago.

LIFE IN PURNULULU

Purnululu National Park is home to over 130 species of birds including the rainbow bee-eater, gray falcon (*Falco hypoleucos*), and Gouldian finch (*Erythrura gouldiae*). Almost 300 species of vertebrates have so far been cataloged, as well as 40 species of mammals—such as the northern nailtail wallaby (*Onychogalea unguifera*)—and numerous amphibians, reptiles, and fish comprising life forms typical of the park's overlapping desert, savanna, and tropical bioregions. Arid land animals found on Purnululu's plateaus include skinks, short-eared wallabies, and monitor lizards, while in its cool valleys field rats, mouse-eared bats, and several varieties of frogs can be found. Vegetation ranges from open forests and woodlands to stunted shrublands dominated by spinifex, acacias, grevilleas, and several species of gum trees, including the silverleaf bloodwood (*Eucalyptus collina*) and the rough leaf range gum (*Corymbia aspera*). The extremely rare *Grevillea miniata* grows only within the confines of the park. The park's northern boundaries reflect a flora typical of monsoonal savannas including *Livistona* palms, orchids, and ferns on the floors of its deeper gorges. Over 650 species of plants, including some 17 species of fern, exist in Purnululu's transitional climate. Its dry season (April to October) sees occasional night frosts in the cooler months and an average temperature of 95° F (35° C), while the wet season (November to March) is extremely hot with an annual rate of precipitation between 20 and 28 inches (500 and 700 mm).

Bungle Bungle Mountain Range, Western Australia

Timor Sea

0 ——— 100 km
0 ——— 100 miles

INDIAN OCEAN

N

Joseph Bonaparte Gulf

Bonaparte Archipelago
Vansittart Bay
Napier Broome Bay
Admiralty Gulf
Montague Sound
Prince Frederick Harbour
Cambridge Gulf
Brunswick Bay
King Edward
Couchman Range
Drysdale
Ord
Princess May Range
Doubtful Bay
Bucconeer Archipelago
Collier Bay
Durak
NORTHERN TERRITORY
Kimberley Plateau
Synnot Range
Isdell
King Sound
Chamberlain
Durack Range
Echidna Chasm
King Leopold Ranges
Osmand Range
Meda
Mt Wells 3,325 ft (983 m)
Ord
Bungle Bungle Range
Mini Palms Gorge
Oscar Range
Geike Range
Cathedral Gorge
Piccaninny Gorge
Piccaninny Structure
Piccaninny Creek
Roebuck Bay
Margaret
Elvire
Fitzroy
McClintock Range
Sturt Creek
Geegully Creek
Christmas Creek
Wolfe Creek Crater
Wolfe Creek
WESTERN AUSTRALIA
Great Sandy Desert

ABORIGINAL HABITATION

The term Bungle Bungle most likely has its root in the misspelling of the common bundle grass found throughout the Ord River region. The word *purnululu* means sandstone in the language of the local Kija Aboriginals.

Aboriginals have had a presence in the Ord River region for the past 40,000 years. Two main groups lived and continue to live here in a typical hunter–gatherer culture, one in the desert and the other occupying the savanna, each with their own language. They regard the landscape as an embodiment of their cultural and religious values.

During the early years of white settlement when massacres of local Aboriginals were common, many fled to the relative safety of the Bungle Bungles, using notched tree trunks to scale its plateaus.

Local Aboriginal communities have borne witness to significant changes to the vegetation within Purnululu National Park since the onset of the pastoral era. Food sources such as root foods, seeds, and lilies that were once found in abundance are now scarce. The introduction of grazing and subsequent soil erosion, together with changes to burning practices, have led to an increased inappropriate dominance of flora including acacias and bunch grass.

Waterholes that once held large fish such as barramundi and rock cod, and reptiles such as water goannas and turtles, are now shallow and stagnant, with few fish or other creatures in them.

ABOVE A compass termite (*Amitermes meridionalis*) mound in Purnululu. These social insects build air-conditioned towers aligned east–west to optimize the inside humidity and maintain the mound's temperature.

LEFT Echidna Chasm is a narrow gorge with walls towering over 300 feet (90 m) that are studded with boulder conglomerates.

Inland New Zealand

New Zealand has a long and varied geological past, due to its location on the Pacific Mobile Belt. This band of volcanic and earthquake activity has given the country distinctive regions of geothermal activity as well as great periods of seismic and volcanic mountain building.

In the wake of this activity, more subdued landscapes such as basins, depressions, and rock units became uplifted and were filled with sediments. Although outwardly less dramatic than other regions, these inland basin areas provide valuable insights into the turbulent past of a country that is continuing to experience high levels of tectonic activity. Inland landform features of New Zealand also include limestone rocks and gravel plains.

THE CANTERBURY PLAINS

Dominated by mountain ranges formed from uplifted land, New Zealand has few flat areas. The

Canterbury Plains, located at the foothills of the Southern Alps in the middle of the South Island, are the largest area of fertile, flat plains in New Zealand. The plains were formed by sediments that were eroded from the Southern Alps and interior ranges then deposited as fans of sand and gravel. The downland area between the foothills and the Canterbury Plains has a base of limestone rocks and marine sands, with some coal and older volcanic rocks. The sloping plains extend to the coast. The soils of the plains consist mainly of greywacke from the mountains and loess, a fine silt that has been windblown from riverbeds. Layers of porous

BELOW The large region of fertile, sand and gravel plains known as the Canterbury Plains has long been used for sheep grazing and the farming of various crops.

gravels, deposited by rivers when the sea was low, alternate with layers of finer impermeable sediment laid down when the sea level was high.

THE UPPER WAITAKI BASIN

With an area of 580 square miles (1,500 km²), the Upper Waitaki Basin, in the center of the South Island, is a typical example of a tectonic-driven, gravel-filled depression. Its features include the clay cliffs near Omarama and the 2,600-feet (800-m) thick uplifted piedmont fan sequences of the Ben Ohau Range. The basin's undulating landscape provides evidence of glacial deposition laid down by Holocene Epoch ice sheets. Its extensive flat plains, which slope toward the basin's outlet along the Waitaki River, overlay older glacial deposits.

The basin's ongoing tectonic activity opens a unique window into four distinct periods of ice advance: the Wolds Advance (135,000 years ago); the Balmoral Advance (60,000 to 70,000 years ago); the Mt John Advance (18,000 to 25,000 years ago); and the final retreat of the ice at the close of the last Ice Age some 8,000 years ago.

THE ROCK UNITS OF NELSON

The rock units to the east and west of Nelson, at the South Island's northern tip, are more varied and representative of a greater number of geological periods than those of any other land district in the country. Broadly separated by the northeastern continuation of the Alpine Fault, the eastern region is characterized by the upper and lower Paleozoic Era greywackes and argillites of the Spencer and Bryant Ranges. To the north, metamorphic rocks and Paleozoic granites dominate.

To the west are New Zealand's largest granite masses, extending along a north–south axis as a series of parallel belts, the longest of which reaches over 120 miles (190 km)—from Westland's Ahaura River to the Victoria Range.

Precambrian greywackes are present in the Paparoa Range, while to the northwest metamorphic and sedimentary rocks of the lower Paleozoic extend southward in a complex series of thrusts, folds, and nappes (large sheets or masses of rock). New Zealand's oldest rock units are found here.

LEFT The volcanic limestone cliffs of Castle Rock, near Christchurch on New Zealand's South Island, are extremely popular with rock climbers. The formation sits at the head of Heathcote Valley.

BELOW Hawthorn thrives on the famous clay cliffs near Omarama in the Upper Waitaki Basin. Here weathering has created elegant pinnacles and ridges separated by deep narrow ravines and canyons. Maoris refer to the area as Paritea, which means white or light-colored cliff.

KOKOAMU GREENSAND FOSSIL SITE

Covering an area of around 1,200 square miles (3,000 km²), the Kokoamu Greensand in the North Otago–South Canterbury region of New Zealand's South Island is an expanse of fine- to medium-grained sandstone dating from the late Oligocene Epoch. A thin, extensive blanket of easily eroded calcareous sandstone, Kokoamu Greensand gets its name and appearance from its deposits of glauconite, a blue-green hydrous silicate.

The Kokoamu Greensand is a fossiliferous Oligocene Epoch rock unit that is rich in vertebrate fossils such as fish and penguins. These are a testament to a time when what is now much of northern Otago was, at that time, submerged.

The nearby Vanished World Trail preserves Kokoamu's fossils, including a partially exposed baleen whale. Kokoamu's Waitaki limestones also contain world-class examples of giant penguins and shark-toothed dolphins. The nearby Elephant Rocks are remnants of 24-million-year-old Otekaike limestones that were uplifted over the last million years and eroded into the beautifully sculptured shapes seen today.

ABOVE Magellanic penguins (*Spheniscus magellanicus*) nest in burrows in coastal Argentina, Chile, and the Falkland Islands. Their colonies can stretch over several miles of coastline.

CENTER Isla Fernandina, one of the Galapagos Islands, Ecuador, has a volcano at its center. As this photograph shows, the island has a black volcanic sand beach.

RIGHT Whitehaven Beach, in Australia's Whitsunday Islands, is one of the most pristine beaches in the world. It is renowned for its miles of pure white silica sand and clear blue waters.

The coast is a broad, distinctive margin with an array of diverse landforms. The meeting of sea and land, as well as the erosive effects of waves, winds, and tides, creates some truly remarkable natural features.

Coastal landforms vary greatly. There can be broad, sandy, or cobbled (shingled) beaches; steep rocky sea cliffs, sometimes with tiny embayments sheltering small curved beaches; tidal flats; or beach barrier islands with their characteristic bars and lagoons.

Waves and currents, produced by tidal activity and river discharge, interact with wind to create these coastal environments. Storm surges and wind-generated giant waves can cause massive erosion. The extent of erosion is determined by several factors: the exposure of the coast to the erosional forces; the strength of currents; and the composition of bedrock. Igneous and some metamorphic rocks are usually more resistant to erosion than sedimentary rocks, such as sandstone and shale. Deposition along the continental margin is the result of the erosion of coastlines and the transportation of sediments to the shoreline by rivers. Coral reefs are also created by deposition. These special coastal landforms occur either as fringing reefs or shelf reefs.

THE ROLE OF PLATE TECTONICS

Coastal landforms can be grouped according to whether they are formed by depositional or erosional processes; for both, water is the main agent. However, where a coast lies relative to its position on a tectonic plate determines just which of these two processes dominates.

The boundaries of Earth's tectonic plates are said to be convergent when plates collide; and divergent when they move apart, or transform as the plates slide past each other. The continents carried by the plates have various types of continental margins influenced by where they are situated in relation to plate boundaries, giving rise to three distinct types: leading edge coastal margins; trailing edge margins; and the marginal seacoasts. The leading edge coastal margins are associated with convergent plates, such as those along parts of the Pacific coast of the Americas. The coastal landforms related to leading edge margins feature rugged and irregular topography, sometimes with high sea cliffs rising from the water line. Erosion dominates these coasts with small pockets of local deposition, such as small beaches in embayments. Large waves, which have been generated because of the deep near-shore water, erode the cliff to form rock platforms, caves, and sea stacks.

Trailing edge margins, the next type, are away from active plate boundaries and so are relatively tectonically stable. The Atlantic side of North America is a typical example of a trailing edge margin. Here, wave action is moderate because of the gently sloping and wide coastal shelf. These are the depositional margins that have little topographic relief and wide coastal plains. Large river systems contribute to the deposition of mud, silt, and sand that form a variety of landforms such as deltas, estuaries, barrier islands, tidal inlets, tidal flats, wetlands, and salt marshes.

ABOVE Tatami Rock Formation, on the southern coast of Okutake Island, Okinawa, Japan, resembles a flat turtle carapace. Low tide reveals the 1,000 or so hexagonal and pentagonal shapes. Fractures in the rock are typical of those created when magma is cooled by the ocean.

LEFT Bonifacio, a French town on Corsica, sits atop a sea cliff overlooking the Mediterranean Sea. The Mediterranean Basin is sited in the middle of a complex group of tectonic plates that continue to slide under one another.

Marginal seacoasts, the third type, are relatively stable because they are protected from open ocean processes, and the smaller waves allow for the accumulation of mud and silt in the coastal zone. Their seaward plate boundary often has island arcs and volcanoes, such as those of Japan. The coastal landforms associated with marginal seas share some of the features of both leading edge and trailing edge margins, although the deltas that form along the marginal seacoasts are often the largest of all; for example, the Mississippi Delta in the Gulf of Mexico. Australia's plate boundaries are all seaward and so the entire coastline is relatively stable, although it does have a huge range of coastal landforms, from the very rugged to the quieter depositional tidal flats and inlets.

Tectonic activity can dramatically alter coasts. Crustal deformation can cause uplift, raising beaches above sea level, or subsidence, so that the depressed land becomes prone to inundation by seawater. Associated seismic and volcanic activity can create new coastal landforms that have not been greatly modified by waves and currents. In a similar way, inundated or drowned basins are protected from marine erosion. Coastal landforms that are not initially shaped by marine processes are known as primary coasts, which differ from secondary coasts that have been altered by exposure to marine processes.

Tsunamis that have been generated by seismic activity are highly erosive and invasive. They flood low-lying land, destroying dwellings and causing loss of life on an enormous scale. The earthquakes that generate them can also cause coasts to rise, leaving raised beaches and other coastal features sitting above sea level.

Climate change can cause sea levels to rise and fall around the world. These global or eustatic changes are due to the amount of ice held at the poles. Earth is currently undergoing a period of global warming associated with climate change, and this appears to be melting an alarming volume of ice at the poles. Should the ice continue to melt, huge volumes of water will be released into the oceans, raising sea levels possibly some tens of feet (several meters), dramatically changing the existing coastlines worldwide.

The last period of glaciation, which peaked during the Pleistocene Epoch, had the opposite effect. Around 18,000 years ago sea levels were approximately 380 feet (115 m) lower than they are today. At that time the island of New Guinea and Australia formed a single landmass, along with Tasmania, Australia's island state, which was then connected to the mainland. At the end of the glaciation, some 10,000 years ago, sea levels rose to around what they are currently, changing the coastlines to close to those we have today.

Isostatic movement within crustal plates can also cause sea levels to change. A good example of this is the uplift that is ongoing along parts of the Scandinavian coast where continental crust is rebounding or uplifting as a result of the removal of massive volumes of ice following the thaw after the last Ice Age. Coastal features caused by the rebound include raised beaches and rock platforms. Features like these are used as evidence for past sea level changes throughout the geological record.

A CLOSER LOOK AT THE COAST

There are a number of different zones to the coastal margin. The shore is that part of the coast between low tide and the highest elevation that is affected by storm waves; and the shoreline is the water's edge, which varies with the movement of the tides. The littoral or intertidal zone is that part of the coast that lies between the high and low tide levels; it is also known as the foreshore, which on some tidal beaches is hundreds of feet wide. Offshore is the zone beyond the breaker line and near shore is the zone that extends offshore to approximately 30 feet (10 m) in depth.

CORAL REEFS

Coral reefs are special coastal landforms that also form through depositional processes. In their case the deposition is not of clastic (rock fragments) origin but biogenic—corals and other carbonate-producing organisms such as coralline algae build them. They usually occur on moderately deep, stable continental margins, either as fringing reefs to the coasts of islands and continents, or as shelf reefs, which lie well offshore. Australia's Great Barrier Reef is a shelf reef, although many parts of Australia's northern coastline also have fringing reefs.

Sometimes, near-shore currents transport sediment to the edge of the continental shelf and down the slope, but most sediment is taken by waves into the surf zone and deposited on beaches. A beach is a feature of the shore. Beaches are formed when waves and currents above and below the shoreline deposit sediments. The foreshore or beach face is continually affected by the swash and backwash of waves, but it is in the shallow waters of the near-shore zone that the sediments are continually worked and reworked. When waves approach the shoreline at an angle, a long-shore current is generated, forming a long-shore drift of sediment along the beach.

Whether waves are constructive or destructive will depend on the shape of the beach and on its composition. Seasonal weather variations change wave patterns and strength so that sediment can either be moved up the beach (constructive) or away from it (destructive). Extreme weather conditions have been known to strip a beach of all its sand, and then deposit it offshore, sometimes in the process exposing long-since buried rock platforms from the geological past.

Several coastal landforms are associated with beaches. Some form coastal barriers to lagoons, ponds, wetlands, and estuaries. Barriers can include spits, tidal flats, islands, and tombolos. The profile of a beach varies with the composition of the sediments. Gravel and shingle beaches are usually steeper than sandy beaches where the backwash of the waves forms a gently sloping beach. On shingle beaches, the swash seeps down and so dissipates the backwash. However, the profile of all beaches varies with wave conditions influenced by tides, weather, and seasons, all of which are capable of causing considerable short-term changes.

Sand dunes are a feature of many coastal environments, particularly in places where sand is abundant. Dunes are a depositional product, mainly caused by wind carrying the sand away from the

shoreline to the foreshore where it is deposited. The formation of sand dunes is determined by factors such as the size of sediment particle, the shape of the coastline and beach, waves and currents, prevailing wind and its strength, as well as the frequency of severe weather events.

The importance of sand dunes is the stability they give to the coastal environment. That stability is only possible when the dunes are left undisturbed or covered in vegetation. In many places, coastal development has been allowed on the dune systems, exposing the area to accelerated beach erosion.

BELOW An eroded sea stack at Muriwai Beach is home for a gannet colony. One of New Zealand's most stunning black sand beaches, Murawai is within Muriwai Regional Park, which covers around 40 miles (60 km) of rugged, windswept coastline.

HOW SEA CAVES, SEA ARCHES, AND SEA STACKS FORM

A. Joints in the coastal cliffs are constantly being widened by the fury of the waves. These slowly enlarge to become sea caves and eventually break right through to another cave to become a sea arch. B. The sea arches grow wider, spanning a larger and larger distance, until they finally become too weak and collapse into the sea. The result is a series of isolated pinnacles or sea stacks C. Wave erosion continues to attack the base of the sea stacks until they, too, finally collapse. With these obstructions out of the way, the sea begins anew its attack on the joints in the coastal cliffs.

A   B

C

Another depositional coastal landform is the delta that develops where a river enters the ocean. Any sediment still carried by the river at this point is deposited as alluvium, comprising mostly silt, mud, and clay particles. Many deltas are roughly triangular and similar to the Δ shape of the Greek letter "Delta." The delta of the River Nile, with its perfect Delta shape, gave rise to the term.

Delta formation depends on the dynamics of waves and of tides and river flow. A delta can take the form of a bird's foot when it is affected less by waves and tides than by river flow that causes it to deposit sediments in a number of lobes formed along its multiple distributary channels.

Sea cliffs are formed by wave and wind erosion of coastal rock. The cliffs usually have steep vertical faces that can be hundreds of feet (meters) high. The undercutting of cliffs by wave action at the shoreline eventually causes slope instability or failure, resulting in landslides, slumping, and rock falls under the force of gravity. The large volume of soil and rock debris produced forms sediments that are eventually transported and deposited on beaches by the action of waves and currents. The type of rock and its geological structures, such as bedding and jointing, determine the rate of erosion as well as the form of the cliff.

Rock platforms commonly form in the inter-tidal zone at the base of sea cliffs by the abrasion of rocks that fall from the cliff face. As weathering and erosion continue, the platform becomes lower and widens as the cliff retreats landward.

Cliffs often develop into headlands that alternate with bays along a coast. In many cases, the fallen rock debris forms scree at the base of the cliff and this gives temporary protection from further erosion, thus reducing the rate of retreat. This, together with a greater resistance to weathering, helps maintain the headland while the less resistant rocks erode more rapidly to form bays. The headlands offer the bays protection from wave action and so beaches are a common feature.

LIVING IN THE COASTAL MARGIN: ADAPTING TO A HARSH ENVIRONMENT

Coastal ecosystems are as varied as the coastal landforms that provide habitats for the diverse biota. Changes in sea levels, climatic variation, and human activities all have an effect on coastal plants and animals. Many organisms have developed remarkable adaptations to withstand extremely hostile environments, such as the pounding waves of rock platforms. Here, barnacles adhere firmly to the rocky substrate, using their thickened shells to withstand the bombardment. Sea cliffs are the sites of nesting birds, while small sheltered coves and bays between headlands offer protection to seals and penguins.

Beaches are relatively unstable environments for the biota living there. Tidal variations, wave action, as well as changes in deposition all create difficulties for living things. Again, adaptations such as burrowing allow worms and small crustaceans to take shelter in the sand while feeding on food brought in by waves. These animals in turn become food for crabs, fish, insects, and birds. Sea turtles also lay their eggs on ocean beaches.

In quieter, shallow waters sea grasses flourish. These habitats are essential for the maintenance of biodiversity because many species of fish and crustaceans rely heavily on sea grass meadows for all or part of their life cycles. This is also the case with mangroves, which thrive along the more sheltered tropical and temperate coasts of the world. Kelp beds provide another habitat for a rich biodiversity. Some kelps in the colder latitudes grow at an astonishing rate of around 12 inches (30 cm) per day to form huge kelp forests up to approximately 130 feet (40 m) in height.

ABOVE The five villages known as the "Cinque Terre" are dotted along 11 miles (18 km) of Italy's Mediterranean coastline. The geology of the Mediterranean Basin has been greatly changed by the erection of buildings along the coast thus altering the natural processes of erosion.

Coastlines of the Arctic

Although both have extensive ice caps, there are major differences between the Arctic and Antarctic circles. The North Pole is located over the ice-covered waters of the Arctic Ocean, whereas the South Pole lies on the massive continent of Antarctica—a landmass so large it occupies much of the area within the Antarctic Circle.

ABOVE Three major caribou herds—the western Arctic, central Arctic, and porcupine herds—migrate to the Arctic annually to calve.

OPPOSITE This aerial view shows the enormous expanse of the Mackenzie Delta in Canada. As it meanders to the Beaufort Sea, the river splits up into numerous smaller rivers and lakes.

BELOW This sea arch is in Alaska's Togiak National Wildlife Refuge. The refuge was set up to conserve wildlife and fish populations and their habitats.

The Arctic Circle has no such continent; instead its circumference is fringed by North America, Asia, Europe, and Greenland—all of which have remarkable coastlines that total approximately 125,000 miles (200,000 km) in length. The circle itself is an imaginary line of latitude at about 66°33'N around Earth. It defines the boundary of the Arctic and marks the start of the area where the sun does not completely set (June 21) or rise (December 22) for at least one day each year.

WITHIN THE ARCTIC CIRCLE

Although many of the Arctic coastlines are covered in ice for most or all of the year, human occupation and activities have the potential to threaten the stability of the area. At the best of times there is a high variability in the rate of erosion and changes, such as those due to seawater encroachment. The Arctic climate maintains permanently frozen soils and rocks known as permafrost. The ground-ice in permafrost in combination with sea-ice has a powerful influence on coastal erosion. Sea-ice can lift frozen sediment and transport it offshore by the action of either waves or tides. Climate change in these high latitudes has already had an impact, causing the permafrost to melt and form large hollows where the meltwater has carried away the thawed soil and rocks.

CLIMATE CHANGE

Because of its sensitivity to the effects of climate change, the Arctic region is a focus of considerable international scientific research. The coastal zone of the Arctic is already experiencing dramatic changes to not only its land and oceans but also its cryosphere or ice cover, and its biosphere. It is predicted that climate change in the Arctic will be more rapid than at the lower latitudes. Already shoreline erosion is one of the more obvious changes, and this is exacerbated by the increase in the length of the summer sea-ice free season. Other factors accompanying climate change include sea-level rise, the increase in frequency and intensity of extreme weather events, and also permafrost melting. In some regions, coastal erosion accounts for the loss of between 3 and 30 feet (1 and 10 m) per year of coastline as the sea transgresses across the land.

The North American Arctic includes the coastlines of Alaska and Canada. Its many islands form the largest archipelago in the world. Over two-thirds of Canada's coastline, with its greatly varying coastal landforms and relief, lies within the Arctic Circle. Canada's great Mackenzie River features one of the world's largest deltas, with its countless channels and islands that spread over 7,000 square miles (18,000 km²) where it meets the Beaufort Sea.

Arctic Russia stretches from the Norwegian border in the west to the Bering Strait in the east. Siberia boasts the widest coastal shelf in the world; it extends 750 miles (1,200 km) from the shore at an average depth of only 300 feet (100 m). Like Canada, this region also has an archipelago that features numerous islands.

The European segment of the Arctic Circle includes Arctic Norway, Finland, and Sweden, including Lappland. Generally speaking, swamps, lakes, coastal flats, and frozen tundra characterize the entire region, with much of its bedrock worn down by repeated glacial activity.

Most of Greenland is covered by an extensive ice cap up to 10,000 feet (3,000 m) thick over its mountainous terrain, with glaciers that carve out a rugged coastline as they reach the ocean.

A HISTORY OF DRAMATIC TECTONIC EVENTS

The Arctic Ocean has been shaped by a sequence of dramatic tectonic events commencing as far back as the Devonian Period, more than 350 million years ago, when the collision of continents and subsequent rifting formed the ocean. Further collision and rifting events during the Jurassic and Cretaceous periods developed a series of submarine ridges, where the continental plates moved apart and molten rock moved in to fill the gaps. These very steep ridges now separate the deep ocean into a number of sub-basins. Some of the ridges are over 900 miles (1,500 km) long and only 650 feet (200 m) wide, and have peaks that rise to within 3,000 feet (1 km) of the ocean surface.

The nature of a coastline is the result of its geology, relief, or topography, and the amount of sediment that is available. The shape or topography of coastal landforms can be a product of geological processes, such as: volcanism and faulting and folding; glaciation; and the erosive energy of ice, wind, and water. Mountainous Arctic coastlines that have been sculpted by glaciers have formed deep steep-sided fjords—characteristic of Norway's coast.

Coastlines of low relief and high sediment input form depositional landforms, such as beaches, long narrow spits, and barrier beaches.

The Arctic coastline is affected by two main environmental variables: relative sea level rise and reduced sea-ice cover. Both variables are able to alter wave energy along the coast, and this, in turn, can cause either flooding or shoreline retreat through increased erosion. Where sea-ice persists for the entire year there is less erosion along the coast. However, the coastal areas that are ice-free for periods ranging from less than a month up to a possible four months are subjected to far greater erosion. This is particularly true when storm surges from extreme weather events remove huge amounts of sediment loosened by permafrost thaw. Such storm events within the Arctic Circle are increasing in frequency as well as severity with the onset of climate change.

Relative sea-level fall occurs when parts of the coast rise in response to the release of pressure from ice sheets melting. The weight of the ice depresses Earth's crust, which rebounds when the pressure is removed through glacial retreat and unloading. This occurs particularly along parts of the Scandinavian Arctic coast and many sections of the Russian and Alaskan coasts.

Relative sea-level change sometimes occurs after massive blocks of ground ice contained in the on-shore permafrost melt and leave huge voids that fill with seawater, causing the sea to encroach onto land. As a result of this, dramatic changes to the extent and shape of the coastline take place. It is a process that is also increasing in frequency as climate change takes hold.

WILDLIFE DIVERSITY ON ICE

Not only does ice incessantly carve coastal landforms, it also acts as a major limiting factor to biological activity. Nevertheless, the ecoregion of the Arctic supports an extensive biodiversity. The entire Arctic Coastal Plain is an important breeding and calving ground for many species.

The Arctic Coastal Plain of Alaska has been declared a region of such high biological significance that the World Wildlife Fund for Nature has included it in the Global 200 Ecoregions. These are the world's 200 most critical regions for the conservation of biodiversity. The Alaskan region is part of the Arctic National Wildlife Refuge of remote Northeast Alaska. The refuge's important role is to protect the full spectrum of subarctic and Arctic plants and animals.

A portion of the Arctic Coastal Plain—threatened by the development of the oil industry in the region—has been set aside as the biological heart of the refuge. Massive herds of caribou calve and feed on the plain. It is also the habitat for many other species such as polar bears, grizzly bears, wolves, wolverines, musk ox, arctic fox, dall sheep, and 160 species of birds. The snow goose *(Chen caerulescens)* breeds in small areas in this ecoregion, which provides its only breeding sites in Alaska.

Typical of some of the Arctic Coastal Plain regions is the tundra vegetation, with its shrubs and sedges. Many areas have marshes, peat bogs, and wetlands, which are all home to a number of migratory birds. Increased coastal erosion that is associated with climate change and human activities in the Arctic threatens a variety of marine, terrestrial, and freshwater habitats.

ABOVE Andrew Gordon Bay is on Baffin Island, Nunavut, Canada. Part of the Canadian Shield, the island is made up of Precambrian Era rocks, and is rich in metals and other minerals.

ABOVE LEFT Part of the Siberian coastline in De Long Strait, opposite Wrangel Island, these craggy rocks provide an ideal nesting place for around 50 species of migratory birds and waterfowl.

ABOVE Larch forest is a common Arctic vegetation. A study by the American Geophysical Union is looking into the effects of global warming on the larch forests of eastern Siberia.

Coastlines of the United States

The east and west coasts of the continental United States have been shaped by varying tectonic and structural features that have led to two very different continental shelves. The Pacific Coast, uplifted and compressed by the subducting Pacific Plate, has a narrow continental shelf and mountain ranges that prevent river drainage and sediment influx. High waves cross this narrow shelf unimpeded with resultant heavy seas dominating.

OPPOSITE The Santa Lucia Mountains form part of the spectacular granite coastline of Big Sur in California, USA. The name comes from the Spanish phrase *el sur grande*, meaning "the big south."

ABOVE Frequent landslides and erosion of coastal cliffs results in large amounts of sediment falling into the ocean at Big Sur. This may threaten nearshore marine habitats, such as that of these bottom-dwelling sea stars.

The Atlantic Coast, by contrast, is moving away from the Atlantic oceanic ridge, causing subsidence of the continental margin. Its broader coastal plains, wider continental shelf, and resultant lower waves are the product of a continuous discharge and build-up of river sediments into the marine environment.

DRAMATIC COASTLINE OF BIG SUR

Along the Big Sur coastline of central California, the rugged Santa Lucia Mountains stretch for around 100 miles (160 km) from south of Carmel to north of San Luis Obispo. The range is 20 miles (30 km) across at its widest point and decreases in elevation from north to south. Descending dramatically from their 5,000-feet (1,500-m) summits only 2 miles (3 km) inland, the Santa Lucia Mountains plunge into the Pacific Ocean. From the air, these mountains are a chaotic mix of canyons and ridges that trend northwest to southeast with the coastline following the same orientation.

Within the Santa Lucia Mountains, the landscape varies from arid desert-like canyons to jagged mountain peaks and serene valleys that harbor a mix of grassland, forest, and shrub communities ranging from pine forests to dense chaparral and groves of ancient redwoods.

Big Sur has a complex geology. A labyrinth of fault lines fractures a mountain range composed of rocks. These rocks range from the remnants of seafloor volcanoes to the steep cliffs of cemented cobbles near Esalen to the gray-layered sediments of Gorda and the light gray granitic rock visible from Highway 1 at Garrapata State Park, 6 miles (10 km) south of Carmel.

SEA STACKS OF OREGON

Haystack Rock, which lies just south of Cannon Beach in Clatsop County, Oregon, is typical of the sea stacks that can be found in abundance along the rugged spectacular Oregon coast.

The 235-feet (77-m) tall basalt monolith is the third tallest landform of its type in the world. It is composed of a mix of sedimentary and volcanic rock remnants of Columbia River basalt and lava flows from the Grande Ronde Mountains that emptied into the Pacific Ocean along ancient lowlands and streambeds.

Once joined to the coast, Haystack Rock and the three smaller sea stacks to its immediate south, known collectively as "The Needles," are the result of millions of years of wave erosion.

EAST COAST BARRIER ISLANDS

Barrier Islands are a common feature along the southeast coast of North America, from the Gulf Coast of Texas to as far north as Long Island, New

SEA LION CAVES

Sea Lion Caves, 11 miles (17 km) north of Florence on the central Oregon Coast, is one of the world's largest sea caves. Set within a 25-million-year-old eroded amphitheater of basalt rock, Sea Lion Caves is 125 feet (38 m) high and stretches about 300 feet (90 m). The cave's principal residents are its 200 non-migratory Steller sea lions, found mostly on the offshore islands and rocks off British Columbia and Alaska. About a thousand reside along Oregon's coastline. Males can weigh as much as 2,000 pounds (900 kg) and are distinguished by external ears that can be closed when entering the water.

York. They originate as offshore bars built up by wave action across a shallow shore, pushing sand toward the shoreline as they break. The resultant undertow then sweeps sand back to accumulate on the developing bar.

Barrier islands are located parallel to the mainland, separated by lagoons or estuaries over gently sloping continental shelves. Factors that determine a barrier island's shape include tidal fluctuations, wave energy, and sand supply. A characteristic absence of healthy dune systems and a paucity of vegetation make the islands susceptible to erosion, leading to apparent migrations as buildups of transported sand cause the backsides of the islands to encroach toward shore.

Of great environmental concern is the growing number of condominiums and tourist accommodation being built on North America's barrier islands.

LEFT Basalt sea stacks called Haystack Rock and "The Needles" can be viewed from the shores of Cannon Beach, Oregon, USA. Migratory seabirds, such as the tufted puffin, black oystercatcher, pelagic cormorant, and others nest on the rock every year.

Cape Horn, South America

There is an old sailors' saying, which goes: "Below 40 degrees there is no law, below 50 degrees there is no God." Cape Horn lies at a latitude of 56 degrees South, and is considered by sailors the world over to be the most dangerous and unpredictable stretch of water anywhere in the world.

OPPOSITE Countless vessels that tried to navigate Cape Horn at the time of the California Gold Rush of the mid-1800s were no match for the huge seas, rocky headlands, and fierce storms they encountered in this wild place. Today, the ocean floor here is littered with wrecks.

ABOVE This almost surreal vision of icebergs is on Palmer Peninsula (Antarctic Peninsula) in Drake Passage. The peninsula was named after nineteenth-century explorer Nathaniel Palmer.

The most southerly of the great capes, Cape Horn remains one of the most isolated and least-studied regions of the world, with numerous questions still to be answered in terms of its oceanography, geology, and marine diversity.

THE GREAT CAPE

Cape Horn is a rocky headland located on the southern end of Horn Island, part of the Tierra del Fuego archipelago. It was named after the birthplace of the Dutch navigator Willem Schouten who was the first to sail round the cape in 1616.

Horn Island is 5 miles (8 km) in length, 1,400 feet (425 m) high at its highest point and set along a northwest–southwest axis. It is the southernmost tip of an extremely diverse collection of fjords, channels, sounds, and islands that make up an ecoregion that stretches north all the way to the mature closed-canopy forests of Chile's Taitao Peninsula. Horn Island is typical in form and vegetation of the other islands that make up the Hermite group. Many species of birds common to Tierra del Fuego are found here, with trees growing only in its deep ravines, sheltered from their harsh environment. Sphagnum moss and tussock grasses dominate the areas of open ground on these islands.

Despite the biota of the region south of the Magellan Straits being extensively studied by scientific expeditions of the nineteenth and twentieth centuries, the unfortunate fact is that much of the region's unique mosaic of intact natural landscapes remains poorly cataloged.

Cape Horn's incessant 100-knot (185-kph) westerly winds, which are the stuff of every sailor's worst nightmare, blow in through the Drake Passage unmodified by the absence of land to the west. The winds track through the passage in summer, while in winter, southwesterly cold fronts can be the harbingers of savage heavy snow squalls.

An absence of weather stations in the Hermite Islands means the only record of rainfall is from an 1882–83 study, which gave the region an average precipitation rate of 54 inches (1,400 mm) and a mean average temperature of 41°F (5°C).

The local Yagan people can trace their origins back to pre-Columbian times. A nomadic people, they inhabit the coastal lowlands and navigate the subantarctic archipelago in search of food. They are recognized as one of the most threatened of Chile's indigenous cultures.

A PRISTINE ECOREGION

The Cape Horn Archipelago is one of the few places left on Earth that remain relatively free of human impact, thanks to their isolation and to the presence of the Chilean Navy reserve.

The archipelago was recently identified as one of the 37 most pristine ecoregions in the world. Its mix of terrestrial ecosystems ranges from deciduous and broadleaf forests set in the midst of an intricate system of glaciers and fjords in the north, through to the southern subantarctic ecoregion of Magellanic Chile. The latter includes Horn Island, with its impressive diversity of non-vascular floral species and bryophytes.

The waters surrounding Cape Horn are also a hotbed of invertebrate and marine mammals. One could see there, for example, black dolphins, killer whales, and minke whales, as well as the Magellanic and rockhopper penguins.

THE MINIATURE FORESTS OF CAPE HORN

The Cape Horn archipelago, and Horn Island in particular, are barren windswept islands that contain few species of trees. The archipelago is, however, home to a bounty of rare and unique mosses that, in 2005, led the United Nations to declare the 12 million acres (5 million ha) of the archipelago a biosphere reserve in the hope of reconciling new economic interests with conservation of its biocultural heritage.

While the diversity of plants and animals decreases as latitude increases, the reverse is the case for bryophyte species such as mosses, with between 5 and 7 percent of the world's mosses, liverworts, and hornworts found throughout the Cape Horn Archipelago.

These bryophytes are often called the "miniature forests of Cape Horn." Scientists and environmentalists are working closely with the Chilean government to promote a greater understanding of bryophytes, as well as training local guides to help them identify and conserve the bryophytes' delicate habitats.

ABOVE Rockhopper penguins *(Eudyptes chrysocome)* live and breed on the rocky windswept islands of Cape Horn. These small penguins survive on a diet of fish, krill, and squid.

LEFT The pristine waters of the Cape Horn Archipelago are home to a variety of marine mammals. Among the not so well known cetaceans sighted here are the goose-beaked whale and Cuvier's beaked whale.

Coastlines of Costa Rica

Costa Rica, at around 19,000 square miles (50,000 km²), is the second smallest Central American nation after El Salvador. It shares borders with Nicaragua to the north and Panama to the south, and sits astride the Caribbean and Cocos plates, a tectonically rich region in the midst of one of the most active volcanic arcs in the world.

ABOVE The scarlet macaw (*Ara macao*) is one of only two macaw species found in Costa Rica. This stunningly colorful parrot is rapidly losing its habitat to deforestation, and is now endangered over most of its range.

It is one of the youngest landmasses on the planet, home to seven of the Pacific Rim's 42 active volcanoes, as well as the 60 dormant volcanoes that dominate Costa Rica's Pacific coast.

COSTA RICA'S TWIN COASTS

Costa Rica's Pacific and Caribbean coastlines are sprinkled with dense tropical rainforests and the most beautiful beaches. The Pacific coast has a varied topography of mountainous peninsulas and scattered narrow beaches, while the Caribbean side is predominantly flat, and characterized by gray and black sand beaches.

Although the Caribbean coastline does lack the extremes of elevation common to the Pacific coast, it remains extremely sensitive to ocean trench earthquakes that have been among the most powerful in Earth's history.

In 1991, an earthquake that occurred not far from the town of Pandora caused Costa Rica's Caribbean coastline to lift by as much as 5 feet

(1.5 m). Coral reefs were thrust above the surface and were reduced to bleached calcareous skeletons.

THE COCOS AND CARIBBEAN PLATES

Costa Rica straddles the Cocos and Caribbean tectonic plates, which are converging upon each other at the rate of 4 inches (10 cm) per year in a classic subduction zone. The Caribbean Plate, forced beneath the Cocos Plate, has produced a deep trench along their boundary, which is slightly offshore and parallel to the coast. The friction created by this convergence produces the magma that fuels Costa Rica's many volcanoes.

The Cocos and Caribbean tectonic plates helped to build Central America and also to shape the Costa Rican coastline. Subduction from the interaction of these plates built an island-arc system, extruding lava and building up landmass, much like the volcanic regions of the Hawaiian islands. As sedimentary material accumulated on top of the Cocos Plate, it was scraped off and piled

onto the margins of the Caribbean Plate. This process led to the production of a volcanic range.

THE OSA PENINSULA

The Osa Peninsula is Costa Rica's southernmost peninsula. Declared one of the "most biologically intense places on Earth" by *National Geographic* magazine, this secluded wonderland on the Pacific coast, just 24 miles (40 km) north of the Panamanian border, is home to rare and endangered species such as the puma, jaguar, and the scarlet macaw. The Osa Peninsula's rocky outcrops are a mid-Eocene to mid-Miocene Epoch montage of basalt, chert, and limestone resulting from the accretion of seamounts. Its extensive wetlands have evolved in partial isolation from the broadly drier tropical Pacific coast.

COCOS ISLAND NATIONAL PARK

Three hundred miles off Costa Rica's Pacific coast lies Cocos Island, first discovered in 1526 by the Spanish explorer Johan Cabecas, established as a National Park in 1978, and listed as a World Heritage Site by UNESCO on December 4, 1997.

The many ridges found on Cocos Island were formed by the alignment of submerged volcanoes that rose from the Cocos tectonic plate between 2 and 2.4 million years ago. Though its geology remains poorly studied, it appears to be a wholly volcanic island made up chiefly of conglomerates, tuffs, and lava flows of labradorite-andesite and related rock types. Its soil is mostly yellow clay, and landslides are common on the island.

The island's irregular coastline is overhung by precipitous cliffs rising from a narrow shore shelf, with an underwater profile of steep shelves with virtually no intertidal zones, all fringed by a shallow submerged coral reef.

Due to its physical isolation and location with respect to oceanographic processes, its 10 square miles (25 km²) are home to a unique and dynamic mix of species. Close to 30 species of fish are endemic to its waters, with an additional 20 species found only at Cocos Island and the Galapagos Islands, 400 miles (650 km) to the southwest.

ABOVE Corcovado National Park is located on Osa Peninsula. The park boasts around 12 miles (20 km) of volcanic black sand beaches. The upland areas of the park were created by weathering and by intensive tectonic activity, which can still be felt there as earth tremors on almost any day.

LEFT This enormous sea arch is part of the coastal landscape of the Gulf of Nicoya. The gulf separates the Nicoya Peninsula from mainland Costa Rica and consists of rocky islands, wetlands, and large mangrove swamps. In March 1990, a large earthquake (Richter Scale 7) occurred at the entrance to the Gulf of Nicoya.

Coastlines of the Mediterranean

Only five million years ago, there was no Mediterranean Sea. Geologically, it is quite recent. Evidence from ancient sediments of the Mediterranean seabed shows that it is made up of layers of salt, which form when seawater evaporates, and layers of desert deposits. So the Mediterranean Basin dried out to the point that it was once a desert. Whether the basin later refilled with seawater slowly or catastrophically is much debated.

BELOW Steep limestone cliffs surround Shipwreck Cove (also known as Navagio or Smugglers Cove) on Zakynthos Island. Sculpted by a series of earthquakes, the island is part of the Ionian Islands chain, which follows the Mediterranean coastline of western Greece.

The meaning of the name "Mediterranean Sea" is "sea in the middle of land." The Mediterranean region gives its name to the prevailing climate, characterized by warm dry summers and irregular rainfall, which is more substantial in the north than in the arid south of the basin.

THE SEA IN THE MIDDLE OF THE LAND

The very narrow Strait of Gibraltar—only 9 miles (15 km) wide with an average depth of 1,000 feet (300 m)—separates the Mediterranean Sea from the Atlantic Ocean. The Mediterranean is also connected through very narrow openings to the Sea of Marmara and to the Red Sea by way of the Suez Canal. Being an almost enclosed sea, the Mediterranean is subject to water loss, through evaporation, that is in excess of top-up, resulting in higher than average levels of salinity.

The coastlines of the three continents that surround the Mediterranean—Africa to the south,

Asia to the east and Europe in the north—have a total length of around 30,000 miles (46,000 km). The continents represent 19 countries and Greece, with its many islands, has by far the greatest share of coastline with a total length of 10,000 miles (16,000 km). The coastline of the Mediterranean Sea is one of peninsulas, bays, and islands with over 50 percent of its sea cliffs typically composed of limestone. Caves and beautiful grottoes filled with clear turquoise water are characteristic of the cliffs, which also have pocket beaches, sometimes shingled, in small coves at the base of the cliffs or in inlets. Apart from the rocky sea cliffs, there are coastal plains that feature wetlands, lagoons, and low-lying beaches, particularly where there is a rich source of sediments from mountainous regions behind the plains.

Another feature of the Mediterranean coastline is some of the great deltas of the world: Europe's Camargue (Rhone Delta), Ebro, and Po; and Egypt's Nile Delta. The greatly varying topographies of the Mediterranean continents also give rise to some of the world's most beautiful and rugged coastal landscapes with their crystal clear turquoise water, creating prized destinations for tourists. Sadly, coastal development that is associated with the rapid growth of the tourist industry has devastated quite a substantial area of the Mediterranean coast and had a major impact on its landscape.

A SEISMICALLY ACTIVE AREA

The Mediterranean Basin is seismically very active; it has a number of active volcanoes, and earthquakes are not uncommon. This is because the basin sits at the center of a complex assortment of tectonic plates that are still actively sliding under one another.

These tectonic processes began about 170 million years ago during the Jurassic Period, with the collision of the African and Eurasian plates and subsequent subduction and rifting. The activity gave rise to the formation of an ocean basin known as the Neotethys. By the Miocene Epoch, the continuing convergence of the plates pushed Africa and Europe together to form the Mediterranean, which was closed off from the Atlantic Ocean.

The collision of the two plates also caused the crust to buckle, pushing up the vast mountain ranges that extend from the Pyrenees in Spain to the Zagros Mountains in Iran. Most of this activity occurred during the Oligocene Epoch, (around 34 to 23 million years ago), and continued through to the Miocene Epoch. Today, many of these mountains give the present Mediterranean coastline its characteristic rocky cliffs. Other consequences of this geologically "young" mountain-building episode include the Mediterranean's few large plains, its poor agricultural land and ports, and harbors enclosed as inlets among the rocky outcrops. The southeast is an exception. Here, along 2,000 miles (3,000 km) of the Libyan and Egyptian coasts, the older Saharan platform is in direct contact with the sea, forming a coastline of relatively low relief.

By the late Miocene Epoch, 11 to 5 million years ago, the Mediterranean completely dried up when the enclosed water evaporated, leaving the layers of dried salt mentioned earlier. At the height of this period of evaporation, the Messinian Salinity Crisis, the Mediterranean Basin would have been an arid desert with small lakes and some rivers

ABOVE Orange and pink porphyritic rock masses and pinnacles tower up to 1,000 feet (300 m) above Calanches, in the Golfe de Porto on the Mediterranean side of Corsica. Erosion has created some odd shapes out of the rock, including a dog's head, a bear, and a tortoise, as well as a formation that has been described by some as a "Moor's head."

CLIMATE CHANGE AND SEA LEVELS

Global climate change threatens to increase the world's sea level. For the Mediterranean this is exacerbated by changes in relative sea level caused by continuing tectonic movements and human influences. Particularly threatened are the deltas of the Mediterranean. Already substantial problems of erosion and flooding are being experienced in several major deltas. In many cases, sediment supply to deltas and other coastal wetlands has been interrupted, causing local subsidence and the invasion of salt water into freshwater wetlands.

@ANUD A squacco heron (*Ardeola ralloides*) nests at the water's edge. This heron breeds around the Mediterranean Basin and the Black and Caspian seas.

QHĪGS Formentor Peninsula, Mallorca, is made of limestone laid down during the Jurassic to Tertiary periods. Mallorca is one of Spain's Balearic Islands, in the Mediterranean between mainland Spain and Algeria.

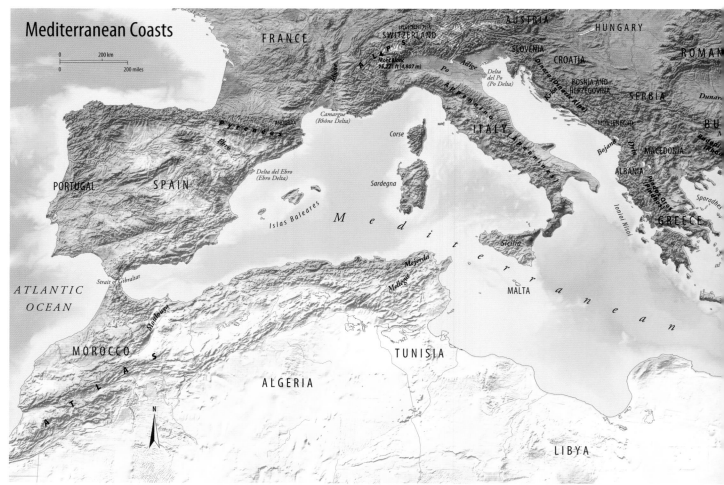

Mediterranean Coasts

running down from the surrounding mountain ranges. The basin refilled from the Atlantic Ocean at the end of the Miocene Epoch when sea levels rose.

Some fossil evidence from the Miocene Epoch seems to contradict the notion that the Mediterranean Basin was a desert. The evidence shows that there was a period when the Mediterranean Basin had a relatively humid subtropical climate with summer rainfall that supported laurel forests. It appears that there was high monsoonal activity around the eastern region that probably peaked during the Messinian Salinity Crisis. The shift to the current Mediterranean climate occurred within the last 3.2 to 2.8 million years during the Pliocene Epoch, as summer rainfall decreased.

A HOTSPOT OF BIODIVERSITY

Its position between the two continents of Europe and Africa, has given the Mediterranean Basin a very high biodiversity. However, Conservation International designates the region as a biodiversity hotspot: although it is rich in biodiversity, many of its species are threatened through loss of habitat. Human activity over a period of 8,000 years has

dramatically altered the habitats of the basin so that a widespread hard-leafed drier vegetation, such as junipers, has replaced much of the earlier covering of the characteristic forests of evergreen oak forests and conifers.

After the Messinian Salinity Crisis, the reflooding of the Mediterranean by the Atlantic was responsible for much of the basin's marine biota. Today, the Atlantic accounts for about 67 percent of the Mediterranean marine species. Coming from the colder and more nutrient-rich Atlantic, the biota has adapted over the past five million years to the lower nutrient level and warmer, more saline conditions present in the Mediterranean Sea. There is a higher diversity of species in the west than in the eastern regions of sea.

Other species have been introduced from the Suez Canal, which from the time it opened in 1869, allowed the passage of the first saltwater migrants to come in from the Red Sea. These account for 5 percent of the total, but up to 12 percent of the southeastern Mediterranean species.

In keeping with its great variety of coastlines is the diversity of Mediterranean species, with many of them being endemic to the region. Of the flora, there are some 22,500 endemic vascular plant species—more than four times the number found in all the rest of Europe. Species inhabiting flat or sandy coasts are different from those associated with the high, steep, and rocky coasts. There are many areas where submerged rocks form a hard substrate that supports a rich variety of benthic communities—those that dwell on the sea floor.

The most significant ecosystems of the coastal areas are the rocky intertidal zones, the estuaries, and the seagrass meadows of mostly *Posidonia oceanica*. The meadows are extremely important ecologically, because of diversity and also because of the vital role they play in providing breeding grounds for many of the Mediterranean's commercial fish stocks.

@ANUD Buildings cling precariously to the craggy limestone slopes on Italy's Amalfi Coast. Underwater researchers have discovered artificial structures beneath the Mediterranean Sea here, including a dockyard that would have been used by sailors many hundreds of years ago and a fourteenth-century seawall.

INSTABILITY AND EROSION

The shoreline of the Mediterranean has been dynamically unstable in recent geologic and historic times. Only around 5 percent of the area of the Mediterranean Basin remains undisturbed. Its geology has been grossly altered by the building of structures along the coast, causing detrimental changes to natural erosional processes. With the region increasing in popularity as Europe's vacation destination, resort development and beach mismanagement, as well as increasing pollution, all contribute to the fragmentation and isolation of many of the populations of threatened species.

Brittany Coastline, France

The Brittany Peninsula in northwestern France is bordered to the north by the English Channel and to the south by the Bay of Biscay. It is characterized by an abundance of rivers that flow out from a ruggedly indented coastline of cliffs, dunes, estuaries, mudflats, and marshlands.

ABOVE The port city of St Malo is situated on the Rance Estuary. The granite wall erected in the Middle Ages to surround the city is made of the same gray granite used at Mont St Michel.

Brittany also has a unique Celtic heritage. Approximately a quarter of its almost 3 million population still speak Breton, an ancient Celtic language similar to Welsh and Cornish, with its customs and traditions practiced mostly in the more isolated west of the peninsula.

BRITTANY'S SEA CLIFFS

Brittany's most lingering images are associated with its sea cliffs. Coastal outcrops of Cretaceous Period rocks border the northern Brittany Peninsula. The *terrain crétacé*—white soft limestone, so beloved of the master impressionist painter Claude Monet —is found around Etretat.

Sea cliffs are plentiful all along Brittany's 750-mile (1,200-km) coastline, with the highest cliffs rising to over 300 feet (100 m) at the Point of Finistère on the Crozon Peninsula, and to almost 200 feet (70 m) high at Cap Frehel and Goelo on the Cotes-d'Armor.

Brittany's sea cliffs are breeding grounds for cormorants, gannets, and other seabirds and are covered in flora resistant to marine environments, such as sea campion and gorse, a spiny evergreen shrub common to Western Europe.

THE CROZON PENINSULA

The coastal scenery of Brittany's Crozon Peninsula is impressive, with its abundance of Paleozoic Era rocks reflecting the different weathering of thermally layered sandstones, mudstones, and limestone lithologies, as well as some complex folding and faulting. A craggy extension of the Brittany Peninsula and the central feature of Finistère's chaotic blend of estuaries and promontories, its mix of hard quartzes and sandstones make it virtually immune from erosion.

Shaped like an enormous stone cross with its "arms," the Crozon Peninsula stretches from the Point of Spain headland to the Cap de la Chèvre. The Crozon Peninsula is located within the Armorique Regional Nature Park, which begins in the Arrée Mountains in the east and continues to the western shoreline of the Crozon Peninsula. The Precambrian Era Arrée Mountains boast some of the oldest geological features in Europe.

WILD AND RUGGED CAP FREHEL

The region of Cap Frehel in northern Brittany covers an area of 740 acres (300 ha), consisting of a wild and rugged coastline covered in heather and gorse over cliffs that rise to 200 feet (70 m) above the foaming waters of the Atlantic Ocean.

Cap Frehel's towering pink sandstone and porphyry cliffs make it difficult to access by boat. Moorlands and marshes cover much of this tiny peninsula's undulating terrain. There are no towns or villages, and only two lighthouses, one of which dates to the seventeenth century.

The nearby Fort la Latte was built atop Cap Frehel's famous clifftops in the fourteenth century on the site of an earlier tenth-century castle. Burned to the ground in the 1500s, it was rebuilt in the seventeenth century by Louis XIV. Sold by its owners in 1892 in a state of disrepair, today Fort la Latte has been restored to its former glory.

THE ABBEY OF MONT ST MICHEL

Mont St Michel is located on the Brittany–Normandy coast and began life as a Benedictine monastery in 966 CE. It is perched on a quasi-island 3,000 feet (1,000 m) in diameter and 260 feet (80 m) high, and is connected to the French mainland by a sand bridge known as a tombolo. Tombolo is an Italian word derived from the Latin for "burial mound." It is a pile of sediment or sand that leads offshore from a beach directly to an island.

Mont St Michel's construction could well be termed arrogant, as it is built on a site that experiences some of the greatest tidal fluctuations in France. Forty-five feet (14 m) separate the high and low water marks here, which can unpredictably shift the surrounding sands.

Mont St Michel underwent several stages of construction with the first serious work beginning under Abbot Hildebert in 1020. In 1170, a new facade was begun on the western elevation, and in 1210, building started on a number of dining and assembly halls, as well as kitchens, cloisters, and a dormitory.

RIGHT On the Alabaster Coast (Cote d'Albatre) at Etretat, is the sea arch "needle," known as the elephant's trunk. It was so named because it looks like an elephant has dipped its trunk into the sea. Alabaster is a fine-grained and compact variety of gypsum.

Coastlines of the United Kingdom and Ireland

The United Kingdom and Ireland possess, in global terms, unusually high tides and a particularly harsh stormy climate. These conditions have combined, over time, to produce a dynamic and varied coastline that includes the spectacular Jurassic Coast.

OPPOSITE From the air, the Cretaceous Period white chalk cliffs of Old Harry Rocks at Hand-fast Point, in Dorset, England, are easily seen. Though there is an ongoing debate about which of these sea stacks is Harry, his "wife" once stood beside him but that stack succumbed to erosion, collapsing around 50 years ago.

Triassic, Jurassic, Cretaceous, and Tertiary period formations are all well represented along this dramatic coastline. They vary greatly in age from the hard Precambrian and early Paleozoic Era examples in Britain's west and north to younger Tertiary Period formations in the southeast.

ENGLAND'S JURASSIC COAST

In 2001, a 100-mile (60-km) stretch of coastline between Old Harry Rocks in East Dorset and Orcombe Point near Exmouth in East Devon became Great Britain's first and only natural World Heritage Site. Known as the Jurassic Coast, its sea cliff exposures provide a continuous sequence of rock formations spanning 185 million years of geologic history along an unspoiled and accessible coast-line of great beauty.

Its landforms include stretches of discordant and concordant coastlines, examples of cove and limestone folding, tombolos, and natural arches that are still the subject of ongoing international field studies. The region gained its name from the richness of its fossil record that was first recognized when the geologist John Ray discovered a wealth of fossils near Lyme Regis in 1673.

Dinosaur footprints, a fossil forest, and huge ammonites represent a fraction of the preserved fossil evidence that can still be found. Within the shale cliffs of Chippel Bay, bone fragments of ichthyo-saurs and plesiosaurs have been unearthed, while ammonites, bivalves, and brachio-pods are common to the limestone and shale Church Cliffs of Charmouth.

A VARIETY OF REGIONS AND ROCK TYPES

Along England's east Devon coast are Permia Triassic period continental red bedrock facie zoic and Cenozoic era stratas are common a middle of the south coast. In the southwest, ian and Carboniferous period shales, slates, a stones are plentiful, while further north, in S older harder Precambrian Era and Cambrian rocks predominate, with Devonian Period o sandstone found in the north around Caithn

IRELAND'S RUGGED COASTAL TERRAIN

The stunning Causeway Coast of Northern is a plateau landscape, with high sea cliffs c

ARCHEOLOGY OF THE DINGLE PENINSULA

No other region in Western Europe possesses the wealth of archeological monuments that can be found scattered across Ireland's Dingle Peninsula (Corca Dhuibhne). Over 2,000 monuments have been uncovered spanning 6,000 years of human habitation, from the first hunter-gatherers of the Mesolithic Period (8,000 to 4,000 BCE) through the Stone Age, Iron Age, and Bronze Age to the early Christian, Viking, and Medieval periods. It is a land of Stone Age passage tombs and cup and circle rock art; of Bronze Age implements and weapons; of the Iron Age hill fort of Cathair Con Ri; and of 70 Ogham stones that preserve the earliest form of Irish writing. The peninsula's isolation also accounts for the presence of over 30 Early Christian monastic sites such as Oileain tSeanaigh, which has remained virtually untouched ever since it was abandoned in the twelfth century.

The rugged sandstone cliffs of Dingle Peninsula—the westernmost point in the Republic of Ireland.

ABOVE The glacio-karstic land-scape of "The Burren" is part of Ireland's coastline. This area is geologically similar to Slovenia's karst region. Bare Carboniferous Period limestone gives The Burren its eerie appearance. Beneath the surface, there are drainage systems, caves, gorges, sinkholes, and seasonal lakes known as "turloughs."

ABOVE Dorset's heathlands are a feature of England's Jurassic Coast. The colorful blooms of heather and gorse are character-istic of the heathlands. They have the ability to cope with the poor acid soil of the area and, like all coastal plants, they are able to tolerate salt.

Tertiary Period basalts and occasionally underlying Ulster white limestone (chalk). Its open indented coastline is home to volcanic vents, faults, and dolerite dikes set among extensive sand beaches and dunes backed by scree and slumped debris. A designated World Heritage Site, the Causeway Coast is in a process of continual erosion. Rotational and planar landslides, rock falls, and the toppling of basaltic columns are commonplace, with cliff collapses threatening the historic landmarks of Kinbane and Dunluce castles.

Four hundred million years ago, Ireland's Dingle Peninsula in County Kerry was a vast inland valley basin bordered by uplands and mountains, whose rivers deposited gravels, sand, and mud that led to the formation of river bottoms, sand dunes, and shallow freshwater lakes. Crustal uplifting led to faulting, erosion, and buildups of skeletal debris produced by mollusks, algae, and coral, followed by further mountain building at the end of the Upper Carboniferous Period 290 million years ago. From 200,000 to 10,000 years ago the peninsula was scoured by glacial ice, which carved out the landscape that can be seen today.

Ireland's coastline is a region of beautiful high ridges and granite outcrops. It includes the Conor Pass, Ireland's highest mountain pass, which at its crest is 1,300 feet (400 m) above sea level.

Coastal Landforms of South Africa

The 1,800 miles (2,800 km) of the South African coast is a mostly rock-strewn, highly exposed, rugged coastal landscape that is battered by high wave energy and constant strong winds. The coastline has relatively few sheltered inlets, coves, or harbors.

BELOW African (or jackass) penguins congregate on Boulders Beach, just a few miles north of Simons Town, False Bay, Cape Peninsula. The immense boulders shelter the cove from ocean currents and large waves.

The South African coast stretches from its desert border with Namibia in the west to Cape Agulhas in the south, before continuing to the northeast along a narrow coastal plain to the border with Mozambique on the shores of the Indian Ocean.

CAPE PENINSULA AND CAPE OF GOOD HOPE
The rocks that form the Cape of Good Hope and the Cape Peninsula are part of the Malmesbury

Group's marine siltstones, which are 560 million years old. The sandstones are identical to those seen in the exposed gradients of Table Mountain.

The fynbos-covered hills of the Cape Peninsula, which forms the western boundary of False Bay, comprise one of the great landmarks of the world. The Portuguese navigator Bartholomew Diaz first rounded its promontory in 1488, and dubbed it Cabo das Tormentas, the "Cape of Storms."

of the warm waters of the time. Weathering of the reef system has formed magnificent gorges and caves that are a huge tourist attraction.

Periods of advancing and retreating sea levels followed the Devonian, the most recent being the rise that followed the end of the last glaciation when the sea inundated about 14 percent of the Kimberley block. It was this sea level change—about 15,000 years ago—that totally altered the coastal outline, giving the region its distinctive ruggedness and forming structures such as the archipelagoes: exposed tops of hills that were once well away from the sea. There are also many drowned river valleys, with the most spectacular forming gorges that run between two ridges of the McLarty Ranges in King Sound, north of Derby. The remarkable 35-foot (10-m) tides in this region surge through the inland passages and gorges, in the process creating treacherously strong and unpredictable currents. Some of the gaps in the sheer cliffs in Talbot Bay are only a few feet wide so that as the tide peaks, the water on one side can be as much as 12 feet (4 m) higher than on the other, forming in effect a "horizontal waterfall."

DISTINCTIVE FLORA AND FAUNA

The Kimberley region, with its huge variety of habitats, has a biota and biodiversity that are quite distinct from elsewhere in Western Australia. The barrier formed by the desert to the south means that the biota is more closely related to that of the northern Top End of Australia. The marine environment is significantly influenced by the warm current that flows from the east through both the Asian and Top End regions.

There are savannas, scrublands, eucalypt forests, rainforest remnants, and mangrove forests that vary across the region according to rainfall and soil types. The mangroves form extensive forests along the tidal flats of the Kimberley coast and support species-rich communities, together with diverse land and marine species that depend on the mangroves.

Recently, the coastal waters have seen a rapid increase in human presence, and the people's various intrusive activities are threatening the marine biota, including a rare dolphin habitat.

ABOVE Tourists come to the Kimberley region to witness an amazing and rare coastal phenomenon: the horizontal waterfall at Talbot Bay in Buccaneer Archipelago. A combination of extreme tidal movements and narrow rock entrances creates these waterfalls in the ocean.

Coastal Landforms of New Zealand

New Zealand possesses a heavily incised coastline of fjords, sounds, harbors, and coves. The coastline stretches from the warm, subtropical waters of the North Island's narrow Aupouri Peninsula to the subantarctic waters of Stewart Island.

BELOW The brilliance of these zoanthids, or yellow sea daisies, shines in a dark area of Riko Riko Sea Cave, Poor Knights Marine Reserve. The reserve also protects Tawhiti Rahi and Aorangi islands, and the sea arches Blue Mao Mao Arch and Northern Arch.

BOTTOM The complex network of sea-drowned valleys known as the Marlborough Sounds was probably formed during the universal rise in sea levels that resulted from the Pleistocene Epoch ice melt. Its geological structure is one of schists and greywackes.

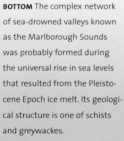

New Zealand's 9,400 miles (15,000 km) of coastline is at its most irregular in the Northland region to the northwest of Auckland. The coastal areas of the South Island are generally less indented; the exceptions being the rugged beauty of Fiordland in the southwest and the picturesque sea-drowned valleys of the Marlborough Sounds along the South Island's northernmost tip.

THE CRENELLATED NORTHLAND COAST

The Northland area extends north from Greater Auckland, and is a sparsely populated region of ancient sand dunes and crenellated coastlines. One of Northland's primary features is the Bay of Plenty coast, a 200-mile (350-km) expanse comprising over 125 miles (200 km) of sand substrate with functioning dune systems.

The number of active dunes in the Northland and across New Zealand has declined by more than 70 percent, from 320,000 acres (129,000 ha) in the early 1900s to less than 97,000 acres (39,000 ha) today. This loss is largely due to the presence of marram grass, an introduced grass species that causes degradation of the indigenous dune flora required for the stabilization and vitality of dune systems.

A rich array of fossil deposits has been found in the Northland region and beneath surface vegetation in Doubtless Bay. Other sites are along the east coast of Aupouri Peninsula and in the karst landscape of the west coast's Waitomo Caves.

Taranaki Sea Cliffs on Northland's west coast, formed by lahar deposits ejected from nearby Mt Egmont, rise to heights of over 650 feet (200 m) above black iron-sand beaches along a 25-mile (40-km) belt of continuous exposure.

UNUSUAL LAYERS OF PANCAKE ROCKS

The distinctive Pancake Rocks lie near the small town of Punakaiki, midway between Westport and Greymouth on the South Island's west coast. The rocks are the eroded remnant of a singular band of Oligocene Epoch limestone that is 165 feet (50 m) thick and 30 million years old.

Originating in the calcareous remains of marine organisms compacted on the sea floor and lifted by seismic action, the heavily eroded limestones of Pancake Rocks have been shaped by a mix of acidic rain, wind, and water erosion. The resultant formation resembles a stack of pancakes dissected to form cliffs, which rise above the Pororari River. This is one of the most-visited of all New Zealand's conservation areas.

RIKO RIKO, THE WORLD'S LARGEST SEA CAVE

The Riko Riko sea cave is located on the northwest side of Aorangi Island in the Poor Knight Islands, just 15 miles (24 km) off the east coast of New Zealand's North Island. Its enormous oval chamber makes it possibly the largest sea cave in the world when measured by volume.

Poor Knight Islands are the remnants of large lava domes composed primarily of rhyolitic breccia and tuff from the Late Miocene Epoch, which run from the Coromandel Peninsula to Northland's east coast. Riko Riko Cave is most likely the result of a gas bubble that emerged in the volcano's cooling lava.

More than 25 sea caves and archways can be found throughout the Poor Knight Islands, including the completely submerged Airbubble Cave.

THE RAISED BEACHES OF TURAKIRAE HEAD

Located on the North Island's southern coastline near the capital Wellington, Turakirae Head is a promontory at the southern end of the Rimutaka Range, where tectonic uplifts during major earthquakes have resulted in the emergence of four Holocene Epoch marine terraces, often called "stepped beaches." The terraces provide us with a complete record of uplift going back 7,000 years.

The most recent event was a consequence of the great Wairarapa earthquake of 1855, which resulted in an uplift of 21 feet (6.4 m). The largest single uplift event occurred between 200 BCE and

Coastal New Zealand

382 BCE, resulting in an increase of 30 feet (9 m). The oldest preserved ridge in the group was raised between 5,400 BCE and 5,100 BCE, and all four terraces tilt relative to sea level.

The Turakirae headland is also the winter home for approximately 500 New Zealand fur seals.

ABOVE LEFT At Pancake Rocks, near the South Island township of Punakaiki, a blowhole bursts through the heavily eroded limestone stacks during a high tide.

ABOVE Beneath the Nullarbor Plain along the southern coastline of Australia, there are large caves with karst lakes. These lakes provide an underground source of fresh water.

RIGHT Visitors are attracted by the stalactites and stalagmites within the Buchan Caves, in the eastern highlands of Australia. The more-than 150 karsts in this region have formed in Paleozoic Era limestone.

The vertical, mist-shrouded, verdant peaks of China and Vietnam, the jagged spire-like islands rising from the turquoise South China Sea, and the subterranean chamber pools lit by shafts of sunlight in Mexico's Yucatan Peninsula all have one thing in common—they are all landscapes that have been etched in limestone rocks.

Karst topography and caves occur wherever areas of limestone are exposed above sea level by tectonic uplift. Limestone is hard and tough, but very soluble compared to other rock types, and is particularly susceptible to slightly acidic rain or groundwater. It does not take very long for surface water to create a typical landscape in limestone: a series of interconnected underground chambers and channelways, through which the water can travel.

Many places across the globe contain karst landscape, often vast expanses of it. The Caribbean Islands, for example, boast large networks of caves and passageways as well as sinkholes and karst formations. In North America, numerous limestone caves and karst systems are now protected within national parks. Australia's Nullarbor Plain is one of the largest areas of karst topography in the world.

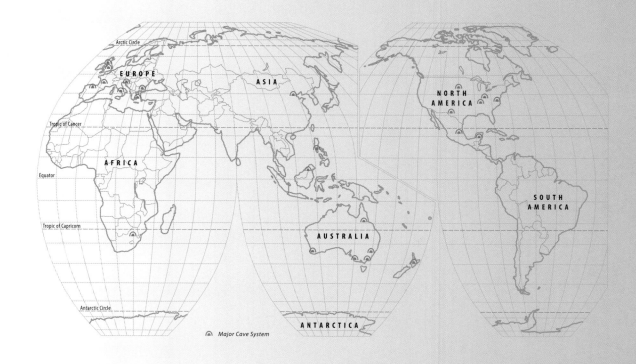

Major Cave System

DRY JAGGED KARST

Slight amounts of acid in rainwater react with the calcium carbonate of the limestone to produce soluble calcium salts and carbon dioxide. Over time, countless raindrops create thousands of tiny pits and fluted rills on the rock surface that slowly become sharper and sharper. Eventually, the rock surface becomes a myriad razor-edged surfaces —almost too sharp to touch and capable of reducing the soles of a sturdy pair of boots to shreds in a matter of hours. These classic limestone dissolution features are known as karren or lapies.

Surface-flowing rivers are very rare in limestone terrains as water quickly disappears through a series of solution-widened joints into interconnected subterranean chambers on its underground journey to the sea. As a result of this rapid disappearance of water, the surface of karst landscapes is usually extremely dry and, even in areas of high rainfall, the vegetation has to adopt strategies to ensure it can conserve water. It is rather interesting to stand underground in some caves beneath the karst surface and see the tree roots growing downward from the surface like long ropes trying to reach the water table. Caves that are located above the water table are called vadose caves. They feature uneroded ceilings and downward slopes.

BELOW THE WATER TABLE

Beneath the level of standing groundwater (the water table) in areas of high rainfall there is a flow of water that works its way outward and downward under gravity toward the sea. This happens in all rock types. Water picks the path of least resistance between the grains of a sedimentary rock, between lava flows, through joints and fractures, or along fault planes. In limestone, however, the results are dramatic as the groundwater slowly turns initially tiny channelways into giant tube-shaped tunnels—huge underground river channels

ABOVE Devil's Ear is part of the Devil Spring system in Ginnies Springs basin, Florida, USA. From deep in this cave-like opening, divers can at certain times look up and see the sun through the tannin-stained river water.

LEFT Extensive caves and above-ground karst features have formed in the highly karstified limestone of the Margaret River region of Western Australia.

that carry fresh water, sometimes miles out to sea to emerge as underwater freshwater springs. These caves that form beneath the water table are known as phreatic caves. Often, such caves are exposed, eventually, after a general lowering of the water table, at which time it becomes possible to explore the caves without the use of scuba gear.

DOLINES AND SINKHOLES

Individual caverns can grow to enormous sizes. If a chamber is located close beneath the surface, its roof will become thinner and thinner. Eventually the roof either dissolves away or, if the cavern is large enough, it can collapse, creating an enclosed crater-like basin called a doline or sinkhole. These can range in size from small grassy depressions to enormous cliff-bounded wells containing entire forests or sometimes lakes. These features are carefully explored as they often provide access points to larger cave systems. On New Britain Island, off the coast of Papua New Guinea, speleologists have explored the Ora Cave System by means of the Ora Doline. Here, a raging subterranean torrent emerges from the side of the heavily forested doline and crosses its floor before disappearing into a cave on the other side.

THE BLUE HOLES OF THE CARIBBEAN

Blue holes are the near-perfect circular sinkholes found particularly in the submerged karst of the Caribbean Sea. Approaching the edge of one of these structures, the light turquoise, shallow waters give way suddenly to a deep indigo blue as the sides drop vertically away into an abyss. The sight of these circular features is even more dramatic from the air, hence the name "blue holes."

One such sinkhole, the Great Blue Hole at Lighthouse Reef, lies 60 miles (100 km) off the coast out of Belize City. It is a very popular dive site. The hole is about 1,300 feet (400 m) across and drops to a depth of 480 feet (145 m). Its walls are lined with ancient speleotherms, and the hole itself is the opening to a system of caves and passageways believed to connect through to the mainland. The stalactites indicate that the cave was air-filled during the last Ice Age, when the sea level was lower. The debris on the cave's floor suggest the sinkhole formed when its roof collapsed. Distinct shelves and ledges carved into the limestone walls and running around their entire circumference show that the sea level did not rise progressively, but in a series of steps.

SPECTACULAR CAVE FORMATIONS

The same limestone that is dissolved to create caves is often re-deposited as glistening formations

ABOVE In the southern part of Palau in the Pacific Ocean there is a group of 200 rounded limestone formations covered in forest. They are examples of "cockpit karst," in which residual hills enclose depressions called dolines or "cockpits."

RIGHT An aerial view of Belize City's Great Blue Hole shows the ink-blue color of its deep center. This almost perfectly circular limestone sinkhole has been declared a natural monument.

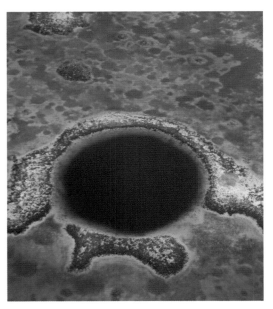

all over their walls, ceilings, and floors, especially in humid, tropical climates. Speleotherm is the general name used for all such cave formations.

Speleotherms grow very slowly; they are often deposited as microscopic crystals, drip by drip, as mineral-charged water falls from the ceiling or runs down the walls of a cave. Pointed stalactites grow downward from the ceiling toward fatter, rounder stalagmites building upward from the floor, drip by drip. Eventually the two touch and fuse to form a single column. Looking at these mighty features, it is sometimes hard to believe that they started life as fragile, thin, hollow proto-stalactites or soda straws.

Wavy or folded sheets of calcite known as draperies or shawls grow from sloping cave walls. These thin rock curtains are translucent and when lit from behind can show a beautiful alternating light- and dark-colored growth banding. Colors vary from dark brown to white, caused by staining of calcite by impurities in the groundwater. Calcite can also build up some formations in the shape of rounded disks that project from a cave wall around a wall rock fracture. These are known as shields and grow at an upward angle, often with their undersides draped with stalactites. Shields are rare, except in Lehman Caves in Nevada, USA.

Helictites are particularly enigmatic speleotherms because they appear to be defying gravity by going neither up nor down. These twig-like growths are like twisted soda straws with a fine central canal. Water appears to travel to the end of the helictite by capillary action allowing it to grow longer in an apparently random direction.

Quite often, small walls or barriers of rock on the floor of wet caves, known as rimstone, trap the water in shallow basins known as rimstone pools. Changing conditions, such as turbulence and the release of carbon dioxide, cause the precipitation of soluble minerals to form the rim.

In contrast, the terms dripstone or flowstone are often used to describe the layered or concentrically banded rocks that are deposited in caves. Secondary cave deposits are mostly calcite (calcium carbonate) but are also commonly a number of other water-soluble minerals such as gypsum (calcium sulfate) or halite (sodium chloride).

CAVES LADEN WITH GEM TREASURES

Occasionally, rare colorful carbonate speleotherms, such as green malachite, blue azurite (copper carbonates), and pink rhodochrosite (manganese carbonate), can be found adorning caves. The caves of Catamarca Province in Argentina are a particularly spectacular example, decorated by pink and white delicately banded rhodochrosite stalactites and

MEXICO'S SACRED CENOTES

The Yucatan Peninsula of Mexico is dotted with deep, circular, well-like sinkholes or collapsed dolines, known locally as "cenotes." These vertical-sided wells, draped with greenery and bottomed by a deep lake, are connected to the Yucatan's network of underground drainage. The Mayan peoples built their towns and temples around these cenotes. They were the wells from which the people drew their vital fresh water, and, as a result of their importance to the survival of the civilization, cenotes attained a religious significance. One of the best known is the Sacred Cenote at Chichen Itza near Merida, which is approximately 200 feet (60 m) across. Here the Maya made offerings of jewelry, clay pots, and figurines to their rain god, Chac, including human sacrifices. Dredging and diving expeditions since the late 1800s have recovered thousands of artifacts, including the bones of about 50 sacrificial victims. The Yucatan Peninsula has some of the world's finest clear-water cavern diving sites.

The sinkhole called Sacred Cenote, at the archeological site Chichen Itza, is 115 feet (35 m) deep.

stalagmites. Most unfortunately, the majority of these rare speleotherms have been removed for cutting and polishing into gemstones.

In Mexico, thick cave deposits are mined for gem and ornamental use. This flowstone material is known as cave onyx or Mexican onyx, and is a delicately banded, pale green to white translucent calcite. Its relative softness makes it a very popular rock for carving.

GYPSUM CAVES AND CHANDELIERS

Caves formed in gypsum are much less common than those in limestone; however, interestingly, in the second-longest cave system in the world, in the Podolie region of the Ukraine, the cave has been created in gypsum. Known as the Optimisticheskaya Cave, it has about 186 miles (300 km) of passageways. Secondary gypsum growths on the walls and ceilings of caves are known as gypsum flowers, while large, spectacular examples are called chandeliers, such as those found hanging from the ceiling of the Lechuguilla Cave in New Mexico. Unlike speleothems, gypsum flowers grow from their point of attachment

outward. Distortion in the crystal lattice causes the
columnar crystals or fibers to curl around as they
grow thus creating the flowerlike effect.

CAVE FOSSILS

Fossils of unfortunate animals that have strayed
into a cave and become lost, or have fallen through
a skylight, are fairly common. Once inside the cave
their remains dry out and are well preserved before
eventually being buried or sealed into the rock
with calcite. In Australia, a team of amateur cavers
discovered well-preserved fossils of giant flat-faced
kangaroos, marsupial lions, thylacines, and wombats
in the Naracoorte Cave below the Nullarbor Plain.

These enormous creatures lived in Australia
during the Pleistocene Era, between approximately
1,750,000 and 10,000 years ago, and then they disappeared. Such finds help shed light on theories as
to why this megafauna became extinct in Australia,
leading scientists to speculate that it was probably
due to over-hunting by early humans rather than
a result of climate change.

The bones of early humans have also been
found in caves where they died or were buried.
Important discoveries, such as the 30,000-year-old
bones from Pestera Muierii Cave, Romania, have
been able to shed light on the course of human
development across Europe. There is much debate

HOW KARST IS FORMED

A. Karst erosion sculpts the surface of a limestone plateau that has been lifted above sea level. The landscape is dry because
all water disappears underground via a series of joints and crevices and travels to the sea in underground rivers. B. Cone-
shaped hills riddled with caves and covered with razor-sharp flutes become prominent as the limestone rocks continue to
dissolve and collapse into sinkholes. Broad valleys become wide flat valleys between the hills. C. The mature karst landscape
is essentially a very broad, level plain that lies close to the level of water table. Small, low, residual limestone hills, known
as "haystacks," dot this flat landscape.

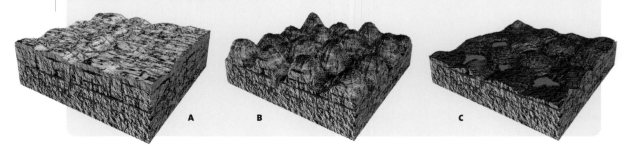

A   B   C

as to whether these early humans may have been able to interbreed with the Neanderthals before this species died out.

MATURE KARST LANDSCAPES

As a limestone landscape continues to erode by widening and deepening of existing sets of joints, a landscape of vertical-sided towers or conical peaks begins to emerge. These are known as tower karst or cone karst landscapes. Some of the most spectacular tower karst can be seen in the Guangxi Province of southern China. Here the magnificent cliffed towers of limestone are so abundant that the name fenglin, or "forest of stone peaks," is used to describe them. At Ha Long Bay on the coast of Vietnam, a rising sea level has flooded a fenglin landscape and created the bay and its 1,600 towering islands. Around the base of each island at sea level there is a deep undercut often big enough for boats to pass under at low tide. This marine erosion notch is caused by increased dissolution of the limestone at this level. It gives the islands a bizarre and dangerously over-steepened appearance.

As a limestone landscape matures, without subsequent tectonic uplift or sea level change, the peaks are dissolved away slowly into a series of remnant rounded or cone-shaped hills surrounded by broad, flat valleys. These distinctive hills are usually riddled with caves and are known by various names in different regions of the world, including pepinos or haystack hills in Puerto Rico, mogotes in Cuba, or hums in Serbia and Montenegro.

ABOVE Carlsbad Caverns National Park in New Mexico, USA, has 113 underground caves that were formed when sulfuric acid dissolved the surrounding limestone. These large caves feature amazing speleotherms.

LEFT This dramatic, vertical-sided formation is part of the mature karst landscape of Coromandel Peninsula on New Zealand's North Island. Many dormant and long-extinct volcanoes can also be seen in this coastal region.

Karst and Caves of North America

Limestone caves and karst systems are found in many parks in the US National Park System, ranging from the 10 or so found in Chesapeake and Ohio Canal National Historic Park to several hundred scattered throughout the extent of Arizona's Grand Canyon.

Kentucky's Mammoth Cave–Flint Ridge system is the longest cave system known. This system has more than 300 miles (480 km) of interconnected passages cataloged so far.

Canadian examples range from Castleguard Cave in Alberta's Banff National Park, with more than 15 miles (25 km) of galleries, to the more than a thousand caves found on Vancouver Island on Canada's west coast.

CARLSBAD CAVERNS, NEW MEXICO

One of the oldest and most famous cave systems in the world, Carlsbad Caverns in northern New Mexico is a collection of 83 separate caves that began life 250 million years ago as a 400-mile

(640-km) long reef in the midst of an inland sea. In 1923, this site was designated a US National Park and became a World Heritage Site in 1995.

Carlsbad Caverns's Big Room measures some 2,000 feet (600 m) in length, with an average width of 600 feet (180 m)—one of the largest single underground chambers in the world, it contains many of Carlsbad's most prominent features, including The Giant Dome and the massive formation known as the Rock of Ages.

Lechuguilla Cave—the fifth longest cave in the world—is the deepest cave in the United States. Not fully explored until 1986, the interior revealed a vast array of rare speleothems, including cave pearls, long soda straws, and enormous gypsum

BELOW Carlsbad Caverns feature numerous spectacular formations. In the "Big Room"—the world's largest single underground chamber—breathtaking sights abound, such as the Rock of Ages, which has formed over many centuries.

chandeliers, surpassing the adjacent Carlsbad Cavern in size, depth, and diversity of formations.

Carlsbad Caverns are also home to a colony of one million Mexican free-tailed bats, a colonial species that feeds exclusively on insects, and which are distinguished by their long narrow wings and thin dangling tails. These bats give birth to their young in June and July each year, before migrating to their wintering grounds in Mexico in October.

LEHMAN CAVES, GREAT BASIN NATIONAL PARK, NEVADA

Despite its name, Lehman Caves is a single cave and is one of the most richly decorated limestone caves in the United States.

Discovered in the early to mid-1880s by miner and rancher Absalom S. Lehman, it wasn't too long before Lehman was charging admission for visitors to tour the cave. In 1922, the cave was proclaimed a National Monument, before being incorporated into the Great Basin National Park in 1986.

The site is famous for its profusion of the rare and unusual structures called shields, which tend to form in caves with highly fractured limestone. Resembling flattened clam shells, they consist of two round or oval parallel plates separated by a thin medial crack. Built from calcite carried by water entering the cave by means of capillary action, the cave's 300 shields are themselves covered in myriad helictites, draperies, and stalactites.

CANADA'S "ISLAND OF CAVES"

Canada's Vancouver Island is the largest island on North America's west coast and home to over a thousand limestone caves. Some 500 square miles

(1,200 km²) of the island is karst, representing the highest density of caves found anywhere in Canada.

Well developed due to high levels of rainfall and a mountainous topography, many of the country's longest and deepest caves are found in the island's north, where mildly acidic run-offs from the surrounding coniferous forests have eroded complex systems into the region's Quatsino limestone formation. Features include the massive sinkholes of Devil's Bath and the Eternal Fountain, as well as numerous karst formations in the Bonanza Valley, and areas to the south and west of Port Hardy.

Mostly hosted in limestone and marble—a metamorphosed limestone—a number of Vancouver Island's caves and karsts occur within the boundaries of provincial parks. The more accessible are Horne Lake, Upana Caves, and Little Huston.

ABOVE Stalactites decorate the surfaces of Carlsbad Caverns. The action of water and calcite produces the spectacular formation of stalactites and stalagmites that have made this cave system world famous.

TOP Emerging as the sun goes down, a cloud of Mexican free-tailed bats fills the skies in their nightly hunt for food. These mid-sized bats have a wingspan of around 12 inches (30 cm), and have a tail that has little membrane covering—hence the name of the species.

MAMMOTH CAVE, KENTUCKY

South-central Kentucky is a region of rolling hills bordered by escarpments and alluvial plains—a community of farmlands interspersed with forested rugged hills over a seemingly innocuous bedrock of limestone and soft sandstone.

ABOVE Crystals of calcium sulfate form these exquisite roseate blooms—known as gypsum flowers. These formations are found in the Crystal Cave section of Mammoth Cave.

RIGHT Sections of Mammoth Cave are so fragile that visitor numbers are restricted. In this section of cave—known as Big Avenue—visitors are limited to 20 per year in an effort to preserve the unique environment.

It is this combination of limestone and sandstone that has enabled the formation of the world's largest system of caves and karsts, with over 300 miles (500 km) of explored caves, spread over six distinct levels ranging in age from Tertiary to Pleistocene.

Set on a broad north-dipping plateau underlain by Mississippian limestone, Mammoth Cave sits on a crossroad of carbonate bedrock stretching from Indiana to Georgia and from the Ozarks to the Cumberland Plateau.

Five of the 10 longest caves in the world are found in the United States, and Mammoth Cave is the longest of them all. Essentially a "dry" cave, Mammoth Cave is more famous for its massive caverns, underground streams, and towering vertical shafts rather than the intricate formations of shields, stalactites, and stalagmites common to the majority of cave systems.

HUMBLE BEGINNINGS

Some 350 million years ago, a warm shallow sea covered much of what is today the southeastern United States. For 70 million years, the warm waters of this sea supported dense populations of organisms, whose calcium-carbonate laden shells became embedded by the millions on the floor of the ocean, eventually depositing over 700 feet (200 m) of limestone and shale.

When the sea levels fell, the limestone became exposed. Tectonic forces acted on the limestone, causing it to buckle, twist, and crack. River systems also began to develop. The core area evolved into a dissected plateau—known as the Chester Upland—comprising a series of sandstone-capped ridges that protected the underlying caverns. This plateau began to rise above what is the present-day Green River, which would eventually erode a channel through the limestone base of the plateau, dissolving the limestone and carving out micro-caverns in the process.

Multiple layers of limestone provided varying paths for water flow. Vertical shafts began to intersect with other passages, adding to the cave's complexity. In time, a profusion of sinkholes developed, allowing water to quickly enter the limestone aquifers. Known today as the Sinkhole Plain, surface streams flow underground to eventually become the subterranean rivers for which Mammoth Cave is justly famous.

MAMMOTH CAVE'S HUMAN HISTORY

The history of human involvement with Mammoth Cave is a long and complex one. The cave's historic entrance was first surveyed in 1798 by Valentine Simons, a miner intent on exploiting the cave's extensive reserves of saltpeter—an ingredient used to make gunpowder.

Remains of Native Americans displaying signs of intentional burial, and dating to pre-Columbian times, have been recovered from Mammoth Cave, with carbon analysis detailing a record of human habitation that stretches back 6,000 years—from hunter-gatherer societies through to sedentary agricultural communities.

Mammoth Cave is the second oldest tourist attraction in America after Niagara Falls, with guided tours first being offered as early as 1816. Nearby, mining pits dating to 1812, and tuberculosis huts from the 1840s still stand near the cave's entrance.

ROCK FEATURE: LIMESTONE

Limestone is a common marine sedimentary rock made of calcium carbonate. It forms when precipitating calcite or microscopic fossils accumulate on the sea floor, or from cemented shell material, or even entire ancient coral reefs. Limestone is immediately identifiable as it fizzes when a drop of acid is placed on it. Limestone displays a variety of attractive colors and patterns, and is often polished and used for decorative purposes, such as tiles, bench tops, and building facades—its use as a building material dates back to ancient times. Limestone will metamorphose into marble when subjected to heat and pressure.

A TREASURE TROVE OF BIODIVERSITY

Mammoth Cave's labyrinth of caves and karsts contain the world's most diverse cave ecosystem, including more than 200 species of invertebrates, 42 of which have adapted to living in total darkness. Its age is such that communities from three distinct karst regions call it home, while its vastness even permits speciation to occur, including more than 40 species of mussels.

Some 350 feet (100 m) below the surface, the River Styx and Echo River—two of Mammoth Cave's underground rivers—are home to a menagerie of invertebrates and insects, including blind and un-pigmented fish and crayfish. Other inhabitants of this underground environment include cave-adapted crickets, ghostly white spiders, and sightless beetles.

Mammoth Cave was named a World Heritage Site in 1981, to preserve the cave system and the scenic river valleys of the Nolin and Green rivers. In 1990, the 53,000 surface acres (21,000 ha) of Mammoth Cave National Park and its underlying cave ecosystem were designated an International Biosphere Reserve.

A number of species have adapted to the environment of Mammoth Caves, such as the near-sighted cricket (top right) and the slimy salamander (bottom right).

CAVES OF YUCATÁN PENINSULA

The Yucatán Peninsula is a flat low-lying plain extending over 76,000 square miles (200,000 km²), taking in the Mexican states of Campeche, Quintana Roo, and Yucatán, and extending southward into northern Belize and northern Guatemala. It is bordered to the north and west by the Gulf of Mexico and by the Caribbean Sea to the east.

BELOW RIGHT Water has long driven the location of human settlement. As Yucatán's cenotes provide a plentiful source of fresh water, they have been a focal point for the establishment of cities and towns since the time of the Mayans.

Karst and caves are common across the peninsula, with a significant feature being the cenotes—from the Mayan word *dzonot*—vertical, subterranean, water-filled shafts and caves containing pools and underground passageways. With little surface water available, these cenotes have been a source of fresh water throughout the region since early times.

A GEOLOGIC OVERVIEW

The Yucatán Peninsula is a vast calcareous platform whose sediments and marine materials date from the late Cretaceous Period. Faults and fractures in its emergent limestone—created by surface tectonics in the Tertiary Period—are responsible for the present-day relief of the peninsula.

The extensive upper limestone outcrops in the north of the peninsula are known in Mayan language as *chaltuns* or "sharp hard stone," which sits over a softer less consolidated limestone known as *sahakab* or "white earth." The soils are generally shallow—the result of weathering of the emergent marine sediments on the limestone platform.

Elevations in the Yucatán region are meager. The topography is characterized by gently sloping hills and shallow valleys and plains over a karstic landscape of wells formed by the deep dissolution of limestone, where natural drainage is conducted primarily via a labyrinth of caves, caverns, and subterranean rivers.

CENOTES—PORTALS TO A MAYAN UNDERWORLD

The limestone of the Yucatán is only a few million years old— geologically young. The peninsula's sparse topsoil means carbon dioxide-laden rainwater has easily been able to penetrate pre-existing faults within the limestone, dissolving it and covering the

Caves and Karst of the Yucatán Peninsula

0 ____ 50 km
0 ____ 50 miles

Isla Contoy

YUCATÁN

Xlacach Cenote
Grutas de Tzabnah
Grutas de Calcehtok
Chelentun Cenote
Cenote Dzitnup o X'keken
Gruta de Balankanche
Nohoch Nah Chich

Gulf of Mexico

Grutas de Lol-tun
Grutas Xtacumbil-Xunan

Sistema Actun Chen
Sistema Ox Bel Ha
Sistema Sac Actun

Isla Cozumel

MEXICO

QUINTANA ROO

CAMPECHE

Caribbean Sea

Uaxactún

Turneffe Island

GUATEMALA   **BELIZE**

N

Lago Petén Itzá
Jobitzinaj   Actun Kan
St Herman's Cave
Chiquibul
Che Chem Hah

BEYOND YUCATÁN—SÓTANO DE LAS GOLONDRINAS

Discovered by cavers in 1967 and named Golondrinas after the Mexican name for the hundreds of thousands of swallows that nest in its cavern walls, Sótano de las Golondrinas, in the Mexican state of San Luis Potosí, is one of the world's most famous caves.

Deeper than the Empire State Building is tall, its entrance is 180 feet (55 m) in diameter and sited directly above the 270,000-square-foot (25,000 m²) floor that gives Golondrinas an inverted cone-like appearance. Its pit measures 1,300 feet (400 m) at its deepest point, and is covered in a thick matting of droppings from the multitude of resident birds. The droppings also combine with rainwater and surface debris to provide a food source for millipedes, snakes, insects, and scorpions.

Cavers willing to descend into Sótano de las Golondrinas—as seen at right—must do so with respirators to combat fungi and bacteria and to reduce the risk of contracting histoplasmosis, a respiratory disease caused by a fungus associated with bird droppings.

Yucatán with sinkholes and caverns roofed with extremely thin layers of surface limestone.

Cenotes are formed when these thin roofs eventually collapse due to erosion, leaving open water-filled holes that represent the Yucatán's sole reserves of fresh water in a flat landscape that is largely devoid of rivers and lakes.

There are over 3,000 cenotes scattered across the Yucatán Peninsula. Many are richly decorated in stalactites, stalagmites, waterfalls, soda straws, and pillars—formed 10,000 years ago during the last Ice Age, when retreating seawaters left behind dry caverns—where the dissolved salts of fresh rainwater began to recrystallize the cave's interiors.

Cenotes also played a significant role in the lives of the Mayan Indians. Ancient Mayan cities, such as Chichén Itzá, were built in the vicinity of cenotes, which not only provided sources of fresh water but also held significant spiritual meaning within the Mayan culture.

Cenotes were believed to be the abode of gods and ancestors, and portals to the Mayan underworld of Xibalba. Their perceived proximity to the divine also saw them used for various sacred rituals and ceremonies, and the discovery of skeletal remains located within cenotes have led some archeologists to speculate that they were also used for human sacrifice and burial.

TOP Yucatán's cenotes have been carved out by penetration and erosion of the underlying limestone by rainwater. The ingress of surface and underground water has filled many of the channels with subterranean rivers.

Karst and Caves of the Caribbean

The islands of the Caribbean are the products of intense geologic activity that has produced myriad caves, passageways, sinkholes, and karstic formations. The region is a karstic wonderland that is home to some of the largest and most diverse caves in the world.

Beneath the surface of many of the Caribbean islands, caves and caverns have been carved out, their surfaces bearing testimony to geological and human history—from the thousands of pictograms and wall paintings that adorn the El Pomier cave in the Dominican Republic to the crystallized limestone of Barbados's Harrison's Cave, and Jamaica's "Cockpit Country."

THE LIMESTONE FOUNDATIONS OF THE TURKS AND CAICOS ISLANDS

A southern extension of the Bahamian island chain, the Turks and Caicos Islands are remnants of continental debris, overlaid with limestone thousands of feet deep, which broke away from Pangea when North America separated from West Africa. Today the plateaus on which they sit rise over 8,000 feet (2,440 m) from the ocean floor below.

The islands are characterized by Aeolian carbonate hills and young unconsolidated dunes of reef-derived carbonate sand near the shoreline. Further inland older hills date to the Pleistocene age and have been lithified into hard Aeolian limestone.

The largest cave network on the Bahamian archipelago can be found on Middle Caicos, where shallow depressions in the limestone fill with salt water high in alkaline content that evaporates, leaving behind gypsum, sea salts, and calcium carbonate—the basis of the island's salt industry until well into the mid-twentieth century.

JAMAICA'S NETWORK OF CAVES

Jamaica possesses the highest concentration of caves per square mile of any place on Earth. Two-thirds of the island consist of a labyrinth of limestone caves that overlie vast deposits of

BELOW The relentless action of the ocean's waves and currents has carved out this sea cave on the island of Dominica, which is a member of the Windward Islands.

older metamorphic, igneous, and clastic sedimentary rocks. Over 1,200 caves, passageways, and sinkholes have so far been cataloged, though less than 5 percent of these have been the subject of any biospeleologic analysis.

The karstification of white and yellow limestone groups in the island's northwestern region—which is known as "Cockpit Country"—began 12 million years ago with the erosion of a faulted limestone plateau that led to the karst towers, deep

depressions, and gently rolling plains that we see today. The region is dominated by "conekarst," where domed hills enclose closed lobed depressions—known as dolines—that drain to nearby aquifers via sinkholes.

BERMUDA'S BEGINNINGS

To the north of the Caribbean islands, in the North Atlantic Ocean, lie the islands of Bermuda, and this island group—numbering around 180 islands—has its share of caves and karst.

One million years ago, when the sea levels around the island of Bermuda were over 410 feet (125 m) lower than they are today, the area of the then Bermuda Platform was 20 times the size it is today. As sea levels rose during this post-glacial era, the encroaching salt waters displaced large masses of fresh water, and Bermuda's complex network of caves and passageways—richly decorated with stalactites and stalagmites—began to form.

Bermuda's oldest Aeolian limestone formation is known as the Walsingham formation. Its limestone overlays a volcanic base of basalt and it possesses one of the greatest concentrations of subterranean caves and fissures in the world. Walsingham Pond is one of many sinkholes in the area, connected to the ocean through a maze of fissures in the limestone and home to over 30 species of sponges, as well as sea squirts and a diverse range of Bermudian algae.

High levels of nutrients are washed in from the surrounding land and provide detritus and other nutrients that support a wide variety of invertebrates, such as filter-feeders that feed off the pond's suspended organic matter.

LEFT Cave diving is one of the many leisure activities that visitors can enjoy in the Caribbean. This cave in the Bahamas is one of many water-filled caves that can be explored.

BELOW Tunicates (*Rhopaleae crassa*), or sea squirts, inhabit the sinkhole known as Walsingham Pond in Bermuda. These unique, usually hermaphroditic, marine creatures have a sac-like structure with two apertures.

BELOW Jamaica offers a wealth of caves for exploration. In ancient times, the local people used the underground chambers as living quarters. Nowadays, the caves act as a tourist attraction and a home to colonies of bats.

RIO CAMUY

The 268 acres (108 ha) of Río Camuy Cave Park represent the third-largest cave system in the western hemisphere, a 10-mile (16 km) long subtropical karst feature carved by the Camuy River in the moist subtropical forests of northern Puerto Rico.

Over 200 caves and 17 entry points have so far been mapped, including the vast 180-foot (55-m) high dry chamber of Clara de Empalme Cave. Experts believe as many as 800 caves are yet to be discovered.

Río Camuy also contains the massive Tres Pueblos Sinkhole. Measuring over 660 feet (200 m) across, viewing platforms around its rim allow views down 360 feet (110 m) to the subterranean Camuy River, the world's third largest underground river.

Karst and Caves of Central and Eastern Europe

With the classical karst in its south, karst making up 43 percent of the country, and 50 percent of its population depending on karst aquifers for their drinking water, Slovenia is the home of karst and caves.

Before the railway linked Vienna to the port city of Trieste in 1857, travelers dreaded the last section of their journey across the barren, rocky, and almost waterless expanse of the karst. They would be surprised to return today to southern Slovenia and see healthy young forest hiding much of the gray rocky ground that caused them so much pain.

Tourists best know the karst for the cave with the train at Postojna and the massive river cave in the Skocjanske World Heritage area. Virtually every type of non-tropical karst landscape—dry valleys, karst plateaus, dolines, giant collapse dolines, caves, unroofed caves, polje, hums, karren fields, stream sinks, and springs—can be found in Slovenia, so scientists use it as a reference area.

Karst landscapes extend through the entire length of the Balkans, including the Dalmatian Coast, and through Greece and Italy. Their primeval forests succumbed to ancient Greek and Roman demands for timber warships, feudal arrangements that effectively taxed trees, and grazing by sheep and goats.

Born in the Mesozoic Tethys Sea, Triassic to Paleocene limestones occur not only in the south, but also throughout Europe. Relicts from the Mesozoic include dinosaur fossils found in ancient filled caves in Slovenia and in the limestone sea floor off the Croatian coast. Great earth movements pushed some of these limestones up to form the pointy white mountains typical of the Carpathians and Alps. Deep and complex alpine caves, such as Dachstein-Mammuthöhle in Austria, occur in these mountains and the associated limestone plateaus.

Older Paleozoic limestones occur in the Bohemian and Brno Massifs of the Czech Republic. Koneprusy (Konepruske) Caves are located in the Bohemian Massif, south of Prague. This area has one of the longest histories of cave development in Europe, with some caves dating back 70 million years. Extensive caves and deep shafts—such as the Macocha Propast (Abyss)—formed in the Moravian karst of the Brno Massif. Tourists

While almost all speleothems are made of calcite, less common and some rare minerals occur in caves. Ochtinská Aragonite Cave in Slovakia contains spectacular deposits of aragonite. The aragonite occurs as shiny needles, spirals, or spiky clusters growing like flowers from the blue limestone walls of the cave's Milky Way Hall. Aragonite has the same composition as calcite, but this rarer polymorph crystallizes in the orthorhombic crystal system. Aragonite stalactites also occur in the Carlsbad Caverns, New Mexico, USA.

Green malachite stalactites occur in caves in the Austrian Alps, the Democratic Republic of the Congo, and Arizona, USA. Malachite is a carbonate of copper and its attractive light to dark green concentric growth bands make it popular as an ornamental gemstone – resulting in many of these cave formations being taken. Together with its blue counterpart, azurite, malachite was once an important ore of copper.

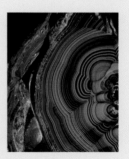

Distinctive banding in shades of green can be seen in this sample of malachite (top). Aragonite often features long needle-like crystals and is named for its place of discovery, Aragon in Spain.

can enjoy the underground boat trip in Punkevni Cave and the large open spaces of Katerinska Cave.

THERMAL CAVES

While sinking surface water formed most caves in Europe, a number of caves in Central Europe were formed by rising thermal water.

The caves of the Buda Hills (Budai-Hegseg), some under residential streets and the castle in Budapest, Hungary, are probably the best known. They consist of mazes and dome-shaped chambers and contain dramatic crystal deposits. Szemlóhegy

and Pálvögyi caves feature large deposits of crystal rafts, cave coral, and complex wall sculpture.

Bubbling mineral water, releasing carbon dioxide, carved out Zbrasov Aragonite (Zbrasovske Aragonitove) Cave and the adjacent deep pit, Hranická Propast, in the western Czech Republic. A computer-controlled pump and sensors keep the atmosphere in the cave at safe levels for visiting tourists. Thermal caves also occur near Lake Baliton in Hungary and near Krakow in Poland.

Exploratory drilling for a mineral deposit in the Rhodope Mountains (Rodopi Planina) of Bulgaria intersected what may prove to be the world's largest single cavern. This enormous cavern measures at least 4,400 feet (1,340 m) in height and has an estimated volume of 8,390 cubic feet (237.6 m^3). The cavity is full of mineral water.

Movile Cave in Romania is another very unusual hot water cave. Air in the cave contains methane and hydrogen sulfide and apart from the top millimeter, the water is anoxic. Life in Movile Cave depends on energy extracted by bacteria oxidizing hydrogen sulfide, not energy from the sun. Movile Cave is the first complex terrestrial environment known to be independent of energy from the Sun.

ICE CAVES

With the right geometry and climatic conditions, ice can build up in limestone caves. Well-known examples are Dachstein-Rieseneishöhle and Eisriesenwelt in the Austrian Alps. The largest body

LEFT Baradla Cave in Aggtelek National Park in Hungary's north is the venue for this musical performance. The cave is a popular venue for musical concerts, offering superb acoustics and an amazing backdrop of stalactites.

GERMANY

P O L A N D

Drachenhohle Syrau

Zbrasovske Aragonitove
Jeskyne

Bozkovske Dolomitove Jeskyne

Jaskinia Niedzwiedzia

Jaskinia Raj

Jaskinia
Lokietka

Jeskyne Na Spicaku

Jaskinia Nietoperzowa

Konepruske Jeskyne

Jeskyne Na Pomezi

Smocza Jama

Optimisticheskaya Pestera

CZECH REPUBLIC

Kri
Pes

Javorícske Jeskyne

Mladecske Jeskyne

Ozernaja Pestera

Konig-Otto-
Tropfsteinhohle

Ve

Katerinska
Jeskyne

Chynovske Jeskyne

Belianska Jaskyna

Dniestr (Dniester)

Grosses Schulerloch

Punkevní
Jeskyne

Sloupsko-Sosuvske Jeskyne

Dobsinska Ladova Jaskyna

Jeskyne Balcarka

Harmanecka
Jaskyna

Demanovska Jaskyna Slobody

Donau (Danube)

Jeskyne na
Turoldu

Bojnicka Hradna
Jeskyne

Jasovska Jaskyna

Ochtinska
Aragonitova
Jaskyna

Jaskyna Domica

Jaskyna Driny

Inn

SLOVAKIA

Rakoczi-barlang

Morava

Eisriesenwelt

Hochkarschacht

Nixhohle

Allander
Tropfsteinhohle

Dunaj (Danube)

Szt. Istvan Barlang

Gassel Tropfsteinhohle

Hermannshohle

Tisza

Lamprechtsofen

Dachstein-
Rieseneishohle

Kraushohle

Neusiedler
See

Budai-Hegyseg
(Buda Hills)

Szemlohegy- barlang

Dachstein-
Mammuthohle

Enns

Mur

Lurgrotte Semriach
Grassihohle

Palvolgyi- barlang

AUSTRIA

Drau

Kateiloch

Loczy-
barlang

Griffener Tropfsteinhohle

Tapolcai-Tavasbarlang

HUNGARY

Obir Tropfsteinhohlen

Balaton

Duna (Danube)

Grotta Nuova

SLOVENIA

Abaligeti-barlang

Pestera Meziad
Pestera Urslior

Mures

Grotta di Oliero

Predjamska Jama
Skocjanske Jame

Postojnske
Jame

Spilja Veternica

Drava

Grotta Gigante

Spilja Vrlovka

ROMANIA

Golfo di
Trieste

Po

Golfo di
Venezia

Spilja Lokvarka

Kupa

CROATIA

Krk

Spilja
Biserujka

Pestera

Reno

Cres

Sava

Dunar (Danube)

Pestera Mujerilor

Appennino

Golubnjaca
Pecina

Dunarea (Dan

Arno

Samograd Spilja

Risovaca Pecina

Rajkova Pecina

Manita Pecina

Cerovacke
Spilje

Petnicka Pecina

Resavska Pecina

Grotta dell'Onferno

Pecina Bijambare

Zlotske
Pecina

Grotta Grande del Vento

Lago
Trasimeno

BOSNIA &
HERZEGOVINA

Stopica Pecina

Pestera Magura

Spilja Vranjaca

SERBIA

Tevere (Tiber)

Lago di
Bolsena

izvor Buna

Stara Planina

Pestera Ledenika

Modra Spilja

Pecina Vjetrenica

MONTENEGRO

Temnata Dupka

Sueva Dupka

BULGAR

La Grotta di Stiffe

Bisevo

Hvar

Skadarsko
Jezero

Mermerna
Pecina

Bacho Kiro

Grotta di
Beatrice Cenci

Grotta del Cavallone

Adriatic Sea

Shpella e Gjolave

Maritsa

Lago di
Bracciano

Drin

Snezhanka Pestera

Grotta di Pastena

Promontorio
del Gargano

Vrelo Spilja

Rodopi Planina (Rhodope)

Dyavolsko Garlo

ITALY

MACEDONIA

Iagodinska
Pestera

Uhlovitsa Pestera

Isola di
Capri

Grotta dello Smeraldo

Grotta di Putignano

Grotta di Castellana

Obridsko
Jezero

Pesterna Crkva Sveti Erazmo

Spilaio Alistratis

Grotta Azzurra

Grotta dell'Angelo

ALBANIA

Pesterna Crkva Sveti Stefan

Spilaio Agios Georgios

Presko-Planina (Pindus)

Strait of
Otranto

Prespansko
Jezero

GREECE

Spilaio Petralona

Samothraki

Grotta delle Meraviglie

Golfo di
Taranto

Grotta Zinzulusa

Kerkira

Aegean Sea

Tyrrhenian Sea

Lesvo

Shpella e Keramoit

Evvaia

Strait of Sicily

Sicilia

Kefallinia
(Kephalonia)

Limni
Trikhonis

Ionioi Nisoi

Katavothres

Limni Melissani
Spilaio Drogkarati

Spilaio Peramatas

Ik

Spilaio Glifada

PELOPONNISOS

Kikladhes

Ionian Sea

Milos

MEDITERRANEAN SEA

Spilaio Andiparo
Andiparos

Malta Channel

Thira

MALTA

**Caves and Karst of
Central and Eastern Europe**

N

0 _____ 250 km
0 _____ 250 miles

BELARUS

RUSSIAN FEDERATION

U K R A I N E

Dnepr (Dnieper)

MOLDOVA

Don

*Azovskoye More
(Sea of Azov)*

Emine Bair Khasar   Mramornaja Pestera

Pestera Movile

Bol'shoy Kavkaz (Caucasus)
Voronya
(Krubera)
Pestera

Black Sea

GEORGIA

Karaca Magarasi

Kelkit

Marmara Denizi

Ayvaini Magarasi   Iznik
Gölü   *Sakarya*

T U R K E Y

Kizil Irmak

Murat

Gediz

Tuz Gölü

Seyhan

Dicle (Tigris)

Inkaya
Magarasi

Aslanli Magarasi   Pinargozu
Magarasi   *Höyran Gölü*

Büyük Menderes   Pamukkale   Insuyu Magarasi

Firat (Euphrates)

Beysehir Gölü

Karain Magarasi   Zeytintasi
Magarasi   Altinbesik Magarasi   Hislayik Magarasi

Spilaio Epta Parthena

Papazkayasi Magarasi

Molla Deligi Magarasi   Damlatas
Magarasi   *Toros Daglari (Taurus)*   Cennet ve Cehennem Cokukleri

Rodhos   SYRIA

ABOVE Explored to a depth of 7,080 feet (2,158 m), Krubera Cave in Abkhazia holds the record as the deepest cave in the world. The pit known as Big Cascade, pictured here, is the largest pit within the cave.

OPPOSITE PAGE The spectacular Plitvice Lakes National Park in Croatia features a karst river basin etched with natural dams that have been created by the deposition of calcium carbonate.

RIGHT Slovenia's historic Predjama Castle is built into the side of a cliff. Beneath the castle lies a cave system, which has a number of features—including the Passage of Names, where the names of visitors dating back to the sixteenth century can be seen inscribed on the walls.

of ice in a limestone cave has a volume of 3,889,275 cubic feet (110,132 m³)—it is not found high in the Alps, but in the World Heritage-listed Dobsinská (Ľadova) Ice Cave in Slovakia at an elevation of only 3,180 feet (969 m).

THE REGION'S LARGEST CAVES

The longest caves in Europe are not in limestone but in gypsum in the Dniester (Dnestr) River region of western Ukraine. These are giant maze caves formed by artesian water rising through joints (cracks) in a horizontal layer of gypsum. Optimisticheskaya Cave, the longest of these caves, is over 186 miles (300 km) long—making it the world's second longest cave.

The search for the world's deepest cave has passed through many places, including the highlands of New Guinea, the Pyrenees, and Mexico. Ukrainian and Russian explorers discovered the first caves deeper than 1¼ miles (2 km) in the Arabika Massif of the Caucasus Mountains in Abkhazia. By September 2006, they had penetrated Krubera Cave to a depth of 7,080 feet (2,158 m).

CASTLES AND CAVES

Caves and karst features have proved useful for castle builders. In Poland, ancient coral atolls have provided the foundations for castles such as Podzamcze. There, the solid limestone rim forms the walls and the lagoon forms the keep.

The most striking use of a cave for a castle is at Predjama (meaning "in front of the cave") in Slovenia. A fifteenth-century castle, Predjama Grad is constructed partly inside a large cave entrance. The cave provided water and a secret entrance for

CAVE BEARS AND NEANDERTHALS

Caves provided shelter for people and wild animals, particularly during the Ice Ages. Cave bears (*Ursus speleaus*) lived deep in caves, navigating through the darkness by scent. Large accumulations of juvenile bones suggest that the infant mortality of cave bears was very high.

Neanderthal cave finds include the original sites in the Neander Valley; the Neanderthal flute discovery from Slovenia, which challenged established ideas about Neanderthal culture; and recently, a Neanderthal footprint from a Romanian cave, dated to between 97 and 64 thousand years.

supplies during sieges. An older castle in the cave was home to the "robber baron" Erasmus, whose siege—and death by cannon fire—in 1484 has become legendary.

KARST LAKES

Flat-floored valleys surrounded by steep hills, called polje, are important features of southern European karsts. These valleys, for example the polje of Peloponesus (Peloponnisos) in Greece, are often the most important agricultural areas in the karst. Polje often flood, forming seasonal lakes. One of the most celebrated is Lake Cerknica in Slovenia. Round residual limestone hills, known as hums, sometimes stand up in the flat polje floors.

Karst lakes can also form by tufa dams growing in streams. Some of the most impressive tufa dam lakes are the World Heritage-listed Plitvice Lakes in Croatia. This is a magical place of mirror-still lakes, tumbling waterfalls, and green tufa barriers. Sixteen lakes have formed along the course of the Korana River. The largest, Kozjak Lake (Kozjak Jezero), is 1¼ miles (2.5 km) long, and 200 feet (67 m) wide at its widest point.

KARST SPRINGS AND STREAM SINKS

While the surface of karst areas is often dry and waterless, large quantities of water lie below. Karst springs are valuable sources of water and provide much of the water supply for towns and villages in the karst, but they are not restricted to dry land. Fresh water springs in the sea occur along the limestone coasts of the Balkans, Greece, and Italy. Some of these springs have large enough zones of fresh water around them to enable pumping of fresh water from the sea. Streams, and sometimes rivers, sink through holes in the ground into caves. These holes are called stream sinks, or ponors. Ponors may be vertical shafts or horizontal cave entrances. In some places, when hydrological conditions

change, water comes out of holes that normally take in water. Two-way holes are called estavelles.

The strangest ponors and springs are located on the Greek island of Kefalonia, where seawater sinks underground on the island's southern shore and emerges, slightly diluted with fresh water, above sea level on the northern shore. Water wheels—the sea mills of Agastoli (Katavothres)—are turned by the seawater sinking through the limestone and also as it flows out to sea on the other side.

LEFT The Greek island of Kefalonia boasts some unusual caves, such as Melissani Cave—known as the Cave of the Nymphs. Brackish water flows through the cave, and sunlight on the water floods the cave with an ethereal blue light.

Karst and Caves of Asia

Asia's rich legacy of limestone karsts is one of its most recognizable natural features, the scattered remnants of an ancient sea that once stretched from Guilin in western China south through Vietnam, Thailand, and on to Borneo and the Philippines during the Permian Period 250 million years ago.

ABOVE The spectacular karst topography of China's Guilin region is laced with waterways and surrounded by lush landscape. The unusual formations bear names that reflect the shapes they resemble, such as Elephant Trunk Hill.

Throughout South-East Asia karst landscapes cover an area measuring some 154,440 square miles (400,000 km²), ranging in age from the Cambrian to the Quaternary. Along the fragmented Sunda Shelf, the difficult karstic terrain has spared the land from agricultural development, instead preserving the natural environment, and presenting an opportunity for biological and ecological research.

THAILAND'S ANCIENT COASTAL KARSTS

Stretching from Tarutao Marine National Park near the Malaysian border in the south to Chiang Mai in the mountainous north, Thailand's spectacular array of limestone karsts were formed when the Indian subcontinent collided with mainland Asia 30 million years ago, rolling southern Thailand and the Malay Peninsula clockwise and rupturing an ancient coral reef that ran for 3,100 miles (5,000 km) from the Philippines to mainland China.

Karst intrusions are most prominent in the southern Thailand provinces of Krabi and Phang Nga. Offshore karst towers emerge from mangrove-fringed tidal flats and the shallow waters of Phang Nga Bay, with Quaternary alluvial and colluvial deposits surrounding inland karst towers.

The coastal tower karsts of Krabi developed in dolomitic and Permian limestone and appear as "peak forest karsts" and "peak cluster karsts." Peak forest karsts are a mix of isolated, mostly tall, cylindrical towers called turm karsts, and moderately steep-sided kegel karsts with elevations ranging from 200 to 700 feet (60 to 210 m) above sea level. Peak cluster karsts range from 800 to 1,300 feet (240 to 400 m) above sea level and exhibit both cone and cylindrical-shaped peaks set on broad limestone plates, with vertical cliff faces along their margins.

Monsoonal rains and persistent wave erosion of the soft sedimentary layers of the karst towers have cut cave chambers and passageways that combine with sparse, dry, limestone-based soils to support a diverse flora and fauna. When the roofs of these

cave chambers collapse they form "hongs"—one of the most intriguing features of karst topography—forming either lagoons or forests depending on whether the cave floor was above or below sea level.

Karst vegetation can withstand extended periods of dryness and an almost soil-free environment. Decomposing leaves create mildly acidic water that further erodes the limestone's calcium carbonate as it penetrates the karst. The leaves also penetrate rock crevasses, providing nutrients for plants such as cycads and bonsai-like palms. Thailand's karsts are also home to several species of figs that provide an abundance of seasonal fruits for animals such as long-tailed macaques. Larger karst systems with as much as 124 acres (50 ha) of continuous forest are known to support gibbons, while smaller karst areas support leaf monkeys whose leafy diets allow them to survive in small range habitats.

VIETNAM'S "BAY OF THE DESCENDING DRAGON"

According to one legend, long ago when Vietnam was under siege from a great enemy, a dragon came down from the hills and came to its aid. Using its tail, the dragon lashed and split apart entire mountain ranges, impeding the advance of the enemy and creating "Vinh Halong"—the almost 2,000 dolomite islets that comprise Halong Bay—Vietnam's "Bay of the Descending Dragon."

Designated a World Heritage Site in 1994, this wonderland of karst topography in Vietnam's northeast is bordered to the west by the Red River delta and to the southwest by the island of Cat Ba. Covering a total area of 600 square miles (1,553 km²), Halong Bay's shallow waters contain 1,969 islands.

Halong Bay's seafloor is submerged karst plains covered mainly in clastic sediments—its islands emerging from a fine-grained Carboniferous and Permian limestone over 3,300 feet (1,000 m) thick—ideal for the development of karst. The bay's geologic substructure is complex, with drowned fault-guided valleys between the islands and bedding ranging from horizontal in its eastern region to small overfolds in the western part of the bay.

Halong Bay's islands are a vast mix of individual karsts and karst clusters. Cluster towers can reach heights of up to 650 feet (200 m), possessing steep cone profiles, minimal lateral undercutting, and few vertical cliffs. The majority of the individual towers reach heights of between 165 and 330 feet (50 and 100 m) with vertical walls present on all or most of their faces.

ABOVE Plants and fruit that flourish in Halong Bay's karst environment provide a food supply for a number of species, including the communities of long-tailed macaques, which rely on fruit for more than 50 percent of their diet.

ABOVE LEFT The breathtaking coastline of Krabi—on Thailand's western coast—features towering karst overlooking the waters of the Andaman Sea.

BELOW Vietnam's Halong Bay is bejewelled with almost 2,000 islands. The waters of the bay are plied by local fishermen as they seek out their catch.

Lateral undercutting of many karsts is evident in a marine "notch" cut into the islands' limestone 10 feet (3 m) above the level between high and low tide. Caused perhaps by the mixing of sea-water and fresh water within the islands' fissures, Halong Bay's well-developed marine notches represent the primary cause of the erosion and retreat of its limestone cliff faces.

The limestone surfaces of Halong Bay's karsts consist largely of jagged open fissures, pinnacles, ridges, and remnant rocks fretted by dissolution that combine to create an inhospitable terrain.

An abundance of lakes exist on many of Halong Bay's limestone islands. Mostly occupying drowned dolines within the karst and known as "hongs" or "rooms," they are mostly tidal, with seawater moving freely back and forth via sea-level caves or inaccessible networks of fissures.

The rugged limestone terrain of Halong Bay's karst towers is home to a wide variety of unique plants and wildlife. A World Bank biological expedition in 1988 found more than 80 previously unknown species, including cave-adapted spiders, millipedes, woodlice, and over 17 species of snails, most of which were endemic to the karsts—making Halong Bay by far the richest habitat for cave snails in South-East Asia.

Extreme fragmentation of the Halong Bay karst may even have led to early speciation, with remarkable similarities existing between closely-related forms of spiders, millipedes, and springtails collected in different limestone units. Botanical surveys have found several species of African violets, previously unknown orchids, *Impatiens*, and even a species of ginger.

PUERTO-PRINCESA SUBTERRANEAN RIVER NATIONAL PARK

The World Heritage Site of Puerto-Princesa Subterranean River National Park is found on the Philippine island of Palawan. Also known as St Pauls Subterranean River National Park, 90 percent of the karst landscape of the St Paul Mountain Range is comprised of sharp karst limestone ridges within a series of eroded limestone peaks running down the island's west coast.

The park's primary feature, however, lies hidden beneath the surface. A colossal subterranean river—the world's longest navigable underground river—flows for more than 5 miles (8 km), rising 3 miles (5 km) southwest of Mt St Paul before emptying into the sea at St Paul's Bay. The river runs past a profusion of stalactites, stalagmites, and several large chambers up to 400 feet (120 m) wide and 210 feet (65 m) in height, with the river's lower portions subject to intense tidal influences.

Puerto-Princesa's tunnels and myriad chambers are home to an abundant population of swiftlets and several species of bats.

ASIA'S LIMESTONE KARSTS UNDER THREAT

Only 13 percent of Asia's karst topography enjoys the luxury of protected status in a region where limestone quarrying is increasing at a rate of almost 6 percent per year. Not even national parks are spared the issuing of limestone concessions, with pressure to mine for limestone used in the production of cement putting these fragile ecosystems and their inhabitants under a great and continuing risk.

More than 30 karst-dwelling species are currently listed by the World Conservation Union as threatened, including the blind cave fish, various mollusks, and rare bats. Animals and plants are also under-represented in lists of endangered species due to data deficiencies and the difficulties inherent in performing biodiversity studies on such difficult rugged terrain.

Since 1988, over 13 new species of fish have been discovered in karst caves throughout South-East Asia, including a blind loach that uses its fins to clamber up the sides of rocks. Sadly, most cave-dwelling karst species such as fish, crabs and crayfish, and various species of karst organisms have not even been fully cataloged, let alone been the subject of any organized programs of threat analysis.

ABOVE Puerto-Princesa Subterranean River National Park features an underground river system within a karst landscape of mountains on the surface and caves below ground.

OPPOSITE Looming out of the ocean, the karst islands of Thailand's Halong Bay—often featuring steep vertical faces—are largely uninhabited.

CAVES OF BORNEO

One of the most extensive and spectacular cave systems on Earth lies beneath the massively bedded Melinau and Subis limestone of Gunung Mulu and Niah National Parks in northern Sarawak, on the island of Borneo.

RIGHT As dusk falls, the skies of Gunung Mulu National Park are filled with nocturnal hunters, as the millions of bats that reside within the cool passages of the caves by day, set off on their nightly hunt for food.

BELOW Limestone "pinnacles" form a contrast to the lush vegetation of Gunung Mulu National Park, as the jagged rock pierces the canopy of the surrounding rainforest.

Gunung Mulu National Park contains the world's largest cave passage, the world's largest river cave passage, and the world's largest cave chamber. The West Mouth of Niah Cave is the world's largest cave entrance, as well as being the site of the oldest known fossilized remains of *Homo sapiens* so far found in South-East Asia, including the famous Niah "Deep Skull."

THE WONDERS OF GUNUNG MULU

Gunung Mulu's Deer Cave is the longest natural cave passage in the world, and measures from 395 to 495 feet (120 to 150 m) in diameter—the product of 5 million years of geological uplifts and the erosion of karst phenomena by local rivers.

Residents of Deer Cave include hairy earwigs that feed on the oils produced by naked bats to protect their skin. Various species of white crab are common, as are terrapins, turtles, flat worms, and the occasional catfish.

After heavy rains, 600-foot (180-m) waterfalls cascade from Deer Cave's roof, and each night over two million wrinkle-lipped bats exit the cave to feast on Gunung Mulu's rich insect population.

Clearwater Cave has been measured at over 60 miles (97 km) in length—making it South-East Asia's longest river cave passage, and the seventh longest in the world. It includes superb examples of fossil passages, curtains, flowstone, and limestone needles set in the midst of a complex labyrinth of side passages.

The Melinau River and its various tributaries bisect Clearwater Cave, and when in flood, some 5,300,000 cubic feet (150,000 m³) of water is carried through the cave. The river's complex alluvial history has defined the formation of Clearwater Cave and adjacent Green Cave, while the river's moisture provides a habitat for some unique plant life including rare orchids, pitcher plants, and photo-sensitive algae, with needle-like formations forever oriented toward the cave's filtered light. Clearwater Cave is also home to the largest recorded windblown stalactite, measuring over 40 inches (1 m)

ABOVE Glow worms—which are really an insect species, rather than a type of worm—are among the inhabitants of Gunung Mulu's caves. They capture their prey—lured by the "glow"—in these sticky silk threads that hang from the cave ceiling.

in height, with stromatolites a common feature at its various cave entrances.

Discovered in 1980 by a small group of English cavers, Sarawak Chamber is the world's largest natural chamber, measuring 2,300 feet (700 m) long, 230 feet (70 m) in height, and up to 1,300 feet (395 m) wide. Measuring some three times the size of New Mexico's Carlsbad Cavern—previously believed to be the world's largest cave chamber—the Sarawak Chamber contains exceptional decorative speleothems replete with aragonite and calcite needles.

NIAH CAVE

The Great Cave of Niah in Niah National Park measures over 25 acres (10 ha) in area, with a history of human habitation stretching back to the late Pleistocene Epoch, 40,000 years ago.

Embedded in 20 million-year-old Subis limestone, Niah Cave is the site of a series of archeological discoveries by Tom Harrisson, curator of the Sarawak Museum in the late 1950s and early 1960s. Over four years, Harrisson and his team gradually uncovered one of the largest stratified cave sequences ever found in South-East Asia, culminating in the discovery of the so-called "Deep Skull," uncovered in a trench near the cave's entrance. Found among a collection of stone implements embedded in charcoal, it remains the earliest evidence of human settlement yet found in Borneo. Pottery used by Mesolithic foragers from the Holocene period—5,000 to 10,000 years ago—as funerary gifts and burial jars were also found.

A passage at the rear of Niah's Great Cave leads to Painted Cave, an ancient burial site whose wooden boat coffins—dating from 0 to 780 BCE—can now be seen in the Sarawak Museum.

Over one hundred paintings, up to 2,000 years old and rendered in red hematite, adorn the rear wall of the cave, depicting spread-eagled human figures representing warriors and hunters as well as snails, snakes, and other animals engaged in various funerary rituals.

ABOVE Serving as a window into Prehistoric times, Niah Cave has revealed evidence of human habitation—some of the earliest recorded—with bones, artifacts, and artwork contained within its interior.

Karst and Caves of Australia

The karst and caves of Australia reflect the unique history of the continent, the last Gondwana fragment to separate from Antarctica. Unlike the northern continents, there are few Mesozoic carbonate rocks in Australia and no high white limestone mountains. The last major folding was in the Carboniferous period and the last major glaciation (except for Tasmania) was in the Permian period.

RIGHT Known as the "City of Craters, Lakes, Caves and Sinkholes," Mount Gambier in South Australia is located on the Limestone Coast. Beneath the city streets lies a subterranean network of caves.

ABOVE Lying some 330 feet (100 m) below the Nullarbor Plain, Cocklebiddy Caves include some of the longest underwater caves in the world.

Australia's terrain is largely flat, generally below 984 feet (300 m) in elevation, and the climate has been arid for much of the last 15 million years. Relief and water, which are two of the driving forces of karst, are in short supply. Due to the combination of low relief and low rainfall, caves—rather than surface landforms—are the defining features of Australian karst.

Australia's deepest caves are found in Tasmania, while Australia's longest cave, Bullita Cave (more than 62 miles, or 100 km long) is located in the Gregory karst of the Northern Territory.

THE NULLARBOR

The Nullarbor Plain is the largest karst area in Australia and one of the largest in the world. The plain stretches across the south of the continent, covering an area of approximately 77,220 square miles (200,000 km²). The flat arid surface of the plain gives no hint that it is a karst landscape until, almost without warning, the ground opens up into a collapsed doline—the entry to the world below.

Large caves with lakes lie below the plain. In the past, Aboriginal people and cattle farmers have used them as water sources, but today the lakes attract cave divers. In Cocklebiddy Cave, divers have penetrated 21,980 feet (6.7 km) underwater.

AUSTRALIA'S LIMESTONE COAST REGION

The Limestone Coast area of western Victoria and South Australia is the Australian equivalent of Central America's Yucatan Peninsula. This large flat area, underlain by Tertiary limestone, is peppered with cenotes. Groundwater flows strongly through the limestone to coastal springs such as Ewen's Ponds and Piccaninnie Ponds. This is one of Australia's most important wine-growing regions, utilizing groundwater from the limestone for irrigation.

A ridge along the northern edge of the karst contains the World Heritage-listed Naracoorte Caves, which contain extensive deposits of Pleistocene Epoch fossils.

Limestone cliffs along the south coast of Victoria are eroded into ravines, bridges, and stacks—forming popular tourist attractions along the coastline, such as the Twelve Apostles.

EASTERN AUSTRALIAN HIGHLANDS

Over 150 karsts, mostly small, have developed in the Paleozoic limestones of the Tasman Fold Belt, extending through the highlands of eastern Australia from Tasmania to North Queensland. In Tasmania, karst is influenced by higher relief, higher rainfall, and the effects of Pleistocene glaciation, while north from Yessabah in New South Wales, the distinct tropical effect has culminated in the development of tower karst at Mitchell-Palmer and Chillagoe. These karsts include tourist localities such as Buchan, Jenolan, Mole Creek, Wellington, and Wombeyan, which are renowned for their impressive displays of speleothems. These karsts are of considerable scientific interest as they contain some of the world's oldest and most complex caves.

WEST KIMBERLEY LIMESTONE RANGES

In the Canning Basin of northwestern Western Australia, fossil coral reefs from the Devonian period are preserved. Caves, natural bridges, and small karst towers feature in this remote area that is now attracting significant numbers of tourists.

PLEISTOCENE DUNE KARSTS

Pleistocene dune limestones occur along the southern and western coastline of Australia. These limestones are highly karstified, with the development of extensive caves and surface karst features.

The caves, such as those in the Margaret River area of Western Australia, contain abundant speleothems, and particularly large numbers of straws.

FOSSILS IN THE CAVES

Local karsts contain fossils of Australia's unique Pleistocene megafauna and of marsupial history dating back into the Tertiary. In 1830, Pleistocene marsupial fossils were discovered at Wellington; significant finds have been made since at Naracoorte and in caves in the Nullarbor region. Much older marsupial fossils occur in caves, paleokarst deposits, and the Tertiary limestone bedrock at Riversleigh in Queensland.

ABOVE Western Australia's Margaret River region boasts a wealth of caves, though few are open to the public. Lake Cave—one of the accessible caves—features limestone formations and an underground lake.

BELOW The limestone cliffs of the Head of Bight edge the Nullarbor Plain. Southern right whales (*Eubaleana australis*) come to the waters here to breed during the southern hemisphere winter.

JENOLAN CAVES

Around 250,000 tourists gasp in astonishment on their way through Jenolan Caves each year, entranced by the stalactites, stalagmites, draperies, and helictites that sparkle in the darkness. While most have some idea that Australian landforms are old, few will realize they are walking through one of the world's oldest and most complex limestone caves.

Jenolan Caves ramify through a narrow band of Silurian Period limestone only 820 feet (250 m) wide and 5.5 miles (9 km) long. The show caves are part of an integrated system that extends for just over ½ mile (1 km) south and for at least 2 miles (3 km) to the north. Cavers continue to find new sections and connections, but the going is hard due to the complex arrangement of the caves.

CAVE LEGENDS AND LOCAL LORE

A narrow winding road from the plateau leads down to Jenolan Caves. The first hairpin bend has Mt Inspiration Lookout, renamed from "perspiration point," where seven-seater tourist cars often lost their brakes in the 1920s. Around the final bend, a wall of gray limestone marks where the road passes through the Grand Archway—a natural cave. Local Aborigines recognized this as one of the holes burrowed by the dreamtime being, Gurangatch. Before the road was built, it took two days to walk the Six Foot Track from Katoomba. Late-nineteenth-century travelers could find their way by following the discarded stalactites. At Jenolan, the stalactites and stalagmites are very close to the path. In 1880, Jeremiah Wilson, the first Keeper of the Caves, began installing chicken wire to protect the fragile speleothems. While there is less wire used today, wire working remains an essential skill for the cave staff.

INSIDE THE CAVES

Caves extend from both sides of the Grand Archway. South of the Archway, there are two groups of dome-shaped solution chambers, called cupolas, connected by large tube-shaped passages that rise and fall in loops and a large breakdown chamber. There are six show caves in the south, Lucas Cave, River Cave, The Pool of Cerberus, The Temple of Baal, Orient Cave, and Ribbon Cave. Lucas Cave includes the Cathedral, a high solution chamber where concerts, weddings, and secret Masonic rituals occur, and the Exhibition Chamber, a large breakdown chamber with the iconic Broken Column. A thin coating of apatite, from bat guano, makes the Broken Column stand out by fluorescing. The River Cave has large still ponds, not a flowing stream. Tourists once negotiated The Pool of Reflections in a punt. River Cave is renowned for two white stalagmites—the Minaret and the Grand Column. The Temple of Baal consists of large cupolas, joined together. Draperies, angel wings, and masses of helictites hang from the walls. A small mass of white clay in the far corner is some of the world's oldest cave sediment, dated at 340 million years. The Orient Cave has high cupolas with draperies and distinctive red flowstone. The smaller Ribbon and Pool of Cerberus caves contain exquisite aragonite helictites.

The northern caves are straighter and more level than the southern caves. Their floors rest on 66 feet (20 m) of sediment deposited by the Jenolan Underground River. Tourists today walk close to the ancient cave ceiling. There are three show caves in the north—Imperial Cave, Chifley Cave, and Jubilee Cave. Imperial Cave runs north along the ancient streamway. White stalactites, stalagmites, and draperies decorate the walls, ceiling, and floors beside the path. Steep stairs lead down to the Jenolan Underground River, a small burbling stream, out of scale with the large passages above. Chifley Cave has a high passage with striking white speleothems, a chamber, Katie's Bower, and a narrow loop through crystal leading back to the Grand Archway. Jubilee Cave takes visitors 1¼ miles (2 km) north from the Grand Archway. This cave shows striking contrasts between fragile white speleothems and the mud and gravel from which these formations grow.

ABOVE A walkway along the banks of the Jenolan River allows visitors to explore the impressive karst terrain from outside the cave system. This river flows underground through the caves in a series of pools and streams.

RIGHT An eastern water dragon (*Physignathus lesueurii lesueurii*) basks in the sun along the Jenolan River just outside the entrance to the caves. These strong-swimming lizards occur in a broad belt along much of Australia's eastern seaboard.

TOP The Temple of Baal is one of the six southern show caves open to the public at Jenolan Caves. Its two massive chambers showcase many flowstones and large shawls. It also contains some of the rare irregular cave formations called helictites.

RIGHT The first glimpse visitors have of Jenolan Caves is when they drive through the impressive Grand Arch, a massive opening leading into the cave system, via a dark and narrow cutting that allows only one car a time to get through.

Karsts of Palau

Gaining independence in 1994, the Republic of Palau is one of the smallest countries in the world. Mostly uninhabited, the 350 islands that make up the tiny island nation are of volcanic and coralline origin, and are home to Micronesia's greatest diversity of terrestrial flora and fauna.

RIGHT Palau's islands arise like verdant pillows in a sheet of blue ocean. The islands, with their underwater wonderland of caves and marine life, are regarded by many to be one of the natural wonders of the world.

Forming the westernmost archipelago of the Caroline Islands, the islands of Palau are exposed peaks of undersea ridges that stretch along a north-south axis for 124 miles (200 km) between New Guinea and Japan, the most physiographically varied archipelago in Micronesia. Seventy-five percent of Palau's land area is found on the island of Babeldaob, a 145-square-mile (376-km²) eroded volcanic island with patches of raised limestone along its northern and southern coastlines.

PALAU'S KARST FORESTS

Palau is perhaps best known for its collection of over 200 rounded towers of forest-covered, karst-weathered, limestone islands, most of which are found within a barrier reef that encircles much of the archipelago, creating a lagoon over 12 miles (20 km) wide. Ranging in height up to 679 feet (207 m), these so-called Rock Islands are found predominantly in a cluster that extends from Koror Island southward toward the island of Peleliu. Remnants of earlier patch and barrier reefs, tidal action has severely undercut these islands at their waterlines. Their porous, fractured, and heavily eroded features make access extremely difficult. Biologically, they are replete with native limestone forests and an array of animal and bird life.

VEGETATION OF THE ROCK ISLANDS

The Rock Islands are home to limestone forests. The coral limestone surface has a thin covering of leaves and humus, which serves as the principal growing medium for the local plant life. The decaying vegetation and humus provide a self-sustaining cycle of nutrients, and this balanced cycle has generated a stable ecosystem, where a wide range of species coexist. While species composition is diverse, it is similar throughout the Rock Islands, and includes many endemic species of palms, forest trees, and understory plants.

LEFT The waters off Palau's coastline feature an extensive reef fringing most of the islands. With colorful marine life, hundreds of species of coral, and wrecks of World War II ships lying on the ocean floor, this region is a favorite with divers.

Some variation is evident from island to island, particularly in the forests of the group's two next largest islands—Mecherchar (Eil Malk) and Ngeruktabel—where stands of old growth trees can still be found in low-lying areas that pro-

vide shelter from typhoons. Remnant coconut palms, once planted for commercial purposes, can also still be found on the islands.

PALAU'S MARINE LAKE

Marine sulfide lakes are a feature of the Rock Islands, and these bodies of water each boast a unique ecosystem, which supports a range of wildlife adapted to the individual site conditions.

The lakes are filled with seawater, with some completely land-locked, being filled by seawater seeping through cracks and fissures in the limestone foundation. This method of entry allows the seawater to fill the lake, but acts as a filter to prevent other sealife entering the lake and disturbing the unique balance. Typical inhabitants of these lakes include fish, jellyfish, and mollusks.

The island of Mercherchar boasts a number of these lakes bearing evocative names such as Jellyfish Lake, Hotwater Lake, and Spooky Lake.

The waters of the well-known Jellyfish Lake are populated by millions of jellyfish—mostly non-stinging—that survive on algae cells attached to their bodies.

ABOVE Beneath Palau's islands lie caves and reefs teeming with colorful marine life. Caves such as Blue Holes attract divers to this largely unspoilt island chain.

Index

Acknowledgments and Credits

The Publisher would like to thank the following picture libraries and other copyright owners for permission to reproduce their images. Every attempt has been made to obtain permission for use of all images from the copyright owners, however, Millennium House would be pleased to hear from copyright owners.

KEY: (t) top of page; (b) bottom of page; (l) left side of page; (r) right side of page; (c) center of page.

Robert R. Coenraads: 332(b)

Grant Dixon: 238(r), 261(r), 391(l)

Fiona Doig: 88(b), 113(r), 435(l), 456(l), 508(l), 509(t), 509(bl), 509(br)

Fredrik Fransson: 117(t)

Getty Images:
AFP: 39(l), 100(l), 115(b), 119(t), 152(l), 182(l), 186(r), 197(b), 214(l), 281(r), 284(t), 494(r), 503(r)
All Canada Photos: Chris Cheadle 205(l); Gary Fiegehen 6–7, 303(c); Chris Harris 16(c); Dean van't Schip 136(b)
Altrendo Nature: 19(b), 26(l), 36(b), 59(t), 151(b), 224–225, 290(l), 371(c), 416(l), 455(l), 456(l), 457(b), 470(b)
Altrendo Panoramic: 92–93(c), 104–105(c), 244–245(b), 380–381(b)
Altrendo Travel: 44(b), 137(b), 486(b)
Amana Images: Satoru Imai/A. Collection 66(b)
Asia Images: Jill Gocher 153(t)
Aurora: 129(t), 175(r), 183(r), 253(t), 459(t); James Balog 197(t); Robert Caputo 282(b); Mario Cipollini 91(b); A. L. Harrington 425(t); Peter McBride 372(t); Alexander Nesbitt 374(b); PatitucciPhoto 229(c); David Stubbs 176–177(b)
Axiom Photographic Agency: Ian Cumming 460(l)
China Span: Keren Su 43(t), 294(l)
DAJ: 262-263
De Agostini Picture Library: 228(l); DEA/Archivio B 128(t), DEA/N. Cirani 87(c), DEA/G.Sioen 14(l), 302(t)
Digital Vision: 429(r), ABEL 48(b); Sylvester Adams 185(l), 370(r); Michael Busselle 352–353(b); Cosmo Condina 449(r); Joe Cornish 345(b); Digital Zoo 342–343(b); Jerry Driendel 460(b); Robert Glusic 454(l); Andrew Gunners 500–501(b), 504(b); Robert Harding 407(r); Adam Jones 37(t); Wilfried Krecichwost 280(l); Sunset Avenue Productions 315(c); Jeremy Woodhouse 280(b), 417(l)
Discovery Channel Images: Jeff Foott 133(t), 345(r), 378(c)

Dorling Kindersley: 43(r), Paul Harris 144(l); Rupert Horrox 499(b); Gary Ombler 22(c), 240(r); Toby Sinclair 290–291(b); Jon Spaull 461(t)
First Light: David Nunuk 171(r), 412(b); Ron Watts 32–33
fStop: Marc Volk 505(r)
Gallo Images: Karl Beath 148(l); Heinrich van den Berg 42(l), 354(r), 404–405, 427(t); Roger de la Harpe 426(b); Lanz von Horsten 355(r); Stefania Lamberti 212(b); Michael Poliza 2–3, 50(b); ROOTS RF collection–Martin Harvey 321(t); Travel Ink Photo Library 355(c)
Gallo Images ROOTS RF collection: Richard du Toit 10–11
Getty Images News: 52(c), 94(l), 101(l), 114(l), 115(t), 183(b), 221(c), 273(b), 297(c); David Goddard 442–443
Getty Images North America: 421(c)
Hulton Archive: 99(b), 294(b), 469(l)
Iconica: Arctic-Images 63(t); John W Banagan 34(l), 439(r); Grant V. Faint 37(b); Frans Lemmens 149(l), 371(t), 375(t), 387(t), 387(t); Bryan Mullenix 415(r); Sergio Pitamitz 141(t); Simon Wilkinson 437(t)
Lonely Planet Images: John Banagan 28(l); Anders Blomqvist 340(l); Graeme Cornwallis 230(b); Trevor Creighton 439(c); Grant Dixon 55(r), 128(r), 326(b); John Elk III 231(b); Jenny & Tony Enderby 476(l); Christer Fredriksson 373(c); David Greedy 179(t); John Hay 296–297(b); Kraig Lieb 135(t); Diana Mayfield 430(b); Wes Walker 300(t), 346(l), 444(r)
Look: Jan Greune 266(l); Andreas Strauss 243(r); Tony Wheeler 463(r)
Medioimages: Photodisc 433(r), 433(b)
Minden Pictures: Gerry Ellis 56(t), 67(rc); Michael & Patricia Fogden 64–65(b), 83(r); Colin Monteath/ Hedgehog House 441(r); Piotr Naskrecki 33(r); Chris Newbert 69(b); Pete Oxford 16(c); Mike Parry 61(b); Richard Du Toit 215(b); Michael Quinton 46(l); Tul De Roy 328(l), 349(l), 459(l); Konrad Wothe 198(t); Norbert Wu 35(b)
National Geographic: 32(c), 39(b), 47(t), 48(c), 51(t), 62(l), 64(l), 86(r), 92(r), 105(r), 112(l), 136(l), 148(b), 151(b), 159(r), 174(b), 176(l), 179(b), 188(l), 199(r), 211(r), 212(l), 215(c), 215(b), 219(r), 233(r), 258(b), 259(c), 259(c), 264(r), 265(l), 271(r), 277(c), 278(l), 313(r), 320–321(b), 337(t), 362(b), 378(b), 387(b), 397(b), 400(b), 410–411(b), 419(r), 422–423(b), 423(r), 424(l), 425(c), 428(l), 428(r), 481(t), 487(t), 501(r), 502(c), 504(r), Stephen Alvarez 345(t), 498(l); James P. Blair 451(c); Sisse Brimberg/ Cotton Coulson/Keenpress 203(b); David Doublett 157(r); Jason Edwards 63(r), 222–223(b); Kenneth Garrett 238(l); Justin Guariglia 295(b); Bobby Haas 62–63(b), 91(t), 128(l), 192–193, 211(l), 340(l), 370(l); Tim Laman 379(c); O. Louis

Mazzatenta 30(b), 268(b); Michael Nichols 287(c); Klaus Nigge 57(r), 353(r), 430(l), 464(l); Richard Nowitz 357(c); Randy Olson 150(b); Carsten Peter 35(t), 106(l), 107(t), 108(l), 116(l), 217(t), 217(b), 493(b); Michael S. Quinton 172(l); Rich Reid 45(b), Stephen L. Raymer 84–85; Jim Richardson 143(l); Joel Sartore 41(l), 279(b), 336(r); Roy Toft 37(r), Gordon Wiltsie 361(t); Steve Winter 180(l), 181(l)
Nordic Photos: Lars Dahlstrom 324(l); Atli Mar Hafsteinsson 170(l), 171(c); Thorsten Henn 23(t); Sigurgeir Jonasson 87(t), 90(b); Gunnar Svanberg Skulasson 20–21
Panoramic Images: 8–9; 45(t), 118–119(c), 204–205(b), 210(b), 234–235(b), 240–241(b), 251(b), 252–253(b), 398–399(b), 454–455(b), 472–473(b), 476–477(b)
Photodisc: 18996 158(b); Sylvester Adams 82(l); S. Alden-PhotoLink 455(r); Michael Aw 511(t); Tom Brakefield 255(b), 459(r); Kent Knudson/ PhotoLink 336(r); Roine Magnusson 185(b); Medioimages/Photodisc 483(c); StockTrek 89(c); Jeremy Woodhouse 354(l), 506–507(b)
Photographer's Choice: Geoffrey Clifford 109(t); Kathy Collins 143(r); Georgette Douwma 29(b), 217(r); Larry Dale Gordon 303(r); Darrell Gulin 220(b); Bruce Heinemann 409(t); Gavin Hellier 408(t); Peter Hendrie 444(b); Bruno De Hogues 240(l) Simeone Huber 499(c); Wilfried Krecichwost 228(r); Mark Lewis 511(b); Ray Massey 40(c); Richard Price 346–347(b); James Randklev 49(r), 304(l); Roger Ressmeyer 121(r); RF/Paul Edmondson 415(r); Steve Satushek 226(t); Kevin Schafer 59(r); Thomas Schmitt 80–81; Kim Westerskov 329(b); Jochem D Wijnands 194(r)
Photolibrary: 112–113(c), 268(b)
Photonica: Theo Allofs 10–11, 40(b), 141(b), 227(t), 438(t), 444(t), 446(b); Jake Rajs 406(b); Jake Wyman 289(t)
Purestock: 414–415(b)
Riser: Astromujoff 60(t), 68(b), 195(c), 216(b), 251(t) 389(t); Jean du Boisberranger 104(r); Paul Chesley 139(b); China Tourism Press 339(r), 358–359(b), 359(c), 478–479; Alain Choisnet 65(c); Daniel J Cox 343(r); Rich Frishman 95(c); Jane Gifford 469(b); Kevin Kelley 113(l); Wilfried Krecichwost 448(l), 477(t); Michael McQueen 266(r); Ted Mead 398(l); Andrea Pistolesi 431(r); Kevin Schafer 67(l), 333(r); Thomas Schmitt 418(b); Harald Sund 135(r), 319(r), 447(c); Darryl Torckler 22(b), 81(b); Richard Ustinich 96(b); Ulf Wallin 134(b); Art Wolfe 314(l)
Robert Harding World Imagery: C Gascoigne 218–219(b); FireCrest 462(b); Robert Francis 411(t); Lee Frost 409(b); James Hager 475(l); Robert Harding 397(t); Gavin

Hellier 147(c); J P De Manne 283(t); Throsten Milse 472(l); Bruno Morandi 248(b); Sergio Pitamitz 52(b); R H Productions 406(r); Roy Rainford 351(b); Jochen Schlenker 155(b); Marco Simoni 140(l), 352(t); Luca Tettoni 129(c); Ruth Tomlinson 350(l), 420(l); D. H. Webster 327(t)
SambaPhoto: Cristiano Burmester 265(c)
Science Faction: Matthias Brelter 49(t); Ed Darack 53(t), 132(b), 304(b); Gerry Ellis 332(t); Stephen Frink 482(b); Fred Hirschmann 15(c), 51(t), 175(r), 450(b), 485(r); Doug Landreth 360(t); G Brad Lewis 83(t), 58(b),159(b); NASA/digital version by Science Faction 110(t); NASA-ISS/digital version by Science Faction 286(b); Dan McCoy/ Rainbow 24(l); Flip Nicklin 277(r); Louie Pslhoyos 26(r), 44(l), 46(b), 269(c); Jim Wark - Stock Connection 160–161; Norbert Wu 295(r), 338(c)
Sebun Photo: Hiroyuki Yamaguchi 445(t); Katsuhiro Yamanashi 184(b)
Stockbyte: Tom Brakefield 286(l), 348(l); Martial Colomb 467(c); Jeremy Woodhouse 382–383(b)
Stone: James Balog 207(r); Tom Bean 270(l), 487(c); John Beatty 82–83(b); Vanessa Berberian 441(t); Gerald Brimacombe 206(r); Carolyn Brown 90(t); Robert Cameron 448–449(b); Kevin Cooley 494(b); Joe Cornish 131(b); Cousteau Society 493(t); Nicholas DeVore 102(b); Jack Dykinga 172–173(b), 341(l); Johan Elzenga 196(l); Tim Flach 347(b); Robert Frerck 348(b); Ernst Haas 413(b); Paul Harris 146(b); William J Herbert 177(c); David Hiser 152(b); Jeff Hunter 38(l); Arnulf Husmo 305(c); Jacques Jangoux 17(r), 276(b); Will & Deni McIntyre 277(b); Eastcott Momatiuk 458(l); David Muench 107(b); Ian Murphy 288(l); David Nausbaum 279(r); Ben Osborne 25(b); Hans Strand 480(r); Darryl Torckler 440–441(b), 506(r); Luca Trovato 507(t); Paul Wakefield 469(r); John Warden 154(l); Stuart Westmorland 300(r); Art Wolfe 107(b), 13–131(b), 238(b), 418(l)
Taxi: Adastra 159(t); Peter Adams 396(b); Gary Bell 493(r); Walter Biblkow 360(l), 428(b), 471(t); Beth Davidow 310(b); Wendy Dennis 382(l); Mike Hill 31(r); Harvey Lloyd 432(b); Ken Lucas 495(t); Keith Macgregor 501(l); Ian McKinnell 1; David Noton 434–435(b); Gary Randall 272(l); Ron & Patty Thomas 227(c)
The Image Bank: Theo Allofs 366(t); James Balgrie 424–425(b); Daryl Benson 27(r), 29(t), 453(r); Walter Biblkow 67(c), 422(l); Carolyn Brown 284–285(b); Angelo Cavalli 358(l); Kevin Cooley 498(b); Stephen Cooper 277(t); Cousteau Society 457(r); Macduff Everton 147(t), 180(b); Andre Gallant 67(r); Sylvain Grandadam 189(t); Peter Hendrie

301(t), 329(t); Gavin Hellier 42–43; Frank Krahmer 484–485(t); Wilfried Krecichwost 485(b); Ted Mead 226(r), 367(c); Michael Melford 313(t), 376–377(b); Bruno Morandi 67(lc); Carlos Navajas 397(r); Daniele Pellegrini 243(b), 385(t); Chip Porter 69(r); Ingrid Rasmussen 142(l); Siqui Sanchez 143(b); David Sanger 231(t); Kevin Schafer 260–261(b); Jonathan & Angela Scott 285(r); Juan Silva 407(c); Paul Souders 61(t); Bob Stefko 364–365(b); Hans Strand 464(r); Travelpix Ltd 126–127; Joseph Van Os 403(b); Michele Westmorland 289(c); Jochem D Wijnands 465(r); Simon Wilkinson 334–335; Winfried Wisniewski 194(t); Art Wolfe 362(l); Gary Yeowell 445(c)

Tim Graham Photo Library: 195(t)
Time & Life Pictures: 206(b), 391(t), 488(l), 489(t), 489(c), 489(b), 492(b)
VisionsofAmerica : Joe Sohm 273(t)
Visuals Unlimited: 97(t), 101(t), 111(t), 234(t), 274(c), 391(r), 488(b); Ken Lucas 35(r); Mark Schneider 41(b); Tom Walker 325(t)
Westend61: Martin Rietze 66(t)

iStockphoto:
Bryan Busovicki 121(r); Alexander Hafemann 388(b); Craig Hansen 108(b); Luca Manieri 244(l); Alexey Poluyanenko 246(b); Vladimir Pomortsev 111(r); Vadim Rumyantsev 119(r); Nico Smit 383(r); Raldi Somers 122(l); Arne Thaysen 383(t)

John I. Koivula:
18(tc), 18(tr), 18(c), 18(r)

David McGonigal:
18(b), 24(t), 31(b), 55(t), 86(t), 116(b), 117(b), 122(b), 123(t), 123(c), 123(r), 125(t), 125(b), 138(l), 145(b), 145(r), 155(t), 156(b), 157(b), 169(l), 169(r), 169(b), 179(r), 186(l), 187(r), 187(b), 189(b), 200(l), 201(l), 201(r), 202(t), 202(b), 203(b), 204(t), 208(b), 209(t), 209(r), 210(l), 219(r), 222(l), 222(r),

232(r), 233(r), 233(r), 235(t), 235(r), 236(b), 237(t), 237(r), 241(t), 242(l), 245(l), 245(r), 246(l), 247(t), 249(b), 250(l), 252(l), 253(r), 254(l), 256(l), 257(t), 257(b), 260(l), 261(t), 274(l), 275(r), 278(b), 296(l), 306(b), 307(t), 309(t), 309(b), 311(t), 311(b), 318(b), 319(b), 320(l), 321(c), 322–323(b), 323(r), 330(b), 333(r), 337(c), 342(l)356(l), 364(l), 366(b), 372(l), 374(l), 376(l), 377(l), 381(c), 385(t), 389(c), 392(b), 393(c), 394(l), 394(b), 399(r), 400(r), 401(b), 402(b), 402(l), 412(l), 413(r), 435(r), 436(b), 450(l), 453(l), 453(r), 466(l), 468(b), 473(c), 474(l), 475(r), 480(t), 481(c), 482(l), 500(l), 509(c), 509(r), 510(l)

Avril Makula
434(l)

NASA:
Earth Observatory/Robert Simmon 271(t); GSFC/METI/ERSDAC/JAROS, and the U.S./Japan ASTER Science Team/Jesse Allen 333(b); Image Science and Analysis Laboratory, NASA-Johnson Space Center 281(t); NASA International Space Station Imagery: 250(b); Robert Simmon/NASA GSFC Oceans and Ice Branch/Landsat7 Science Team 403(t); NASA/GSFC/LaRC/JPL, MISR Team 363(t); NASA/GSFC/METI/ERSDAC/JAROS, and U.S./Japan ASTER Science Team 298–299; NASA/USGS 301(b), 368–369; NASA Visible Earth 218(l), 223(t), 249(t) 386(l); NASA Visible Earth/ Jacques Descloitres, MODIS Land Science Team 283(b); NASA Visible Earth/USGS EROS Data Center Satellite Systems Branch as part of the Earth as Art II image series 395(r); USGS EROS Data Center 267(b), 390(b)

Armstrong Osborne:
494(l), 495(c)

The Art Archive:
Global Book Publishing 506(l)

UC Santa Barbara Department of Geography:
Jeffrey J Hemphill 89(b)

Chapter Opener Images:
Creek Brennisteinsoldukvisl, Landmannalaugar, Iceland: Getty Images/Nordic Photos: Gunnar Svanberg Skulasson 20–21

Aerial view of Blue Hole, Lighthouse Reef, Belize: Getty Images/Photographer's Choice: Thomas Schmitt 80–81

Aerial view of Mt St Helens, Washington, USA: Getty Images/National Geographic: Stephen L. Raymer 84–85

Basalt Formations, Giant's Causeway, County Antrim, Northern Ireland: Getty Images/The Image Bank: Travelpix Ltd 126–127

Aerial view of Grand Prismatic Spring, Yellowstone National Park, USA: Getty Images/Science Faction: Jim Wark - Stock Connection 160–161

Massed Flamingoes on Lake Bogoria, Kenya: Getty Images/National Geographic: Bobby Haas 192–193

Aerial view of Swiss Alps, Switzerland: Getty Images/Altrendo: Altrendo Nature 224–225

Aerial view of Iguazú Falls, South America: Getty Images/DAJ 262–263

Satellite view of Chapman Glacier, Ellesmere Island, Nunavut Territory, Canada: NASA/GSFC/METI/ERSDAC/JAROS, and U.S./Japan ASTER Science Team 298-299

Aerial view of Dead Horse Point, Utah, USA: Getty Images/The Image Bank: Simon Wilkinson 334–335

Aerial view of dunes in the Namib Desert: NASA/USGS 368–369
Rock Arch Formation, Namibia: Getty Images/Gallo Images: Heinrich van den Berg 404–405

Aerial View of the Seven Sisters, Sussex, England: Getty Images: David Goddard 442–443

Seven Star Grotto, Guilin, Guangxi, China: Getty Images/Riser: Chinese Tourism Press 478–479

Front Cover:
Creek in Iceland: Getty Images/Iconica: Arctic-Images

Back Cover (from top to bottom)
Satellite view of Chapman Glacier, Ellesmere Island, Nunavut Territory, Canada: NASA/GSFC/METI/ERSDAC/JAROS, and U.S./Japan ASTER Science Team

Aerial view of Grand Prismatic Spring, Yellowstone National Park, USA: Getty Images/Science Faction: Jim Wark—Stock Connection

Seven Star Grotto, Guilin, Guangxi, China: Getty Images/Riser: Chinese Tourism Press

Aerial View of the Seven Sisters, Sussex, England: Getty Images: David Goddard

Creek Brennisteinsoldukvisl, Landmannalaugar, Iceland: Getty Images/Nordic Photos: Gunnar Svanberg Skulasson;

Aerial view of Blue Hole, Lighthouse Reef, Belize: Getty Images/Photographer's Choice: Thomas Schmitt

Aerial view of Iguazú Falls, South America: Getty Images/DAJ